THE ATMOS BOOK OF PIPELINE SIMULATION

Moemen Metwally, AMIChemE

Application Support Engineer, Atmos International

Dr. Jason Modisette

Chief Scientist, Atmos International

Editors

Andy Hoffman

Dr. Jun Zhang

First published 2022

1st edition

Authors Moemen Metwally and Dr Jason Modisette

Edited by Andy Hoffman and Dr Jun Zhang

Cover design by Sam Edge

ISBN 978-1-7397829-2-4

TABLE OF CONTENTS

PREFACE BY JASON MODISETTE

I would first like to thank my co-author Moemen Metwally, who wrote the lion's share of this book. I also want to thank Atmos International for supporting and encouraging me during this process. Finally, I would like to extend my sincerest gratitude to my father, Jerry Modisette, without whom I would not be in this industry. He was one of the first people in the world to use digital computers to simulate pipelines and was a leader in the pipeline software industry for its first several decades. Parts of this book are based on my lectures at the Pipeline Simulation Short Course offered by the Pipeline Simulation Interest Group (PSIG). Those lectures were in turn partly based on my father's lectures when he taught the same course.

Very few people go to college to learn pipeline simulation; possibly nobody does. Everybody currently working in the pipeline simulation field came from either some more general field like civil engineering - and maybe had one class on flow in pipes - or (as in my case) came from other fields entirely. Numerical electronic structure calculation taught me how to solve partial differential equations with computers, but that was about the only overlap my later education had with my actual career. And as we wrote this, I've seen news stories about universities shutting down their petroleum engineering programs due to lack of interest, so this problem is only going to get worse.

And yet, pipelines aren't going away any time soon. The energy transition is likely to first shift much of the world from coal to natural gas, then perhaps to natural gas with some hydrogen in it, a topic we cover in this book. More pipelines are going to be built; also, existing ones are going to be regulated more tightly, requiring more leak detection and more safety studies. So the world is still going to need pipeline software for a long time to come, even for hydrocarbon systems – and there will always be water pipelines.

This book is intended to be the book I wish I'd had when I started in the field. It covers how oil and gas pipelines work in practice, or at least such aspects as are relevant to someone writing, installing, or using simulators. We tried to present how our company's simulator, Atmos SIM, works, as well as some of the simple equations that one could solve in a spreadsheet to confirm the accuracy of all aspects of any gas or liquid pipeline simulator. In addition to the mathematics and physics of pipe flow, we also present pipeline terminology, and the reasons pipeliners do things the way they do.

Of course, to some extent this book is also an advertisement and companion text for Atmos SIM, the simulator / optimizer we developed during my time here at Atmos. There are multiple approaches to any problem solved by a simulator; this book tries to provide an overview of the possibilities, but in each case, it mostly concentrates on the approach we actually used in Atmos SIM and tries to provide justification as to why that approach is the best. We also discuss how to best use a simulator, not just how to write one: how to validate a model, how to tune it up for best accuracy, problems to watch out for, and how to diagnose the causes of misbehavior.

Note that this is not intended to be a complete textbook on fluid flow, so we have deliberately omitted older flow equations, equations of state, and such that we consider not to be accurate enough for modern use.

PREFACE BY MOEMEN METWALLY

In this book we address a set of pertinent discussions of interest to pipeline simulation professionals. To our knowledge these have never been compiled before in a single convenient place. We present the industry's answers to a number of big questions that often come up, reflecting the state of the art. Many significant improvements and major advances are happening in pipeline simulation, and we are excited to share that story.

Nobody wakes up one day and, unprompted, asks themselves: how rough is a steel pipe? This book strives to *contextualize* what pipeline simulation is all about. It isn't intended to inform experts with decades of experience about their operation – they live and breathe it every day, it is we who learn from them. Rather, it presents an outline of this broad domain in a manner accessible to anyone who finds themselves interested, passing on the spark which ignited our curiosity. Pipeline simulation is a fascinating, interdisciplinary enterprise. For this reason, we've taken it upon ourselves to explain what it is we do: the application of *simulation* to *pipelining*. Regrettably, both of these terms have long been impenetrable to most people, not because they are uninteresting or irrelevant, but because they have been perceived as extraordinarily complicated.

This is unfortunate; the general public are an important stakeholder in what we do. They should have an insight in plain terms into what exactly it is we are up to. They hear nothing from us, it seems, and only hear about us when there is an incident, spill or dispute. How can something so innocuous as a hollow cylinder be so controversial, powerful and indispensable all at once? Like a knife, a pipeline can be a tremendously useful implement in our world, safely and reliably transporting water to homes, crude oil to refineries and jet fuel to airports. Pipelines enabled the entire gas industry, an energy carrier to domestic boilers and power plants, and a feedstock for a multitude of further industries. When abused, a pipeline might indeed earn itself a poor reputation, but a pipeline remains indispensable when it is used properly. It is critical infrastructure; the routes taken by the UK's *Government Pipelines and Storage System* are top secret.

Figure 1 – Channels (even covered channels) and ducts (even cylindrical ducts) are *not* pipes

Unlike *channels* open to atmospheric pressure, or *ducts* for conveyance over short distances, *pipelines* are capable of operating at significant pressures. Other geometries are used in select applications up to a few atmospheres, but it is a circular cross-section that offers the optimal strength to withstand intense pressurization. So a pipeline is always cylindrical, characterized by a diameter. Pipelines are really rather interesting in and of themselves. Like any good infrastructure, when a pipeline works well, it is almost invisible to the general public, and so goes largely unappreciated. When it goes wrong, hard lessons are learnt. As an avid reader, I've spent many an hour shuffling through books, new and used, high-street and online. There's almost nothing for the ordinary reader to dive into the fascinating world of fluids, at the heart of engineering.

We are afforded an opportunity to tap into the spirit of popular science and technology that's become part of today's zeitgeist. To make pipeline simulation accessible to a broader audience, we share thought-provoking ideas on energy, fluids and pipelining – nuggets of knowledge gained from picking someone's brains on some pressing matter; observations we happened upon whilst troubleshooting; a mishmash of intrigue compiled into a digestible form. Pipelines at their core are surprisingly intuitive, and modelling them is frankly not that hard. Engineers just often work in silos! The underlying mathematics should never hinder us from talking about what we do: mathematics is inevitable, beautiful and wise; the language of nature, no less. Yet it has an unparalleled ability to bore and intimidate, so where possible we shy away from heavy mathematical derivations.

Pipelines let us make sweeping generalizations and simplifying assumptions. Their fluid mechanics are far more straightforward than almost any other application of fluids one might come across. They motivate us to use advanced tools at our disposal, including control theory and thermodynamics, at a level that is quite accessible. They can naturally teach people to be better engineers.

If there is an omission, suggestion or section to be rephrased or expounded to better convey these ideas, we know from experience that our global community of pipeliners will be as forthcoming as always in thinking it through together.

On a personal level, thanks are due for the long hours spent on this book by kind experts, most of all my incredible co-author Jason Modisette, and the countless experts at Atmos International who embody passion for pipelines. It is their incredible insights and technical discussions we present here. Many people are working hard on each and every aspect of this book – and have gladly and willingly shown me their world, even from other organizations. Tommy Isaac helped me navigate the vast, treacherous waters of hydrogen. Every chapter was written with someone's kind input. Most of all, my team around the world, with sincere, gracious support from my directors, managers and colleagues. Our customers, project teams, development experts – and everyone else who makes Atmos International – are solving so many interesting problems, and are always so keen to share their expertise and ideas.

To all those who have supported, inspired and taught us throughout our careers, this book is dedicated to you. This book is the brainchild of Andy Hoffman. Thank you Andy, for having faith in us and guiding us throughout the process.

CHAPTER OVERVIEW

In the first three chapters, we examine pipeline simulators, the universal conservation laws they rely on, and how a solver uses numerical methods to apply model equations. In chapter 4, we consider shut-in pipelines to focus on density and equations of state, discussing their significance and their implications. In chapter 5 we look at steady states, flow equations, and how they help us to streamline a pipeline's design. In chapter 6, we discuss the thermal model, and ask whether a familiar approach – the overall heat transfer coefficient – always gives the right answer. Chapter 7 looks at how temperature is affected by pumps and compressors, calculating their power and performance, and briefly run through a basic sizing of their drivers.

Chapter 8 looks at pipeline operations, the full range of transient scenarios to be investigated, and how a simulator can perform a host of useful offline studies. Chapter 9 devotes special attention to rapid transients, control logic and surge analysis. Chapter 10 looks at the simulation of leaks and ruptures in offline scenarios, and the performance of model-based leak detectors and locators in online scenarios.

Chapter 11 then shifts our attention to real-time models deployed live on site, working in tandem with auxiliary and supporting software. We examine their automatic tuning. In a section on trainer-trainee applications, we explore how an offline constrained scenario can mimic a real system by feeding online meter-based scenarios in real-time, in order to improve and assess an operator's response. Chapter 12 looks at how an online live system has been supported by look-aheads of many kinds, for forecasting and prediction. New, exciting opportunities are facilitated by the advent of recent technologies: fast, concurrent and advanced look-aheads. We look at several kinds of optimizer, what they have been able to do for pipeline operations, and whether the promise of a true operational optimizer is mere hype or a feasible and emerging reality.

In chapter 13, the final chapter, we discuss a set of features that our industry urgently needs: taking simulators beyond single phase flow. We touch on a few reasons why running a multi-phase model can be challenging, and alternative approaches to deal with phenomena such as slack-line in liquid pipelines or liquid drop-out in gas pipelines.

At the end of this book, we summarize some of its key points. This main book focuses on the task of pipeline simulation. It is complemented by a number of appendices in a separate volume, to allow the reader to dive straight into the technical details with no distractions. The appendices offer a background on the energy sector, pipeline history, a basic primer on pipeline fundamentals, hydrogen and a quiz to challenge the reader.

1 – MODEL-BUILDING: WHAT IS A PIPELINE?

Figure 2 – An offshore pipeline approaches a station

"Horses for courses" – British proverb

1.1 – START WITH WHY: THE USE CASE

Offline studies: running virtual experiments

When I pick up a book about the birth of modern science and engineering, and compare it to a typical undergraduate degree, I can't help but ask: where's the sense of wonder? Has the sheer volume of terminology and concepts around our subject matter become a hindrance, *"missing the woods for the trees"?* Simulation rekindles our spirit of curiosity, of tinkering and experimentation. The best way to understand a system is to run a few experiments. Build a set of scenarios, and let the physics play out, probing and comparing the results to explore the system's behavior in each case. Previously, this was done using clever apparatus, scale models, and observational rigor. Indeed, the world's best and brightest still work in this way. To investigate fluid mixing, multi-phase flow, and countless challenging phenomena, impressive large-scale experimental rigs are still built today to validate models.

Thankfully, the 21st century curious mind need not build these models up from first principles. Software is available off-the-shelf, answering all manner of interesting questions at the click of a button. We simply set up a scenario, launch a run, and watch a simulation unfold faster than any pioneer could have ever imagined.

Today, it is tempting – dare I say *enjoyable* – to build a numerical model and harness computing power to solve it using a physical simulator. Physical and empirical equations play out discretely, hidden from view, in a well-choreographed performance culminating in solutions to the problem at hand. This happens so elegantly it goes almost entirely unnoticed. By tinkering we see for ourselves *why* fluid behavior can be counter-intuitive, or at least surprising: *"careful analysis of experiment naturally inspires the inquiring mind to synthesis and design".* [23]

These experiments can be turbo-charged and taken to their limit when conducted on properly validated simulation packages. Engineering teams critically examine simulated results, comparing them against independently calculated expected values. It thereby proves itself as a trusted tool in our arsenal, for tackling all manner of tasks.

In the design phases, it allows optioneering by exploring various approaches to meet the design intent. We evaluate how profitable and flexible each proposed solution would be, helping a team rapidly reach informed decisions. Hydraulic studies cover a wide range of scenarios in these design phases:

- Flowsheeting, laying out the network and finding operating conditions
- Specification, selecting appropriate pipes or equipment from vendors
- Control philosophy, defining a narrative for how best to operate it

In the operational phase, simulation helps the diverse experts that comprise an operations team with their routine and ad-hoc tasks, including:

- Troubleshooting, identifying issues and overcoming them
- Forecasting, anticipating disruption and avoiding it
- Operations planning, especially modifications to fluids and flows
- Expansion studies, assessing the retrofits involved in a proposal
- Operator training, demonstrating their competency to regulators
- Optimization, finding the best strategy or debottlenecking capacity

Online simulators: monitoring in real-time

The same model we assemble for offline studies simulations forms the foundation of an online simulator deployed live on site. This monitors the pipeline in real-time as it operates. Depending on the application, it is either a real-time transient model (RTTM) or a real-time incompressible model. We'll come back to both in detail in a later chapter. They might serve many purposes.

A simulator supports problem solving at all stages of a pipeline lifecycle. From the front-end design of a new pipeline, through the detailed design phase which refines and re-examines these designs, all the way through to supporting the operations team day-to-day. To properly harness its power, an engineer's skill and expertise are indispensable. By critically examining *how* we simulate, we ensure all our results are reliable, and draw the right conclusions.

1.2 – GOOD MODELS: WHAT TO CONSIDER

A graphical view on canvas

The simplified pipeline diagram

Although some systems are as simple as a single-inlet single-outlet (SISO) pipeline, the topology of a network is rarely so straightforward. In the oil & gas sector, upstream gathering lines and commercial gas distribution grids are particularly extensive. Even midstream transmission infrastructure typically has branches, mid-line offtakes, and all manner of complexity. In the water sector, scores of pipelines are even more complex.

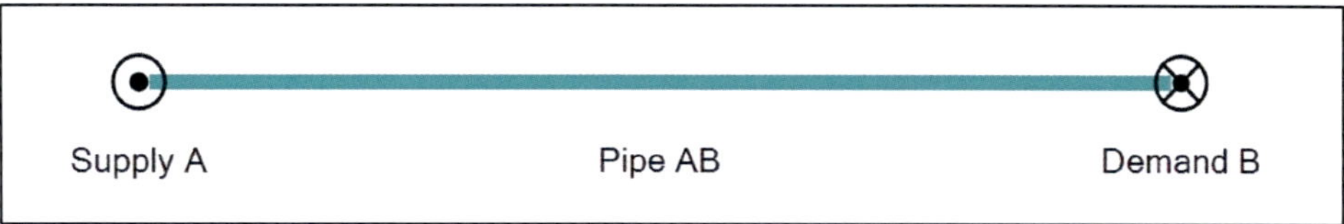

Figure 3 – A single-inlet single-outlet (SISO) pipeline displayed on a graphical canvas

Building up a schematic drawing of a pipeline is an insightful exercise in and of itself. We simplify valves in series, for instance, representing them as one model valve. A *simplified pipeline diagram* (SPD) simplifies *piping & instrumentation diagrams* (P&ID) but keeps more detail than the blocks of a *process flow diagram* (PFD). We abstract the important features for our simulation, keeping just enough detail to describe fluid behavior throughout the various pipeline operations. Atmos SIM lets us build intuitively on a visual canvas by drag-and-drop, allowing us to encapsulate detail within stations.

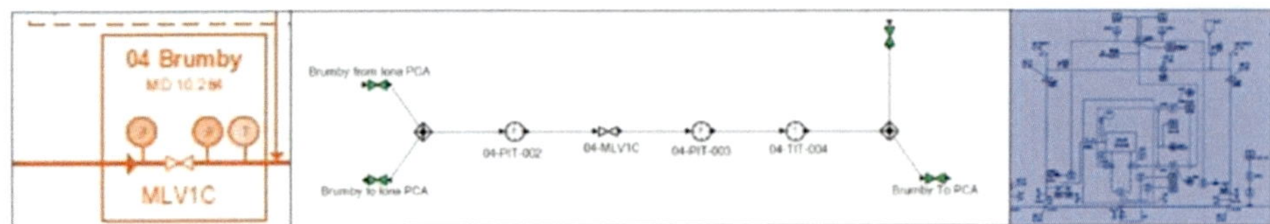

Figure 4 – A high-level abstraction as a schematic block on a *process flow diagram* (PFD, left), alongside a suitable *simplified pipeline diagram* (SPD, middle) configured in Atmos SIM, and a detailed *piping and instrumentation diagram* (P&ID, right). All represent the same station.

Experienced engineers build simple models

"I apologize for such a long letter; I didn't have time to write a short one." – Mark Twain

A fresh engineer is often inclined to argue the case for *not* simplifying a model. Simplifying takes thought, and the engineer can mess up, oversimplify and get the wrong answer without realizing. It's easier they just simulate everything! We can tell a fairly untrained, lazy or bad engineer to *"just configure everything on this diagram in the model"*. But the best model is not one including every single component of the real pipeline. It is expensive to build and maintain due to the amount of data involved, and the number of constraints to define, implement and tune. And it is as complicated to operate as the real network! [29] The trouble is: adding detail doesn't necessarily improve the *accuracy* of a model, even if it increases its *precision*. All we know for sure is that detail slows down the model solution, and makes it more complex. This introduces more places for something to go awry and makes troubleshooting it more of a hassle.

Experienced engineers build simple models. They establish exactly what the purpose of the model will be, decide on its key features, and on that basis add only the level of detail that is appropriate. Their efforts are focused on defining functional requirements and satisfying them, not throwing in all known data. This allows them to spend their time on other more subtle model parameters. Starting from the simplest possible arrangement (single-in single-out), they build up the pipeline from the ground up. Once they are happy with the simple model, according to the principles we discuss later in the chapter on validation and accuracy, only then does an experienced engineer start adding complexity. They build it up in reversible steps (a reliable undo button is a must). And at every step, they validate again, going back and amending when necessary. [29]

Crucially, they stop as soon as it is fit for purpose. The use cases for a pipeline simulator are well-defined, and different users have very different priorities.

The humble pipe itself

What makes a pipe a pipe?

Though every pipe is unique, a pipeline is built to specifications which strive for predictable consistency. A pipe's chief characteristics for pipeline hydraulic modelling are *inner diameter* and *wall roughness*. Other factors are lumped into roughness or into a separate hydraulic efficiency. This seems a bold move, but unlike pipework on a plant, a pipeline is straight and *"uneventful"*. Tees, bends and fittings are of no consequence in all but the most elaborate stations.

Pipe parameters also vary along the length of a pipeline. These parameters affect the operating pressures and flow rates as calculated by our simulator. The behavior of a fluid in a pipe is modelled using a set of *pipe-leg equations*, including a suitable *flow equation* capturing the effects of friction. A long section throughout which there is no change in characteristics, metering or tie can be treated as one single *"model pipe"*. Despite consisting of many individual pipes in reality, this proves to be a reasonable abstraction to keep a model simple. More detail can always be added by splitting a pipe within the model into several separate pipes if it turns out a section is somehow special.

Length, elevation and inclination

Figure 5 – A pipeline heads offshore. It is uncertain whether the subsea portion is still buried.

Some pipelines are flat. Others traverse treacherous terrain. One crosses the Andes, following the route of the infamous *"death road"*! Another has a vertical riser climb a sheer cliff face, deep beneath the North Sea. Terrain makes a pipeline's *length* rather involved. Once its *"right-of-way"* is determined, the length of a section with varying elevation is expressed equivalently in one of two separate ways: either the *chainage*, or the *horizontal distance* (like in an aerial view) ignoring elevation. Pipeline simulators expect users to input the chainage distance. Elevation may be defined above any meaningful datum, not necessarily *above sea level* (ASL).

Pipelines are typically constructed from segments each measuring 10 to 15 meters in length, with no internal bends. Some pipelines have a slope exceeding a 20% grade. Some connect with vertical risers. The slope of a pipe is defined as the *"rise over run"* – the change in elevation (Δz) divided by the horizontal distance its length covers (Δx):

$$\sin[\theta] = \frac{\Delta z}{\Delta x}$$

Pipeliners prefer to talk about angle of inclination (θ) rather than the elevation gradient ($\Delta z/\Delta x$). We substitute that into equations whenever there is an elevation term at play.

Lack of knowledge of true elevation profile can be a challenge. Some pipelines have detailed elevation profiles recorded by tools traveling through the pipe, but common practice used to be a survey along the route, with a large number of points that don't actually reflect the true elevation profile. Elevation profiles should be filtered to remove excess points, while preserving all high and low points. Local peaks and troughs are important for modelling phenomena such as slack in liquid lines. Localized sag-bends and over-bends might exist in a pipe network but be missed by an elevation profile.

How much the elevation profile and the inclination it implies actually matters depends on the application. For an incompressible liquid with no slack, all that matters is elevation at each station suction and discharge, and at each batch interface. In a gas, elevation is generally less important as the density is relatively low – a gas model might remain accurate even if an elevation is wrong by 20 meters. Conversely, elevations are important to model how fluids flow through pipelines in other contexts (e.g. slack flow).

Routes and groups

Often, we need to retrospectively modify an aspect of our model's topography. A branch or a split is added here, an extension there, or a correction elsewhere. It is useful, then, to think of the total length of a pipeline as not being fixed in stone: even if the line is built and operating, it could change in future! Atmos SIM lets a user set the properties of a *route* via an overarching *group*, applying elevations etc. with reference to an *origin*, allowing us to manipulate the model items at will without pointless rework.

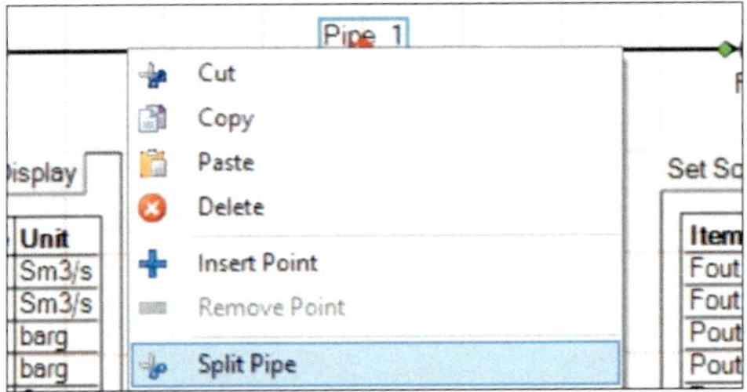

Figure 6 – Splitting a pipe in the Atmos SIM user interface.

Which model pipe objects constitute a physical pipeline route does not matter, so long as all properties are as specified at a given distance along. Distances on pipelines are referred to by milepost or kilometer markers (MP or KP). It is also helpful to view results along routes which overlap with others, or view the results of any combination of non-contiguous pipes as an inventory group.

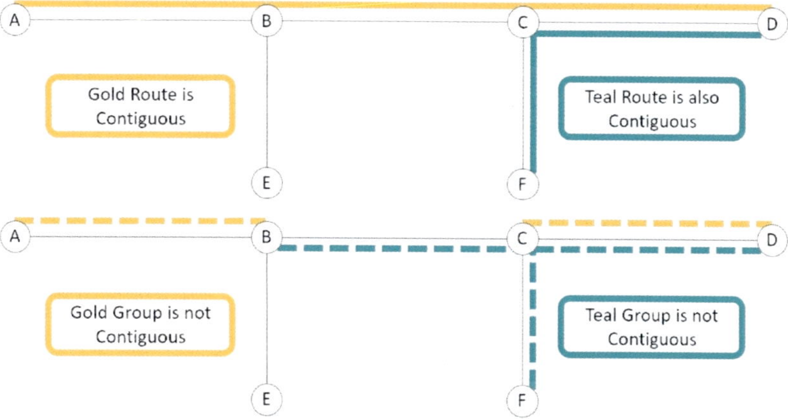

Figure 7 – A contiguous route (top) is *unbranched* and *connected*, else it's a group (bottom)

Inner diameter, nominal diameter and wall thickness

A prime consideration for engineers analyzing a pipeline is diameter (D or ∅). Once diameter is set, everything is based on that. If throughput is to be higher than expected, and a bigger diameter is needed, this incurs a significant cost as it entails re-work by design engineers, including specifying duty points on each pump or compressor, selecting control valves to order, and sizing motors. Seemingly inconsequential work is interdependent: small changes might force the team to start over! [45]

What size pipe is capable of handling twice the flow rate of an 8-inch schedule-80? Cross-sectional area (A) varies with pipe diameter as $A = \pi \cdot D^2/4$. At first glance it looks like the answer is 25% larger: 10-inch. But a ten-inch pipeline doesn't actually carry twice as much as an eight-inch line! In reality it's a lot more complicated than that. Again – for every complex problem there is an answer that is clear, simple, and wrong. In a rough-pipe turbulent regime, pressure gradient of an incompressible liquid scales with flow-rate-squared divided by fifth power of diameter.

Thicker walls linearly increase design pressure, offsetting the reduced design pressure inherent with a larger diameter. As each pipe size offers a choice of thicknesses, a pipe is usually referred to by more than just its stated *"diameter".* For historic reasons, this may be defined either as its outer diameter (*"OD"*, i.e. the outside diameter or external diameter), or as its inner diameter (*"ID"*, i.e. the inside diameter or internal diameter). Which of the two is referred to depends on the size of pipe in question, and the system of reference for pipe sizes. In Europe, this is expressed as a nominal diameter (*diameter nominal* DN). In the Americas, it is expressed as a *nominal pipe size* (NPS) or *nominal bore* (NB) – and that standard uses inner diameter for pipes of twelve inches or smaller (12", 10", etc.) and outer diameter for pipes larger than that (14", 16", etc.). A pipe's diameter must be accompanied by the pipe's *schedule* which lets us look up its wall thickness, allowing us to deduce whichever of the two diameters is unknown (outer or inner) by adding or subtracting twice the wall thickness from the other one.

Roughness and hydraulic efficiency

A newly manufactured might be as smooth as 1 micron, whereas an old, rusty pipe might be 1 mm in roughness. A typical pipe roughness is 0-10 microns (coated pipes) and 10-20 microns (uncoated pipes). What does roughness even mean? It's nominally the typical or characteristic height of material jutting out of a pipe's inner wall, but that must be some sort of global average over regions with very different roughness. How the material was manufactured affects what *kind* of roughness is at play. But how much of a pipeline's *"roughness"* comes from the welding bead every 15 meters? Those beads are much bigger than the true nominal roughness of any pipe material.

There are people who spent a lot of effort and attention on verifying how rough a steel pipe actually is, especially in relatively *"clean"* natural gas service. The manufacturing process, as well as the fluid being conveyed, both have some effect on the true roughness. Pipeline simulators generally overestimate roughness to account for other inefficiencies as pressure and flow data cannot usually distinguish between the two, and this usually serves our purposes reasonably well. On long-distance pipelines, roughness is typically far more significant than the effect of bends and fittings.

A pipeline section's hydraulic resistance evolves over time during operation. Some simulators *learn* efficiency whilst a model is deployed live on site, whereas others capture the same effect by adjusting roughness as a proxy. Notwithstanding this learning, for offline studies, it's important to reasonably *calibrate* these tuning parameters based on a stable averaged dataset. Gas pipelines always operate in the rough-pipe turbulent flow regime, so tuning them on efficiency or roughness is more-or-less equivalent. But for liquid pipelines, efficiency and roughness are not interchangeable. In laminar flow, roughness has no effect on friction. In the smooth-pipe turbulent regime, roughness has a limited effect. Efficiency, on the other hand, always applies regardless of flow regime, so *efficiency tuning* can still be used on liquid pipelines to compensate for effects – such as incorrect batch properties – that cause pressure gradient to change, whereas roughness might not work in those scenarios.

Material of construction

High-pressure gas transmission pipelines are mainly made out of steel because of the high pressures at which they are operated. Low pressure gas distribution pipelines are made out of cast iron, fibrous cement, PVC-enriched polyethylene (PE) or steel. The oldest cast-iron or fibrous cement pipelines are being replaced across Europe.

Figure 8 – Exposed high-density polyethylene (HDPE) pipeline in the Australian outback
GordonJ86 / Wikimedia Commons / CC BY-SA 4.0

Compatibility of insulation might affect the risk or rate of corrosion, but we won't delve into any materials science here. To quote the brilliant chemist Andrew Szydlo: *"it's all rather complicated, and I'm not sure I fully understand it myself"*. The compatibility of fluids and construction materials is a specialist subject beyond our scope, but it is worth being aware of an interesting pitfall: some gases pass right through solid materials. For example, rubber balloons filled with carbon dioxide don't stay inflated for long. Similarly, plastic pipework isn't well suited for carbon dioxide service. Nevertheless, in many applications, plastic pipework such as polyethylene is being preferred over carbon steel. As pipeline simulators, the characteristics that concern us about the material of a pipe are its elastic modulus (Y), density (ρ) and thermal conductivity (λ). In certain applications, we might also be interested in other properties such as how it's restrained.

Burial, lagging and insulation

For a river crossing, each pipe section might be coated in a layer of concrete. Good knowledge about the properties of layers surrounding a pipeline is important if we need a transient thermal model for the ground and for the fluid in the pipeline. For our purposes, we consider three widely used terms – lagging, cladding and insulation – as synonymous. In reality, it is typical to see several sorts of layer, such as a layer that is intended as thermal insulation, and cladding intended to weigh a subsea pipeline down.

Figure 9 – A buried pipeline might be exposed to air only for a crossing, like this pipe bridge

If we are building a thermal model of the pipeline environs, as we often must, when the pipe is buried, we need to know if the surrounding soil is clay, loam, or sand, and we might not be certain. If it is exposed, and contains slow-moving or shut-in fluid, the unknowns of ambient weather are influential. An anecdote pipeliners share is that when a pipe is completely exposed, they can see the effect of a cloud passing overhead!

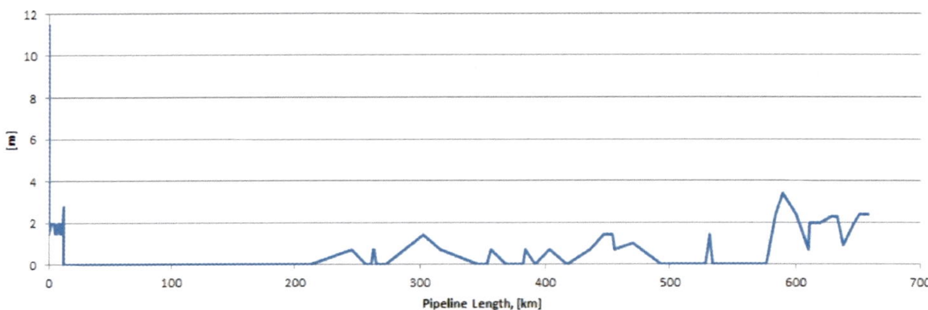

Figure 10 – A typical profile of a subsea pipeline's burial depth might change over time, introducing substantial uncertainty to the thermal properties of a given section of pipe.

Levels of detail: stations and equipment

Abstracting a station

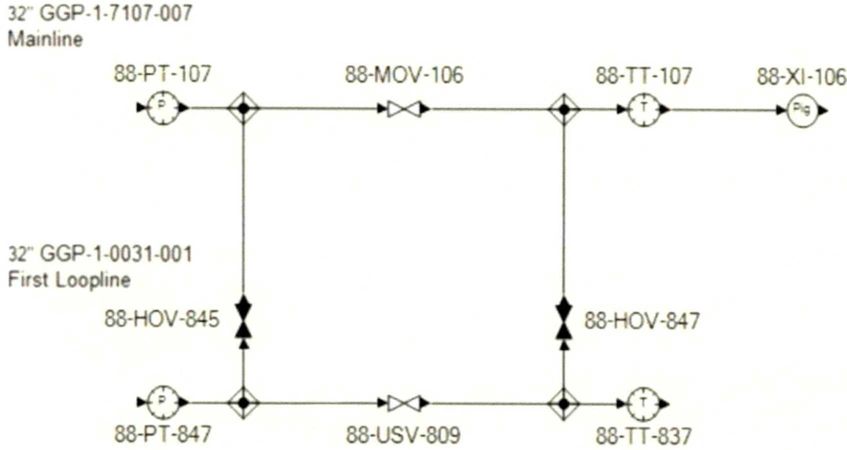

Figure 11 – A typical block valve station (BVS) with a mainline and a loopline. Here both cross-over valves are closed. There may also be inlet and outlet valves on each of the two pipelines.

Modelling stations is a specialist endeavor. It involves fast transients and advanced control logic, each an entire field of expertise in its own right, with its own dedicated professionals and software. Atmos SIM can capture all kinds of detail within stations, but we should abstract enough behavior to capture just enough of a station's characteristics for our model, and no more. Only a select few types of equipment affect how a pipeline operates. Although it is useful to say if it is a *"block valve station"* (BVS), a *"compressor station"* (CS) or a *"pressure limiting station"* (PLS), for instance, each pipeline station is unique. Not every station fits into a neat pre-defined category. Inlet streams bringing in supplies and outlet streams taking off deliveries may coexist, or a bi-directional pipeline might operate in reverse flow. Moreover, conceptualizing stations in this way shouldn't lead us to think of them as merely a pipeline's entry and exit points.

They also serve as inspection points, metering points, and / or isolation points. Some act to pressurize or depressurize the fluid using banks of pumps or compressors, and regulators. Some allow us to insert, hold or retrieve solid contraptions called pigs. Pigs are tracked as they move through a line, so operators follow their progress and prepare for their arrival at receiving stations. A station may contain complex arrangements of block valves, pig launchers, holders and receivers. Stations on gas lines might have filters, or slug catchers to remove liquid. Some have tanks or pressure vessels, heaters (on liquids) or coolers (on gas lines). And in most cases, they will have some metering.

Adjacent items

Pipelines may contain lots of equipment laid out within stations. If we include thousands of real pipes in our model, when going to that level of detail doesn't impact our results, we would needlessly slow down our simulation. It has been best practice to ignore a short section of pipe, within a station say, using a *"zero-length"* connector that Atmos SIM calls a *virtual link*. As computing power is less of an issue now, there are other advantages attained if we instead have our simulator link items using short *mini-pipes*.

Junctions: ties, mixers and splitters

A junction, alternatively called a *connection point* or a *tie-in point*, can link a number of adjacent pipes or items to one another, acting as a *mixing point* or as a *splitting point*.

When a junction connects only two pipes, we want it to act just like a knot in a pipe. How do we generalize it if it connects more than two pipes? There should be a single pressure at the junction. All inflows and outflows should total zero.

What about temperature? As we only apply temperature boundary conditions at the upstream end of a pipe, we only set the temperature on the *outflowing* branches of a junction. By performing an energy balance, we deduce this outflow temperature (T_{out}) from the temperatures of incoming streams (T) and their respective heat capacities (c_P) and mass flow rates (Q_{mass}):

$$T_{\text{out}} = \frac{\sum_{\text{in}} Q_{\text{mass}} \cdot c_p \cdot T}{\sum_{\text{out}} Q_{\text{mass}} \cdot c_p}$$

Situations can arise where we don't know in advance which pipes are flowing in and which are flowing out. Having a numerical simulator deduce which of several situations is taking place and handle it gracefully is an involved affair.

Mixing rules can also in themselves be quite involved, and for many fluids are empirically derived. For example, special rules deduce the viscosity of a mixture of two liquids defined by bulk parameters rather than composition. Taking a simple average weighted by flow rates would yield a poor estimate.

Point resistances

If a fluid must pass through a component such as a sharp bend, its resistance might significantly affect the fluid's flow rate. The component is said to impart a pressure drop across it, much like a length of pipe does. But the component has no significant length and thus almost no inventory hold-up, so it can't be abstracted as a pipe. We instead represent it as a *point resistance*. Equipment not explicitly modelled can be abstracted in this way. Filters, for example, are a resistance that varies over time as they become accumulated with material.

If we may pause for a gentle digression on two concepts:

- *discharge coefficient* (C_D) is the ratio of real flow rate to ideal flow rate
- *flow coefficient* (C_V or K_V or A_V) relates a pressure drop to a flow rate

Those unfamiliar with these should bear in mind they are not synonyms for the same concept! Moreover, though many in the industry use it as such, C_V (often written *"CV"*) is merely the algebraic symbol of flow coefficient – it is not a unit of measurement!

Figure 12 – An elbow in a pipe incurs losses that can be modelled as a point resistance

A flow coefficient relates the volumetric flow rate of water through a component to any given pressure drop applied to it. The units used might be any of the following.

Units of flow rate Q	Units of pressure P	Further notes	Symbol in use
Gallons per minute GPM	Pounds per square inch PSI	At a reference temperature of 60°F	C_V, as used by liquid pipeliners
Standard cubic feet per minute SCFM	Pounds per square inch PSI	At a reference temperature in °R	C_V, as used by some gas pipeliners
Pounds per hour lb/hr	Pounds per square inch PSI	Via a multipliers and a density in units of $\mathrm{lb/ft^3}$	C_V, as used by other gas pipeliners
Cubic meters per hour m^3/h	bar	5 to 30°C	K_V
Cubic meters per second $\mathbf{m^3/s}$	Pascal Pa	This definition gives SI-units of area ($\mathrm{m^2}$)	A_V

Table 1 – A number of different units and symbol conventions are used for flow coefficient

The simplest of these is the final definition listed in the table, which just uses SI-units to relate the actual volumetric flow rate (Q) to fluid density (ρ) and pressure drop (ΔP):

$$A_V = Q \cdot \sqrt{\frac{\rho}{\Delta P}}$$

For a liquid, correction factors (s) relate our pipeline fluid to clean water, as listed in the Crane Handbook no. 410, [24] based on actual volumetric flow rate (Q):

$$\Delta P = s \cdot \frac{Q}{C_V}$$

For a gas pipeline, one of the typical definitions is in terms of mass flow rate (Q_{mass}) and density (ρ). The following conversion factor, for example, relates pressure drop (ΔP in imperial units of psi) to mass flow rate (Q_{mass} in $\mathrm{lb/hr}$) and density (ρ in $\mathrm{lb/ft^3}$):

$$\Delta P = \frac{Q_{\text{mass}}}{4010 \cdot C_V \cdot \rho}$$

Bends, diameter restrictions, and so forth are usually described in this way. We can also describe a short pipe in the same way, if the fluid inventory within it is negligible.

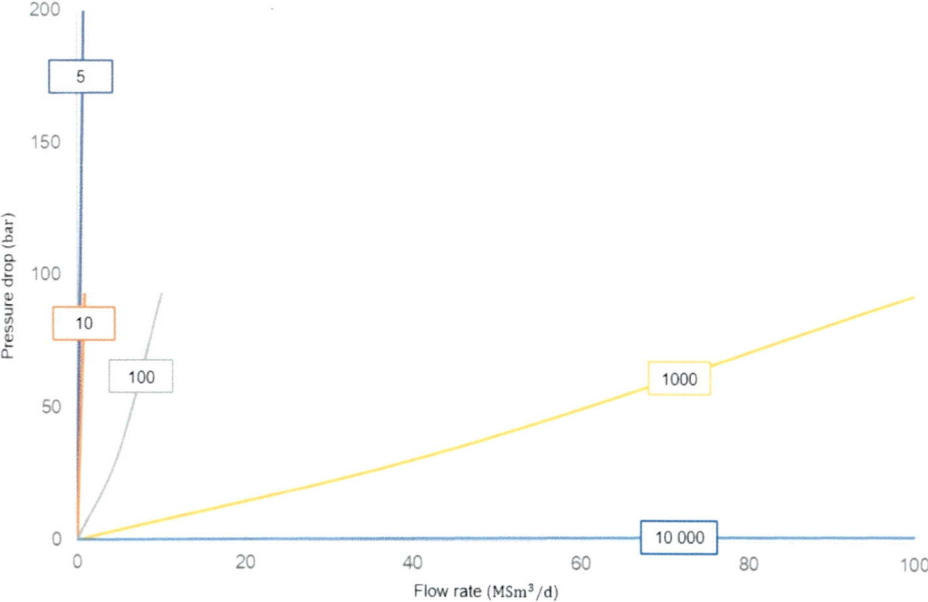

Figure 13 – A point resistance imparts a pressure drop dependent on the flow rate, so we can draw a graph plotting the curve for each value of flow coefficient (5 / 10 / 100 / 1 000 / 10 000)

Conversely, it is sometimes useful to express the resistance of an item of equipment in terms of an *equivalent length of pipe* (L). This is common engineering practice, even when the component is not actually a pipe. For an actual volumetric flow rate (Q) a length of straight pipe has a friction factor (f) defined:

$$Q = f \cdot \frac{L}{D} \cdot \frac{v^2}{2}$$

Note that the following combination is just a multiplier on the flow velocity (v) squared:

$$K = f \cdot \frac{L}{D}$$

This combination can be empirically determined for fittings, usually applying only in the turbulent flow regime. The equivalent length of an elbow might be estimated as around:

$$L = 10 \cdot D$$

And the equivalent length of a check valve might be estimated as something like:

$$L = 100 \cdot D$$

Block valves

Block valves (often BV) are like an on/off switch. They are present in most stations, allowing sections of pipeline to be shut or isolated (*sectionalizing*) or flows to be redirected (*bypass*). Block valve stations are spaced along a line, with special arrangements for emergency shut-down procedures. The pipeliner ensures each pipeline section can be isolated and safely drained (*blowdown*) quickly in case of a leak or malfunction. This is also essential for maintenance, repairs and non-routine operations such as the recovery of stuck pigs. In a gas pipeline, valves are typically spaced more closely than is needed in a liquid pipeline. [51] Pressure transmitters should be placed on both sides of a valve. One reason is so that if the valves are shut and the pipeline is isolated into sections, we have a pair of indicated pressures located at both ends of each section, allowing us to detect when a leak occurs and deduce its location.

It should be noted that *"a valve is not enough"* to be sure that a section is truly isolated, but a simulator can only believe what is told. In the field, one cannot necessarily assume that a valve status is accurately representing what the valve is doing at all times, or that a valve which is apparently closed is actually providing isolation. Best practice for high-pressure applications includes concepts such as double-block-and-bleed, as well as a number of relevant safety standards that vary by country. But to abstract to the level of a *simplified pipeline diagram*, we ought to represent block valves in a row with one valve.

What happens when a valve fails to operate as instructed? Does it fail-open or fail-shut? A contributing factor to the disaster that killed thousands at Bhopal was a valve failing *open* when it ought to have failed *closed*. It's worth double checking the terminology on every project to be sure. An actuated block valve may either be a *motor operated valve* (MOV), which is always fail-last-position, or it may be a *remotely operated valve* (ROV), which may be configured to be fail-safe.

As a rule, a valve is selected to have a smaller orifice diameter than the pipeline it is installed within. An exception is if pigging is required, a topic we discuss later. To allow a pig through, a full-bore (or *"full-port"*) ball valve is selected. Whatever the valve's *"open"* diameter, as a valve is closing over time it has a smaller orifice than the pipeline. The effect of the shrinking orifice and ensuing head loss on fluid velocity is not linear throughout a valve closure. Each valve type behaves differently. A gate valve, for example, is nearly closed (only 2% to 5% open) before generating enough head loss to significantly decrease velocity. This has vital implications.

A pipeline simulator abstracts valves as the industry does, by characterizing them using a *valve map*. The model item imparts a variable point resistance, varying based on its position and the line's actual volumetric flow rate (Q). For liquids, this is equivalent to a flow coefficient (K_V) which we might express as:

$$Q = K_V \cdot \sqrt{\Delta P / \rho}$$

And for gases, it is equivalent to flow coefficient (K_V) defined a little differently:

$$Q = K_V \cdot \sqrt{P_{\text{up}}^2 - P_{\text{dn}}^2 \div (\rho \cdot T_{\text{up}})}$$

These are a good approximation. The real equations used in a simulator and in the Crane Handbook also include corrections for non-ideal-gas behavior and other effects. Each valve conserves flow through it, whilst enforcing that flow coefficient at its position by introducing pressure drop. Across a fully open valve, we want virtually no pressure drop, so we may assign a very high flow coefficient on the order of $C_V = 100\,000$. An open block valve is captured as offering a very small resistance rather than zero resistance in this way as the latter approach can get the simulator into trouble in situations where, say, there are two open block valves in parallel. Besides, a real valve always imparts some pressure drop; manufacturers boast about their valves offering the highest flow coefficient in the industry, but none claim a valve offering infinity.

A closed valve enforces zero flow as a constraint on both sides, essentially breaking up the pipeline at each closed valve. In a simulation, this can leave a region between two closed valves without any equations for pressure or temperature. A transient step can maintain previous-step pressure and temperature, but what do we do for an initial steady state? Either we ask the user about each of these regions, or we fall back on some reasonable assumption. We could assume, say, that the pressure is equal to whatever it is at the other side of the valve.

Check valves

A *check valve*, also called a *non-return valve* (NRV) or sometimes a *one-way valve*, allows flow through only in one direction. For this purpose, a diaphragm valve might be selected, so that it opens on reaching a certain pressure drop. It must be suitably sized for its application: knowing the line size is not enough. An oversized check valve delivers full flow at too little pressure drop. In operation, it would repeatedly open and close, a cycling phenomenon known as chattering, fluttering or hunting. We end up with poor control and early actuator damage. An undersized check valve delivers full flow rate at too much pressure drop, and would leave us with insufficient flow rate during operation.

Inside the simulator's solver, a check valve works exactly like a block valve. What is special about them is that outside the time-step over which we perform a calculation, a check valve must open and close appropriately. We want it to go to a closed position on reverse flow, ideally on some non-zero threshold of reverse flow. And we want it to move to an open position on upstream pressure exceeding downstream pressure. If a check valve moves during the course of a time-step, we must cut that step short to when the valve moves, else we can expect to see a small spurious surge that might become a nuisance.

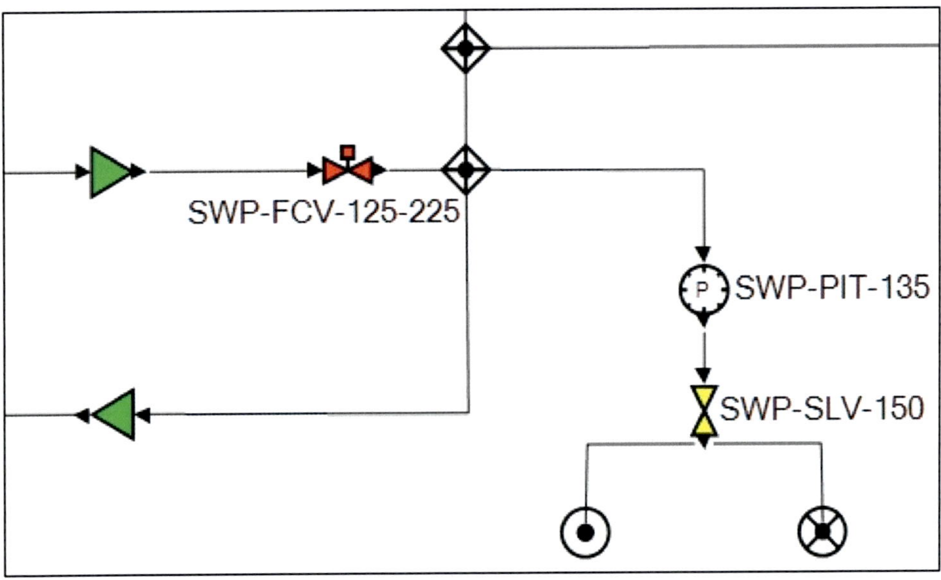

Figure 14 – Block valves, check valves and regulators in an Atmos SIM station.

Relief valves

A relief valve follows a curve, opening only if pressure exceeds a set threshold. Thereafter there are two possible approaches available to model its behavior.

A timed-style approach moves the valve from its original fully closed position to a fully open position over a user-configured duration, with this movement beginning when the relief pressure is reached.

An over-pressure-style approach works as follows. The more pressure rises above the relief threshold, the more the relief valve opens, until at some higher pressure threshold, the relief valve is at a fully open position.

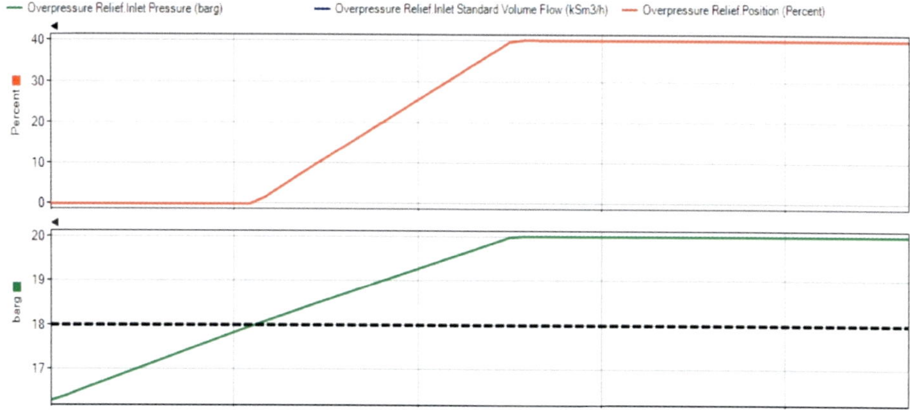

Figure 15 – A trend over time of Atmos SIM simulating an over-pressure style relief valve.

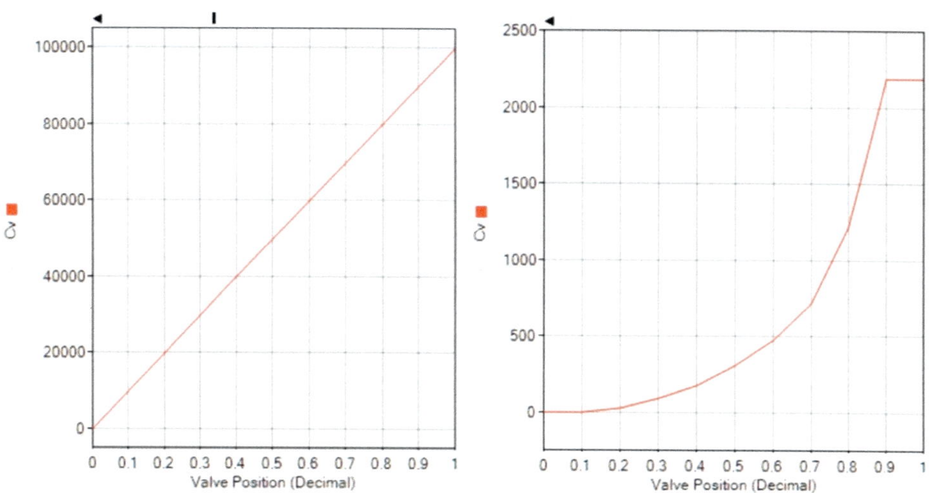

Figure 16 – Valve maps plot flow coefficient (C_V) against valve position for an ideally linear valve (left) and for a regulating control valve (right).

Regulators

A *regulator, throttling valve* or *control valve* (CV) is not intended for isolation, and should not be used as such. Above and below certain position thresholds (for example 95% and 5%), the status of a regulator can be considered fully open or fully closed – the manufacturer often recommends a *soft minimum* to prevent it from closing, because exceeding that position leaves it prone to fluttering. Unlike block valves whose status can only ever be *"open"*, *"closed"* or *"transitioning"*, a regulator's position is adjustable. By adjusting its position fraction, a regulator enacts the control philosophy by *regulating* the flow rate, inlet pressure or outlet pressure. The setpoint it controls is a *boundary condition* (BC) for our set of model equations.

Regulators are automatically controlled. This can either be achieved using a simple mechanical device, like the pressure regulator used for propane on a barbeque grill, or, in the case of modern pipelines, they can be controlled by a small computer that implements sophisticated control strategies designed to meet setpoints with fast response times with minimal overshoot and oscillation.

A regulator is a versatile model object. In reality it is a class of physical devices, deployed in many versatile ways. A *back-pressure regulator* may be placed at the inlet of a delivery point. On a liquid line, a control valve might be installed at the outlet of a single-speed pump to govern outlet pressure or flow rate. On a gas line, one of these devices may be tasked with adjusting how much flow is recycled through a compressor, while another controls mainline throughput.

In Atmos SIM we configure a regulator in an offline scenario by directly instructing it to constrain flow or pressure (ideal control) or by implementing logical control blocks, which we shall discuss later. Regulators controlled by a logical scheme could – rather than adjusting their position based on the immediate conditions directly at their inlet or outlet – instead refer to pressure at any (sensible) metered location. It is up to the user to configure a regulator so that its stated goal is actually reasonable and achievable!

In an online scenario, a pipeline simulator deduces the position of a regulator from a pair of pressure meters located either side of it.

Idealized control: constraints in offline scenarios

When we add a regulator valve to a model, and run an offline scenario, Atmos SIM assumes it can perfectly and immediately hit its setpoint – a regulator's setpoint being its user-specified pressure, flow rate or position. To attempt a solution, Atmos SIM tries to enforce that setpoint as a hard *constraint*. If we've specified a pressure as that regulator's setpoint, it then back-calculates what fraction open the valve should have for the regulator to hit that pressure.

This approach is an idealization of how equipment is truly controlled. A real regulator behaves in a slightly different fashion, gradually responding to changes in setpoint according to some control scheme with complex circuitry. That circuitry is embedded in what is referred to as *control logic*.

A regulator probably has – at the least – a PID controller * that strives to hit the setpoint gradually. That gradual control philosophy means some period of time will pass until the setpoint is actually achieved; the regulator is heading for that same position deduced by the idealized control, but it takes a while to get there. Moreover, a regulator governed by a real control scheme can overshoot. It can oscillate. It can even go unstable, if its controller is poorly configured. A user who wishes to model all of that in Atmos SIM can do so.

But for most applications, the user doesn't need that level of detail. So it is a convenient feature that we can just throw in a regulator, for example, specify its setpoints without worrying about the specifics of the real logical scheme, and have Atmos SIM assume that the overall system is set up so that it works perfectly. That is the spirit in which idealized constraints enforce user-specified setpoints on equipment used to operate a pipeline. It makes offline scenarios an easy, quick and powerful way to simulate many useful pipelining operations.

---

* Standing for *"proportional – integral – derivative"*. We revisit its logic later.

Pumping and compression

Pumping or compression stations impart mechanical energy to the fluid to sustain its flow. This sophisticated equipment intimately affects a pipeline's operation, so we shall dedicate a detailed chapter to discuss how to model pumps and compressors, as well as how they work in series and parallel, how their performance is intertwined with the wider pipeline system, and the control strategies employed for their operation.

If a pump or compressor controls on a flow or pressure setpoint then we include that boundary condition as a model equation, just as we do with the setpoint of a regulator. The performance calculations describing their power draw and outlet temperature are a surprisingly involved affair, even in steady state. A fixed-speed pump or compressor may control on speed or power, for example. The appropriate curve is taken from the map and applied as a boundary condition within our model equations. As solvers do not lend themselves well to handling curves with irregularities or bends, we use a polynomial cubic fit of the user-specified pump or compressor map.

That performance map is dealt with quite carefully, to avoid a few pitfalls that might offend a solver. One of these is that the fan-law approximation of a map becomes singular as the pump or compressor's operating speed approaches zero. In that case we must use another setpoint until it hits its minimum speed of flow, head or otherwise. This kind of question makes simulating the start-up and shut-down of pumps and compressors particularly tricky, and we visit that in our chapter on transient scenarios.

Figure 17 – An electrical motor driving a centrifugal pump at the start of a 28-inch oil pipeline

Heaters and coolers

Heat exchangers might be installed to meet the needs of the pipeline. These are categorized as *heaters*, typically seen on liquid lines, or *coolers*, typically seen on gas pipelines. Heat exchangers are analyzed via well documented methods: *log mean temperature difference* (LMTD), or if enough information is not available to do so, via the *"effectiveness"* or *number of thermal units* (NTU).

Gas compression necessarily heats the gas quite a bit, so it often requires substantial cooling to be provided. Hot gas can potentially damage equipment, is more expensive to compress, and incurs higher frictional losses in the line. Surprisingly, it is often more economical to compress gas through a series of compressors if it is cooled between stages with *"inter-coolers"*, despite those inter-coolers themselves drawing significant power. On some pipelines, there are shell-and-tube liquid-cooled heat exchangers, but more common are air coolers. Those may be simple pipes with fans blowing air across them, the fans spinning as fast as needed to maintain a set desired temperature until the limiting duty (power) is reached. In Atmos SIM models, we abstract this as a minimum temperature constraint in tandem with a maximum power constraint.

As a gas progresses along a pipeline, the head drops, so the pressure drops and thus the density drops. This means that along a pipeline, a flowing gas *speeds up* in its velocity (a consequence of conservation of mass) and cools down (a consequence of energy pulled out of intermolecular attractive forces). We can avoid letting the temperature become so cold that equipment freezes by being careful with our operations. A gas pipeline won't freeze in states that are reasonably steady, but if we blow a gas pipeline down, that blowdown operation might freeze it unless we are careful to perform it gradually enough.

On a viscous liquid line, a heater might be needed because at lower temperature, the liquid becomes too viscous and head losses become impractical. Some refer to this informally as *"improving flowability"*. Heaters are modelled in the same way as coolers within a pipeline simulator. In any pipeline system, the rate of heat loss is proportional to the difference in temperature between a fluid and its immediate environment. We want to warm it up to the requisite temperature, but no more than that. Doing all the heating at the discharge of a station would be inefficient, as it would oblige us to heat a liquid much more than necessary to accommodate these thermal losses. So instead, many crude oil pipelines in cold climates use trace heating, or include that provision in plans for how to operate at lower flow rates. By installing trace heating – essentially, running electric current through electrical resistors – all along the outside of a pipeline until the next station, less heat is lost to the environment, and so less power is wasted.

Storage tanks and pressure vessels

Storage facilities for liquids are tanks, arranged in *tank farms*. Atmospheric liquid tanks hold strategic supplies, or inventory as an intermediate buffer. An accumulator terminal might decouple a multi-product pipeline from variations in upstream supply and downstream demand. This lets us run the pipeline to optimize product batching, and to minimize the amount of contaminated fluid, called *interface*, *inter-mixture* or *transmix*, created at batch interfaces. A tank farm is also where a typical batch is headed, as a depot for onward distribution.

These tanks might be gargantuan, holding quantities on the order of 20 to 80 thousand tons apiece. The tank farm might consist of dozens of these tanks, storing millions of barrels. Storage of this kind warrants specialized modelling and engineering that lies well beyond our scope. Nonetheless it is useful to make a basic model available to our simulator, as we can readily relate level and pressure, provided we accept this is only a reasonable approximation. Tanks might also impose some limits on the rate at which they may be filled, to avoid a situation in which a fast-flowing hydrocarbon splashes so much that it generates sparks.

The level of a volume of liquid within a tank is modelled as a straightforward identity relation, which is a fair approximation. The true inventory ought to be monitored and controlled by specialists, using dedicated software involving strapping tables and all the bells and whistles. Many safety implications come out of that specialist software: consideration is given to venting and proper operating procedures, to avoid inadvertently creating a vacuum by filling up too quickly.

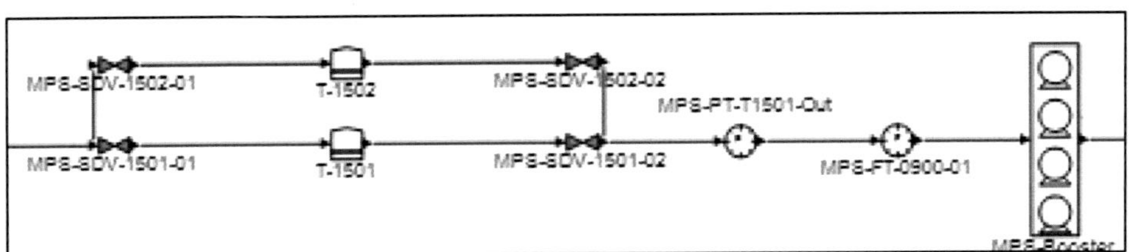

Figure 18 – A tank farm and a pump bank with valves and metering inside part of a station along a crude oil pipeline, as displayed on the Atmos SIM GUI.

Not all liquids are stored in tanks. For example, the United States strategic oil reserve is partly held in the caverns of what was formerly an underground salt mine. Such facilities hold millions of barrels of liquids in long-term storage.

Figure 19 – A cylindrical tank for storing liquids and pressurized spheres for storing gas

Gas is stored in spherical pressure vessels which some refer to as *spheres*. A gas pipeline's scope, which some call a project's battery limits, is unlikely to include storage facilities; all pipeliners need is the incoming pressure. Gas is difficult to store in large amounts, but not impossible. Long-term gas storage is already being done in underground mines, caverns or salt domes. Natural gas can instead be refrigerated and thereby liquefied. Liquefied natural gas (LNG) is then stored in special liquid tanks. The purpose of this inventory can include strategic storage to accommodate variations in demand across entire seasons. Depending on the geography of consumers, this is relied upon to support domestic heating and electric power generation. Gas storage can be viewed in a pipeline simulation model as being equivalent to a pipe that is blocked at one end and which has an incredibly large diameter. In reality, there are characteristics to salt caverns, depleted wells, and so forth which warrant a far more detailed discussion; we won't be covering those here.

A compressible fluid such as liquified petroleum gas (LPG) has a liquid level, in equilibrium with its vapor. Unlike an atmospheric tank, the headspace is at the vapor pressure, so LPG is stored in spheres or in cylindrical *bullets*. Interestingly, none of these fluid storage facilities typically maintain anywhere near the same operating pressures seen within a typical transmission pipeline. In the case of gases, this invites us to appreciate the utility of *linepack* used by operators of high-pressure gas pipelines, a concept which is instrumental in their daily operation and which we shall explore later.

A liquid tank or gas storage facility is described by an ordinary differential equation (ODE) in time. This is because flow rate into or out of a storage facility depends on the pressure at the bottom of the vessel, but that pressure in turn depends on the amount of fluid in the vessel, which varies as the time integral of the flow rate. It is therefore solved as an integral part of the set of equations being computed by the solver.

Metering

Metering of pressures, flow rates and temperatures might be marked with different terminology. As they communicate with the wider world, they are properly referred to as *"transmitters"* – we mark pressure transmitters as PT, temperature transmitters as TT, and flow rate transmitters as FT. There is also a host of other instrumentation, but these three parameters get us surprisingly far, so they are where our primary focus lies. We look at how to interpret these devices in our chapter about simulators deployed in real-time, *Live on Site*.

Arrangements of valves and metering can be complex, not only for redundancy, calibration and maintenance, but also for all kinds of historic and circumstantial reasons peculiar to every unique pipeline. Missing, unknown or bad metering can pose a significant challenge for a project team commissioning an online pipeline simulator.

Dealing with meter issues is a large part of any online model installation. It can constitute the lion's share of the workload. To a first approximation that is what the teams delivering projects in a business like Atmos actually do day-to-day!

1.3 – GREAT MODELS: WHAT TO IGNORE

> *"Not everything that can be counted counts, and not everything that counts can be counted."* — Einstein

Assumptions are not cheating

Having established what we care about, we decide on suitable simplifying assumptions. Certain details might well exist, but in the grand scheme of things many details have little significant impact on a pipeline's overall behavior. It is good practice, then, to simplify the problem at hand by making reasonable assumptions *wherever possible*. A great model omits whatever can be considered negligible, whilst still satisfying the specification we set out to meet.

Constructing a well-engineered simulation model is as much about what we leave out as it is about what we include. After all, omitting some aspect altogether eliminates the possibility of getting it wrong. Even if a quantity is correct, leaving it out of a model where it serves little to no useful purpose allows us to get on with the task at hand with minimal distraction. If we do eventually find we need to include it, then – and only then – should we add it. We will, after all, have demonstrated that it's truly indispensable!

> *"All physical laws are a mathematical abstraction of observed reality. We choose a suitable level of abstraction to represent the behaviors we are interested in. Too many assumptions means the model doesn't represent the behaviors we are interested in, but this is easily fixed by revising the model. Too few assumptions leaves us with a complex, multi-equation model with numerous adjustable parameters. If things don't work, it becomes hard to identify where the problems lie. Too many assumptions is easier to fix than too few."*

A simpler model with fewer parameters offers numerous advantages over needless complexity. Considering fewer mechanisms facilitates the configuration. Solving fewer equations speeds up a solver. Having fewer parameters makes troubleshooting easier.

Simulators can attain incredibly accurate results even with a remarkable degree of simplification and abstraction – the skill here is in making all the right assumptions, without missing any important parameters or details where they do actually matter.

Parameters entirely beyond our scope

There are many aspects of a pipeline which are important for, say, its structural integrity but which are beyond the scope of our work in modelling their hydraulics. The designer's operating philosophy imposes constraints for most and lowest allowable operating pressure, for which components are rated as a system, but we do not include each and every component in our model. This doesn't mean that they aren't important: just that they aren't within our scope.

Some pipeline design considerations are beyond the scope of our simulation model, including important structural engineering questions such as:

- How loops accommodate the expansion of pipe materials
- How hydraulic surge conditions affect structural supports
- How an earthquake would impact the pipeline

A typical pipeline simulator doesn't model plumes, combustion, brittle transition of steel, gas rupture, or countless other phenomena that may be of interest to some pipeliners. All of these are features one might conceivably want in a pipeline model, but the higher priority has been to focus on the bread and butter of pipeline design and operation.

The reason we do this is again that best practice is for a model to only describe what it must describe, and go no further, eliminating complexity where we can.

Assumptions for pipeline hydraulics

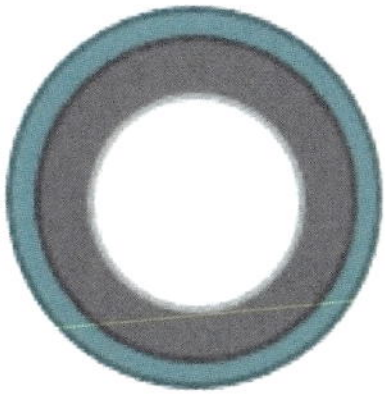

"All models are wrong, but some are useful."

– George Box

What a pipeline is and what it is not

Having assured ourselves that sound abstractions are a crucial element of any good pipeline simulation model, we choose what to ignore with as much care and attention as when we choose what to consider.

Compared to, say, process plant, a pipeline is largely devoid of bends and fittings, so special attention is paid to long, straight stretches of pipe. This affords the modelling engineer an opportunity to greatly simplify their model. Acknowledging that a pipeline is different to pipework, we can make many exceedingly convenient assumptions, modelling the underlying physics of our hydraulic and thermal model in a way that is tailored for pipelines.

Process pipework has different priorities to pipelines. Pipelines allow us to make many simplifying assumptions.

Assumptions are flexible

In the following subsections, we list common assumptions that are sometimes useful when modelling a pipeline. These are not set in stone, and should not be seen as universally applicable. A list of assumptions is presented here to say: *"we might find this helpful in many applications"*, rather than being a guide for what is always the right way to approach a situation. In fact, many of these assumptions are useful to start with, but Atmos SIM will relax them in favor of more advanced representations of real phenomena.

Let's say it's a continuum

There are more molecules per mole than any human intellect could fathom; it's impossible to follow each one, so we consider the average *"bulk"* properties of a fluid. We say it's a *"splodge"* of shared overall properties – ignoring the fact that fluid is composed of discrete molecules that collide with one another and solid objects, and considering only its overall average fluid properties and operating conditions. This is equivalent to considering those properties and conditions to be well-defined at infinitely small points, varying continuously from one point to another.

Let's say it's one-dimensional

The Navier-Stokes equations are so notorious that there is a million-dollar prize for making progress *towards gaining insight* explaining turbulence! [25] Thankfully the mathematical open questions of whether the equations solve concerns three–dimensional flows. Flow in our pipelines can be treated as effectively *"one-dimensional"* in almost all circumstances. This simplifies all the conservation laws. Assuming that flow is *"straight"* in the sense that bend forces acting in the pipes are negligible, and average flow direction aligns with the pipe's axis. There are indeed no sharp turns in most pipelines; the radius of a typical bend is very gradual. This doesn't mean ignoring pipe diameter, the effect of surrounding ambient conditions, or elevation – we just take the physical properties to be constant across a pipe, for a given distance along it, so the only dimension we must think about is *"along the pipe"*.

The one-dimensional assumption lets us consider properties and conditions to be constant across the pipe's cross-section. That is never actually true unless flow is stopped, but it is a reasonable approximation. When a fluid is flowing, a layer adjacent to the pipe wall always has zero velocity, famously called the no-slip boundary condition. *Average* velocity is taken across the cross-section.

Laminar flow is not at all one-dimensional: it has a parabolic velocity profile across the pipe. But in laminar flow, we can ignore the radial dimension from consideration by hiding it wrapped up within the friction factor.

This assumption holds true when flow is in a laminar regime, except locally around bends and diameter changes, effects which we model as a resistance if they become significant overall in a station. But when flow is in a turbulent regime, the one-dimensional assumption is plainly false! We get away with it nonetheless, by employing a trick: we lump all non-one-dimensional effects, such as vortices in the flow, and call those losses *friction*. Variations across the pipe, turbulent vortices – indeed all non-one-dimensional effects – are wrapped up into this empirical friction factor.

We do get our comeuppance for making this particular assumption when we come to attempt a proper thermal model and find that it doesn't quite work especially if flow is laminar, but we'll cross that bridge when we come to it.

Let's say it's single-phase

Fluid completely fills the pipe's cross-section, and is entirely in a single-phase; no column separation, liquid drop-out or any kind of multi-phase phenomena. We end the book by talking about what it takes to model beyond single phase.

Let's say friction dominates losses from fittings

A special aspect of long pipelines is that they must primarily overcome *friction*. This is a peculiarity of flow in long pipelines that means we think more about the roughness of a wall and the viscosity of a fluid. Unlike flow through *"normal"* pipework, in pipelines we can ignore losses across most fittings altogether – especially as bends, tees, etc. are gradual, and are comparatively few and far between. If there's a particularly large station, we lump in a simple resistance.

Can we say heat transfer is lumped?

The thermal properties of the fluid, pipe and surroundings are distributed in space. Could we consider them to be discrete lumps? One *"lumped"* element represents a uniform thermal capacity at some temperature, and another *"lumped"* element represents a uniform thermal resistance. This abstraction of thermal properties has a strong circuit analogy (elucidated in the appendices), but we'll see where it falls short when we discuss the thermal model in detail. In pipeline simulation it is often a poor assumption: it is never recommended online and rarely in an offline scenario. Atmos SIM has a more advanced thermal model.

Further assumptions

A pipe is assumed to be rigid, so definitely not a hose. For low-pressure gas distribution networks, the material of construction might be a plastic such as polyethylene of various grades, which challenges this assumption. Many also assume a pipeline has a constant cross-section except at changes in nominal diameter, but as we shall discuss shortly, Atmos SIM captures pipe expansion.

We also make decisions during a simulation about assumptions to enhance performance. One is to use a simplified elevation profile for our calculations. Then, we make a second pass to put the detail back in – the effect of elevation. This lets us calculate the effect of an elevation profile that changes every few meters, for instance, without saddling the model with knots every few meters. Also, it's important to remember the straight pipe sections welded together to make a pipeline are generally 15 or 20 meters long, so even if a survey shows elevation points every few meters, a pipe is unlikely to be bent at those points.

Figure 20 – Pipelines are made from sections of certain standard lengths for ease of transport

Assumptions for incompressible liquids

An incompressible approach is useful for some online simulations live on site. In this assumption, density is always constant throughout (for the same batch, at least), and frictional losses are deduced by the hydraulic profile in a manner that keeps the friction factor constant between each pair of pressure meters. To calculate flow rate, it follows a user-defined fiscal meter exactly, decoupling the effect of pressure from flow altogether. This leaves the task of enacting corrections to inline batches at the operations team's disposal. It estimates the pressure profile by assuming constant friction factor throughout each pressure-meter bounded interval, and a simplified form of the momentum equation suitable only for incompressible liquids in steady state. For most products and water pipelines, this gives a fairly accurate pressure profile almost all the time.

An exception is when significant pressure surges move through the line, but those don't last long. A model with this level of accuracy – say, *"almost always within a bar or two of reality"* – is far easier to configure than a more detailed model that correctly reproduces all the transient effects. It may be good enough for our application – if it is, there's certainly no need to spend valuable time configuring more detail.

1.4 – INSTRUCTING A PIPELINE SIMULATOR

Scenarios and libraries in an Atmos SIM project

An Atmos SIM project consists of much more than just an underlying physical model. It includes scenarios and libraries. It also records preferences for, say, which set of units to use for displaying results. And it incorporates a data manager saying which SCADA tags to look for, where to look for them, what validation to perform on them, which output tags to log, and where to do so.

A *scenario* is a set of instructions for a simulation engine to apply over the course of a run's duration. There are one or more *scenarios*, which describe how conditions like setpoints change over time as well as some of the details of the simulation to be performed. A great many scenarios are often set up and repeatedly modified within an overarching project, to capture all kinds of conditions, operations and simulation approaches. In Atmos SIM, some item properties are applied across an entire project, while others vary scenario-to-scenario. This is done in a convenient, logical and consistent way. A project includes universal data that applies to all scenarios, capturing the infrastructure of the pipeline itself. Elevation data, how pipes interconnect with station equipment, and the diameter and wall thickness of every pipe are all project-wide parameters, that transcend all scenarios.

For each type of equipment, pipe, unit, and so forth, a set of pre-configured standard objects of that type are available in a global library. This is a handy repository of reference materials and equipment to facilitate building a model in the early stages, until information is available for a custom project library. That project library may contain time-profiles for nominations, batch plans, etc.

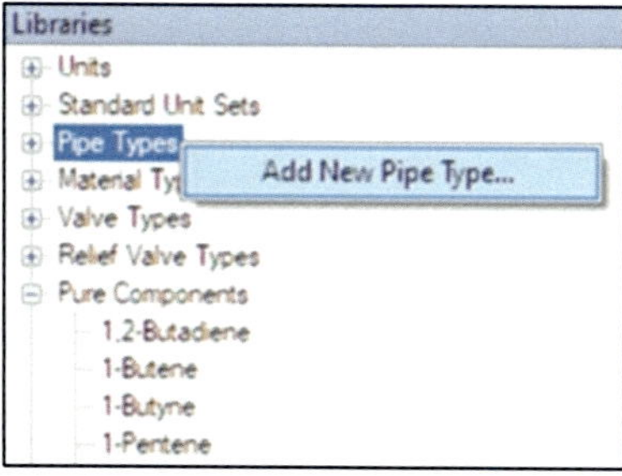

Figure 21 – Adding a new pipe type to the project type libraries within Atmos SIM.

Simulation approach and run-type

A simulation scenario can be requested to run in one of three ways:

- Finding a *steady state*, as a stable snapshot at a certain instance in time, perfectly balanced so no quantities evolve over time
- Sprinting *continuously*, advancing steps as fast as possible throughout a pre-defined duration in its entirely
- Progressing in *real-time*, advancing steps second-by-second in parallel with the true clock – useful for live monitoring, or to mimic a pipeline for training

Figure 22 – A simulation can be run to steady state, continuously, or in real-time.

Any scenario we request works by one of three principles:

- an offline model (based on constraints)
- an online transient model (based on meters)
- or an online incompressible model (based on meters)

This means a simulation in progress can be thought of as being in one of six categories.

	Offline scenario ***"on constraints"***	**Online scenario** ***"on meters"***
Steady state run	Ignores all meters, one stable snapshot	Only follows meters, one stable snapshot
Continuous run	Ignores all meters, fast for a set duration	Only follows meters, fast for a set duration
Real-time run	Ignores all meters, second by second	Only follows meters, as fast as new data is available

Table 2 – Offline scenarios and online scenarios can be simulated to solve as steady state, and progressed as fast as possible or in real-time depending on the use case.

Offline scenarios are based on exactly satisfying the most restrictive of a set of constraints on pressure, flow, etc. – constraints that are specified by the user at each relevant item (supply, demand, regulator, pump, or compressor). Engineers use this sort of offline study to perform all sorts of investigations.

Online scenarios are based on live meter data, which is used to deduce what's happening throughout the pipeline. This is how simulation is deployed on site.

a) *Real-Time Transient Model* (RTTM) scenarios estimate the most likely state, combining statistics with physics to find a consistent set of pressures, temperatures and flow rates. This is the only live approach that works for a gas pipeline, and it is needed for some liquid pipelines too.
b) *Real-Time Incompressible Model* (RTIM) * scenarios for certain liquids exactly follow a specified fiscal flow meter. It estimates pressure profiles for monitoring, but flow rates are entirely independent. Liquid pipeliners interested in batch tracking are well served by this approach.

All of the above approaches to simulation are perfectly valid. Each satisfies a different purpose. A well-built underlying pipeline model in a versatile simulator like Atmos SIM can readily accommodate any of these approaches to meet the task at hand.

Scenarios can be thought of as scenes playing out,
each using the underlying pipeline as their film set.

In Atmos SIM, all simulation approaches and run types share the same project file. This means that a project team can build and calibrate a model offline before deployment to site. It also means the internal *state* of a model deployed online, matching live metered site data, can be saved, exported, then imported to an offline scenario. This can be done manually for ad hoc studies or it can happen automatically as scheduled *look-aheads*. It is also useful to do the opposite: have a *replay*-run sprint through a historic dataset originally logged in real-time.

* I coined this term because it's useful to package an incompressible-mode batch tracker complemented with a simplified hydraulic profiler for pressures.

Offline simulation using constraints

The state of a *constrained* segment in an offline simulation is completely and equivalently determined by applying any of the following boundary conditions:

- Supply pressure and demand pressure ($P \rightarrow P$ mode)
- Supply pressure along with demand flow rate ($P \rightarrow F$ mode)
- Supply flow rate along with demand pressure ($F \rightarrow P$ mode)

The reader may be wondering why there is no flow-flow constraint here, to fill out the set. In steady state, a flow-flow controlled line has no unique solution. Countless pressure combinations would satisfy the flows. Pipelines, especially gas pipelines, operate briefly on flow-flow control if we have a history of what they were doing just before, but eventually linepack changes enough to trigger a control valve at one end switching over to a min. or max. pressure constraint. The simplest model we can build is a straight pipeline, across flat terrain, with constant diameter, from a supply at the same pressure as its demand. When a pipe goes from, say, atmospheric pressure at one end to the same pressure at the other end, across no elevation changes throughout, the fluid it contains – unsurprisingly – sits still and doesn't flow. This is a valid, interesting scenario. To instruct a pipeline to perform it, we input its constraints.

Interestingly, a stopped pipeline isn't exactly the same as an isolated one. An isolated section is said to be in shut-in if its ends are blanked with flanges or closed with block valves. But flow might also cease in a non-shut pipeline.

A constraint can be instructed to vary over time. A dynamic simulator progress from an initial state to take subsequent *transient* steps, executing a continuous run which progresses as quickly as it can. This describes pipeline operations.

Live deployment and monitoring on site

We are sharing so much intimate knowledge about pipeline simulation as this industry expertise has never been collated in such a comprehensive readable format, but most of what is written here is already freely available to whoever seeks it out. What really defines a successful simulation project is the support of a dedicated, capable and tenacious team. The value we add to our projects is our experience negotiating countless hurdles that present themselves as we engineer, commission and maintain models simulating real pipelines. Live on site, we are paid to build real-time systems that can keep the most challenging raw data processed into reliable simulation results!

2 – THE SIMULATION ENGINE

"Our brain simulates reality. Our everyday experiences are a form of dreaming... mental models, simulations; not the things they appear to be." – Morpheus, the Matrix

A pipeline simulator is special. We don't just say that because we like pipeline simulation, it is actually true in a technical sense. By focusing on the pipeline domain, it is an ultra-specialized simulator – because it is tailored for pipelining, it has a significant strategic edge over generalist process simulators. Many of the latter simplify pipes beyond what is reasonable. Some even ignore pipes altogether!

A pipeline simulator includes features relevant to a real pipeline, such as a fully transient thermal model capturing the physics of temperature. By ignoring irrelevant factors, it directs efforts to improve the product against validation datasets. By virtue of this focus, pipeline simulators are strong performers.

Unique features of Atmos SIM's state-of-the-art simulation engine include advanced implementations of an adaptive spatial mesh, moving knots for composition tracking and a maximum likelihood state estimator interpreting meters live on site. Longstanding experience, specialization and continual improvement through dozens of projects has equipped us with the confidence to implement the right *heuristics* within our proprietary algorithms. By meticulously incorporating countless tolerances and decades of experience, the Atmos SIM engine strikes a balance for accurate, swift simulations that perform in anger.

2.1 – THE CONSERVATION LAWS

Universal laws and proprietary equations

Physics does not have proprietary equations. However, simulators perform their calculations to apply these physical equations in proprietary ways. [26]

In the next few sections, we ask a few questions. Why do we often see different forms and expressions of the same universal conservation laws? How do we implement conservation laws and other model equations as computational algorithms? How are initial conditions used to cold-start a scenario and advance a simulation? And what scenarios actually benefit from automatic adaptive time-stepping and knot-spacing?

Control mass or control volume?

One can approach fluid mechanics from either of two equivalent perspectives.

Lagrange's frame of reference follows a specific clump of *material* as it moves along as a *system*. Those working with fluids say *"system"* with this particular meaning: a parcel of fluid, i.e. a *control mass*. This is a useful way to think when we perform a force balance. In contrast, for any given portion of a pipeline, Euler's *laboratory* or inertial frame of reference watches what flows through a *control volume*. Thinking the Eulerian way lets us express the fluid's thermodynamic state with *operating conditions* at a certain point in a pipeline.

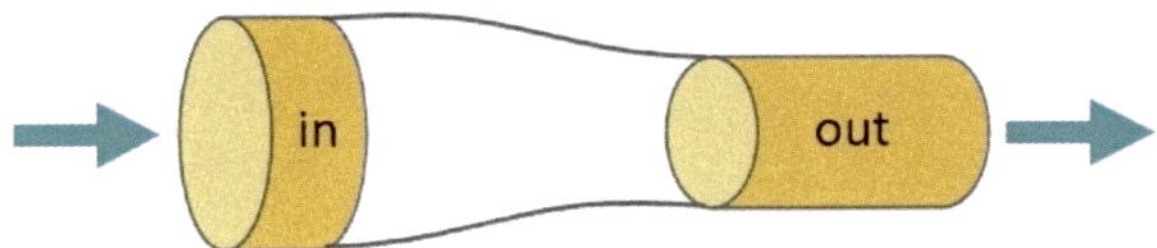

Figure 23 – Instead of tracking a fixed control mass (the Lagrangian *"system"*) of flowing fluid, we look at whatever happens to be present within a *control volume* (the Eulerian perspective).

For this reason, Atmos SIM and most other commercial pipeline simulators use the Eulerian control volume approach.

Continuity equations

An equation for every conserved quantity

The core of any dynamic physics model is a *continuity equation*, balancing a *conserved quantity* over a *control volume*. A basic continuity equation always takes the following form, intuitively describing an accumulating rate of change:

$$\text{Accumulation} = \text{Inflow} - \text{Outflow} + \text{Generation} - \text{Consumption}$$

Process engineers refer to these continuity equations as mass, energy, and momentum balances when applied respectively to those three conserved quantities. The accumulation is set to zero if we are solving for a steady state. In transient conditions, there might be accumulation (along with dissipation), as well as generation (along with consumption). A conservation law holds true for mass, energy and momentum. Applying it to mass, energy and momentum separately provides the basis for a pipeline simulator.

Mechanics, kinetics, statics, dynamics and steady states

Across all of *mechanics*, fluid mechanics included:

- *Kinetics* describe how fast something moves, as opposed to *statics*
- *Dynamics* describe how fast something changes, unlike *steady states*

These definitions are carefully phrased. A *kinetic* system could well be *steady*.

Each continuity equation is applied over a *control volume* holding the quantity being balanced. It's the rate of change of the conserved quantity within this volume (namely, the *accumulation* term) which describes our *dynamics*. The other terms are modelled by *kinetic* expressions, such as a mass balance for our flow rates and an energy balance for our heat transfer.

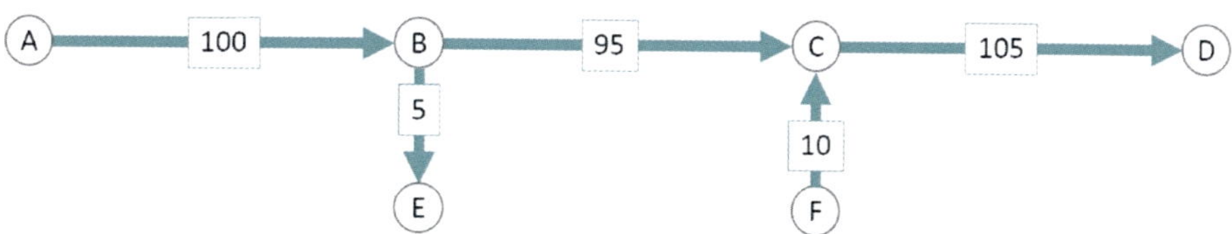

Figure 24 – A pipeline in a steady state: flows are perfectly balanced, unchanging over time.

The general conservation law in one dimension

Pipelines are long and straight. We can assume fluid properties are constant across a pipe's radius (r), and velocity (v) is directed along the pipe's axis (x). This lets us draw up a schematic of a distance element along a pipe as follows.

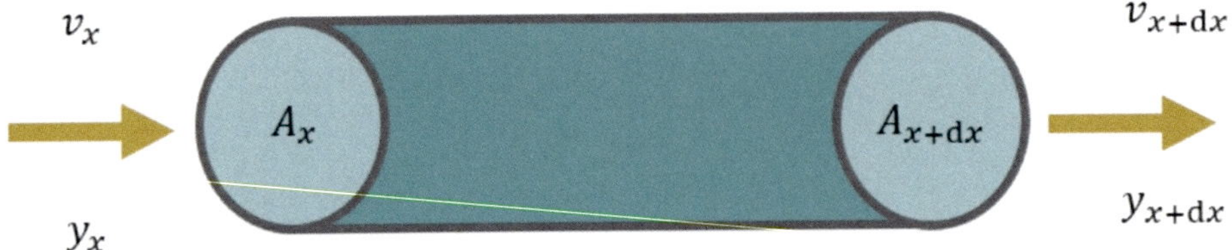

Figure 25 – A conserved quantity (y) flowing through a one-dimensional pipe element.

We place it at a point (x) and say it extends for an incremental length ($\mathrm{d}x$). We consider the density of each conserved quantity per unit volume (which we call y in this general equation, to be substituted with the appropriate quantity later). If, for the moment, we just look at the part of the conserved quantity (y) that is carried along with the fluid – and not the part of the conserved quantity that might in some manner pass through the pipe wall – we can make a general statement for all of mass-density, energy-density and momentum-density alike. The inflow and outflow can be expressed in terms of fluid velocity (v) and cross-sectional area (A) as follows:

$$\text{Inflow} = y_x \cdot v_x \cdot A_x$$

$$\text{Outflow} = y_{x+\mathrm{d}x} \cdot v_{x+\mathrm{d}x} \cdot A_{x+\mathrm{d}x}$$

Here we use subscripts to indicate the *location* (x) at which the quantity is evaluated, and we use superscripts to indicate the *time* (t) at which the quantity is evaluated – whatever quantity (y) might happen to be. So from the definition of a derivative: *

$$\text{Accumulation} = \text{Inflow} - \text{Outflow} = -\frac{\mathrm{d}(y \cdot v \cdot A)}{\mathrm{d}x} \cdot dx$$

The rate of change of the amount of conserved quantity (y) contained within our control volume ($y_x^t \cdot A_x^t \cdot \mathrm{d}x$), at any instance in time (t), is to a first order approximation:

$$\text{Rate of Change} = \frac{\text{Amount at } (t + \mathrm{d}t) - \text{Amount at } (t)}{\mathrm{d}t}$$

$$= \frac{y_x^{t+dt} \cdot A_x^{t+dt} - y_x^t \cdot A_x^t}{\mathrm{d}t} \cdot \mathrm{d}x = \frac{\mathrm{d}(y \cdot A)}{\mathrm{d}t} \cdot \mathrm{d}x$$

* We dropped the generation and consumption terms in this special case of what a fluid carries along with it as it flows, but they might make an appearance if momentum and energy are exchanged between the fluid in a pipeline and the surrounding environment.

Thus a conservation law can be written for any conserved quantity expressed as a volumetric density (y), flowing through a one-dimensional pipe element:

$$\frac{\mathrm{d}(y \cdot v \cdot A)}{\mathrm{d}x} + \frac{\mathrm{d}(y \cdot A)}{\mathrm{d}t} = 0$$

This holds true as long as there is no other way for the conserved quantity (y) to get into or out of the stretch of pipeline. This holds true for mass ($y = M/V = \rho$), unless there is a leak in that pipeline. For momentum and energy we will need additional terms.

Does a pipe's cross-sectional area change over time?

We say cross-sectional area (A) is a function of distance (x) as well as time (t):

$$A_x^t \text{ or } A(x,t)$$

This is because a pipe is constructed of solid material, like steel or plastic. As it is stressed by differential pressure, or warmed and cooled by temperature, a pipe expands and contracts. So as fluid pressure and temperature varies over space and time, so does a pipe's inner cross-section. When does this matter?

Pipe expansion can be substantial in gas and liquids pipelines alike. While pipe diameter may only vary by 1% or less, the fact that the pipe is slightly flexible has an exaggerated effect on the speed of sound (c), with several implications. Firstly, a pressure surge will be modelled to propagate at slightly different speeds to reality, meaning that if we ignore pipe expansion the arrival time of a surge will be off. And secondly, the relationship between a transient pressure surge and the associated flow rate change is linearly dependent on speed of sound, so ignoring pipe expansion will for example introduce a significant error in peak pressures calculated in surge studies.

A good simulator accounts for how a pipe's cross-sectional area changes over time, not as a matter of pedantry or an obsessive extra, but as a key feature essential for maintaining accurate results in those scenarios where it matters.

The mass equation

A mass balance on our control volume

Conservation of mass dictates that the mass of fluid ($M = \rho \cdot A \cdot \mathrm{d}x$) contained within a control volume ($A \cdot \mathrm{d}x$) changes over time (t) based on the incoming and outgoing mass flow rate (Q_M) expressed in terms of velocity (v) and density (ρ):

$$Q_M = \rho \cdot v \cdot A$$

We set our conserved quantity (y) to be mass per unit volume (M/V) – the volumetric mass density, usually just called the density (ρ):

$$\frac{\partial(\rho \cdot A)}{\partial t} + \frac{\partial(\rho \cdot v \cdot A)}{\partial x} = 0$$

As we said, the inner cross-sectional area (A) varies with operating conditions. The mass equation we stated enforces conservation of mass (M). It dictates that total mass convected in and out of the control volume must accumulate within it. Pipeliners refer to whatever fluid goes missing as *"unaccounted for fluid"*.

One point to note here is that conservation of mass doesn't make any blanket promises about any other descriptor of flow. So we can't say something like standard volumetric flow rate (Q_{std}) is universally conserved in all scenarios. If incoming fluid composition differs to outgoing fluid composition, *shrinkage* may cause the accumulating standard volume not to balance inflow minus outflow.

A species balance on our control volume

Another point to note is that a sort of corollary of the mass equation applies conservation of mass not to total mass but to a certain species, say methane specifically rather than the entire mixture making up a flowing natural gas. This is known to process engineers as a species balance, and along with the mass, energy and momentum balances, is treated as another balance equation.

A species balance is indeed not only valid but indispensable when we perform manual calculations. However, if we attempt to deploy it within a pipeline simulator as a method of *composition tracking*, it turns out to be an inadequate approach, introducing unacceptable levels of numerical dispersion.

The momentum equation

A general form of the momentum equation

Now let's take our conserved quantity (y) to be the density of momentum ($\rho \cdot v$) on a volumetric basis. The conservation equation for this quantity is considerably more complex than the equation we derived for mass. Let us first present the momentum equation then delve into the reasoning behind each term in the following few sections:

$$\frac{\partial(\rho \cdot v \cdot A)}{\partial t} + \frac{\partial(\rho \cdot v^2 \cdot A)}{\partial x} + \frac{\partial(A \cdot P)}{\partial x} + A \cdot \rho \cdot g \cdot \sin[\theta] + A \cdot \frac{\rho \cdot f \cdot v \cdot |v|}{2 \cdot D} = 0$$

In addition to quantities used in the mass equation, we now see the pressure (P), gravitational acceleration (g), the angle of inclination (θ) of our stretch of pipe capturing the effect of any change in elevation, the friction factor (f) and the inner diameter (D).

We can simplify this a bit. We use the product rule to make the first two terms:

$$\frac{\partial(\rho \cdot v \cdot A)}{\partial t} + \frac{\partial(\rho \cdot v^2 \cdot A)}{\partial x} = v \cdot \frac{\partial(\rho \cdot A)}{\partial t} + \rho \cdot A \cdot \frac{\partial v}{\partial t} + v \cdot \frac{\partial(\rho \cdot v \cdot A)}{\partial x} + \rho \cdot v \cdot A \cdot \frac{\partial v}{\partial x}$$

Because of the mass equation:

$$\frac{\partial(\rho \cdot A)}{\partial t} + \frac{\partial(\rho \cdot v \cdot A)}{\partial x} = 0$$

We can say that the first and third term of the aforementioned equation cancel:

$$v \cdot \frac{\partial(\rho \cdot A)}{\partial t} + v \cdot \frac{\partial(\rho \cdot v \cdot A)}{\partial x} = 0$$

This allows us to state the first two terms of the momentum equation as follows:

$$\frac{\partial(\rho \cdot v \cdot A)}{\partial t} + \frac{\partial(\rho \cdot v^2 \cdot A)}{\partial x} = \rho \cdot A \cdot \frac{\partial v}{\partial t} + \rho \cdot v \cdot A \cdot \frac{\partial v}{\partial x}$$

By substituting the mass equation and dividing through by the area (A), we get:

$$\rho \cdot \frac{\partial v}{\partial t} + \rho \cdot v \cdot \frac{\partial v}{\partial x} + \frac{1}{A} \cdot \frac{\partial(A \cdot P)}{\partial x} + \rho \cdot g \cdot \sin[\theta] + \frac{\rho \cdot f \cdot v \cdot |v|}{2 \cdot D} = 0$$

Though the above analysis is instructive, in a real pipeline simulator we might choose our parameters more carefully to be mass flux ($\rho \cdot v$) and mass density (ρ).

Forces are flows of momentum

The first term in the momentum equation looks a lot like the *"mass times acceleration"* part of Newton's Second Law ($F = m \cdot a$), and indeed the last three terms do represent forces due to the pressure gradient, gravity, and friction – respectively. Since we wrote them on the same side of the equality as the mass-times-acceleration term, they are actually the negative of the forces. Like mass, momentum flows with a fluid. And like mass, momentum is convected along with the fluid; but these forces provide additional momentum flows into or out of the fluid element.

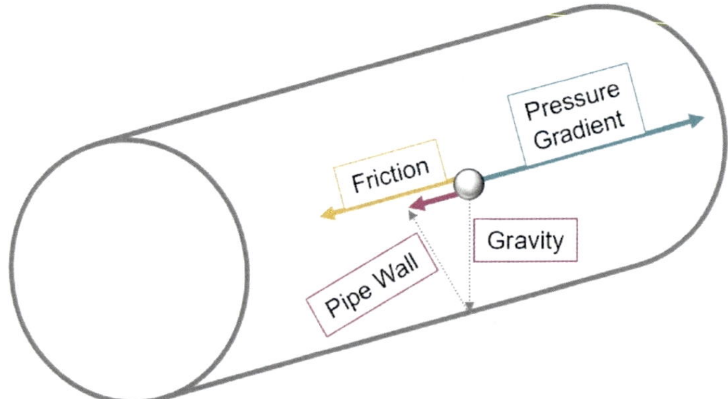

Figure 26 – Forces acting on a fluid element due to pressure gradient, gravity and friction. In the case of steady incompressible flow, the net force (magenta arrow) would be zero.

The force of pressure

At each end of a fluid element (upstream and downstream), there arises a force due to pressure, amounting to pressure (P) times the inner cross-sectional area (A):

$$F_{\text{up}} = P_x \cdot A_x \text{ and } F_{\text{down}} = -P_{x+\text{d}x} \cdot A_{x+\text{d}x}$$

Therefore across the section of pipe, the net force is:

$$\sum F = F_{\text{up}} + F_{\text{down}} = \frac{\partial(A \cdot P)}{\partial x} \cdot \text{d}x$$

Our derivation of the conservation laws has been in terms of quantity-per-unit-volume, so we also need to convert this to a force-per-unit-volume. To do this we just divide through by our infinitesimal control volume ($A \cdot \text{d}x$), and then this becomes:

$$\frac{\sum F}{\text{d}V} = \frac{F_{\text{up}} + F_{\text{down}}}{A \cdot \text{d}x} = \left(\frac{1}{A}\right) \cdot \frac{\partial(A \cdot P)}{\partial x}$$

The force of gravity

Our elevation term is time dependent, not because we want to simulate earthquakes, but because density evolves. The force of gravity acts on mass – a certain mass ($\mathrm{d}M$) of fluid is held within the volume of an increment of pipe and is dependent on density:

$$\mathrm{d}M = \rho \cdot A \cdot \mathrm{d}x$$

Gravity exerts a vector force acting downwards, but in a single-phase flow, the part of the gravitational force that is perpendicular to the pipe wall is opposed by the force of the wall pushing on the pipe. So we are only concerned with the scalar component of gravitational acceleration acting in parallel to the pipe's axis. Thus, a horizontal pipe across flat ground isn't affected by the gravity term, and an inclined pipe is affected as a function of its angle of inclination (θ) above the horizontal (i.e. the elevation change).

$$F_{\text{grav}} = \mathrm{d}M \cdot g \cdot \sin[\theta]$$

$$= \rho \cdot A \cdot \mathrm{d}x \cdot g \cdot \sin[\theta]$$

Converting this to a force per unit volume:

$$\frac{F_{\text{grav}}}{\mathrm{d}V} = \rho \cdot g \cdot \sin[\theta]$$

The force of friction

All non-one-dimensional effects such as turbulence are wrapped into a frictional force, acting to oppose motion. Darcy friction factor (f) is defined as *"whatever it must be for the following equation to describe the force of friction"*:

$$\frac{F_{\text{friction}}}{\mathrm{d}V} = -\frac{\rho \cdot f \cdot v \cdot |v|}{2 \cdot D}$$

This is proportional to the negative of flow velocity squared ($- v \cdot |v|$) and thus always acts to oppose motion, always having the opposite sign to the direction of flow velocity. This particular functional form was chosen because it was observed that friction factor does not vary too much with changing flow rate and pressure.

A simpler momentum equation for steady liquid flows

A less compressible fluid has a higher speed of sound, so after an operational change hydraulic transients will settle at a new equilibrium quicker than in a compressible fluid. This means that a pipeline carrying an almost-incompressible liquid will often operate jumping from steady state to steady state. For many online real-time applications, we get away with relying on a simpler physical model which completely discards all gradual time variation, instead assuming everything jumps immediately to a new equilibrium whenever conditions change in the pipeline. We take the flow velocity (v) to be constant over space and time, so the momentum equation simplifies down to the following, where D is inner diameter and angle of inclination (θ) tells us the change in elevation:

$$\frac{\partial P}{\partial x} + \rho \cdot g \cdot \sin[\theta] + \frac{\rho \cdot f \cdot v \cdot |v|}{2 \cdot D} = 0$$

Density (ρ) is constant throughout except at batch interfaces, and flow is always steady. Standard flow rate (Q_{std}) is thus constant, as are velocity (v) and friction factor (f). In an online model, we usually have pressure meters available to measure the inlet and outlet pressures (P), and a flow meter lets us deduce the flow velocity (v). We know what batches are present and the elevation profile. Using those boundary conditions, we solve for friction factor (f), which we assume to be constant along the line. Once we have calculated the friction factor, we can compute the full pressure profile $P(x)$.

Having thrown out the time-derivative term, this approach effectively solves each successive model step in a hydraulic steady state. This is known as a *"succession of steady states"* (SSS). In addition, it takes advantage of knowing the pressure at both the upstream and downstream ends, coupled with the other approximations described, to avoid having to solve any partial differential equations (PDE's) – or even simpler ordinary differential equations (ODE's) – at all. This makes it extremely fast and robust.

A succession of steady states can be used as a standalone *"simple hydraulic profiler"*, without having to build and simulate a model for transients or compressibility. Despite being mathematically so simple that it can be set up on a spreadsheet, it is surprisingly accurate. For a steady flow of liquid we can expect accuracies within 10 psi ($< 1\text{ bar}$) everywhere, provided thermal effects are insignificant – which excludes some pipelines such as crude lines in cold climates – and provided hydraulic surges can be ignored.

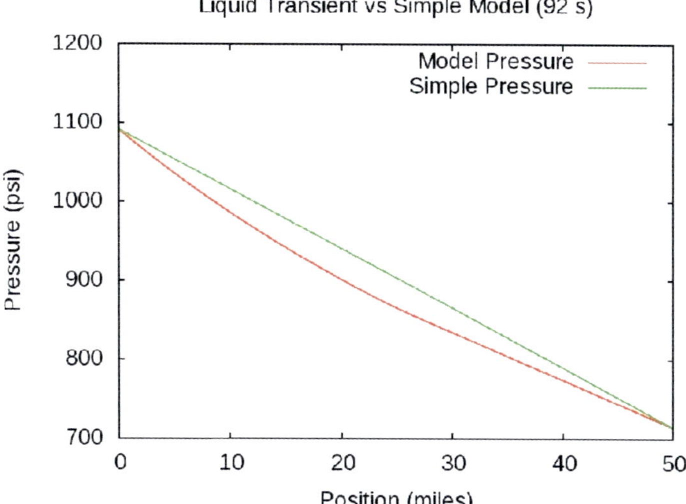

Figure 27 – Two hydraulic profiles, compared following a rapid rise in inlet pressure. The *"simple hydraulic profiler"* doesn't capture transient effects, going straight to a steady state. For the duration of this transient, the true hydraulic profile is seen only in a fully transient model.

During many routine operations on pipelines carrying liquids with low compressibility, those assumptions generally hold true and so we can use an incompressible model. Of course, we must remain cognizant of the limitations of this approach. In formulating a simpler equation, we assumed a *steady state*. If transients occur – and surges do in reality – the results obtained by a succession of steady states will be far off the mark. For those situations, we need a hydraulic model fully capturing the transient physics.

The energy equation

Forms of the energy equation

Several textbook authors ignore the energy equation and neglect heat transfer. They implicitly assume temperature to be constant – but in pipelines it usually is not. Change in temperatures can cause significant variations in the quantities of interest in pipeline simulation, such as flow rate and linepack. [21] The energy equation can be derived by setting out conserved quantity (y) to be the energy density (e):

$$\frac{\partial(e \cdot v \cdot A)}{\partial x} + \frac{\partial(e \cdot A)}{\partial t} + \text{other energy flows} = 0$$

So far we mostly worked in either total quantities or else quantities defined per-unit-volume. We can also talk about quantities-per-unit-mass, called *"specific quantities"*. The specific energy of a fluid has three pieces: the specific internal energy (u), specific kinetic energy ($½ \cdot v^2$), and specific gravitational potential energy ($g \cdot z$). To convert a specific quantity to a per-volume quantity we multiply by mass density (ρ). The energy per unit volume is a conserved quantity we can plug into our general conservation law:

$$e = \rho \cdot u + ½ \cdot \rho \cdot v^2 + \rho \cdot g \cdot z$$

This gives us a form of the energy equation that is pleasingly intuitive, which we again show here first in its complete form before justifying each of its terms:

$$\frac{1}{A} \cdot \left(\frac{\partial(e \cdot A)}{\partial t} + \frac{\partial(e \cdot A \cdot v)}{\partial x} + \frac{\partial(P \cdot A \cdot v)}{\partial x} + P \cdot \frac{\partial A}{\partial t} \right) + 4 \cdot U \cdot \frac{T - T_{\text{ground}}}{D} = 0$$

The final term, representing heat flow into or out of the pipe through the wall, is written in terms of an *overall heat transfer coefficient* (U), abbreviated *OHTC*, acting on the *outer diameter* (D) of our pipe element. Here T is fluid temperature and T_{ground} is the *"ground"* temperature, which we define in the case of a buried pipe as the temperature of the soil at the burial depth of the pipe and at a far enough horizontal distance away from the pipe that it isn't affected by the pipe. For unburied subsea pipes, this would instead be the ambient temperature of the seawater.

We again cancel out the part proportional to the mass equation, as we did with the momentum balance:

$$\rho \cdot \frac{\partial(u + ½ \cdot v^2 + g \cdot z)}{\partial t} + \rho \cdot v \cdot \frac{\partial(u + ½ \cdot v^2 + g \cdot z)}{\partial x} + \frac{1}{A} \cdot \frac{\partial(P \cdot A \cdot v)}{\partial x} + \frac{1}{A} \cdot P \cdot \frac{\partial A}{\partial t}$$

$$+ 4 \cdot U \cdot \frac{T - T_{\text{ground}}}{D} = 0$$

Further simplification is possible, but we stop here. A solver can happily deal with this form. We want to avoid writing an equation with large terms that are very similar in magnitude but opposite in sign, so that they sum to a much smaller value. If we end up with a form of the equation that looks simple on paper, but where the first term is typically a million ($1\,000\,000$) and the second is typically negative a million and one ($-1\,000\,001$), then we introduce potential numerical issues. Removing, say, terms proportional to the momentum equation is possible and simplifies the energy balance on paper, but it would give rise to this exact problem, so we resist that temptation.

What the terms mean

The energy density (e) that is conserved includes internal energy (u) capturing thermal and intermolecular potentials, plus the kinetic energy and a gravitational potential term:

$$\underbrace{e}_{\text{Energy density}} = \underbrace{\rho \cdot u}_{\text{Thermal and intermolecular}} + \underbrace{\tfrac{1}{2} \cdot \rho \cdot v^2}_{\text{Kinetic}} + \underbrace{\rho \cdot g \cdot z}_{\text{Gravitational}}$$

We saw on the previous page that further terms represent energy flows other than convection: *work done* and *heat transfer*. Adjacent fluid does work by compression or expansion, the pipe does compression and expansion work as it contracts and expands with changing temperature and pressure, and heat is transferred through the pipe wall.

Dealing with internal energy: thermodynamic identities

So far we have been talking about the energy bound up in heat and intermolecular forces as the *"internal energy"* (u). But we can't measure this quantity, so we need to write the energy equation in terms of quantities we *can* measure, like temperature (T). To deal with internal energy we first ought to talk about enthalpy (h), because some handy thermodynamic identities exist which can get us from enthalpy – which we also can't measure – to quantities that we finally *can* measure.

If we look at some volume of fluid present in the world, say inside a pipeline, it has an amount of internal energy which represents the heat content and *"spring compression"* of intermolecular repulsion. But that fluid also takes up a chunk of space, and if it wasn't there, some of the adjacent fluid would immediately expand into the space it occupies. Enthalpy is the internal energy in the fluid plus the work needed to push that adjacent fluid out of the way to make room for the fluid's presence, a function of pressure (P) and density (ρ). Mathematically we write this as follows – in specific quantities:

$$u = h - \frac{P}{\rho}$$

The *chain rule* for derivatives states that if specific enthalpy (h) can be written purely as a function of pressure (P) and temperature (T), which it can, then its derivative in terms of some other variable y – for instance distance (x) or time (t) – can be written:

$$\frac{\partial h}{\partial y} = \left(\frac{\partial h}{\partial T}\right)_P \cdot \frac{\partial T}{\partial y} + \left(\frac{\partial h}{\partial P}\right)_T \cdot \frac{\partial P}{\partial y}$$

These derivatives of enthalpy, as it happens, correspond to real quantities that can be measured! The specific isobaric heat capacity (c_P) – i.e. constant pressure – is defined:

$$c_P = \left(\frac{\partial h}{\partial T}\right)_P$$

And the thermal expansion coefficient (α) is defined:

$$\alpha = \frac{1 - \rho \cdot \left(\frac{\partial h}{\partial P}\right)_T}{T}$$

Introducing these two definitions into the preceding equation:

$$\frac{\partial h}{\partial y} = c_P \cdot \frac{\partial T}{\partial y} + \frac{1 - T \cdot \alpha}{\rho} \cdot \frac{\partial P}{\partial y}$$

Taking our independent variable (y) to be distance (x) or to be time (t), we finally have everything we need to calculate the variables within the energy equation:

$$\frac{\partial h}{\partial x} = c_P \cdot \frac{\partial T}{\partial x} + \frac{1 - T \cdot \alpha}{\rho} \cdot \frac{\partial P}{\partial x}$$

$$\frac{\partial h}{\partial t} = c_P \cdot \frac{\partial T}{\partial t} + \frac{1 - T \cdot \alpha}{\rho} \cdot \frac{\partial P}{\partial t}$$

Now we can work all the way back to internal energy (u) by considering how it changes over space ($\partial u/\partial x$) and time ($\partial u/\partial x$). For a change with respect to either variable (y):

$$\frac{\partial u}{\partial y} = \frac{\partial h}{\partial y} - \frac{\partial (P \cdot \rho^{-1})}{\partial y}$$

$$= \frac{\partial h}{\partial y} - \rho^{-1} \cdot \frac{\partial P}{\partial y} + P \cdot \rho^{-2} \cdot \frac{\partial \rho}{\partial y}$$

$$= c_p \cdot \frac{\partial T}{\partial y} - T \cdot \alpha \cdot \rho^{-1} \cdot \frac{\partial P}{\partial y} + P \cdot \rho^{-2} \cdot \frac{\partial \rho}{\partial y}$$

So we can calculate how internal energy at a certain moment varies along the pipeline:

$$\frac{\partial u}{\partial x} = c_p \cdot \frac{\partial T}{\partial x} - T \cdot \alpha \cdot \rho^{-1} \cdot \frac{\partial P}{\partial x} + P \cdot \rho^{-2} \cdot \frac{\partial \rho}{\partial x}$$

And we can calculate how the internal energy at a certain location changes over time:

$$\frac{\partial u}{\partial t} = c_p \cdot \frac{\partial T}{\partial t} - T \cdot \alpha \cdot \rho^{-1} \cdot \frac{\partial P}{\partial t} + P \cdot \rho^{-2} \cdot \frac{\partial \rho}{\partial t}$$

Work from fluid compression

The energy flows in the *"other energy flows"* term of our energy conservation equation are positive when directed out of the control volume, so we want to compute the net work done by the fluid on the environment as it is pushed into and out of the control volume. This work is force times velocity at each side, and force is pressure times cross-sectional area. On the upstream side, at x, the work done by the fluid on the environment is negative, and on the downstream side, at $x + \mathrm{d}x$, it is positive, and so:

$$\text{net work done by fluid} = P_x \cdot A_x \cdot v_x + P_{x+\mathrm{d}x} \cdot A_{x+\mathrm{d}x} \cdot (-v_{x+dx})$$

We convert this to a per-unit-volume basis by dividing by unit volume $(A \cdot dx)$:

$$\text{net work done by fluid per unit volume} = -\frac{1}{A} \cdot \frac{\partial(P \cdot A \cdot v)}{\partial x}$$

Since in a level pipe the pressure gradient $(\partial P/\partial x)$ is negative in the direction of flow, this correctly results in positive work being done by the fluid against friction.

Work from pipe expansion

The work done by the fluid on the pipe wall in the process of pipe expansion depends on the operating pressure (P) and the volume (V). Using the same reasoning as above, we express it as a rate of work done per unit time (w_{pipe}):

$$w_{\mathrm{pipe}} = P \cdot \frac{\mathrm{d}V}{\mathrm{d}t}$$

The volume is the product of the cross-sectional area (A), which changes, multiplied by the length increment $(\mathrm{d}x)$, which is assumed not to change. So:

$$w_{\mathrm{pipe}} = P \cdot \mathrm{d}x \cdot \frac{\mathrm{d}A}{\mathrm{d}t}$$

Dividing by our control volume $(A \cdot \mathrm{d}x)$ converts to a per-volume rate of work:

$$\frac{w_{\mathrm{pipe}}}{V} = \frac{P}{A} \cdot \frac{\mathrm{d}A}{\mathrm{d}t}$$

This is the work done on the pipe wall by the fluid as the pipe expands or contracts. Then all we need to do to include this effect in our model is to express the cross-sectional area (A) as a function of pressure and temperature. We will do this in the chapter on the thermal model.

Heat flow through the pipe wall

Newton's Law of Cooling says that the heat flux (q/A) from conduction is proportional to the temperature difference driving it. In this case that is the temperature difference between the bulk fluid temperature – which is assumed to be constant across the pipe cross section, including immediately adjacent to the inner side of the wall – and the *"ground temperature"* (T_{ground}) as seen far outside the pipe. That gives us the form:

$$q/A = U \cdot (T - T_{\text{ground}})$$

Where we have introduced an as yet undefined constant of proportionality (U) in the equation. The surface area (A) of a control volume of unit length $(\mathrm{d}x)$ is proportional to its circumference:

$$A = \pi \cdot D \cdot \mathrm{d}x$$

So the rate of heat transfer (q) is:

$$q = \pi \cdot D \cdot \mathrm{d}x \cdot U \cdot (T - T_{\text{ground}})$$

And the volume (V) of the fluid element within the control volume is:

$$V = \frac{\pi \cdot D^2}{4} \cdot \mathrm{d}x$$

Therefore the rate of heat transfer per unit volume (q/V) is:

$$q/V = \frac{4 \cdot U}{D} \cdot (T - T_{\text{ground}})$$

The overall heat transfer coefficient (U) – abbreviated OHTC – which we used in this expression is a concept that warrants its own discussion. We shall return to examine it more closely in the chapter on the thermal models used in pipeline simulation.

2.2 – SOLVING THE UNSOLVABLE

Why we need a solver

"As soon as an Analytical Engine exists, it will necessarily guide the future course of the science. Whenever any result is sought by its aid, the question will then arise: By what course of calculation can these results be arrived at by the machine in the shortest time?" – Charles Babbage

The aforementioned set of conservation laws for mass, momentum, and energy are partial differential equations (PDEs). Together with fluid property correlations, friction factor and other equations describing the performance of equipment such as pumps and compressors, these make up a complete, solvable system. By finding a solution of our equations at steps in time, the simulation advances.

But this set of equations has non-linearities even in its algebraic equations. For example, the friction term in the momentum equation is non-linear, as is the equation of state. If we are careful in our choice of variables, the *mass* equation can be linear, and most of the *momentum* equation can be linear, but its friction term has a velocity-squared term in it and there is just no way to make that linear. * Once we throw the *energy* equation into the mix, there is potentially all sorts of non-linearity hiding in there. We only have a linear system for some crude pipelines while they are in laminar flow, but most pipelines need to solve non-linear systems.

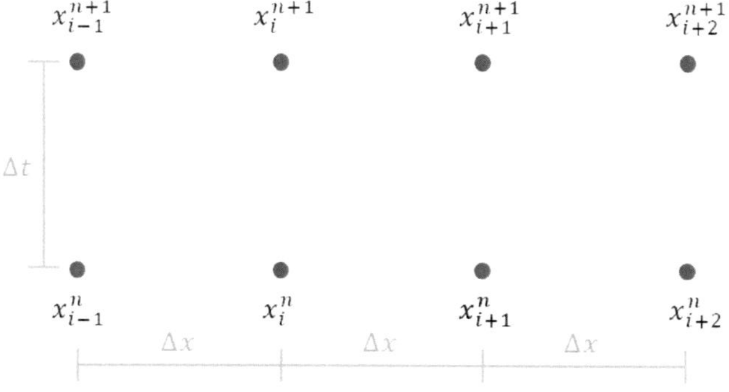

Figure 28 – A discrete grid in space (knots denoted with subscript i) and time (steps denoted in superscript n). Distance is on the horizontal axis; time is on the vertical axis.

* Without losing the linearity of other significant terms, at least.

A computer can't directly solve non-linear equations. A simulator employs a solver to approach the task *numerically*. Our system is continuous in time and space. We can't directly represent continuous functions or differentiate them in a discrete way, so we discretize our system, making it algebraic so digital computers can solve our equations. This entails getting rid of every derivative in space or time ($\partial P/\partial x$, $\partial P/\partial t$, etc.).

There are lots of choices around how we discretize. We might turn $\partial P/\partial x$ into some form akin to $(P_{i+1} - P_i)/(x_{i+1} - x_i)$ for example. Together, those choices constitute the *numerical scheme* of a solver. A user does not usually consider which of these is being used when deciding whether to use a pipeline simulator. But the choice a pipeline simulator opts for entails consequences. We need accurate, stable, fast-solving simulations. Not all solvers were created equal. Famous approaches for Navier-Stokes solvers include Galerkin and finite element methods. Most are overkill in a *"one-dimensional"* pipeline. Almost all commercial pipeline simulators have used one of three approaches:

- An upwind explicit finite difference method
- The method of characteristics
- The box scheme, an implicit finite difference approach

Despite our omission of more complex options, this section is still inevitably quite technical, so the casual reader is welcome to skip it. In practice, a typical user of pipeline simulator software does not consider the method by which it solves its equations, unless conducting a careful surge analysis for instance.

In the 1980's, a study compared simulating a gas pipeline using three such schemes: the implicit box scheme, an explicit scheme, and the method of characteristics. It found all three were accurate. [27] An implicit box scheme is usually the fastest way to achieve a given accuracy.

The principle of finite differences

Most modern commercial simulators use finite difference solvers. This is a family of methods which share in common an underlying principle. They assume derivatives can be approximated by algebraic equations embodying differences between adjacent groups of points on a mesh, to thereby numerically approximate any derivative term. At the limit approaching an infinitesimal step in time ($\Delta t \rightarrow 0$), this becomes an exact approximation:

$$\frac{\mathrm{d}w}{\mathrm{d}t} \approx \frac{\Delta w}{\Delta t}$$

By rearranging we advance whichever variable (w) we need to a new value:

$$\Delta w = \Delta t \cdot \frac{\mathrm{d}w}{\mathrm{d}t}$$

$$w^{\text{new}} = w^{\text{old}} + \Delta w$$

We can use this finite difference method to solve for steady state by setting time derivatives to zero. We can use it to solve for transients: there are several valid options.

Why not use an explicit scheme?

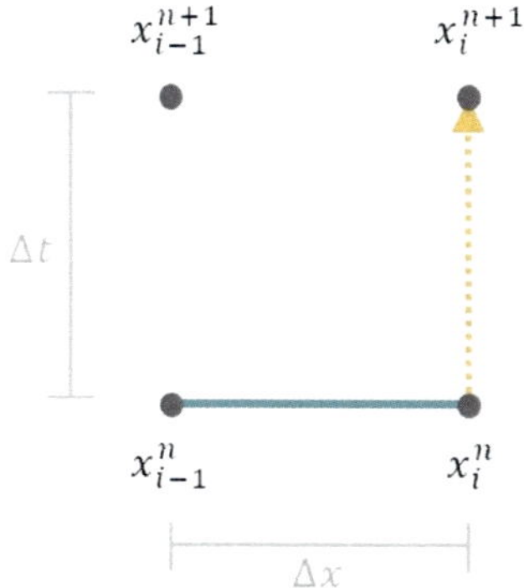

Figure 29 – An upwind explicit scheme. A horizontal bar connects points used to evaluate spatial derivatives. The vertical dashed arrow represents where the time derivative is evaluated. Time advances upwards. The horizontal direction represents distance along the pipe.

An explicit scheme that seems to be easy at first glance is *upwind differencing*. We solve for one knot's new state based on its old value along with the old value of the preceding knot (the one lying *upwind*, relative to flow direction). The reason the upwind knot is chosen is that it intuitively makes sense that the fluid flowing into the knot influences what is happening inside the knot on the current step; the fluid that has already flowed away downstream doesn't seem relevant. And indeed, upwind differencing as this approach is called tends to produce stable solutions, while the converse (which would be called downwind differencing) does not. This converts a distance derivative ($\partial w / \partial x$) into a difference between known values as follows:

$$\frac{\partial w}{\partial x} = \frac{w_{\text{this.knot}}^{\text{old}} - w_{\text{upwind.knot}}^{\text{old}}}{\Delta x} = \frac{w_i^n - w_{i-1}^n}{x_i - x_{i-1}}$$

These *old* values are known from the last solved state, so we can immediately compute the new value! For non-derivatives, like those space-derivatives, this explicit method advances our variable (w) to a new value based only on its known old value at step-start.

A time-derivative becomes a *difference* over time. We only refer to the new value in computing our variable's time-derivative ($\mathrm{d}w/\mathrm{d}t$). This replacement of the exact time derivative ($\mathrm{d}w/\mathrm{d}t$) with an expression relating step-start and step-end quantities is equivalent to Taylor expansion of the underlying function $w(t)$ about the center of the step to second order in time.

Each time derivative is evaluated at the middle of the time-step, then Taylor-expanded to first order to approximate the true derivative ($\mathrm{d}w/\mathrm{d}t$):

$$\frac{\partial w}{\partial t} = \frac{w_{\text{this.knot}}^{\text{new}} - w_{\text{this.knot}}^{\text{old}}}{\Delta t} = \frac{w_i^{n+1} - w_i^n}{t^{n+1} - t^n}$$

We can't immediately solve this, because it has an unknown ($w_{\text{this.knot}}^{\text{new}}$). But we plug it into any partial differential equation to convert it into an algebraic equation. Thus our PDE's become a set of algebraic equations equal in number to the number of unknowns (the new values of our variable). That is solvable for the next state!

The rate of change of velocity (v), for instance, ends up being a function of the known *old values* of velocity, density, and so forth:

$$\frac{\mathrm{d}v}{\mathrm{d}t} = \text{function}\left(v^{\text{old}}, \rho^{\text{old}}, \frac{\mathrm{d}\rho}{\mathrm{d}x}, \ldots\right)$$

By independently solving all along the pipeline, knot-by-knot, computing each new value in turn, we thereby advance the simulation by one step in time.

An explicit approach seems easy to implement as it requires no matrix solver. It gives us one equation per knot for the new values of pressure, flow, etc. If we just solve each grid point's equation then we're done. Sadly, there is a catch.

This explicit method introduces an inaccuracy by including only first-order terms in its linearization of the spatial derivatives; the higher-order terms it ignores become a new kind of error. This isn't necessarily a problem with all explicit schemes; predictor-corrector methods can be accurate to higher order.

The real downside with any explicit scheme is that it is subject to a fundamental limit which might require the size of our steps in time to be inconveniently short, significantly limiting the fastest speed at which we are able to simulate.

The Courant limit on time-step

Every explicit scheme's major drawback is that for all but the very shortest time-steps, it is liable to becoming utterly unstable. An explicit scheme is bound by a stringent limit, beyond which that instability ensues. That limit is the Courant limit on time-step,[*] a ratio of knot spacing to speed of sound (c):

$$\max(\Delta t) < \frac{\min(\Delta x)}{c}$$

For a particular method, the limit of stability might be half that, or some other coefficient of order 1. This is *exceedingly short*. For 1 km knots in water it's already less than 1 second, shorter than the scan rate we can expect from a typical SCADA system. For a 30-meter length of pipe transporting water within a station, the Courant limit is *20 milliseconds*, and even shorter if we want to break that pipe up into pieces to capture any internal dynamics!

That short a time-step is not necessarily impracticable, modern computers can possibly still use it. But it would mean a simulator can't ever run at a long time-step. Most applications of a pipeline simulator involve long periods of running where no violent hydraulic events are occurring. During those periods, fairly long time-steps would be accurate enough, so if we used an explicit model those would be a lot slower. Each individual explicit time-step is generally very fast, faster than an implicit method for that step, but still slower than taking longer steps.

We could attempt to conceive some tricks for dealing with this limit. Perhaps our whole system takes these milli-second steps somehow, or we invent a way to advance different parts of a pipeline at different time-steps within the same simulation. This would be as cumbersome and as complicated as it sounds.

Some explicit methods are used in some circumstances, but as a blanket approach to pipeline simulation, it's not an adequate choice. It all gets rather complicated; explicit schemes don't turn out to be so *"simple"* after all!

---

[*] Or CFL, after Richard Courant, Kurt Friedrichs and Hans Lewy, all leading German mathematicians who fled the Nazis to the United States in the 1930's.

The method of characteristics

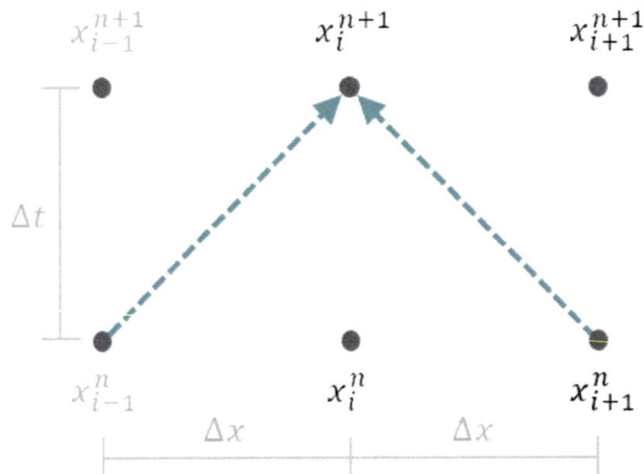

Figure 30 – The stencil of the method of characteristics. It calculates a new value based on *characteristics* (teal lines) from old values of the *adjacent* knots.

The method of characteristics (MoC) is easy to set up and solve, making it a favorite in textbooks. It assumes transients of pressure and velocity propagate along a pipeline at a constant speed of sound (c). On an $x - t$ grid, these transients take independent paths called *characteristics* $x(t)$. We locally linearize our conservation laws and model equations, then decompose the linearized equations onto the *"characteristics"*. Along these trajectories certain combinations of fluid properties remain constant. On that basis, we deduce *"the new value here"* ($w_{\text{centre}}^{\text{new}}$) based on *"old values elsewhere"* ($w_{\text{upwind}}^{\text{old}}, w_{\text{downwind}}^{\text{old}}$ etc.). Because waves can propagate upstream and / or downstream, those other locations lie both upstream and downstream. This converts the partial differential equations (PDEs) varying in both space and time to ordinary differential equations (ODEs) varying only in time. It allows us to algebraically calculate our unknown variables at points where one characteristic curve intersects with another.

Doing this at each knot, solving its algebraic equations, and moving onto the next, we cover the whole pipeline, thereby advancing to the next state.

The method of characteristics is easy to implement for a simple pipeline. In some circumstances, it introduces no numerical dispersion, and preserves wave behavior, making it very accurate for scenarios with fast, brief transients. It is therefore a preferred method for performing surge analysis studies such as those investigating a pipeline during an emergency shut-down.

But this method also suffers drawbacks. It requires the ratio of knot-spacing to time-step to *equal* speed of sound (c) everywhere and at all times:

$$c = \frac{\Delta x}{\Delta t}$$

This is just the same Courant condition that limits explicit models. It severely restricts what we can do with knot-spacing and time-step, a major challenge on a real pipeline. It's tricky to make it work with variable speed of sound, batch interfaces, and the energy equation, alongside other issues. [26] It is always complicated to combine it with equations for equipment. And, again, it only works with short time-steps.

Nonetheless, the Method of Characteristics is used in some commercial pipeline simulators.

The box scheme

The principle of a box scheme: implicit finite differences

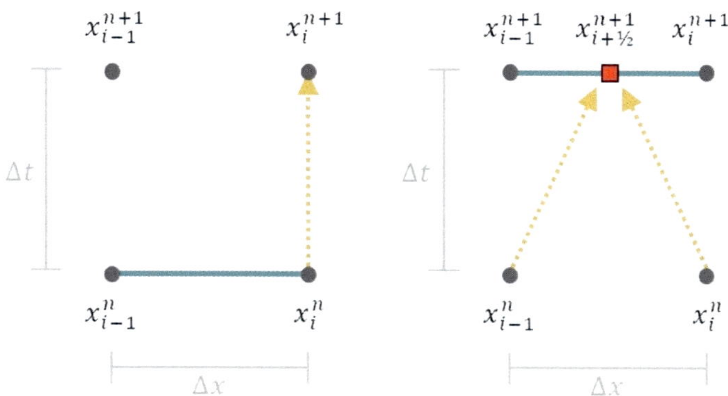

Figure 31 – An explicit (left) versus implicit (right) discretization schemes of a finite difference solver, progressing a step based on old or new spatial differences.

A major advantage of implicit schemes is their stability at time-steps longer than the Courant limit. This has made implicit schemes the standard approach in modern pipeline simulation software.

For reasons that are not straightforward, several pipeline simulators use a relatively obscure method which the basic books of numerical tricks and recipes omit to mention, which came to the industry from river simulation. The *Preissmann box scheme,* [28] *box method* or *collocation scheme* is an implicit finite difference approach. It is centered about each cell in space. Distance derivatives involve two adjacent knots. The conservation laws are applied somewhere in the center of spatial cells.

The box scheme itself can be implemented in different forms. An explicit box scheme would apply its equations (conservation laws etc.) at the *start* of a time-step – though this is not stable, so the box scheme cannot be used as an explicit approach. A fully-implicit box scheme applies its equations at the *end* of a time-step. A semi-implicit box scheme applies them somewhere in the middle of a time-step. Fully- and semi-implicit schemes alike are all referred to as *implicit*.

The box scheme is widely used in the pipeline industry: it is reasonably robust, and, for an implicit method, it is easy to implement. The premise is that it's more accurate to take some average over a step's duration rather than just the old values. So equations are written in terms of not just distance $x_{i+½}$ but also time $t^{n+\theta}$ where theta (θ) is a custom parameter adjusting how far into a step we take this average. The four corners of our box is made up of the old and new values of two adjacent knots: $(x_i, t^n), (x_{i+1}, t^n), (x_i, t^{n+1})$ and (x_{i+1}, t^{n+1}). This relates two new values ($w_{\text{left}}^{\text{new}}$ and $w_{\text{right}}^{\text{new}}$) to one another as well as to two old values ($w_{\text{left}}^{\text{old}}$ and $w_{\text{right}}^{\text{old}}$).

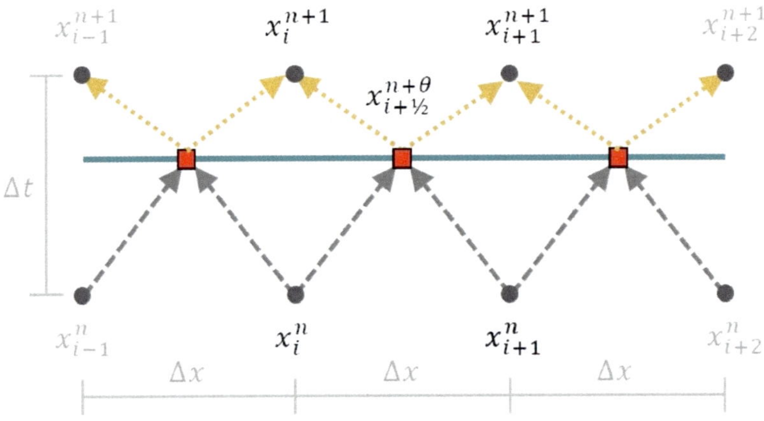

Figure 32 – Stencil for a box scheme: equations involve both points at step start and both points at step end. This is equivalent to Taylor expansion about some point in the step center, depicted here as the small square.

The governing partial differential equations (PDEs) are discretized about a special point (indicated in red). It lies half-way between the two spatial knots, and *part*-way into the step in time. The arrows from the bottom show the dependencies on adjoining variables. The arrows to the top indicate the use of a linear solver, which at once solves the *entire pipeline's* equations to advance the state. The choice of how far into a step we place this point is described by a parameter denoted *theta* (θ), being the fraction of the time-step about which we discretize. We examine how to choose this value shortly.

Distance and time derivatives via the implicit box scheme

To illustrate how the method works, for whatever variable (w) we need, let's discretize the distance derivative ($\partial w/\,\partial x$). We find that this implicit box scheme ends up *coupling* the equations at adjacent knots as follows:

$$\frac{\partial w}{\partial x} \rightarrow (1-\theta) \cdot \frac{w_{right}^{old} - w_{left}^{old}}{\Delta x} + \theta \cdot \frac{w_{right}^{new} - w_{left}^{new}}{\Delta x}$$

$$\left(\frac{\partial w}{\partial x}\right)_{i+½}^{n+\theta} = (1-\theta) \cdot \frac{w_{i+1}^{n} - w_{i}^{n}}{x_{i+1} - x_{i}} + \theta \cdot \frac{w_{i+1}^{n+1} - w_{i}^{n+1}}{x_{i+1} - x_{i}}$$

$$= (1-\theta) \cdot \frac{w_{i+1}^{n} - w_{i}^{n}}{x_{i+1} - x_{i}} + \theta \cdot \frac{w_{i+1}^{n} + \Delta w_{i+1} - (w_{i}^{n} + \Delta w_{i})}{x_{i+1} - x_{i}}$$

$$= \frac{w_{i+1}^{n} - w_{i}^{n}}{x_{i+1} - x_{i}} + \theta \cdot \frac{\Delta w_{i+1} - \Delta w_{i}}{x_{i+1} - x_{i}}$$

Unknowns at the upcoming time-step (w_i^{n+1}) are expressed in terms of known values at the present time-step (w_i^n), as well as other *unknowns*: neighboring and related values at the next time-step (w_{i+1}^{n+1}, etc.). An implicit finite difference solver evaluates the time derivative equalities mid-step. So the time derivative is:

$$\left(\frac{\partial w}{\partial t}\right)_{i+½}^{n+\theta} = \frac{w_{i}^{n+1} + w_{i+1}^{n+1} - (w_{i}^{n} + w_{i+1}^{n})}{2 \cdot (t^{n+1} - t^{n})}$$

$$= \frac{\Delta w_{i} + \Delta w_{i+1}}{2 \cdot (t^{n+1} - t^{n})}$$

Solving the "box-set" of equations

An explanation to bear in mind here is skimped on in textbooks. Until the first time they actually try to implement one, people don't realize the big, single, important practical difference between an implicit scheme and an explicit scheme. Unlike an explicit scheme, an implicit scheme doesn't just produce one equation per step-end value. Instead it answers our question with a challenge. An implicit scheme gives us a big set of linear relations *between* the step-end values. This requires a simultaneous solution of all the equations at every box. This means there is additional work for implicit schemes: we use another linear solver (matrix methods) to solve that *"box-set"* calculating step-end values.

Our set of equations has partial differentials and non-linear algebraic terms. We want to feed them into a *linear solver.* [*] To do so, we must first linearize them in space and in time, converting them into a linear system of equations. We already have a way to do this: a Taylor expansion to first order. Linearizing the implicit box equations is a bit complex. Equations are written at a *non-grid point* ($x_{i+½}^{\tau+\theta}$), so the linearizing happens there, and all the corners of the box contribute. The *"linearized box-set"* ends up being very big, and very sparse.

A lot of authors, particularly before the 1990's, concluded *"... and therefore it's impossible to use implicit schemes because inverting that huge matrix is too slow."* But this is not quite right. Practically, it is faster to invert that big, sparse matrix than it is to generate it. Sparse matrix solvers are available off-the-shelf as a third-party package. Today they are *blindingly* fast, their solution time scaling linearly with the size of the matrix. The time spent solving these sparse linear systems of equations is a small fraction of the total time Atmos SIM spends in pipeline simulation; just generating the matrix elements prior to solution is considerably slower than actually solving.

---

[*] There are non-linear programming solvers available capable of handling sparse sets of quadratic equations. So we could alternatively make our model equations quadratic (*quadratizing*?). This is an interesting approach, but sparse quadratic solvers are quite a bit slower than sparse linear solvers, so it may be impractical. But then again, that is exactly what people used to say about implicit schemes in general!

A theta to balance stability against numerical dispersion

To numerically evaluate any derivative in time, we do a linear expansion. * Most pipeline simulators simply perform this linearization about step start. Atmos SIM usually solves with a *fully implicit* discretization, but can be alternatively configured to use a more accurate but slightly less stable scheme.

How do we choose the best theta (θ) based on which our implicit box scheme conducts its expansion? This choice affects the weighting of the box scheme's artificial dispersion term, as $(\theta - 0.5)$. But usually, people intuitively think of it as indicating which point during a time-step we are discretizing around. We typically use the box scheme with theta just above midway ($\theta\ =\ 55\%$ or so).

Choosing *step-start* ($\theta = 0\%$) gives us an *explicit* scheme, an unstable choice. In fact, any theta below the midpoint ($\theta < 50\%$) is unstable and can't be used.

Choosing a time-centered step-midpoint ($\theta = 50\%$) is intuitive, and indeed it is *"most accurate"* in a sense, to a *higher order*. Unfortunately, that choice, too, turns out to be borderline unstable: during hydraulic transients oscillations will appear, and they then won't grow but they also won't ever go away, so the true solution will always have significant oscillations superimposed on it. For the implicit box scheme the stability requirement is supposedly $\theta > 50\%$, but we observe that $\theta =\ 51\%$ or so is still bad. It's technically *"stable"* in that a sudden change causes oscillations which eventually die out, but it is a *long* eventually!

The im"licit box scheme is bound by a further requirement for stability we haven't yet encountered in other methods. At a very short time-step, we get oscillations in the thermal model. So when an online model is deployed live on site with such a theta, it fills our pipeline with non-physical oscillations. Even $\theta = 60\%$ might not be great if we have, say, a lot of rounding error in our meters, even sensible rounding, such as a pressure to the nearest $0.1\ \mathrm{bar}$.

At the half-way point ($\theta = 50\%$), errors are not just proportional to time-step, but rather to the time-step *squared* $(\Delta t)^2$. This is great: halving a time-step cuts errors by three quarters! But this theta is unstable, as we said. 55% of the way into a step, or thereabouts, is more stable, and nearly has that desirable averaging property. Using

---

* We also need to linearly expand in order to average a non-derivative quantity over a time-step, but we will not discuss that here.

something like $\theta = 55\%$ for near-second-order accuracy in time makes errors proportional to a little over time-step *squared:*

$$\text{error} \propto (\Delta t)^2 + k \cdot \Delta t$$

Discretizing about step-end * ($\theta = 100\%$) introduces significant numerical dispersion but is very stable at long time-steps, unlike other values that may be prone to producing oscillations in the results. At *infinite* time-step, it becomes equivalent to solving for a *steady state*! But it causes the worst numerical dispersion, washing out crucial physical details of pressure waves, fluid composition fronts and temperature fronts as they traverse the pipeline.

Pausing to examine the difference between a step-end scheme ($\theta = 100\%$) compared against a step-center scheme ($\theta = 50\%$) we see a sort of diffusion:

$$\frac{\partial f}{\partial t} + k \cdot \frac{\partial^2 f}{\partial x^2} = 0$$

That is what theta (θ) does as it becomes greater than 50%: it diffuses. Sharp non-physical oscillations quickly die, but so do sharp fronts which smear out somewhat as a non-physical artefact.

In practice, the value of theta is a matter of judgment and experience. Selecting something like 55% or 60% is more accurate than 100%, but 100% is the most stable. The typical user is not even conscious of numerical dispersion. Most people except seasoned power-users are less offended by sharp fronts than they are by oscillations.

When does numerical dispersion become untenable?

For almost every operational scenario, a pipeline simulator gets by just fine with a fully implicit theta ($\theta = 100\%$). Even surge studies are concerned with the magnitude of a pressure surge: the highest pressure reached anywhere. They don't care how *fast* the surge traverses the pipeline, which is where a fully implicit theta ($\theta = 100\%$) causes the worst inaccuracy. The highest pressure is calculated correctly despite the velocity of the surge wavefront being wrong, because it occurs after the wavefront passes.

When we come to fast transients for other purposes than *"highest pressure"*, that's where we might run into meaningful inaccuracies introduced by numerical dispersion. Consider a scenario investigating an *emergency shut-down* (ESD). We want to determine how an emergency shut-down valve (an ESV) should be tuned. It is triggered by conditions involving both the magnitude of a pressure surge from a rupture (or one

* Some say *"fully implicit"* here, and *"semi-implicit"* for the others, but strictly, any scheme involving forward-time variables outside derivatives is implicit.

that *shouldn't* trigger it, say from a compressor trip) and the *sudden-ness* of the surge. So here we care about how long the wavefront takes to pass, which is wrong at this theta-setting due to dispersion.

If the waveform of a surge is a step function, a fully implicit theta doesn't degrade the height – the pressure will be quite accurate well in front of the step and also well behind the step. But if the surge is a sudden spike, then this theta setting *does* degrade the height. However this is not typical in pipelining.

It might be relevant if we wish to set up the coefficients on a PID controller. That depends on detailed device characteristics at a level of accuracy that we probably don't know until we are there in the field. Another pipeline phenomenon generating rapid spikes is the collapse of a vapour pocket (a slack bubble). A pressure spike would arise that – with this theta-setting – would appear to go away in the simulator much faster than it did in real life. And finally, if we are using our pipeline simulator for leak detection, it is important to capture the surges as accurately as possible on the time-scale that we are detecting the largest leaks. In some cases this might necessitate the use of a value of theta below a hundred percent ($\theta < 100\%$).

Why the box scheme is often the best

The implicit box scheme has several advantages that make it a favorite today.

It is stable at longer time-steps, and supports adaptive knot-spacing, making it faster-solving. For a given accuracy this makes it faster than a second-order explicit scheme like Lax-Wendroff or Method of Characteristics for the sort of problems we solve, which tend to have relatively long runs on large pipelines.

It has a parameter that we can tune to trade numerical dispersion (reduced accuracy) for increased stability. It is easy to couple to boundary conditions and equipment equations, unlike method of characteristics and second-order explicit methods. Its only downside appears to be perceived complexity of initially implementing it compared to the other two *"simpler"* methods we mentioned here, but in fact once implemented this implicit box scheme is incredibly flexible, and does not need modifying as a developer extends the simulator to support new features.

The only exception, perhaps, is that during fast transients, there is an inevitable degree of numerical dispersion associated with any finite difference method (implicit or explicit), so for the special case of a simple (one in-one out), unbatched liquid pipeline undergoing an active surge, the method of characteristics may be worth considering. Even in that special case, the method of characteristics solver can only model surges with no dispersion provided all of the conditions are perfect. If we introduce a thermal model; or a variable speed of sound due to changing temperature, or a pipe whose length isn't exactly an even number of timestep-times-soundspeed intervals; or anything that violates the conditions in which the method of characteristics' equations were derived, we must introduce some sort of hack to make it solve. Those hacks invariably introduce numerical dispersion back into the problem.

In almost all circumstances, the implicit box scheme has proven to be the best suited to our application. As with anything in life, there's a caveat, but it is one we can happily live with.

The caveat with an implicit box scheme

All box-type schemes (except for explicit methods) come with a caveat. They have a limitation nobody talks about: the energy equation's calculations are prone to oscillations at very short time-steps. We assumed all variables – pressure, temperature, and so forth – always change linearly over a knot ($\mathrm{d}x$). But this assumption doesn't always hold true for the energy equation.

When a big pressure change occurs whilst we're taking a short time-step:

$$\mathrm{d}t \ll \frac{\mathrm{d}x}{c}$$

Happily, we never encounter such an exceptionally short time-step in practice. Unfortunately, there's a second condition, concerning changes in temperature, and this second condition is *frequently* encountered in practical scenarios:

$$\mathrm{d}t \ll \frac{\mathrm{d}x}{v}$$

Therefore, changes in temperature simulated with short time-steps could give rise to oscillations, but only in the *thermal* model. It's not really *"instability"* in a true sense as those oscillations do not balloon. Rather, oscillations arise whose magnitude is unacceptably enormous *from the outset*. They do die off eventually, but we can't trust the results of our model until they do. So we don't say that it's *unstable*, but rather it is *prone to oscillations* at short time-steps.

To avoid these oscillations, if a rapid change occurs, in Atmos SIM we used to just add extra short knots at the end of a pipe in the variable mesh. That worked in getting accurate results, but it slowed us down quite a bit, especially in situations where the user didn't care about the temperatures in that area until they started oscillating and became clearly wrong. Situations like a stopped lateral pipe hydraulically connected to a flowing main line: every time a surge came through in the main line, a tiny bit of flow entered the lateral, at a very slow velocity, and so those oscillations would appear.

Atmos SIM deals with this by decoupling pipe-end temperatures from knot temperature, using a pipe-end temperature buffer. This is a zero-length knot whose thermal mass is significant enough to prevent oscillations. When we change the temperature of incoming fluid, heat goes into this buffer and its temperature couples to the first knot temperature, rather than a naïve implementation of the energy equation directly setting first-knot temperature. These buffers are only used when the incoming flow velocity gets small enough that thermal oscillations are a danger.

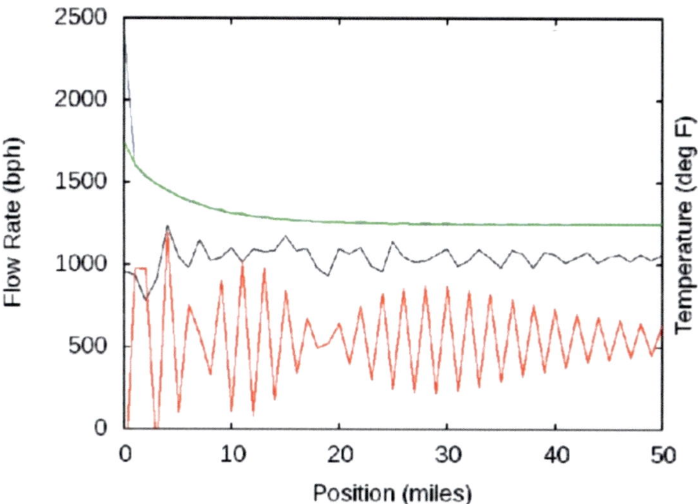

Figure 33 – With a non-explicit scheme, a brief time-step, such as the 1 second used in the scenario shown, causes thermal instability. When original flow rate (green) is ramped up over one second, temperature starts oscillating (red) in a manner that does not reflect any true physics going on. This oscillation is not *truly* unstable, because it eventually dies down to lesser amplitudes (black), but this happens very slowly indeed.

Explicit methods	
Easy setup	Only first-order accurate Short time-steps and knots Introduces dispersion
Method of characteristics (MoC)	
Doesn't require a linear solver Preserves fast transients Accurate surge analysis	Only first-order accurate Short time-steps and knots Time-step determines knot spacing * Knots might not divide into length Difficult to add batches
Box method (implicit)	
Second-order accurate Stable for long time-steps Stable for variable knots Fast and accurate for most use cases	Involves sparse matrix Challenging to debug Introduces dispersion

Table 3 – Advantages (left) and drawbacks (right) of the three methods usually chosen for the solver of a pipeline simulator

Example: discretizing the mass equation

Let's see a discretization scheme in action by discretizing the mass equation.

$$\frac{\partial(\rho(P,T)\cdot A(P,T))}{\partial t}+\frac{\partial \dot{M}}{\partial x}=0$$

Take our independent variables to be pressure (P), temperature (T), of which density $\rho(P,T)$ and cross-sectional area $A(P,T)$ are functions. Rather than, say, velocity, let's choose our third independent variable to be mass flow rate (indicating it $\dot{M}$ instead of the usual Q_M for convenience). Using a semi-implicit box scheme, let's discretize the

* This means that the length of every stretch of pipeline in our network is unlikely to divide exactly into this pre-determined knot spacing. It also means that if we vary the time-step, we must vary the knot spacing accordingly. If interpolation is used to work around this, we lose this method's chief advantage: the property of avoiding dispersion.

distance derivative of mass flow ($\partial\dot{M}/\partial x$) by expressing it in terms of a knot in space (i) and a step in time (n):

$$\frac{\partial\dot{M}}{\partial x}=\frac{\dot{M}_{\mathrm{i}+1}^{n+½}-\dot{M}_{i}^{n+½}}{x_{i+1}-x_{i}}$$

We approximate the value at the time-step's midpoint ($n+½$) to be half of the step-start (n or old) and step-end ($n+1$ or new) values:

$$\frac{\partial\dot{M}}{\partial x}=\frac{\dot{M}_{\mathrm{right}}^{\mathrm{new}}+\dot{M}_{\mathrm{right}}^{\mathrm{old}}-\dot{M}_{\mathrm{centre}}^{\mathrm{new}}-\dot{M}_{\mathrm{centre}}^{\mathrm{old}}}{2\cdot\Delta x}=\frac{\dot{M}_{i+1}^{n+1}+\dot{M}_{i+1}^{n}-\dot{M}_{i}^{n+1}-\dot{M}_{i}^{n}}{2\cdot(x_{i+1}-x_{i})}$$

Taking step-midpoint, however, is numerically unstable. Evaluating the distance derivative ($\partial\dot{M}/\partial x$) at any chosen fraction (θ) of a time-step looks like this:

$$\frac{\partial w}{\partial x}\rightarrow\frac{\theta}{\Delta x}\cdot\left(\dot{M}_{\mathrm{right}}^{\mathrm{new}}-\dot{M}_{\mathrm{left}}^{\mathrm{new}}\right)+\frac{(1-\theta)}{\Delta x}\cdot\left(\dot{M}_{\mathrm{right}}^{\mathrm{old}}-\dot{M}_{\mathrm{left}}^{\mathrm{old}}\right)\cdot\frac{\partial\dot{M}}{\partial x}$$
$$=\frac{\theta\cdot\dot{M}_{i+1}^{n+1}+(1-\theta)\cdot\dot{M}_{i+1}^{n}-\theta\cdot\dot{M}_{i}^{n+1}-(1-\theta)\cdot\dot{M}_{i}^{n}}{x_{i+1}-x_{i}}$$

All fractions greater than 50% ($\theta>0.5$) are stable. Fractions very close to 50% cause persistent oscillations. To balance stability and accuracy we choose 55%. The next-step terms ($n+1$) are unknowns, and every other term is a known constant or a known step-start value of a variable, so this can be solved.

Having dealt with that distance derivative, let's move on to the other term of the mass equation, the time derivative ($\partial w/\partial t$). Discretizing, say, velocity (v) gives us:

$$\frac{\partial v}{\partial t}=\mathrm{function}\left[(v+½\cdot\Delta v)\quad,\quad\rho\quad,\quad\frac{\partial\rho}{\partial x}\quad...\right]$$
$$=\mathrm{function}\left[\left(v+½\cdot\frac{\partial v}{\partial t}\cdot\Delta t\right)\quad,\quad\rho\quad,\quad\frac{\partial\rho}{\partial x}\quad...\right]$$

As the time-derivative of our variable ($\mathrm{d}v/\mathrm{d}t$) appears on both sides of this equation, we cannot directly solve it. This is just as true if we discretize density times area ($\rho\cdot A$):

$$\frac{\partial(\rho\cdot A)}{\partial t}=\frac{\rho(P_{i+½}^{n+1},T_{i+½}^{n+1})\cdot A(P_{i+½}^{n+1},T_{i+½}^{n+1})-\rho(P_{i+½}^{n},T_{i+½}^{n})\cdot A(P_{i+½}^{n},T_{i+½}^{n})}{\Delta t}$$

We approximate the knot-midpoint ($i+½$) terms as an average of the spatial knot-start (i) and knot-end ($i+1$) values. This equation is non-linear in our variables, so we Taylor-expand it in these variables:

$$\dot{M}_{i+½}^{n+1}\ ,\ P_{i+½}^{n+1}\ ,\ T_{i+½}^{n+1}\ ,\ \dot{M}_{i}^{n+1}\ ,\ P_{i}^{n+1}\ ,\ \text{and}\ T_{i}^{n+1}$$

We get one such equation for each knot throughout the pipeline. And that is how we handle the mass equation.

The same discretizing and linearizing-by-Taylor-expansion is also done for the momentum and energy conservation equations. This gives us a linear system:

$$J \cdot (\vec{y}_{j+1} - \vec{y}_j) = \vec{b}$$

Finally, this is what we solve! The dimension of this matrix might be 100 000+ for a large pipeline. It is very sparse; most rows have no more than six non-zero elements. Although solving a non-sparse linear system is slow, because its run-time is proportional to the third power of the number of equations, solving sparse systems (like this one) is rapid. Run-time is directly proportional to the number of equations. We spend longer evaluating the Jacobian (J) than actually solving the equations.

Equipment equations and singularities

Most junctions, valves, and equipment are, as we saw earlier, described by straightforward *algebraic* equations. For example, the pressure drop across a resistance or an open valve is described by its flow coefficient (C_V). We evaluate those at the end of each time-step. Others are described by more involved equations. A storage facility requires an ordinary differential equation; that too must be thrown into a matrix for a solver to calculate *during* every step. Tracking the angular momentum of a centrifugal compressor's impeller lets us model trips and the emergency shut-down requirements of a station very well. This too would add an ordinary differential equation to the matrix. However, we don't really need to model compressors so closely in a pipeline simulator, and nobody seems to do this for pumps.

It is the equations describing equipment which often inadvertently make our set of equations singular, more so than the pipe equations. For example, two pressure meters might be configured as zero resistance and attached in parallel. In this arrangement, pressure is over-constrained, and flow split is under-constrained, so the matrix solver will fail! Another example is two flow meters in series, both attempting to set flow rate. If they attempt to set the same value twice, this is still a singularity. Such singularities can be most inconvenient when we're seeking a steady state! How do we avoid them?

We could opt for an independent leg model rather than a single network model. In the independent leg approach, we don't model non-pipe equipment at all. But this is only feasible in an online model with limited goals, such as leak detection or batch tracking. Instead, what we might call the *mini-pipe* approach places short pipes between every non-pipe object, which is actually realistic, and was only avoided in the past as it was perceived to hinder model performance. It should be noted that any approach reliant on heuristics for catching problems can only be perfected with ample experience.

2.3 – LET'S GO! THE INITIAL STATE

How a simulation begins

When we instruct a simulation to commence, the first message we see on the status bar is a tantalizing one: *"initializing"*. Why do we have to wait for this?

Whether we're setting out to perform an offline study or kick off a live system, a simulation starts at an initial state and can only proceed on that basis. *Initialization* is the act of finding that initial state. Ideally, we have a historic saved state to hand, so a simulation can *warm-start*. That saves most of the hassle we otherwise have to go through in this part of the book. If we don't know the history of a pipeline segment, we have no saved state to start from, and we don't have data to give us clues about what the pipeline is doing, so we start from nothing. This is known as a *cold-start*, as it presumes no prior knowledge of conditions.

It is common practice to deduce an initial state for a cold start by seeking a state in which all conditions are steady. As we said earlier, by steady we mean that all flows are in perfect equilibrium, so nothing varies over time. We call this the steady state. Finding a steady state is essential for a simulator to progress subsequent steps in every scenario, be that a live real-time scenario deployed on site, a sprinted continuous run, or any simulation at all. How do we solve for a steady state?

Our first guess at a steady state

Initial constraints in an offline scenario

Constraint-switching is how solvers traditionally approached finding a steady state to cold-start an offline scenario. How are those constraints switched?

A simulator first attempts to solve steady state using the *initial constraints* specified by the user. It then checks if all *enabled constraints* are satisfied. If they are, great; we have our steady state solution! But oftentimes, the user-specified initial constraints are inconsistent with one another, or with a further enabled constraint. We cannot allow a constraint violation to occur, else we would be answering a different question to the one the user has asked the simulator.

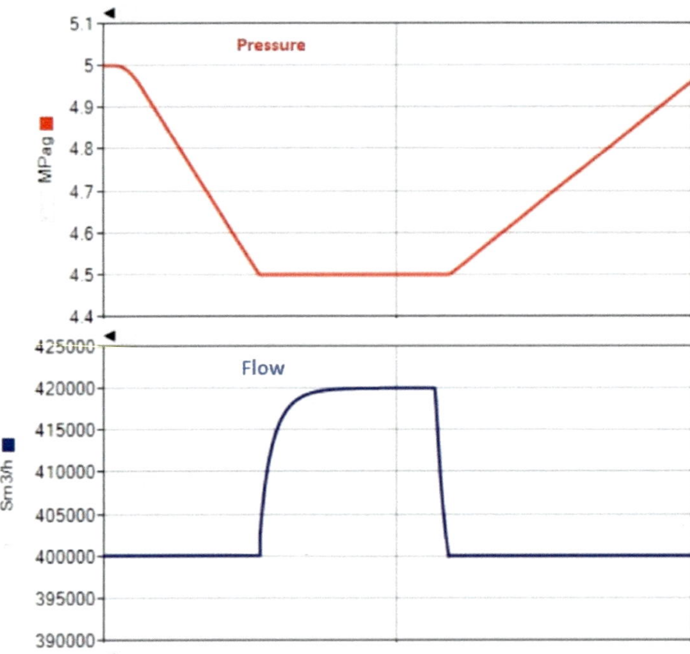

Figure 34 – Here, an initial constraint (minimum flow) controls until the item switches to another enabled constraint (minimum pressure). It then controls on that for a while then switches back.

If there are violated constraints in our first guess, the simulator picks the *"worst offender"*, switches that item's constraint, and makes another attempt. It keeps track of constraint-combinations it has tried as it goes. So far, so good.

When a simulator is switching constraints to find an initial set that solves steady state, it might inadvertently create regions without a pressure constraint. This would be impossible to solve, so heuristics are put in place to prevent it from choosing such impossible guesses. Further complications then crop up. For example, if pumps or compressors are in parallel, their constraints must be kept compatible. A set of constraints might be chosen causing flow to reverse in a region of the pipeline so the solver must be prepared to do something sensible in that case. Even more heuristics.

If the simulator finds that switching the worst violator has it repeat a guess, stepping around in a circle, it gives up and tells the user that it *"failed to solve"*. It can't exhaust every possible combination of constraints as that would entail making an unfeasibly large number of attempts. This usually doesn't happen; if it does, our best bet is to look at the model result messages and see if we can do away with one or more of the constraints it switched to while seeking a steady state. If we really need all of them, we might remove one that we need, solving to steady, then bringing that state back to the desired value in transient.

Cold-starting in an online scenario

Our constrained approach only gets us anywhere in an offline scenario. In an online scenario using meters, we don't have setpoints. To interpret metered data, we use a state estimator and consider it a minimization problem. We discuss state estimators at length in the chapter about simulators live on site.

Any steady guess we make is necessarily quite far off the true state of a typical online pipeline. This means that even after we solve for a feasible steady state, it won't be close to the transient reality of what's happening. We must wait for a *characteristic period* before results reach good agreement. How long that wait lasts depends on the *scale time* of the parameter that's not yet aligned; pressure, composition, linepack, etc. If we're interested in tracking batches in a liquid line or composition variations in a gas line, it's particularly bad to cold-start the linefill. The simulator could be significantly different from reality for days: the characteristic period it takes composition to move through the line. So a good feature is to couple a hydraulic steady state with known linefill, a feature which lacks a distinctive name. One might call it a *"lukewarm start"*.

Three ways to find a steady state

By applying formal equations with zero time derivatives

One approach is to seek a steady state by directly attempting to solve formal equations, with time-derivative terms set to zero. Atmos SIM can do this because it formally handles every equation, providing it in a solvable form either as the steady-state version or the transient version, using the same code. For historical reasons, other commercial simulators could not do this, because their steady state model was an entirely separate piece of software to their transient model, solving slightly different equations, so a steady state solution wasn't truly equivalent to their transient model with time derivatives set to zero.

This approach – solving formal equations with zero time-derivatives – is *fast-solving*, but it does present its own challenges that must be overcome.

To implement it, we typically seek a global matrix solution. Doing so obliges us to *iterate* in order to *linearize* the model equations to self-consistency. This procedure might require many iterations, often on the order of ten.

By integrating the momentum and energy equations

For sufficiently simple pipeline networks, we could alternatively integrate the momentum and energy equations from the upstream end of the pipeline to the downstream end. This involves iterating not for linearization but because we need values for everything (pressure, flow, temperature, etc.) in order to solve. But a guess only gives us boundary conditions for only one or two of those; we can't physically constrain pressure and flow together at the same point. So we use the one we're given (say pressure), make a wild stab in the dark at the other (flow), integrate to get the first parameter at the other end (pressure at B), and compare that with the measurement or constraint there, then refine that guess for flow, and repeat.

Then, challenging situations arise as we try to solve steady state for more complex network topologies, especially if there are loops – which in pipeline terminology refers to parallel pipelines or multiple paths that flow can take from A to B.

By advancing transient steps until they don't change

What commercial simulators other than Atmos SIM do, as of time of writing, is attempt *transient* steps until they find one that happens to converge. This approach, *"run-transient-until-convergence",* can be slow. Which of the two approaches is preferable? As with many things, it depends.

This approach – advancing transient steps until nothing changes – does offer an advantage. It involves less code, because the simulator just runs the same transient step in such a way that, eventually, it ends up finding a steady state.

Using the full transient model equations, it makes a first attempt, then advances to the next transient step, again and again, until all variables cease changing. This *transient initialization* progressively gets steadier every step, advancing them through *"time"* until the time-derivative terms all become zero. By this method, a steady state is reached – and we forget about all the steps we took before reaching it.

That steady state solution is then what we can take as the initial state for a cold start.

Barriers to steady state convergence

Only an advanced simulator solves these reliably

Much as a take-off is one of the riskiest parts of a flight, an attempt to initialize is often the instant when a simulation is most likely to go awry. Atmos SIM automatically does a lot for the user here. A shabby simulator does not. Here are some ideas to throw at a simulator to see if it converges to a steady state.

What to do about a bad initial guess

Firstly, if our first guess is *"too wrong"*, a solver isn't starting from *"close to the solution"* as it would on a transient step. So a steady step is forced to navigate a non-linear solution space – for example the compressor speed setpoints, the flow coefficient of valves and resistances, and the thermal model as a whole. Although this solution space is always non-linear, a transient step doesn't have us venture too far into it, so locally it's more-or-less linear. A steady state's initial guess is nowhere near the converged solution. Flow might even reverse, sending successive-linearization algorithms running off in the wrong direction.

We address this first problem by employing simplified forms of the most non-linear equations. These more-linear forms are used until the solver gets close to the right solution, at which point it switches back to the full forms. We might switch all pumps or compressors from controlling on a speed or power setpoint to a pressure setpoint. It might converge an isothermal model first, then throw in the energy equations once it has pressures and flows about right. This is where much of the effort of making a commercial simulator goes!

If we can (correctly) make all of the equations linear, we can solve no matter how bad our initial guess. If we could even just give all the equations a *linear form* equivalent to the first-order Taylor expansion around the true steady solution, we can solve no matter how bad our initial guess. So Atmos SIM tries to do this with the initial approximations. For instance, consider a pump or compressor which is controlled by a discharge pressure constraint, but the user tells us to initially control on the *speed* constraint. That discharge pressure constraint is probably not far from the correct pressure, so we switch this item to control on discharge pressure, getting this equation:

$$\text{discharge pressure} = \text{discharge pressure constraint}$$

This is linear in the model variables, unlike the non-linear speed constraint:

$$\text{speed} = \text{that computed for flow and suction and discharge pressures via the map}$$

What to do about physically impossible initial constraints

Secondly, a steady state imposes stringent restrictions on what the solver can get away with. For example, in transient we could control on flow rate at both ends of a pipe, and it will just pack or unpack over that step. In steady state this situation would leave us with a singular set of equations; unsolvable. Atmos SIM addresses this second problem with input-checking code, walking the network seeking issues of that nature and switching constraints as needed.

What to do about missing initial constraints

A third question is what to do about shut-in sections whose pressure is unmetered. Atmos SIM opens all check valves when attempting to solve steady state, obliging a simulator to perform more constraint switches, as closed check valves need to be re-closed. But if check valves are incorrectly closed, this measure lets the simulator make a reasonable guess based on the pressure at the other side of the valve. Another way around this problem would be to identify shut-in sections and ask the user what pressure should be in each one. If we work around a shut-in section, we set flow to zero, pressure to a default, and each pipe's temperature to the ground temperature.

What to do about physically unlikely initial constraints

A set of constraints might be chosen that produces choked flow somewhere in a gas pipeline. These are almost certainly not what the user intended; nobody runs pipelines in choked conditions in steady state. Furthermore, choked conditions are more liable to create difficulties in solving on the next iteration. This condition can inadvertently arise when a simulator is trying to control on various combinations of constraints, whilst attempting to solve a steady state. It is hard to avoid encountering some unlikely combinations of controlling constraints. What matters in this circumstance is to survive and switch away.

What to do about equipment laid out in inconvenient ways

Say there are two check valves in series. If flow reverses, they both close. Without special handling, this causes the model to fall over. Because if there is no pipe in between the two check valves, then the pressure in there is unknown and the flow is over-constrained. Two different equations set flow rate to zero, and no equations involve pressure. That gives a singular matrix. To deal with that kind of challenging situation we need a gargantuan number of heuristics covering every fathomable special case and arrangement of equipment.

2.4 – TRANSIENT STEPS

A transient step in various scenarios

As a simulation progresses beyond the initial steady state, a number of challenges arise. We delve into how a simulation progresses over time in an offline scenario in the *Operations* chapter, and in an online scenario in the chapter titled *Live on site*. Just as there are potential barriers to convergence in steady state, there are also difficulties that present themselves during a transient step, and through practical experience a mature pipeline simulator implements measures to remain accurate and robust. There are distinct challenges associated with liquid models or gas models, in offline scenarios or online scenarios, which would be impossible to summarize here.

Offline Liquid	**Online Liquid**
Offline Gas	**Online Gas**

Figure 35 – Atmos SIM is capable of handling four different applications of pipeline simulation within the same engine by employing the appropriate heuristics and features to each scenario.

To illustrate this, let's examine one feature and where it helps. It is possible to link every model item using artificial *mini-pipes*. This doesn't help fix singularities in steady state, as each mini-pipe still can't pack or unpack yet, but it does help in solving a transient step. It means that if, for example, a liquid is subjected to flow imbalance for any reason, the model won't crash. Instead, its pressure shoots up fast, causing something to switch constraint, as fluid in the pipeline cannot sustain that flow imbalance.

The settle-in period for online models

Another issue that emerges motivates a feature in Atmos SIM known as the *settle-in period*. If the pipeline is transient but we cold-start, the steady state is necessarily quite far off the true state of the pipeline, because a cold-start is completely steady whereas the pipeline is transient. This manifests as pressures and flows not matching all the meters particularly well. If we then take a transient step, the model sees that if it creates some big transients, it gets the meters to match reasonably well! If we let it go ahead and do that, it gets itself into real trouble in the next few time-steps. But the state estimator is blissfully unaware of the havoc its present decisions will wreak a few steps into the future, and that is fair enough because it doesn't even know what the SCADA pressures and flows are going to be!

The crux of the matter is: a solver cannot handle this impossibly quick transition from initial steady conditions (that aren't happening) to true transient conditions (that are). To relieve the state estimator of this burden, we define a *settle-in period*. Atmos SIM adds offsets to every pressure and flow meter reading, to get them to exactly agree to whatever solution it came up with for steady state. Then it gradually turns these offsets down to zero over an hour or so, while the model builds up to aligns itself with the correct transient state.

How long after initializing do we reach agreement?

After the initial settle-in period, we still aren't quite done yet.

A steady state is used in live simulators as a quick way to find a rough first estimate, but we must then wait for some characteristic period until the simulator comes gradually into alignment, eventually matching the reality of our transient state. In fact, even an outdated saved state might be quite different to reality.

A feature some simulators offer is to *"build-up"* by effectively starting at some moment in the past, simulating that as an initial steady state, then sprinting forward to reach the latest transient state at the present instance in time. This is important in a gas pipeline, as the steady state linepack might be misleading.

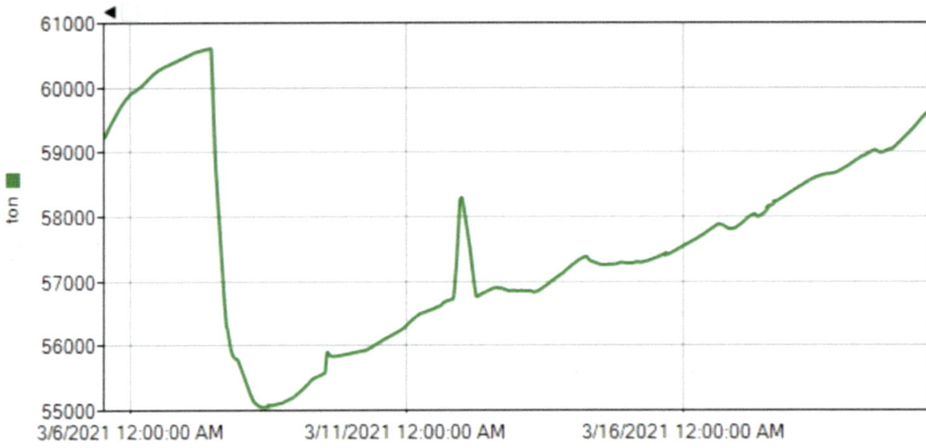

Figure 36 – The linepack in a gas pipeline varies over time. A gas pipeline's state at an instance in time is transient. *"Build-up"* prepares the transient state *"now"* based on recent historic data.

2.5 – An adaptive mesh in space and time

Staying quick, robust and accurate

A robust simulation engine is coded to overcome a series of challenges, often around numerical stability when solving challenging fluid equations. Each software vendor employs their own proprietary tricks. It usually boils down to heuristics. Balancing accuracy and performance relies on any number of tweakable tolerances identified and honed to perfection through experience, running many years of site data through the engine both *on-site* and in *replay*.

One question is how to choose a suitable mesh. We chop the pipeline up into *knots* forming a spatial mesh, only performing calculations at the boundary of each interval, and progressing the simulation every *step* as a temporal mesh.

Evenly-spaced knots and steps are adequate for simulating a single-inlet single-outlet pipeline. Keeping life simple works fine there. But when we simulate the gas distribution network of a megacity, or the water transmission network of an entire country, both of these simple approaches come into question. For a simulator to be *accurate*, shorter steps in time and tighter knots in space are indispensable near a transient. For a simulator to be *fast*, it should avoid overly short steps and overly tight knots when they're not needed.

For every application, there exists some optimal time-step. The best step duration varies depending on what the pipeline is doing at each given moment. What fewer people talk about is that there is also a complementary range of suitable *knot-spacing*. This depends on several factors, the simplest being what operation is happening at this instance in time near every given *location*.

As of time of writing, most pipeline simulators can automatically adapt their own *time-step*, but not their knot spacing. The novice user is still forced to pick that up as a new skill of their own accord, and a seasoned expert is none the happier to be burdened with such mundane tasks. At best, it's a finicky mildly annoying barrier to entry. At worst, it's another pitfall we might stumble into.

Although it's useful to make levers available at the user's discretion, it's not nice to distract them needlessly when they're not interested. These issues have nothing to do with pipelines, and it would be nice if the pipeliner didn't have to think about them. Choosing a knot spacing is not the kind of question that gets engineers up in the morning. Manually selecting a mesh is *work*! In practice, people just pick a knot spacing for an entire simulation and move on.

Yet we know that mesh requirements change over time, and differ in each segment of a pipeline! An unsuitably fine knot-spacing, like an unsuitably short time-step, wastes our time by slowly performing pointless calculations.

Moreover, how tight a knot-spacing we need depends on the chosen *time-step*, a time-step that itself varies in offline scenarios.[*] If knots are spaced out too coarsely, we end up with avoidable inaccuracy, bringing a risk of *instability* that might wreak havoc. A poor solution of this kind contains oscillations that are in no way innocuous. Such artefacts might stick out as obviously non-physical, or might look seemingly plausible despite being very seriously wrong.

Even if an unsuitable knot spacing doesn't cause instability, knots or time-steps that are too long introduce *truncation error*, limiting the accuracy of our solution and possibly entirely hiding specific results of interest to the engineer.

For a user to focus on the actual tasks they set out to do, we want an adaptive mesh that automatically adjusts both time-step and knot-spacing together. Ideally, adaptive techniques add mesh points when and where needed, get rid of them when and where they aren't, and guarantee some specified accuracy.

Adaptive spatial meshes have been widely used in simulators of reservoirs and pipe stress for a long time, solving partial differential equations (PDE's) in more than two or three dimensions. Those domains are too complex to get away with throwing too many knots at the problem. Whereas pipeline simulators, merely solving PDEs in *"one-dimension plus time"*, grew accustomed to waiting patiently for needlessly slow simulations to complete. So when we equipped Atmos SIM with an adaptive spatial mesh, we were trailblazers starting fresh.

---

[*] An offline scenario benefits a lot from varying its time-step, but an online scenario generally runs at the SCADA time-step. No accuracy is gained at shorter time-steps than that, and possibly excepting the largest and most complex pipeline systems there is usually no need to run at longer time-steps.

Adaptive time stepping

Shorter and longer steps

We ask our simulator to execute its algorithms once every discrete *time-step*, interpolating to fill in the gaps in between. This assumes nothing of interest happens until the next step, so this interval must balance several factors. Steps must be brief enough to capture ongoing effects in our pipeline. The question of finding an optimal time-step is unimportant in online real-time scenarios, because in those, steps are taken at the SCADA scan interval if at all possible.

For the constrained scenarios used to conduct offline studies and look-aheads, a time-step that is too brief will be simulating too often, keeping us waiting for pointless computations. So simulators offer *adaptive time-steps*. An algorithm decides how frequently to simulate, taking coarser steps during stable pipeline conditions. A typical algorithm computes how non-linear the results of the last three time-steps have been, and adjusts the next step duration accordingly.

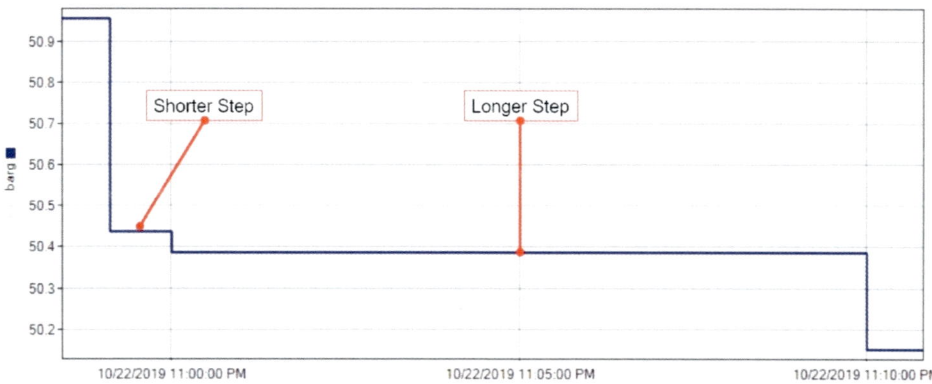

Figure 37 – An adaptive step algorithm has Atmos SIM vary its step interval between user-set limits on briefest and longest step. Shorter time-steps are taken during periods of non-linearity.

The step is reset to its shortest minimum if the inputs change in a non-linear fashion, or if some other non-linear behavior happens – for example: a regulator switching from controlling on pressure to flow control. Otherwise, longer steps are taken up to some maximum. In this way, steps adapt to strike a balance during the course of an elaborate scenario.

Here is an example of adaptive time-step in action. It is not the exact system used by Atmos SIM, but gives an idea. Say we set the shortest allowable step to 1 second, and the longest to 60 seconds. The simulator initially attempts using the minimum: it takes two 1-second steps. Then it re-simulates the same time interval with a single 2-second step, and compares the results. If the longer-step yields results close enough to the first attempt, that is to say a 2-second step gives us the same answer as does a 1-second step, we might get away with even longer. So it again extends step-length, making a third attempt with a step 4 seconds long. If these results differ to those of the preceding second attempt, we decide to use the second attempt's step-length (2 seconds) for the subsequent step and continue on the basis of those results. The procedure of trying several intervals to identify the best is repeated every step. If results deviate, the simulator takes a briefer step intervals, else it tries for a longer step interval. This might sound like it needlessly incurs computational overhead, but it turns out to be worthwhile. Adaptive time-step has been substantially demonstrated to simulate rapidly, with robust accuracy.

This is all premised on a user having chosen their allowable limits on briefest and longest step-interval to be suitable in the first place. Too brief a longest step slows a simulator down without improving accuracy. Too long a briefest step ruins the accuracy of our results. When a user sets up a new scenario, determining these limits is a matter of judgment. It is advisable to try various values and inspect results. If briefer step intervals give the same results, our chosen briefest-step is fine. If longer steps yield results of poor resolution, resembling square waves, our longest-step can't be pushed any further.

Traditionally, it is only upon fully completing a given step's calculations that a simulator may move onto the next step. This is because every step in a fluid's story depends on its present state, and that in turn depends on the history of everything that led it up to that point. For a long-duration scenario capturing many days or weeks, it was not unheard of until very recently for users of offline-study simulators to wait *overnight* for their run to reach completion.

Stepping in real-time

When simulating a live online system (a real-time model RTM), the chosen time-step is not varied in the same way as for offline scenarios. Instead, the aim is to process as much incoming data as possible. Computational power permitting, it seems tempting to have steps happen at the SCADA scan rate, also called a refresh rate or a polling rate. * However, data today might be coming in *"too often"* – in the sense that steps are limited by what a simulator can keep up with, to solve a step before the next step is due. Typically a project engineer would set steps to equal some integer multiple of

---

* Although on a gas pipeline, whether there is any tangible benefit at all to simulating more often than every 15-30 seconds or so is possibly up for debate.

SCADA scan rate. People want to avoid handling *"superfluous"* data if there's no tangible benefit.

Moreover, SCADA might be performing a *"scan"* according to one of several principles, encountering issues of its own. Flaws in the incoming data are to be expected. Even with auxiliary software supporting our simulator by handling these incoming data flaws, certain steps are bound to be numerically intensive, requiring more iterations to converge. It is inevitable when running for months on end that the engine might take too long to solve a step. The system must gracefully handle falling behind. It recovers using a feature called *catch-up steps*, which we detail in the chapter on real-time transient models live on site.

Gas transients occur gradually, so it is meaningful to configure step-intervals for a gas pipeline to happen every fifteen to thirty seconds or so. It remains possible to capture many longer-term effects of a gas pipeline relatively well using intervals as long as *ten minutes*! This makes it possible to run the full duration of a *"look-ahead"* scenario simulating days and weeks of pipeline behavior within a few minutes. If we're smart, simplified liquid pipelines could be amenable to that same treatment. Peering into the future in that way is a feature known as *look-aheads*. Many look-ahead simulations can be launched alongside a real-time simulation. It presents opportunities we look at later.

Adaptive knot spacing

Why adaptive knots matter

Having looked at the simulator *temporal*, we now look at the simulator *spatial*. Say we have a 2 000 km pipeline, but for some brief period we're interested in something happening along a certain stretch. To maintain such a tight knot spacing throughout the pipeline for the duration of the entire simulation would be untenable, and pointless where and when nothing of interest is taking place. How do we home in on a section of pipeline when something interesting is happening in the vicinity of a few hundred meters? Atmos SIM offers *adaptive spatial knots*, deciding how closely to position knots to keep results accurate within a tolerance, whilst maintaining computational speed. This is a feature glossed over by many, and it can be quite a handful. Because it offers us several major benefits, we afford it special attention in the following discussion.

Simply put, a coarser knot spacing is automatically used in pipes when their conditions are stable, and a finer spacing is applied to affected sections when transient effects are taking place. The primary purpose of this is not so much to solve faster, but to boost accuracy when needs arise, thereby avoiding instability. Nobody wants to see *"failed to converge"* during their simulation! A good starting point is a coarsest knot spacing of

1 000 meters, and a tightest knot spacing of 1 meter. For a look-ahead to simulate even faster, a coarsest knot spacing of 5 000 meters may be acceptable for faster solution.

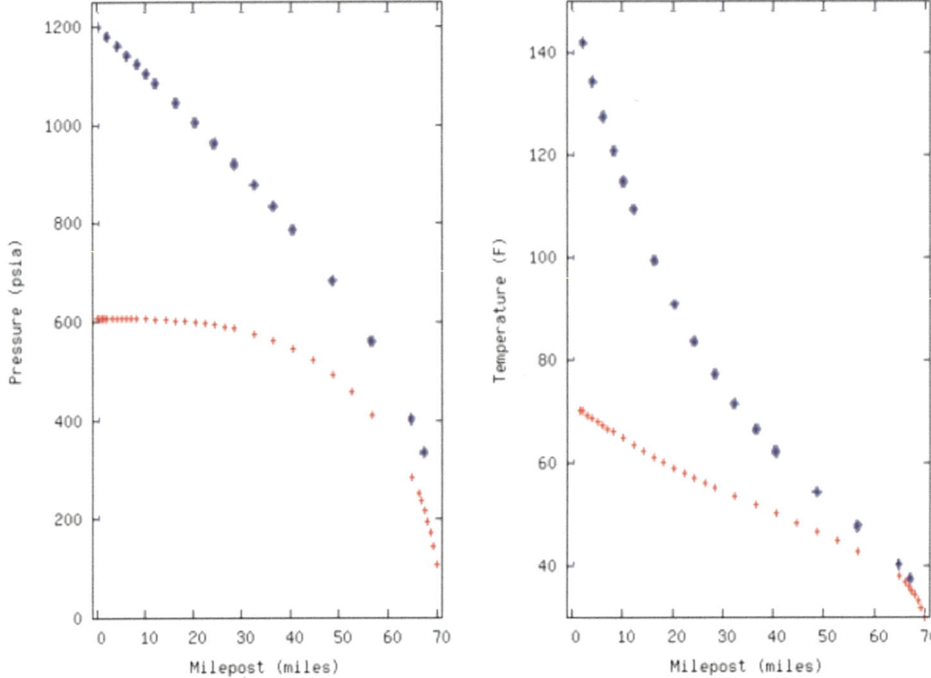

Figure 38 – Knots are automatically spaced to capture features of the pressure profile and temperature profile along a pipeline, following a valve closure event. This spacing dynamically varies throughout the scenario, so knots are only dense during transients. Shown are knots spaced during steady conditions (blue) and transient conditions (red).

The adaptive spatial mesh algorithm can be used in all scenarios. It computes the non-linearity of three adjacent points. On that basis it varies the knot spacing, splitting knots in half, or merging adjacent knots back together. The ability to refine down to a detailed mesh in one part of the pipe, while using coarse mesh elsewhere, is a powerful feature. The biggest benefit we've seen for using an adaptive mesh is that the user need not think about it, making for a seamless plug-and-play experience that remains robust against instability. Our investigations and practical experience demonstrated that one set of mesh parameters works reasonably well for a wide variety of transient scenarios.

There are a great many complications to any algorithm of this nature as deployed to a real system. One of these is how to deal with an elevation profile that has changes at closer intervals than the knot spacing we would like to use. Nonetheless, at its core, Atmos SIM's adaptive knot scheme works according to straightforward principles. It reliably produces sensible meshes placing denser knots at active transients and sparser knots elsewhere. Let's see how.

Heuristics versus self-shadow

How do we choose as few knots as we can get away with while staying accurate? A perfect scheme for adaptive knot spacing adds only those knots that are needed and only those, removing them as soon as they're no longer needed. Unfortunately, no such scheme seems to be available off-the-shelf. We had to look into it extensively and essentially invent one for our purposes.

There are two broad approaches to mesh refinement.

The straightforward approach is comparing results between a solution with the current mesh and a denser mesh. Run each time-step once with the current mesh and once with a coarser mesh, then see if both solutions agree within some tolerance. If results differ, add extra knots. *"Trying again"* in this way runs the model twice, an approach called a *self-shadow-mesh*.

The more challenging approach is to use what people usually call *heuristics*. A few commercial pipeline simulators using adaptive meshes since the early 2000's rely on heuristics of various kinds. While it is easy to get heuristics that work very well for steady conditions, the challenge is to get heuristics that are generally applicable to all the different situations the scheme might encounter.

A good set of heuristics offers the promise of a model that would run at about 150% the speed of one using a self-shadow mesh, because it avoids having to repeat the model solution. That's a significant speed improvement. It must conduct its duties in such a way that we are certain that it is *really* adding enough knots. And it must avoid being overly conservative by identifying exactly where a refined mesh is needed to maintain solution accuracy; adding more knots in all circumstances drowns out any speed gained in the first place.

The curvature of the friction term in the momentum conservation equation is a particular challenge. Rapidly changing values in that term seem to give simpler heuristics the most trouble.

Tighter knot spacing might improve run-time

It might seem surprising that in some cases, subdividing a stretch of pipeline into more tightly spaced knots actually improves the run-time of a simulation. This can happen if there is a sudden requirement for a very dense mesh at the onset of a transient. If that requirement goes unmet, instability or unphysical oscillation might develop. That slows down the solution of subsequent steps.

Judging if a mesh is useful

The big question here is where to add knots. We must keep *truncation error* introduced by the coarseness of our knot spacing within specified limits. In steady state, we keep re-running the model until the mesh is seen to be sufficiently refined. If we are only interested in simulating steady states, simply using a fairly tight fixed knot spacing is probably a reasonable approach: even with a *very* tight fixed knot spacing, a steady state takes only a few seconds, or possibly a few minutes at worst on an extensive pipeline network. Transient scenarios are long enough that the time spent solving this steady state is a tiny fraction of the overall run time. So our focus is on transient steps.

Atmos SIM is unique in employing this sort of adaptive mesh, so it is worth giving some proof that it works. In the next sections, we will look at a few examples of Atmos SIM's adaptive spatial mesh in action. Rather than just explaining what Atmos SIM does, we shall examine the results of some numerical experiments that led us here. A reader who already knows why Atmos SIM uses this approach may skip the next two sections!

Adaptive mesh schemes are judged by how their speed and accuracy compare to baseline fixed meshes. Finest-mesh results are assumed correct as a basis of comparison – something like a 12-foot knot spacing. We define accuracy as the worst disagreement between adaptive and finest-fixed results, in terms of pressure, temperature, or flow rate, sustained along a distance of 0.1 miles. The mesh refinement scheme needs to be given some equivalence between the different quantities driving it: such as pressure, flow rate, and temperature. This choice is a matter of taste, driven by what we consider to be a pressure error *"equally as bad as"* a degree of temperature error. One might choose to define equivalence as being something like 1 psig, 1°F, and a standard flow rate of 30.5 MMSCFD for gas or 22.6 BPH for liquid.

First we compared the various fixed-knot-spacing cases to determine the effect of knot spacing on speed and accuracy. Then we performed similar runs for adaptive cases using various different mesh expansion schemes. We compared execution speed of an adaptive mesh with a fixed mesh producing the same accuracy. These comparisons involved test cases including gas blowdowns, liquid surge scenarios, and the propagation of temperature fronts.

Demonstration: smooth gas transmission

A 12-inch 50-mile gas pipeline runs from a 1 000 psig source to a 500 psig delivery. Both ends of the pipe are pressure-controlled. It is run for one minute at these steady conditions, then the source pressure is increased smoothly to 1 100 psig over a minute. It is finally run for another hour, at 10-second time-steps. This test demonstrates why we can't simply measure truncation error by taking the difference between our shadow solution and the regular solution at shared mesh points.

In this gas line, to accurately represent the significant pressure drop at the downstream end, a fine mesh is needed in that region. But because we impose downstream pressure as a constraining boundary condition, downstream pressure in both solutions is forced by that constraint to be exactly the same, whatever the knot spacing.

Another reason we don't get away with this simple measure of the truncation error is that in steady state, flow rate is necessarily constant along the pipe's length.

Instead, it's better to take as our basis of estimating truncation error the *slopes* of pressure, temperature, and standard flow rate profiles over each knot.

Demonstration: liquid temperature front

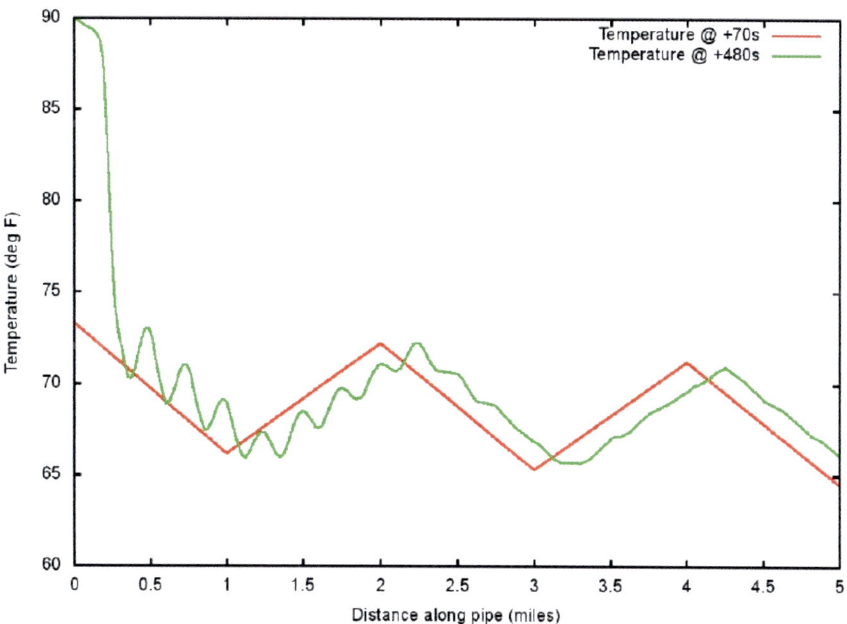

Figure 39 – A thermal transient causes instability which is then locked in.

A 12" diesel pipeline 50 miles long runs from 1 000 psig to 700 psig. Initially the incoming product is at 70°F. Over a minute its temperature is raised to 90°F. The temperature front is then allowed to move down the line, simulated with 1-second time-steps. After two hours the temperature front has disappeared, due to transient heat loss to the much cooler inner layer of soil. This scenario tests the stability and numerical dispersion of thermal fronts. We ran this model without allowing any transient step reruns, for testing purposes.

The line showing temperature was taken on the first step that the incoming fluid's temperature changed. Knot spacing is large, so non-physical oscillations arise, locking in this instability. The plot of knots was taken a few minutes later, when warm fluid moved down the line. Although there's a coherent temperature front, oscillations from that first step of inadequate mesh persist.

Atmos SIM features special measures to avoid this kind of instability. It is a particular problem for temperature as it propagates slower than pressure or flow.

Demonstration: liquid valve closure

A 12" diesel pipeline 50 miles long runs from 1000 psig to 700 psig. After an initial steady state, this shut-down scenario closes its valves over 1 second, at both ends of the line. This is simulated for 12 minutes at 0.1-second steps. It tests the stability and numerical dispersion of hydraulic fronts. The shorter time-steps are chosen to avoid numerical dispersion even at a very tight mesh.

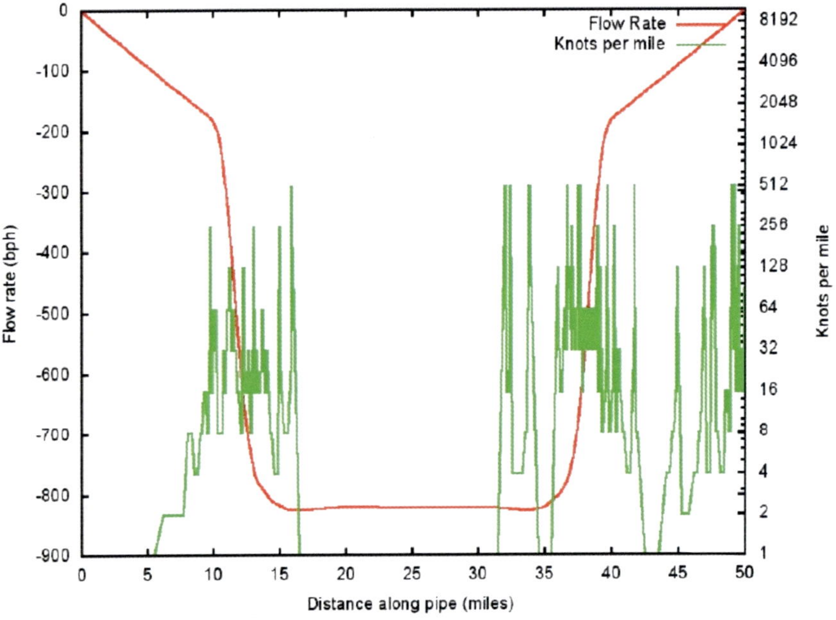

Figure 40 – Knot density throughout the pipe during a liquid pressure surge scenario, taken after the surges have bounced up and down the pipeline a few times.

The plot of flow rate is negative because at this point in time, flow is sloshing backwards. Knots are seen clustering around places where the flow profile has complicated structure. A cluster of knots is also visible at the downstream end of the pipe, because there is a little temperature instability at that end. It's only at the 0.1°F level, but with the tolerances used in this run, that was enough to hang onto a dense mesh there to represent those oscillations. The tolerance chosen here needs adjusting.

The knot density is automatically adjusted to follow interesting goings on within the pipeline, and is automatically returned to an efficient spacing when the goings on have ceased. In the following figures, we look at a few snapshots of a surge event. The knots *"densified"* to capture hydraulic features as they traversed the pipeline. The same adaptive knot algorithm can handle all kinds of transient scenarios, including those in which the thermal effects are of interest not just the hydraulics.

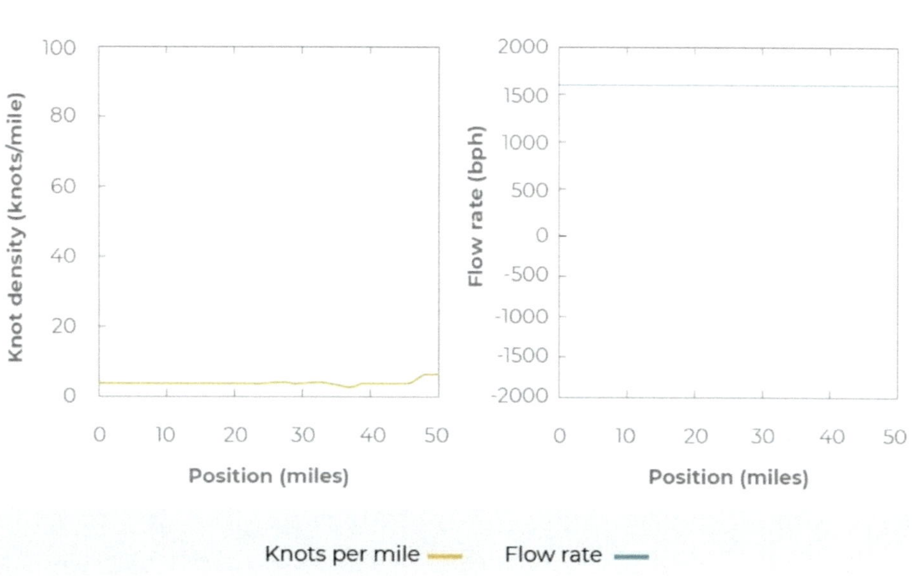

Figure 41 – Knot density throughout a pipeline whilst in steady flowing conditions.

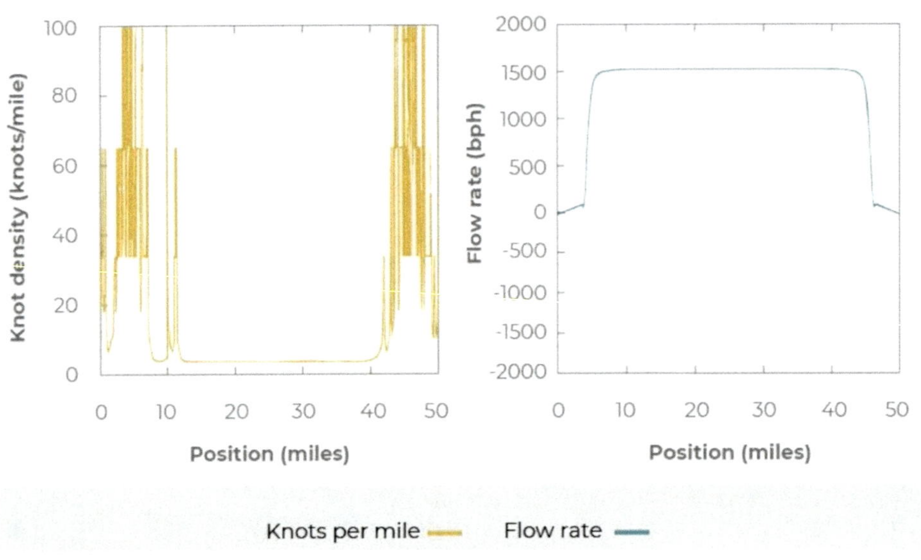

Figure 42 – Knot density throughout a pipeline during a surge event after a shut-down.

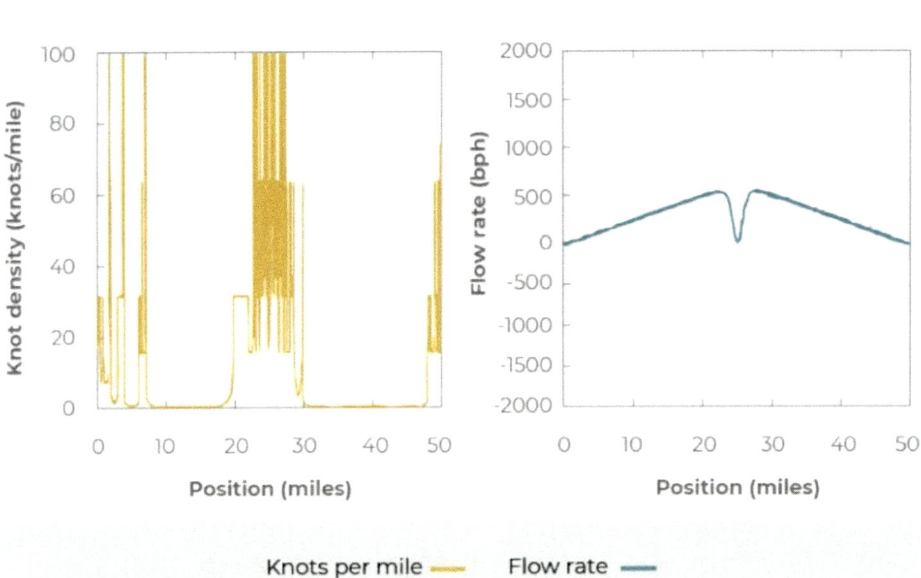

Figure 43 – The reflected surge wavefront makes its way back along the pipeline.

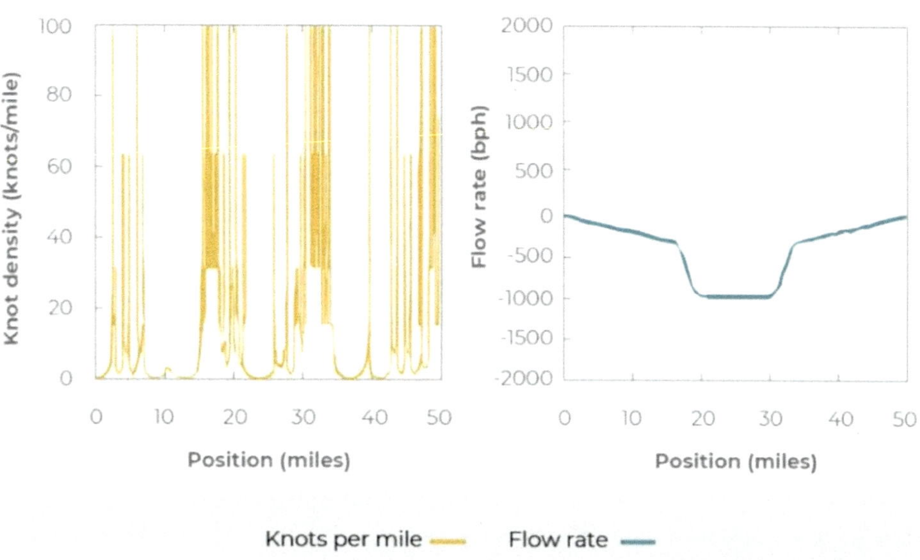

Figure 44 – Knots are placed densely only at locations where something is changing.

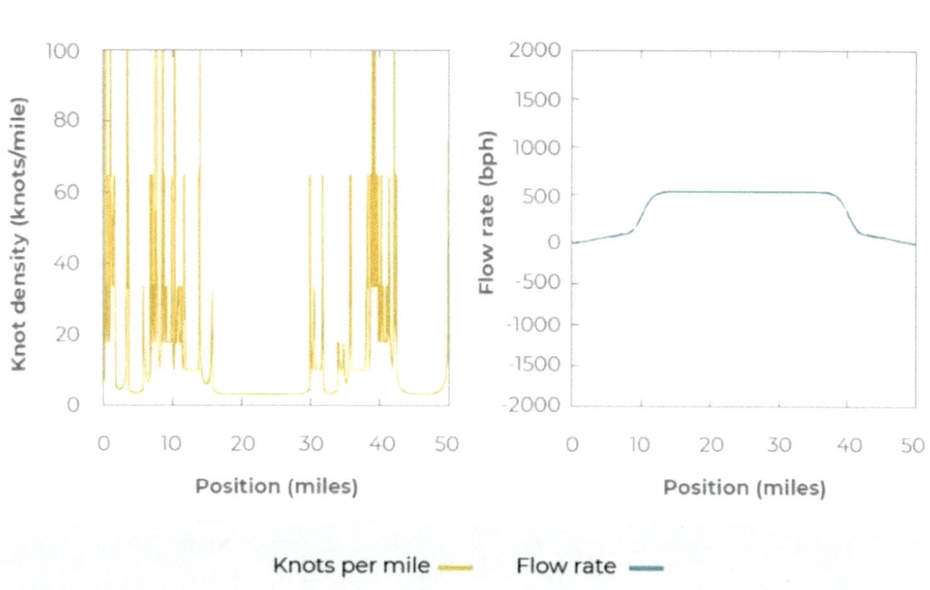

Figure 45 – Gradually the surge flow rate attenuates. Knots return to the usual density.

An adaptive mesh for challenging scenarios

Experience with adaptive meshes in the field have brought up some surprising problems. Adaptive spatial meshes in Atmos SIM are capable of handling many advanced scenarios, by incorporating a whole range of innovative features.

Adaptive knots must suit adaptive steps

Ensuring our adaptive knot spacing always suits the time-step is challenging, because the time-step itself is varying via its own adaptive algorithm.

Fluid and pipe properties vary

All automatic mesh adaptation schemes have in common that they only function inside regions of pipe with constant properties. Our discussion so far has assumed our fluid remains the same everywhere throughout the pipeline. What happens when the situation is more complicated? At a batch interface or a diameter change, the mesh adaptation on one side must be performed independently of that on the other side. Atmos SIM uses moving knots to handle moving batches or composition fronts.

Plotted results feel inaccurate

The box discretization scheme, widely used in our industry, produces a suitably accurate solution with a coarse knot spacing – in many cases 10 miles or more. If a good mesh adaptation scheme sees 10-mile knots to be adequate, then 10-mile knots are what we get! But such a big mesh spacing might *seem* inaccurate when its results are plotted as line segments connecting the knots, because interpolation errors can visually appear to be substantial.

In practice an easy way around this is to maintain a mesh spacing of no more than a couple of miles. For very long pipelines where diameter doesn't change often, this may be inefficient, making fancier schemes for interpolating plotted results tempting. However, arbitrarily fitting curves to make results look more aesthetically pleasing by smoke and mirrors is undoubtedly asking for trouble.

Elevation profiles are detailed

Pipeline operators usually know elevation profiles in great detail. In liquid lines especially, it can be important to consider all this detail in our model. However, if too much detail is thrown directly into an adaptive knot spacing algorithm, it becomes bound to an extremely fine mesh. Burdened with thousands or tens of thousands of knots, the solution would be needlessly slow. Instead, details of the elevation profile can be ignored at first, then reapplied afterwards, with a transformation of the pressure variable for the effect of density (ρ), gravitational acceleration (g) and elevation (z):

$$P \to P + \rho \cdot g \cdot z$$

This allows the normal adaptive mesh algorithm to be used. But this *wouldn't* work if density changes rapidly with pressure, say in a natural gas liquids (NGL) pipeline nearing its vapor pressure, or where there's slack in a liquid pipeline, unless special measures are taken so elevation detail is kept where needed.

Choked flow needs special handling

In general, flow can only choke at the end of a pipe, or at a diameter change. In the event of choked flow, the model physics aren't going to be right unless the simulator explicitly handles the shock. A choke condition would otherwise cause an adaptive scheme to add very many mesh points in a futile struggle to converge within its accuracy at the end of that pipe, yet it will never succeed. Atmos SIM detects choked flow conditions at orifices or ends of pipes, applying a formal choking condition. This lets us handle blowdown scenarios.

Choking is a particularly difficult condition to handle gracefully because when we do explicitly handle it, we are adding flow constraints to the middle of our pipe, which in steady state makes it really easy to inadvertently create a singular system. But hold on: scenarios that solve steady state to a choked condition seem rather exotic, right? Well; it's not that clear. It might have gotten there by some mode switches that the user didn't foresee. Maybe they weren't even right; but the user needs to be able to debug his config, so we really can't ever afford to have it just die! The simulator should always come up with a sensible solution so the user can see what is going on that they don't want.

Tiny flows into stopped laterals

Sometimes the adaptive mesh chooses dense meshes, drastically slowing the simulator down, to capture legitimate physics that nobody is interested in.

An example of this is very slow flow into a pipe where the fluid's equilibrium temperature (approximately the ground temperature) significantly differs to that of the incoming fluid. If the flow is slow enough, the fluid might reach equilibrium over just a few feet; this will necessitate knots of that size. This can happen easily with a stopped lateral branch that is still hydraulically connected to the main line: as the pressure in the main line goes up and down due to hydraulic transients, tiny amounts of hot fluid can be pushed into a stopped lateral, causing an explosion of tiny knots in the lateral, added to adequately model the rapid heat loss of that barely-moving fluid.

We deal with this in our simulator by automatically identifying the condition and handling it outside the normal model. In the absence of an automatically adaptive mesh, this might cause some temperature oscillations to arise in the lateral branch, but that's about it.

Case study: an adaptive mesh for gas transmission

This natural gas pipeline is hundreds of kilometers in length and a meter in diameter, delivering a daily capacity of 1 560 TJ. It transports *coal seam gas* from upstream facilities across a remote hinterland to an LNG export processing facility. One crossing portion extends from the mainland to an island. The pipeline connects to several laterals, and to another network via a spur.

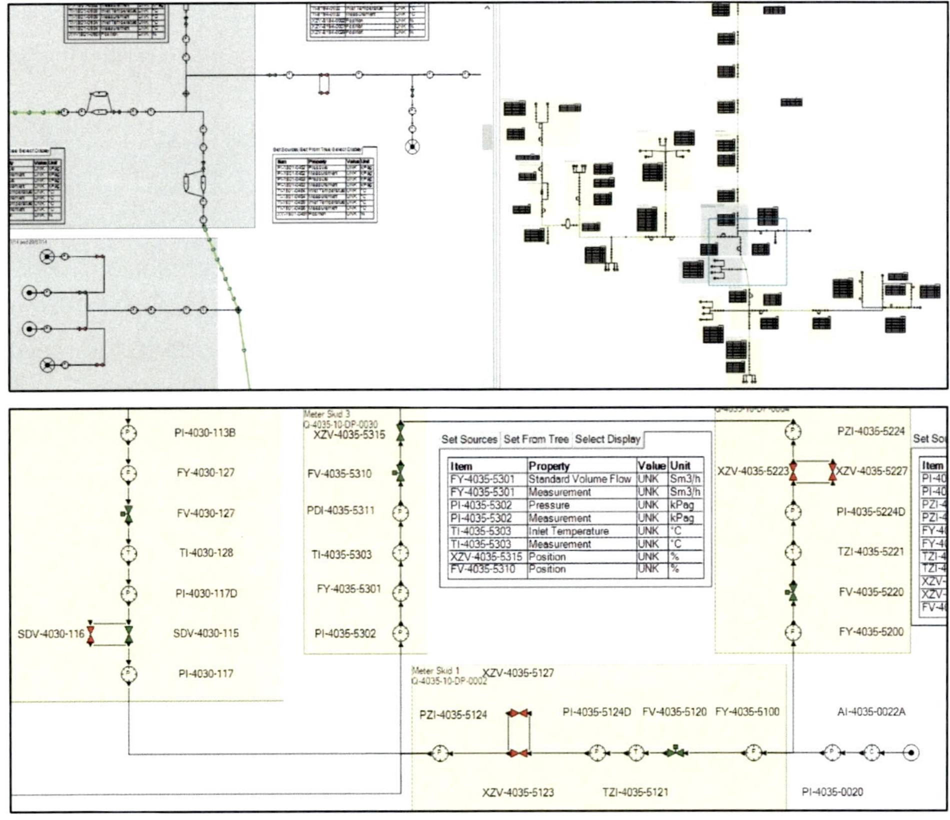

Figure 46 – Part of the gas pipeline (left) alongside a wider network overview (right) and a closer view of a station (bottom).

This particular network is only hundreds of kilometers in length, not thousands, so it is still not inconceivable to simulate it using fixed knot-spacing and fixed time-stepping, albeit slower than the performance would be with Atmos SIM's adaptive mesh. Taking care of the knot-spacing and time-step automatically respects the user by trying not to waste their time, but it is still ultimately an optional feature in this particular case. This is why not every pipeline simulator offers an adaptive mesh algorithm.

In other networks, it is harder for a user to attempt setting an appropriate knot spacing by splitting a pipeline into regions: an adaptive mesh is *indispensable*. For example, there is a pipeline with a tremendous temperature gradient in a riser for 1 km, as very hot gas comes out of a production platform, and then much less change in temperature along the sea floor for hundreds of km. It is technically possible to handle that manually by tweaking the meshes. But for simulating extensive networks in all the full complexity of their online and offline scenarios, any fixed knot spacing the user chooses to reasonably capture what a pipeline is doing expends effort on millions of pointless calculations.

With Atmos SIM we don't need to think about exactly where and when tighter knots and shorter steps are needed: as it is all done automatically. The same algorithm seamlessly handles online scenarios, offline studies and look-aheads by adapting knots and steps to always stay fast solving and accurate.

3 – A SIMULATOR'S PERFORMANCE

±0.1%?

"The trouble with the world is that the stupid are cocksure and the intelligent full of doubt."
– Bertrand Russell

3.1 – CERTAINTY, ACCURACY AND SENSITIVITY

Where do uncertainties arise in simulated results?

Accuracy doesn't concern one specific aspect of a simulator or a model. It is a function of the overall end-to-end system, including the meters themselves. [29] Design parameters might be uncertain, or might be varying over time. Some degree of *uncertainty* or *error* is bound to come from all of the following: [26]

- Input measurements, including lags and skews in time
- Pipe parameters, such as wall roughness and inner diameter
- Thermal properties such as ambient temperature and soil conductivity
- Fluid properties, such as viscosity, heat capacity and even the density itself which is subject to the uncertainty of the chosen equation of state
- Model approach, such as whether to use a transient ground thermal model
- Numerical errors within the calculations of the simulator's solver
- Empirical correlations in the model equations, such as friction factor
- The state estimator interpreting meter data into boundary conditions
- Flaws in the incoming data: each meter itself, data conversion, or polling

Conducting a *sensitivity analysis* shows us the potential impact of any of the above, or several at once. What's the impact of changing a parameter? How does that affect a model? Can we quantify how it propagates to the results? [29]

Flaws in the incoming data

Analyzing data from the field is quite a specialist affair. Any comparison of model vs meter data should bear in mind that metered data is subject to its own inaccuracies. [29] Every meter will be subject to its own combination of:

- an *offset*, which may be referred to as *bias*
- and a level of *imprecision* known as *repeatability*

We might also need to consider other factors, like analogue-digital conversion. This all means there's an inevitable degree of noise in a practical engineering calculation. The relevant standards outline a few terms with some concrete definitions. API 1149 defines the *precision* of a value and its *bias* with reference to that true value. The accuracy or error of any measured or calculated value is a function of both of these.

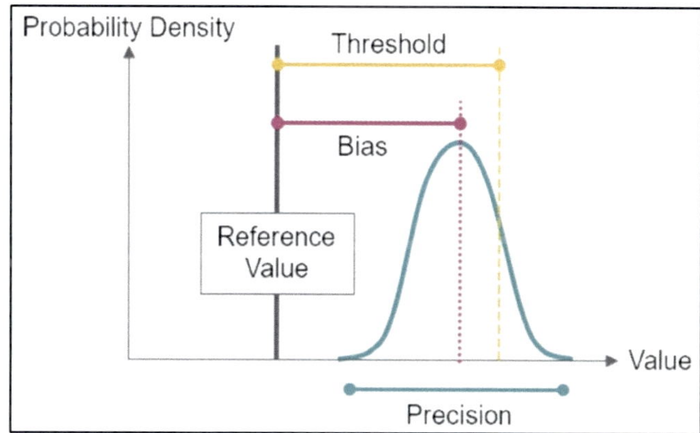

Figure 47 – Precision and bias (API 1149) [30] This probability function may not be normally distributed if it includes the effects of analogue-to-digital conversion.

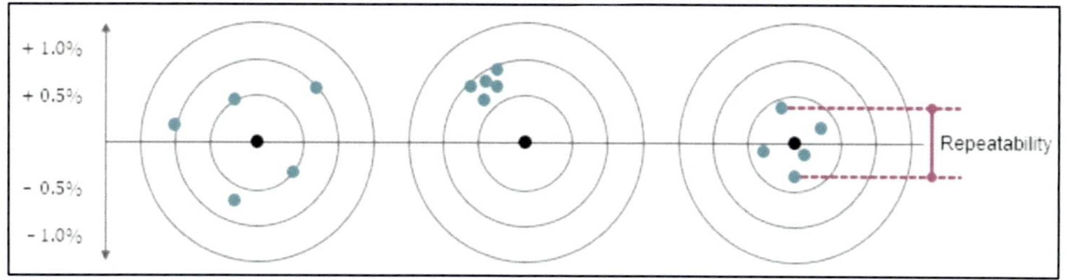

Figure 48 – Repeatability and what it means to be accurate. True value is at the center. Imprecise and unbiased (left). Precise and biased (middle). Precise and unbiased (right). [31]

How accurate do we need to be?

A simulator might always reach the *correct* result, but how *accurate* does it need to be? The answer depends on what a model is going to be used for. The purpose of an application dictates its required accuracy, speed and complexity. Different applications of the same system each have their own requirements. The required accuracy must be balanced with the cost of gathering data, building a model, and maintaining it in operation. [29]

In a batch tracking or composition tracking application of an incompressible liquid, operating conditions of pressure and temperature are only important to scheduling personnel to the extent they influence flow rates and flow paths. Those practical considerations take precedence over exact accuracy. [29]

A simulator deployed for operator training is again more concerned with other matters than absolute accuracy.

In a forecasting look-ahead or predictive what-if analysis on a gas pipeline, the nominations or actions driving a simulation are themselves subject to massive uncertainty. A short-term forecast of average upcoming supplies and deliveries might easily be off by 10% or more. Results are only expected to give a ball-park estimate of future conditions. A look-ahead user expects quick estimates, not exact accuracy. [29]

In many design applications, the engineering team is initially using the model to assess options and is bound to re-check the final detailed calculations as a matter of due diligence, and then apply generous safety margins to boot. While a typical design user does need a certain level of accuracy, at some point beyond that he would prefer fast solutions to further significant digits. For offline studies, scenarios should be convenient to launch in large numbers, with easy modification of the underlying pipeline itself, each progressing of its own accord recreating the behavior of control systems and routine actions by operators. Many long-duration scenarios need to be run to investigate the many possible alternative options. A transient model might be used to identify additional capacity in a complex pipeline network. Some studies extend over *decades* of operation, and steady state models might be best suited for those.

Then, when the options of interest have been selected, they can be simulated again. Finally, designers always apply *safety factors* in various guises. The point to appreciate is that for offline studies, the challenge isn't just accuracy, but launching scenarios and analyzing results manually or via an optimizer. [29]

So who *does* need obsessively accurate simulations?

There seem to be two chief applications crying out for ever-tighter accuracy: model-based leak detectors, and real-time modelling in a broader sense. How accurate do users of those applications expect their simulated results to be, and has our industry been meeting those expectations?

Model-based leak detectors typically use a *real-time transient model* (RTTM) driven by pressure meters to calculate a packing rate, which is then compared with the packing rate reported by flow meters to look for a leak. If the line is apparently unpacking faster than the flow meters say it is, that signals a leak. But to find a small leak, or usefully locate any leak, a high level of accuracy is needed from both the physical model itself and the instrumentation driving it. A real-time model deployed live on site must calculate operating conditions very accurately throughout the pipeline. If it is used for composition tracking, for instance, miscalculating the quality of an arriving fluid at a given moment could disrupt the receiving station. And those very accurate calculations must happen reliably within a mere few seconds (the time-step). A very accurate model is quite useless in this application if it can't keep up in real-time. [29]

We are also now asking: how accurate must a simulator be for an optimizer to be useful at aiding daily operations? For an *operational* optimizer to become a reality, doesn't it have to give us any meaningful recommendation within a pragmatic margin of error? If our accuracy is within 1% of true pressure, and our optimizer is recommending us to change our pressure by 0.5%, is that recommendation of any use whatsoever? We are already within tolerance of the optimal operation! Actually, it might still be worthwhile, as there's a good chance an optimizer still gives us a better solution than we have now. Probably the same issues generating the 5% cost error in our baseline case compared to operation of the real pipeline are generating the exact same 5% error in the optimizer solution for the same reasons. So if the optimizer gives us cost savings of 2%, there's a good chance that if we follow its recommendations 2% savings are realized.

One might finally ask: how accurate must we be to achieve that most elusive of aspirations: fast, accurate simulation of pipeline flows beyond single phase?

The success of emerging innovations in pipeline simulators, namely fast operational optimizers and fast simulation of flows beyond single phase (including a *"slack"* model), hinges on the accuracy of the end-to-end system (meters, model, and auxiliaries). This intimate reliance stems from the value of an application being how well it anticipates a systems behavior, with certain systems being utterly *chaotic* in a mathematical sense (butterfly effect).

If we're serious about accomplishing great performance in a given application, we should think rationally about where uncertainties lie in our system.

How accurate can we feasibly be?

Accuracy is hard to quantify. It's possible to get good steady state results in a badly set up model whose transient results are atrocious. And it's possible to do good surge studies with a model whose steady state results are mediocre.

There is arguably no one *inherent* level of inaccuracy. If we reduce the size of our mesh, taking shorter steps and tighter knots, the same engine simulating the same model would be more accurate. Another consideration is that if a significant transient occurs, transients may cause significant short-lived local inaccuracies: certain localized spots end up being less accurate than the rest.

Moreover, as much as it sounds like a get-out clause, a real-time simulator's accuracy strongly depends on how good the pipeline's instrumentation is. How far apart are meters for pressure, flow rate, and temperature? [26] Is ambient temperature metered? Is quality reliable or does it drop intermittently? Do co-located meters agree? Are all meters well-calibrated? Do we meter the fluid properties?

If the model is accurate but the meters are not, how are we meant to validate it? It is difficult to say what the *"true"* value is in a real-time model. Which meter do we trust, and to what extent? True value is almost never known, as inputs driving the model are inherently inaccurate. *"Simulation accuracy should not be viewed as an absolute measure of quality of results, but as a link in the overall process of establishing confidence in the results."* [29]

Even on a pipeline with perfect metering, almost every model is set up such that every once in a while, it might produce dreadful results localized in some region along the pipeline, due to some specific configuration error in the local vicinity, or nonconvergence due to local conditions, or some other reason. There are so many *unknowables* prior to model-building that something like 1% accuracy is only achievable at all with extensive online tuning.

All of the above caveats haven't stopped vendors issuing out specific figures, and customers demanding that those figures be fulfilled. Some figures might be issued with the best intentions, but end up being almost meaningless. If a tickbox exercise needs to be fulfilled at some point, someone along the way might well be tempted to make something up, or at least agree to it in haste. Some figures feel suspicious. 0.06% sounds like it has a lot of significant figures; let's say 0.054% - even better! Whatever figure we encounter, we must bear in mind that it refers to an accuracy *given whatever choices they made*.

Even if we are *"100% sure"* about a figure, it must certainly only apply within certain reasonable bounds. What if a pipeline is vastly oversized, or flowing very slowly? What if its flow rate fluctuates suddenly and dramatically? What does a steady operation even mean in a gas pipeline that packs and unpacks?

Is there a thorough, systematic way to check a whole pipeline model for sensible accuracies? Or do we just wait until someone reports an obvious issue and work back hoping to happen upon some source of gross inaccuracy? A good development team writes something akin to unit tests in a codebase; automated tests that examine a pipeline model for wrong-seeming behavior.

One possible cause of inaccuracy is the adaptive mesh algorithms. One might reasonably say that by reducing the mesh size, we can get agreement with theoretical results to better than 0.1%. In practice we are not going to have meters and physics that are accurate within 0.1%, so this tells us that the numerical method and the simulator are not becoming a major source of error.

Other potentially inaccurate parts of a pipeline simulation engine include:

- the correlation (such as Colebrook-White) used to calculate friction factor
- the equation of state (such as Peng-Robinson) used to calculate density
- the numerical solver itself

There are also *"rounding errors in physical unit conversion; rounding errors in the actual computation; residuals from numerical matrix inversion, etc."* [29]

The following really ought to never be a source of inaccuracy. In a mature simulator, they are never the culprit, and only constitute significant sources of error (insofar as data-spoiling bugs are concerned we assume there are none):

- the data manager
- the state estimator
- the learner (online tuning)
- the results logging and display

Last but not least, inaccuracies might also creep in during the phases of a project's execution, which in many cases span many months and years.

In general, we can get a liquid pipeline model within a 1 psi (0.07 bar) average error most of the time, via a combination of manual offline tuning (*calibration*) and automatic online tuning (*learning*).

Gas pipelines are harder to simulate accurately. It is doable for a pipeline simulator to be better than 0.5 bar. If we aren't hitting that, the reason is typically that the model needs to be better tuned by learning or by calibration.

So how accurate should we strive to be? In practice, there is so much uncertainty around what bends and fittings there are within a station, and their overall effect is so insignificant compared to frictional loss along a long-distance pipeline, that worrying about factors for losses across every valve etc. – as listed in the CRANE handbook for instance [24] – is not a good use of a project engineer's time.

Does tuning improve accuracy?

Simulators have a lot of tweaks available to the user. Even if a simulator's physics are wrong, an engineer might be able to make it reproduce the behavior of a pipeline reasonably well by adjusting tuning parameters. We use multiple test sets for checking our models and only use some of them for parameter fitting. Complicated models might have so many parameters that they can be fitted to almost any set of data. The only real test is to use clean datasets that haven't been used for parameter identification. [29]

The catch-all term *"tuning"* is used by different people to refer to a number of quite distinct tasks. The first is the creation and configuration of a pipeline network model for the simulator to simulate in the first place, and we touched on the principles of that task in the first chapter: *"Model-building"*. The adjustment of that model's parameters before launching a simulation is referred to as *calibration*, and is detailed in the chapter titled *"Steady states"*. We also take a close look at adjusting parameters automatically during the course of a simulation while it is running – a feature we may refer to as *learning* – in a later chapter, *"Live on site"*, when we talk about real-time transient models (RTTM).

3.2 – TROUBLESHOOTING

What do we do if something goes wrong?

It is not typical, on the whole, that the physics, numerical algorithms or simulator are grossly inaccurate. Most issues are caused by bad input data, and though it's easy to shrug off *"user error"*, one should never jump to blame the user for bad inputs! The good news is, we can do something about this. [29]

Although isolating the cause of bad behavior in a model has developed a reputation as being something of a dark art, there are loose rules of thumb that come with experience. For example, if we see oscillations, they might be due to some combination of the numerical solver and the adaptive mesh algorithm.

3.3 – VALIDATION

More than checking a few results

Engineers are critical thinkers who don't accept a result until they somehow cross-check it, understand where it has come from, and why it is the way it is. To be worth its salt, any simulator had better be giving us the right answer. The unglamorous word experts use to express this important idea is *validation*.

Once a simulated model earns their trust, only then can it prove its real value. Particularly if a team chooses to operate closer to physical or commercial limits, a simulator and a model must be subjected to rigorous validation. This demonstrates it works to a satisfactory standard. To meet professional and legal requirements, it might even be necessary for qualified engineers to re-examine simulated results by verifying them using their own detailed calculations.

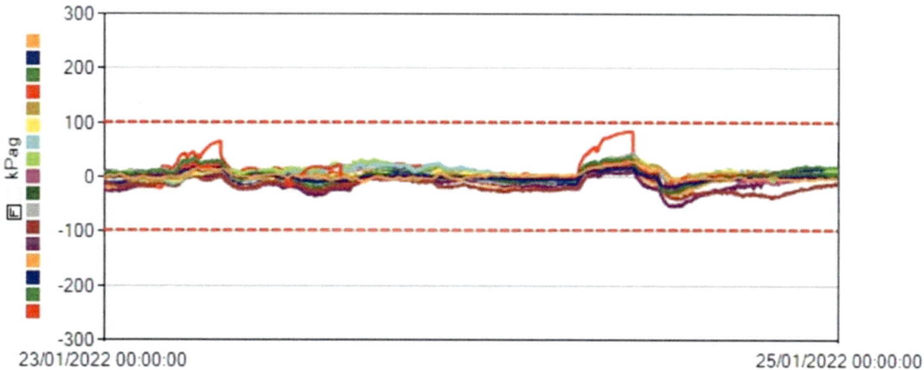

Figure 49 – A trend of the *differences* between modelled and measured pressure at each meter indicates this simulation remained accurate within 100 kPa throughout two days of transients

It is reassuring that models have been successfully used to simulate pipeline behavior for decades now. So although we open up a Pandora's box of possible sources of inaccuracy whenever we closely examine some part of a pipeline simulation system, ultimately the industry's practical experience in the field proves our accuracy has been fine, for all the applications we've been serving so far. [29]

Deciding *how* to validate a model, and establish its accuracy, is a hurdle in itself. Validation can be done by comparing simulated results against analytical results that are more thoroughly explained in later chapters in this book. It can also be done by subjecting the model to any of a number of reasonable tests:

- *Analytical solutions* for special cases of shut-in, steady, or transients
- *Numerical approximations* manually confirmed using spreadsheets
- *Common sense* ball-park figures, reasonable profiles, mass balances
- *Solving again* in the same pipeline simulator with a more refined mesh
- *Solving independently* using different vendors software packages
- *Field tests* using historic datasets to verify what happens in reality on site

We use all the above validation approaches, but the best way is to review real site data.

This may all seem daunting or obsessive at first, but we should not be oblivious to the importance of validation. It isn't merely a means to an end to quantify the accuracy of a model or simulator. Validation is important on its own merit. Thinking about validation raises some poignant questions. How do we rigorously quantify the uncertainties within our results? Can we formally bound the prerequisites of each band of accuracy, caveated with the practical circumstances within which it can be absolutely guaranteed (1% during transients of this size, 0.5% if steady as defined thus, etc.)? If we commit to being accurate to X% in Y conditions, then on what basis are we saying that? And if a pipeline simulator falls short somewhere, what would constitute *"best efforts"* here?

It is almost unheard of for a vendor of simulation software to properly qualify their claimed accuracy and performance with rigorous metrics. This can become a point of contention, as software vendors and project teams are accustomed to selling their product or project as nothing short of *perfect*. That is plainly untrue. No meter is perfect, so there is no way a simulator's results will be perfect, just on that premise alone. Arguments can ensue about who can be *"best in class"*, maybe. What matters is that a model and simulator together meet a sensible threshold of end-to-end accuracy, enough to reliably be fit-for-purpose.

Validating against manual calculations

There isn't always an analytical solution to check what we've built. That is why we resorted to a solver to perform numerical approximations in the first place. Apart from special cases, equations governing variation in pressure, flow and temperature in a pipeline cannot be solved analytically: there is no attainable true solution. Luckily, there happen to be a fair number of special cases.

The whole enterprise of validating a model against hand calculations hinges on being able to isolate separate aspects of model behavior. If we can't isolate behavior, we cannot deduce *why* a model isn't giving us a number we expect.

We might propose that any simulation engineer validates a model in this order:

- At a supply with known pressure, temperature, and composition, validate the reported fluid properties
- At a different point during a shut-in condition, check the fluid properties using the equation of state to calculate the density
- In an isothermal frictionless steady state, check flows versus pressures
- In an isothermal steady state with friction, check the friction losses
- In a thermal steady state with friction, check all the results again
- In an isothermal fast-transient surge event, check versus Joukowsky
- In a thermal fast-transient surge event, check versus Joukowsky

A simulation engineer should be at ease setting up spreadsheets to conduct these calculations, and check them manually by hand *("back-of-the-envelope calculations").*

Shut-in checks

Test one: fluid properties and equation of state

We can validate fluid properties using hand calculations by looking at any points where the pressure, temperature, and fluid composition are known.

Simulated results can be checked against property correlations and in some cases against physical equations too. An easy validation check in a gas pipeline is to compare the reported standard density for a given composition against the ideal gas law, accounting for its molar mass. Supercompressibility and density at pipeline conditions can be checked using the equation of state.

Steady state checks

There are advantages to validating steady state results before any transients:

- The quantities to be verified are simple numbers such as pressure at the end of a pipe, rather than plotting trends over time or plotting profiles over distance. Hand-calculating a trend is cumbersome – an evolving profile even more so.
- There is no packing for us to take into consideration. Even on a liquid pipeline, there is enough packing during transient operations to cause us trouble if we are trying to validate results to several decimal places.
- Because there is no packing, flow rate is therefore the same along a pipeline, making splitting points, mixing and balances easier to calculate analytically.
- The heat transfer calculations are simpler than during a transient. At steady state, the heat transfers are balanced at an equilibrium, so a familiar easier approach can be used – the *overall heat transfer coefficient* (OHTC).

Checking a liquid pipeline's flow at a steady state is relatively straightforward. In the chapter on steady hydraulics, we will walk through one common way to *"check"* a simulated liquid flow rate result – deduced from inputs like pressure drop, fluid properties and pipe dimensions – against engineering calculations.

For a gas pipeline, manual calculations even at steady state are more involved.

Test two: isothermal frictionless steady flow

We should be aware of the limitations of the manual calculations we employ to check against. The only pipeline fluid that truly approaches incompressibility is water, and the only one approaching a laminar regime is cold, slow crude oil in small-diameter pipes. Nonetheless, these remain useful cases for validation.

Saying a pipeline is frictionless is never true to reality. Because of the way the Moody chart works, being log scaled, even a minuscule roughness is still going to produce friction in far turbulent flow – a friction significantly different to zero. Before asking a pipeline to flow, we can run a stopped-flow scenario with an upstream pressure constraint and a downstream zero-flow constraint. We check that the downstream pressure is a value we expect. If it is, the elevation profile and density are correct. If it isn't, either the elevation or the density might be to blame. We use high enough pressures to avoid encountering slack here.

Then the simplest scenario is arguably steady, frictionless and isothermal. A frictionless assumption incurs no pressure drop due to friction. This is only reasonable at low flow rates where the main factor is gravity due to elevation. An isothermal assumption should isolate hydraulic calculations from thermal calculations. Atmos SIM offers an isothermal setting for testing purposes maintaining all pipes and fluids at a set temperature everywhere at all times. If a pipeline simulator does not offer this feature, we assign the overall heat transfer coefficient a high value, and set the ground to the desired temperature. We initialize the model at nominal operating conditions, then make changes in each of the inputs. When the response reaches a steady state, we check those simulated results at the end of the pipeline match the results expected via manually calculated balances. In a steady state, terms with a time derivative go to zero, by definition. Isothermal frictionless flow reduces to the original Bernoulli's equation for an incompressible fluid (see the Primer appendix). We can check liquids with that.

Gases are harder. Technically, if a gas flows in an isobaric (constant pressure) or an isochoric (constant volume) process, that is to say its pressure and volumetric flow rate do not change simultaneously, then a gas flow could be considered incompressible below 30% of sound speed. But maintaining gas density to be constant is quite an unrealistic assumption, and *"frictionless"* gas flows do not really happen at all.

Test three: isothermal steady flow of a gas with friction

Are the frictional losses right using the same upstream pressure and downstream flow boundary conditions? If not, we know the elevation part and density and fluid properties are right from test one and test two, so we check Reynolds Number, and friction factor if possible as those must be what's wrong. We need a broader framework than the basic form of Bernoulli's principle to deal with most liquid lines. We certainly need more sophisticated techniques for meaningful calculation of steady flow along a gas pipeline.

The solution is in an excellent paper from the Pipeline Simulation Interest Group (PSIG) conference proceedings that is the best modern reference on pipeline model validation, *"On Simulation Accuracy"* by Lagoni & Barley. A steady-state isothermal ($T = \text{constant}$) flow of gas with a constant supercompressibility (Z) and molar mass (m_r) in a horizontal rigid pipeline of fixed inner diameter (D) with constant friction factor (f) follows this flow equation relating pressure drop ($\partial P/\partial x$) to fluid density (ρ) and flow velocity (v): [29]

$$\frac{\partial P}{\partial x} = -\frac{(\rho \cdot v)^2 \cdot \rho}{\left(\rho^2 - \frac{(\rho \cdot v)^2}{c^2}\right)} \cdot \frac{f}{2 \cdot D}$$

This does have a textbook analytical solution, albeit in an inconvenient form: [29]

$$\frac{1}{2} \cdot \left(\frac{m_r}{Z \cdot R \cdot T}\right) \cdot \left(P^2 - P_{\text{upstream}}^2\right) - (\rho \cdot v)^2 \cdot \ln\left[\frac{P}{P_{\text{upstream}}}\right] = -(\rho \cdot v^2) \cdot \frac{f}{2 \cdot D} \cdot x$$

Test four: thermal steady flow of a gas

It is possible to directly calculate downstream temperature and account for its effect on pressure profile if we maintain the original set of assumptions: ideal gas, rigid horizontal pipe, steady state flow, constant fluid properties other than the density, and constant friction factor. From the same paper (*"On Simulation Accuracy"* by Lagoni & Barley): [29]

$$\frac{\partial P}{\partial x} = -\frac{\frac{4 \cdot U}{D} \cdot \left(T - T_{\text{ground}}\right) \cdot \rho \cdot v \cdot \frac{\partial \rho}{\partial T} - \frac{f}{2 \cdot D} \cdot (\rho \cdot v)^2 \cdot \left(c_P \cdot \rho - \frac{(\rho \cdot v)^2}{\rho^2} \cdot \frac{\partial \rho}{\partial T}\right)}{c_P \cdot \left(\rho^2 - \frac{(\rho \cdot v)^2}{c^2}\right) - \frac{(\rho \cdot v)^2}{\rho} \cdot \frac{\partial \rho}{\partial T}}$$

$$\frac{\partial T}{\partial x} = -\frac{\frac{(\rho \cdot v)^4}{c^2 \cdot \rho^2} \cdot \frac{f}{2 \cdot D} + \frac{1}{(\rho \cdot v)} \cdot \left(\rho^2 - \frac{(\rho \cdot v)^2}{c^2}\right) \cdot \left(\frac{4 \cdot U}{D} \cdot \left(T - T_{\text{ground}}\right)\right)}{c_P \cdot \left(\rho^2 - \frac{(\rho \cdot v)^2}{c^2}\right) - \frac{(\rho \cdot v)^2}{\rho} \cdot \frac{\partial \rho}{\partial T}}$$

Where all symbols are as in the previous test. The steady state assumption allows us to use the overall heat transfer coefficient (U) – the OHTC – to describe the rate of heat transfer between the fluid in the pipeline and the surrounding ground. There is also now a dependence on speed of sound (c) and specific isobaric heat capacity (c_P).

If each equation is applied sequentially to short intervals, for instance in a spreadsheet, this can even solve for the pressure profile or temperature profile while relaxing several of the original assumptions, by accounting for varying fluid properties and a non-ideal gas. It is already getting rather unwieldy to validate by manual calculation at this point.

Dynamic transient checks

Once we are satisfied that steady-state behavior is sufficiently accurate, only then can we deal with the dynamics. It is wise to look at the timescale and shape of the responses and ask whether they appear reasonable. But we can do more than that. Checking transient operations and dynamic behavior without plant data is *not* impossible. There are rules of thumb interspersed with good engineering judgment that help us check key model results against the profession's wider experience. This is often neglected; many engineers had previously assumed it's impossible to thoroughly validate transient results.

For example, we can check the results from the following transient scenarios:

- A wavefront at the front of a disturbance moves at the speed of sound
- How long a blowdown operation takes to drain a section of the pipeline
- The magnitude of upsurge and downsurge due to a sudden transient

Test five: an isothermal surge

How big a change in flow accompanies a change in pressure, or vice versa? We make a pressure surge upstream and watch it propagate. The isothermal speed of sound (c_T) is calculated via the fluid's *isothermal* bulk modulus (B_T) and density (ρ) as follows:

$$c_T = \sqrt{B_T/\rho}$$

If this is a liquid described by a bulk modulus equation of state, isothermal bulk modulus is a known input parameter for the fluid. For an ideal gas, the isothermal bulk modulus is identical to pressure ($B_T = P$). But for a real gas the isothermal bulk modulus must be calculated – numerically is fine – from the density at two points, using its definition:

$$B_T = \rho \cdot \left.\frac{\mathrm{d}P}{\mathrm{d}\rho}\right|_T$$

The Joukowsky equation gives an accurate short-term transient response. Change in head (ΔH) is calculated from speed of sound (c_T) – just calculated with an isothermal assumption – with change in fluid velocity (Δv) and acceleration due to gravity (g):

$$\Delta H = -\frac{c_T \cdot \Delta v}{g}$$

Multiplying head by density gives pressure. So the magnitude of a pressure surge (ΔP) also depends on the fluid's density (ρ):

$$\Delta P = -\rho \cdot c_T \cdot \Delta v$$

Test six: an adiabatic surge

Now let's test the hydraulics with the transient thermal model enabled. A pressure surge travels quickly enough that it is almost entirely unaffected by heat conduction through the pipe wall, so we can say that this process is adiabatic. Any process that is reversible and adiabatic is – by definition – isentropic. We initiate a pressure surge at the upstream end and watch it propagate along the pipeline. Are waves observed to propagate at the speed of sound (c_s) as calculated via the *isentropic* bulk modulus (B_s)?

$$c_s = \sqrt{B_s/\rho}$$

Isentropic bulk modulus (B_s) can be found from isothermal bulk modulus (B_T):

$$B_s = \left(\frac{c_P}{c_V}\right) \cdot B_T$$

Via the same Joukowsky result, change in head or pressure on an isentropic basis is:

$$\Delta H = -\frac{c_s \cdot \Delta v}{g} \quad \text{i.e.} \ \Delta P = -\rho \cdot c_s \cdot \Delta v$$

Sharpness of a traveling discontinuity smears out over time

The aforementioned paper by Barley & Lagoni identifies a problem: *"the sharpness of [a] discontinuity smears out over time – this effect (known as numerical diffusion) is more apparent with finite difference methods than it is with the method of characteristics."* [29] This too was the subject of a Pipeline Simulation Interest Group (PSIG) conference paper published in 2021 by one of the authors of the present work.

Validated with field data, an idealized compressor trip scenario was considered as a test-bed for examining the evolution of a surge pressure wavefront as it moves through a gas pipeline. A second scenario, a surge like one from a pipe rupture, was simulated by causing a sudden drop of 16 bar over a period of 0.1 second. This produced gas velocities just short of the speed of sound at the point of rupture. The test pipeline was long enough that moving at 370 m/s, over a simulation lasting 80 seconds after the initial time of trip or rupture, surge effects did not reach the far end of the line. Scenarios were run for an isothermal case and for a near-adiabatic case, with three different numerical schemes to determine if dispersion is driven by a physical effect, or if it is merely a numerical artefact.

It was found that manually adjusting simulated results can match the shape of a real surge wavefront. We discuss this in rapid transients chapter.

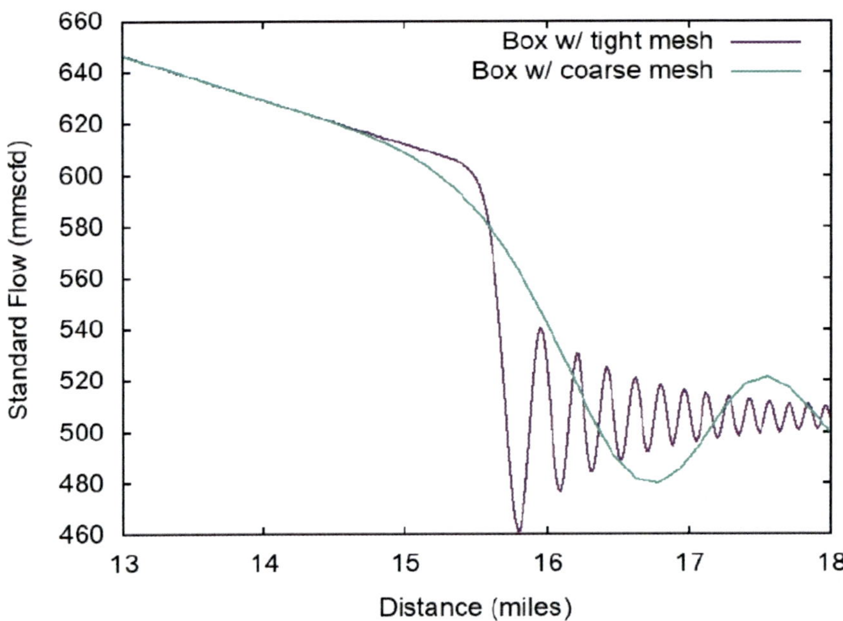

Figure 50 – A scenario in which a tight mesh has more oscillations than a coarse mesh. This paper examined the numerical dispersion of the wavefront using different solvers. [32]

Validating against common sense

Any discrepancy between inflows and outflows should be accounted for as an accumulation or depletion of inventory. As we discussed earlier, provided the fluid does not experience shrinkage or expansion on blending, conservation of mass is equivalent to conservation of standard volume. If the composition of a gas is constant it's also equivalent to do this check in units of combustion value.

Test: inventory and conditions changing over time

This test can be adapted for liquids, we present gas pipeline nominations here.

A clever way of checking our inventory calculation is to simulate a scenario applying time-profiles of flow rate and composition that result in a balanced mass entering and departing over some extended length of time. For example, run the same balanced nomination cycle (on a daily basis, say).

We repeat this cycle, many times in succession, for some long period (say a month), such that every repeating-interval during our simulation, all conditions should be exactly as they were initially. The process should be repeatable, returning the pipeline to its original state, so we expect simulated conditions, inventories and so forth reflect that reality. So we check that:

- all results stay consistent, repeating exactly after the balanced period
- inventory changes to exactly reflect the incoming and outgoing flows

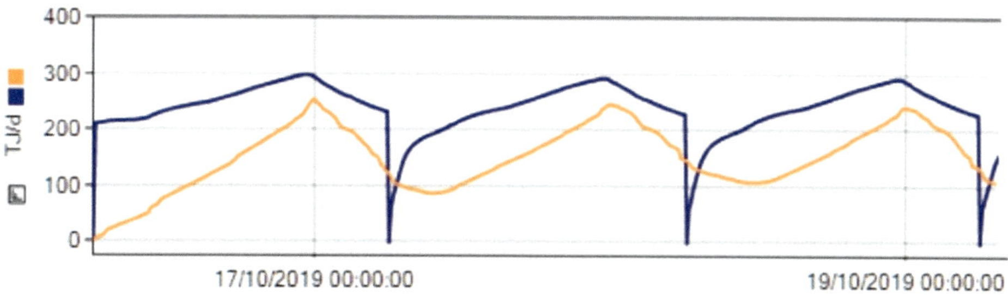

Figure 51 – A rolling-average logic block implemented within Atmos SIM to calculate energy flow rate over a gas pipeline's preceding 24 hours, resetting daily.

Validating against other simulations

Using a tighter mesh in the same simulator

Using a tighter mesh – *"grid refinement"* – gives a solution that is presumably more accurate than a standard mesh or an adaptive mesh. We don't use the tightest possible mesh in all circumstances as that would be needlessly slow. We can compare the relative error in the recorded downstream pressure and temperature for grids spaced at various other intervals against the tightest. [29]

Using a different vendor's simulator

Although it is reassuring to see several independent calculations agree, it is not impossible for several vendors to have made the same false assumption. So validating software against other software isn't as bulletproof as it appears.

Nevertheless, until we have historic site data once the pipeline in question is commissioned, and once the team ensures two separate vendors simulators pass their *"manual"* validation for the simple scenarios, then the best anybody can do for the more complex scenarios is to validate against other software.

Validating against site data

Acceptance tests and historic datasets

Once we've made internal checks on the model, we need real process data to build confidence in a model by verifying its predictions on site. If the pipeline has not yet been built and commissioned, we don't have that. Historic datasets gathered from presumably similar pipelines only take us so far.

Software to be deployed live on site is handed over by a project team according to set procedures. Milestones, typically a *factory acceptance test* (FAT), then an *integrated factory acceptance test* (IFAT), and at the end a *site acceptance test* (SAT), may be years apart. This presents a challenge in project delivery. *

Before it is fully operational, a pipeline system logs data whilst it waits in shut-in. Are the reported fluid properties as expected by the model's calculations?

* Some add a *model acceptance test* (MAT), inspecting the model in isolation.

When the pipeline starts flowing during commissioning activities, we have our first opportunity to validate model performance against field process data from site. When a historic dataset for the pipeline in question is available, we can replay it to see how the simulator performs in anger. We start validating a model by simulating it during a *steady operation*. Errors in our steady-state predictive ability may be due to errors in parameters or invalid assumptions: [29]

- What is the impact of a bad model parameters (e.g. thermal model)?
- Can the overall system handle bad data?
- Can the simulator accurately guess the value of a missing meter?
- How true to reality is the estimated inventory?
- Do unmetered offtakes from a shut-in section trigger leak alarms?

The tasks once site data become available are not only about validation, but also updating model parameters, recalibration and setting the learnt tuning. Many model properties can only be set by an engineer looking at a live system.

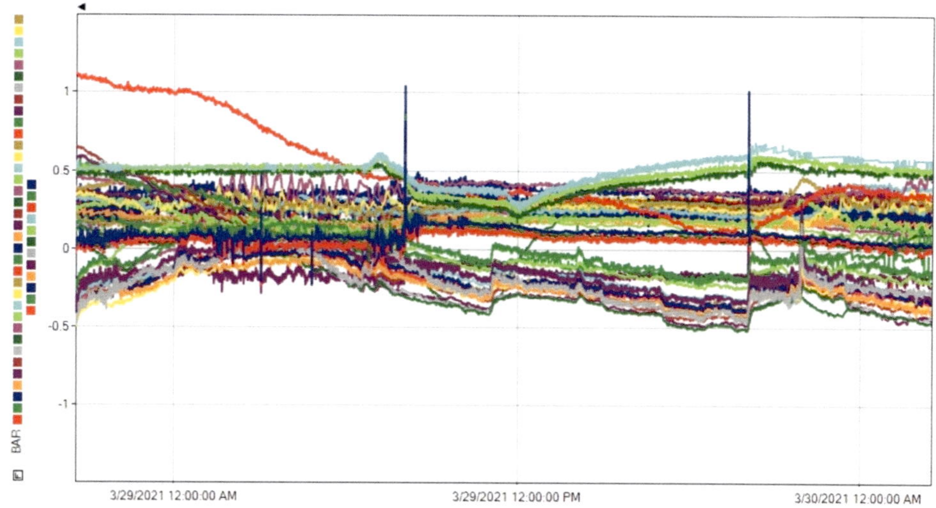

Figure 52 – Model versus meter differences on a live gas pipeline, broadly within ± 1 bar. Two spikes are visible, due to flaws in the incoming data from metering.

3.4 – CASE STUDY: SOUTHERN GAS PIPELINE

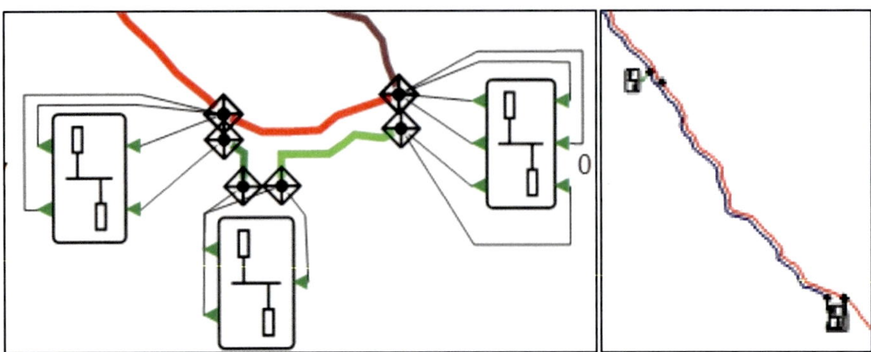

Figure 53 – The southern gas pipeline system in Atmos SIM. Stations are connected by several routes in much of the network (left). Much of the mainline consists of twin parallel pipes (right).

The main pipeline traverses 680 km across several jurisdictions. For approximately half of this it consists of twin 14-inch pipes, shown in blue and red on the right of the figure. The remainder is a single 18-inch diameter pipe. There are two compressor stations located on this pipeline, equipped with units each rated at 4.5 MW, roughly equivalent in power to thirty family sedans. The system connects to other pipelines, and via laterals to several gas-fired power stations as well as to a gas production facility. Atmos SIM was instructed to run two simulations: one with full metering (A), and another with a missing pressure meter such that a mid-line station became unmetered (B). The value at that point simulated by A should already be very close to the metered value. If the underlying model is well configured and well calibrated, then scenario B should match the simulated value of scenario A very closely, only being a little further off the metered value. Indeed, this is what we see, as illustrated in the real trend shown below. The nearest meters are approximately 90 km upstream and 60 km downstream.

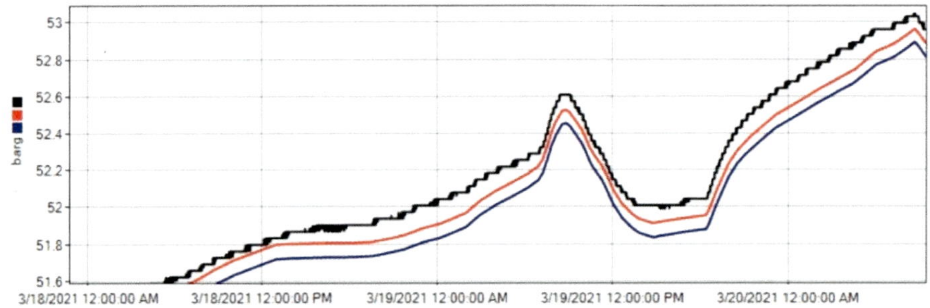

Figure 54 – The pressure at a mid-line station on a gas pipeline, as measured by a meter (top), is close to the fully-instrumented simulated value (scenario A – middle – within 0.2%) and the simulated value with that station's metering removed from the model (B – bottom – within 0.4%).

4 – SHUT-IN PIPELINES: SITTING STILL

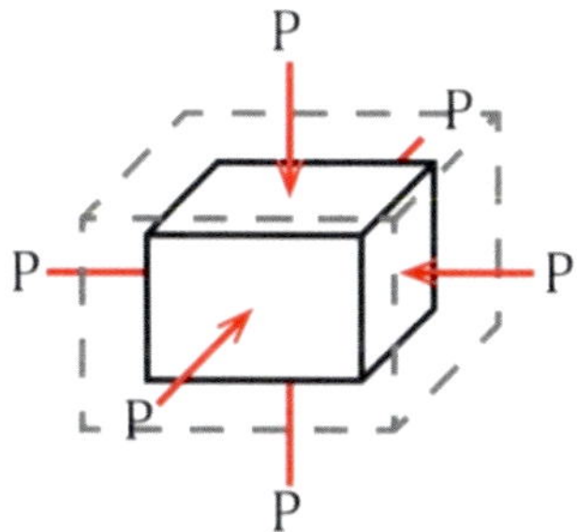

4.1 – DENSITY IN A CLOSED SYSTEM

We discussed the basic premise of pipeline fluids and flow in the basic primer earlier, treading cautiously enough not to drown. Now we'll dive in for a swim. By shutting-in a pipeline, we create a closed system, not an isolated one. Material can't enter or leave, but heat certainly can. Crude oil expands a lot when it's warmer, five times as much as water does. Fewer molecules of oil occupy a given volume on a hot day. Because of expansion and contraction, a compressible fluid *"tries to flow"* if temperature changes. It won't stay stopped unless we prevent it from moving by closing valves, *"shutting it in"*. Saying a line is *"in shut-in"* isn't exactly equivalent to saying that it is *stopped*. Even if we close all isolation valves when we stop a flowing liquid line, the liquid continues to shift around for some period while it cools, unless the whole pipeline is perfectly level.

A shut-in pipeline at a true steady state is in perfect equilibrium with its surroundings so line temperature at every point equals the environment's ambient temperature. In this chapter, we grant ourselves the assumption that the shut-in pipeline is truly at steady state. In reality, it is rare for any pipeline to be truly at equilibrium, unless nothing changes for days, and the day/night weather cycle is uneventful, or the pipeline is buried very deep underground.

4.2 – THERMAL EXPANSION: LINE TEMPERATURE

A change in temperature (T) causes the density (ρ) of a fluid to change by a certain proportion. This is described by a thermal expansion coefficient (α), in units of inverse temperature. If we multiply a temperature change by the thermal expansion coefficient, we get a number with no units (such as 0.01), meaning that the fluid expands by 1%. So if there is no change in pressure:

$$\Delta\rho = \alpha \cdot \Delta T$$

In an almost incompressible liquid, this thermal expansion coefficient is an input, used by the bulk modulus equation of state to calculate the density. Crudes and liquid products typically expand by around 0.1% per °C. Conversely, in a compressible liquid or a gas, the thermal expansion coefficient is deduced *from* our relevant equation of state, alongside other results such as density. Thermal expansion of compressible fluids varies significantly with conditions.

But what do we mean by *"compressible"* and *"incompressible"* exactly?

4.3 – COMPRESSIBILITY AND BULK MODULUS

We carefully define compression as "*contraction caused by pressurization*". That is to say, a contraction caused by a cooler temperature is <u>not</u> what people working with fluids call compression, though both manifest as higher density. * As such, terms such as *incompressible* specifically refer to how *pressure*, not temperature, affects a fluid. If the pressure changes but temperature doesn't:

$$\Delta\rho = \kappa \cdot \Delta P$$

This equation defines the *compressibility* (κ) of a fluid. By convention, liquids are described instead by a *bulk modulus* (B), defined as its inverse ($B = 1/\kappa$).

We said that as temperature in a pipeline falls at constant pressure, a liquid's density increases. And as its pressure falls at constant temperature, liquid density decreases. But the average density in a shut-in pipeline cannot change over time, because the line contains the same mass (M) of fluid while it is shut in, and the pipeline has the same volume (V) which essentially doesn't change (it does expand or contract, but barely). The density therefore, being mass divided by volume, must be constant during shut-in!

Yet we know heat is transferred out of a warm liquid pipeline by conduction. So what is happening here? The temperature falls, and to keep density constant, pressure also falls. This can produce a significant depressurization in a shut-in liquid pipeline while it cools down over hours or days. Estimated depressurization due to falling temperature:

$$\Delta \mathrm{P} = B_T \cdot \alpha \cdot \Delta \mathrm{T}$$

For a typical crude oil pipeline in shut-in, this might amount to around 100 psi per degree Fahrenheit of cooling. This depressurization can easily lead to the formation of vapor pockets at the high points of the pipeline, as we shall discuss in a later chapter.

---

* Unfortunately, few people use punchy terms to distinguish thermal expansion (due to change in temperature) from an isothermal expansion (due to pressure). Some people may prefer to call the latter change in density a decompression.

4.4 – EQUATIONS OF STATE

There is no universal formula for density

A good simulator, like a good engineer, obsesses over getting density right. A bewildering array of *equations of state* relate a fluid's density, pressure and temperature. If we know the fluid's constituent breakdown, some equations of state describe the effect of composition if it changes. Many fluid properties can be derived from some equations of state by standard methods, not just density; they give compressibility factor, speed of sound and phase envelope.

There is no universal equation of state. All of these are empirical, with caveats and provisos. At our discretion we must select one to suit our particular fluid across the full range of conditions. Using several within the same model can be useful, if very different fluids are batched for example, or if some extreme transient takes the same fluid *"out-of-bounds"* for a preferred equation of state.

More formally, an equation of state expresses any thermodynamic property in terms of two others. Choosing pressure and temperature along with density, we note that this trio of quantities are all *thermodynamic properties.* In technical terms, they are *state variables.* That is to say: their value depends only on the operating conditions at a point, not the process by which a fluid reached them. This sounds like nit-picking, but it implies the choice of parameters is up to us.

What are our independent variables?

Any one of our trio of quantities – pressure, temperature and density – can be calculated if we know the other two. Intuitively, we think of pressure and temperature as our prevailing conditions, so we expect it is simplest to find density from the others as known parameters:

"Density at a pressure and temperature" $\rho(P,T)$

But it turns out many equations of state are inconvenient *"that way round".* To avoid inverting an equation of state, a simulator could equally choose to instead take pressure as its unknown:

"Pressure at a density and temperature" $P(\rho,T)$

It's not just equations of state that are written as functions of density $P(\rho, T)$, rather than of pressure $\rho(P, T)$. Density is also the star of the show in the *conservation laws*, appearing directly in conservation laws, more so than pressure. Pressure does not even make an appearance in the mass equation!

We talk in terms of pressure because it's intuitively easier to tell if a pressure is right. This isn't to be dismissed: reality-checks aid in troubleshooting. But density instead of pressure as our independent variable is an equally valid choice to work with. Choosing density has the advantage of equation of state being explicit in density, and also that the conservation equations are more linear when we write them in terms of *density* rather than in terms of *pressure*:

$$\rho(P, T) \leftrightarrow P(\rho, T)$$

We could also use specific enthalpy or internal energy instead of temperature:

$$h \leftrightarrow u \leftrightarrow T$$

For the same reason – convenience in the physical underlying equations – a simulator might choose to use the mass flux (mass flow per unit area; Q_{mass}/A) rather than the velocity (v), actual volumetric flow rate (Q) or standard volumetric flow rate (Q_{std}). Just as a volume (V) can be converted to or from a standard volume (V_{std}) via the fluid's standard density (ρ_{std}), or to or from a mass (M) via the fluid's density (ρ), so can flows:

$$V \leftrightarrow V_{\text{std}} \leftrightarrow M$$

$$v \leftrightarrow Q \leftrightarrow Q_{\text{std}} \leftrightarrow Q_{\text{mass}}/A$$

In fact, other sensible choices are possible to form our trio. We pick any two of these quantities to be the unknowns that we are going to solve for. This technical detour is the explanation for why we call this equation *"the equation of state"*. Whatever simulators do under the hood, all results should drop out of the calculations with correct values. People intuitively think in pressures (P), temperatures (T) and densities (ρ), so we stick with that trio in this book.

The most modern, complex, and accurate equations of state such as AGA-8 for natural gas, Span-Wagner for CO2, and GERG-2008 for almost everything, are actually written in terms of the Helmholtz free energy. Derivatives of this quantity not only return the density but also the specific isobaric and isochoric heat capacities (c_P , c_V), so this lets these equations include correlations for both of those important quantities. Viscosities (μ) still remain outside the purview of equations of state.

The bulk modulus equation of state for liquids

An *"ideal liquid"* is one that is completely incompressible. But even the most incompressible liquid there is, water, changes in density *ever so slightly*, albeit by a vanishingly small amount. To calculate the pressure drop that any other liquid undergoes as it flows through a pipe, we must express its *density*. For *moderately* (not *very*) compressible liquids, we adopt the same approach.

Liquid properties are observed to vary with conditions by simple relationships. At low pressures, a cooler liquid is linearly denser, linearity with temperature holding valid almost universally. The only exception in pipelining may be water below 4°C, which gets *less dense* as it cools for the same reason ice floats.*

> At low pressures, saturated liquid density varies linearly with temperature: throughout a remarkably wide range of temperatures, a cooler liquid gets proportionally denser.

At a more intense pressure, an almost-incompressible liquid is only slightly denser. Crude oils and liquid products exhibit this behavior, and their composition is unknown, so they are typically described by bulk parameters, rather than a breakdown of their constituents. To calculate their density, we use an equation of state based on constant *bulk properties.* This *bulk modulus equation of state* relates the mass density † (ρ) to the operating pressure (P) and temperature (T) relative to their values at standard conditions (ρ_{std}, P_{std}, T_{std}):

$$\rho = \rho_{\text{std}} \cdot \exp\left[-\alpha \cdot (T - T_{\text{std}}) + \frac{P - P_{\text{std}}}{B_T}\right]$$

Pressure dependence is captured by an isothermal bulk modulus (B_T) which is large for a nearly-incompressible liquid (typically measured in gigapascals) and assumed constant. As we said earlier, some pipeliners choose to express this same equation in terms of the reciprocal of bulk modulus: the isothermal compressibility (κ_T). Many real liquids have a very large bulk modulus, but none have a truly infinite bulk modulus.

---

* The other exception is extremely cold liquid helium turning into a superfluid.

† The bulk modulus equation of state is written directly in terms of the mass density, not the mole-basis density. We do not know a molecular weight for bulk modulus fluids because we generally don't have any idea of the exact composition for something like a crude oil or a liquid product.

As we discussed earlier, this sort of approximation is the basis for Atmos SIM's incompressible online model. However, it is not possible to make a transient hydraulic model coupled with the assumption that the fluid is absolutely incompressible, so even for water we need to use something like this Bulk Modulus equation of state. This remains a very good approximation for certain liquids, such as water, whose density for practical purposes doesn't vary with pressure.

We should, however, bear in mind that its density *does* vary with temperature: for example, the warm water heated by a riser from a natural gas platform will itself rise due to buoyant forces as it's less dense than the colder seawater above it. This temperature dependence is captured by a thermal expansion coefficient (α) associated with the fluid, and it is assumed to stay constant.

A simpler linear form approximation to the bulk modulus equation of state is often used, referred to as the *"ideal liquid"* or *"slightly compressible liquid"* equation of state just like the original exponential form:

$$\rho = \rho_{\text{std}} \cdot \left(1 - \alpha \cdot (T - T_{\text{std}}) + \frac{P - P_{\text{std}}}{B_T}\right)$$

Provided the thermal expansion coefficient (α) is small and the isothermal bulk modulus (B_T) is large, it gives us the same answer as the original, to several decimal places. We avoid this minor inaccuracy in Atmos SIM by avoiding the approximation altogether and implementing the original exponential form. A thermodynamic identity relates the specific isobaric (c_P) and isochoric (c_V) heat capacities to these parameters:

$$c_P - c_V = \frac{T}{\rho} \cdot B \cdot \alpha^2$$

For water at $T = 293.15\ \text{K}$,
$\rho = 1\,000\ \text{kg} \cdot \text{m}^{-3}$
$B_T = 2.2 \times 10^9\ \text{m}^2 \cdot \text{N}^{-1}$,
$\alpha = 2.14 \times 10^{-4}\ \text{K}^{-1}$,
so $c_P - c_V = 29.5\ \text{J} \cdot \text{kg}^{-1} \cdot \text{K}^{-1}$
and $c_V = 4186\ \text{J} \cdot \text{kg}^{-1} \cdot \text{K}^{-1}$
therefore $\gamma = c_P / c_V = 1.007$

The ratio (γ) of isobaric (C_P) to isochoric (C_V) heat capacities is an important quantity:

$$\gamma = \frac{C_P}{C_V}$$

Its square root ($\sqrt{\gamma}$) is the ratio between speed of sound in an isothermal model (c_T) and in an isentropic – i.e. reversible adiabatic – model (c_s):

$$\sqrt{\gamma} = \frac{c_T}{c_s}$$

The latter is what we see in real pipeline, and therefore what we expect to see from a good thermal model. A sound wave propagates so fast that there is not enough time for noticeable heat to be conducted anywhere, so it is adiabatic.

Water has a very high specific isochoric heat capacity (c_V) and a low coefficient of thermal expansion (α), so its ratio of heat capacities (γ) has a value of about one. The ratio of heat capacities is greater than that for typical pipeline liquids.

For gasoline at $T = 293.15\ \mathrm{K}$,
$\rho = 750\ \mathrm{kg \cdot m^{-3}}$,
$B_T = 1.3 \times 10^9\ \mathrm{m^2 \cdot N^{-1}}$,
$\alpha = 9.5 \times 10^{-4}\ \mathrm{K^{-1}}$,
so $c_P - c_V = 458\ \mathrm{J/kg/K}$,
and $c_V = 2220\ \mathrm{J/kg/K}$
therefore $\gamma = 1.207$

A widely applicable standard for calculating the density of liquid hydrocarbons is API 2540. If a typical liquid hydrocarbon, such as those batched in multi-product pipelines, becomes 12°C cooler, it will become around 1% denser.

Use the bulk modulus equation of state
to calculate the density of crude oil.

Equations of state for compressible liquids

Liquefied petroleum gas (LPG) is a massively compressible fluid. The bulk modulus approach, or any modification of it, just won't do here: the isothermal bulk modulus and thermal expansion coefficient are not constant for different pressures and temperatures. Instead, we opt for using *"gas-like"* equations of state, taking an entirely different approach thereafter. If this seems surprising at first glance – well, the clue is in the name!

All of thermodynamics doesn't boil down to the simple relationships describing nearly-incompressible liquids, as the field was developed to describe gases. In the spirit of a great model ignoring everything it can get away with, the watchword is simplicity. We only resort to equations of state such as *BWRS* for liquids that are so compressible as to be inadequately described by the foregoing bulk-modulus approaches.

Equation of state		**Validity**
Constant bulk modulus	Parametric approach	Most common approach for liquids. Not accurate for compressible liquid e.g. LPG.
Aalto [33]	Component-defined	For lighter slightly compressible liquids.
Benedict Webb Rubin Starling (BWRS)*	Component-defined	Gas, liquid, supercritical. Used for compressible fluids.
Groupe Européen de Recherche Gazière (GERG)*	Only 18 selected components	Gas, liquid, supercritical, vapor-liquid equilibrium. Very accurate. Slow to solve.

The forms of some of these equations of state can pose quite a challenge to performance of a solver. Those marked with an asterisk (*) in the table for instance are *transcendental* equations which can only be inverted numerically. But if we work in density and temperature, we don't ever *need* to invert them.

Equations of state for gases

The simple assumption we made about our fluid being incompressible is reasonable enough for some liquids, but grossly unsuitable for all gases. For a highly compressible fluid or a gas, density * is calculated at a given set of pressure and temperature conditions via an equation of state. All equations of state for gases derive conceptually from the famous ideal gas law relating pressure (P), volume (V), moles (N) and absolute temperature (T) by way of a universal gas constant (R):

$$P \cdot V = N \cdot R \cdot T$$

Which we can write equivalently in terms of molar mass (m_r) and density (ρ):

$$\rho = \frac{M}{V} = \frac{m_r \cdot N}{V} = \frac{m_r \cdot P}{R \cdot T}$$

The basic essence of this statement is reasonably intuitive. The density of a gas varies with operating conditions: warmer gas at low pressure is less dense.

> A hot air balloon flies because the heated air in the balloon is lighter than the cooler ambient air at the same pressure, but only up to a maximum altitude. Above that ceiling, the difference in density reduces because of the lower pressure. Eventually the heated air gives too little buoyancy to lift the basket and balloon.

The ideal gas law assumes we are far from the critical point, which means it applies only at relatively low pressures and relatively high temperatures. It is also only valid for a relatively hot gas at low pressures. If those conditions are satisfied, it holds true. In a cold, pressurized pipeline, PV is certainly *not* NRT!

Standards define a custody transfer range corresponding to the conditions under which natural gas is traded: negative 3 to positive 56°C, up to 120 bar. In this range, the chosen equation of state must be accurate enough to predict both the density (ρ) and the speed of sound (c) within 0.1%. How do we relate such significant changes in pressure, temperature and density?

---

* Noting that *"density"* in any equation of state for gases or component-defined fluids is mole-basis density, converted to mass density via the molar mass.

A real gas occupies a different volume than the ideal gas law would suggest. Its density varies in a way that *itself* varies in turn. This deviation from ideal gas behavior can be calculated by using a cubic equation of state (such as the Peng-Robinson equation). *

The *supercompressibility* (Z) of a fluid, also called the Z-factor, is defined as:

$$Z = \frac{P \cdot m_{\mathrm{r}}}{\rho \cdot R \cdot T}$$

We can convert between actual volume (V) and standard volume (V_{std}) using this factor evaluated at operating conditions (Z) and evaluated at standard conditions (Z_{std}), with reference to standard pressure (P_{std}) and standard temperature (T_{std}):

$$V = V_{\mathrm{std}} \cdot \frac{P_{\mathrm{std}}}{P} \cdot \frac{T}{T_{\mathrm{std}}} \cdot \frac{Z}{Z_{\mathrm{std}}}$$

There is yet another level of abstraction. The *supercompressibility factor* (F_{PV}) is defined as the square root of the ratio between the supercompressibility at standard conditions (Z_{std}) to its value at the conditions of interest (Z):

$$F_{PV} = \sqrt{\frac{Z_{\mathrm{std}}}{Z}}$$

Some seem to apply a different definition again, further thickening the plot!

Many different equations of state are used in the gas industry. They are typically expressed in a form such that the pressure is calculated from a density as an input parameter, so an easy way of validating an equation of state is to take a reported density from a pipeline simulator and see if the reported pressure agrees with the pressure calculated by the equation of state (rather than calculating the density). † The properties of a gas depends on its composition, so a simulator defines a gas as a *mixture* of known composition, composed of components in known proportions. The engine then refers to default library parameters for each of its components, and the chosen equation of state deduces the mixture's overall properties. Physical chemistry dictates the components combine to a mixture whose properties aren't straightforward!

---

* The old-fashioned way was to use a generalized compressibility chart.

† Another way of validating a pipeline simulator properly implements an equation of state is to check the reported *standard density* against the ideal gas law. All gases are ideal at low pressures and moderate temperatures – this is true of standard conditions.

How accurate is each equation of state in terms of gas density, compressibility factor and speed of sound?

Equation of State	Validity
Soave Redlich Kwong (SRK)	Gas, or supercritical.
Peng Robinson (PR)	Gas, or supercritical. Good phase envelope.
Benedict Webb Rubin Starling (BWRS)	Gas, liquid, supercritical. Tuned coefficients. Popular for gases and compressible liquids.
Groupe Européen de Recherche Gazière (GERG)	Only 18 selected components. Gas, liquid, supercritical, vapor-liquid equilibrium. Very accurate, but slow to solve.
American Gas Association (AGA-8)	Only 21 selected components available. Slow to solve.
Generalized Soave	Gas, supercritical.

This list isn't exhaustive. Of the listed equations of state, ethylene is handled accurately only by GERG. Atmos SIM also provides a bespoke equation of state especially for ethylene which is thought to be even more accurate.

Even if an equation of state has rules allowing the user to add new components not originally covered by its formulators, doing so generally degrades its purported accuracy. So what should someone in a rush use as a *"default"* to fall back? One equation of state commonly chosen as a default is Peng-Robinson, but in practice, each gas pipeline engineering team has its own preference. SRK, BWRS and AGA-8 are all widely used, and some seem to be preferred in certain countries. GERG's equation of state is the most recent (2004), and it is renowned for its accuracy, but its form is intensive to compute, so it isn't commonly chosen for real-time applications, nor for any application where fast solution matters. The hydrocarbon results available at NIST WebBook, a handy online reference for pure-substance fluid properties, are mostly from the GERG equation of state.

4.5 – PRESSURE, TEMPERATURE AND INVENTORY

A fluid's density varies with both pressure and temperature. Accordingly, the same pipeline's actual volume * contains different amounts of fluid depending on conditions all along the line. These line conditions are affected both by the ambient environment, and how we are operating our stations. For this reason, pipeliners rarely describe the quantity of a fluid, especially a compressible fluid, by its *actual volume* (V). Instead, they refer to inventory as a *standard volume* (V_{std}): what the volume *would* be if those same molecules of the fluid were at the pre-defined standard temperature (T_{std}) and standard pressure (P_{std}). Most pipeliners prefer standard volume rather than mass.

We may be inclined to think there is some kind of universal rule that pressure and temperature somehow *result* in a density. But this is not the case. Any two imply the third of the trio, but *which* two are fixed depends on the scenario. In a shut-in condition, the locked-in fluid cannot reduce in density: the same amount of mass (and moles) occupies the same volume. If it gets warmer, the aforementioned thermal expansion doesn't manifest as a fall in density, but as a rise in pressure. We should prepare for the fluid to exert this extra pressure on a warmer shut-in pipeline; relief valves are used only as a last-ditch resort.

Even as fluid sits still in the pipeline, a suitable set of steady shut-in scenarios can help us explore how pressure, temperature and inventory are intertwined:

A. What is the pressure at a given inventory and temperature?
B. What is the inventory at a given pressure and temperature?
C. What is the temperature at a given pressure and inventory?

For a scenario or a pipe group, a basic reality check calculates what inventory corresponds to a global *maximum pressure*. This allows for a quick comparison of inventory against a maximum allowable inventory. Similarly a global *minimum pressure* can be used to compare the present inventory against minimum allowable inventory.

---

* Pipeline volume is almost exactly fixed except for pipe expansion and contraction.

4.6 – Dead-ends and extended shut-in

A pipe that has been blocked off with a flange to seal it (*blinded* or *blanked*) is not the only sort of dead-end. Branches closed off for extended periods by rarely-used valves can be just as dangerous.

Water is present in trace amounts in many oil and gas streams. It can collect in dead-ends and freeze, breaking the pipe. Or corrosive materials can dissolve in the water and breach the pipe by corrosion.

Atmos SIM calculates the inventory within a dead-end, accounting for all the conditions and the elevation profile.

5 – STEADY STATES: FLOWING FOREVER

Figure 55 – Water flowing from an outlet at steady state.

"You must spend money to make money." – Plautus

5.1 – AN INTRODUCTION TO STEADY STATES

What is a steady state?

Think of something a pipeline does. If it changes over time, we won't be talking about it in this chapter. Not just yet. Looking at how a pipeline (or *anything*) behaves in a condition which remains completely unchanging over time seems like a bit of a nothing. Surely, it isn't what happens in practice, so why are we even bothering with it? Today's heavyweight dynamic simulators brag about their ability to handle transients, why don't we jump straight to those?

When we think about *"something that is happening"*, it is tempting to jump ahead and not consider what happens when nothing is *changing*. This helps a fresh engineer hone their skills, getting to grips with how fluids flow through pipes, offering a novice learner of fluids, thermodynamics and heat transfer an easy way to see these toolkits in action. Steady states are the basic foundation of hydraulics. But this is just a happy coincidence. It's worth looking at a steady state before deep-diving into transient behavior because they tell us a great deal about the system we are dealing with. With steady states, hydraulic engineers conduct countless studies, drawing useful conclusions. So much so, that some teams don't use transient simulators at all. Simply put, steady states are in fact, *useful in their own right*.

A steady state describes a physically viable, *stable* way to operate the pipeline. Inlet and outlet flows all balance perfectly, a state of affairs which can persist forever. Although nothing in this kind of arrangement is *changing* over time, this doesn't mean nothing is *happening*. Material and heat is flowing steadily. In a steady state, flows of material and heat perfectly balance in equilibrium. It can theoretically persist forever. Flow rate is driven by the mechanical energy of the fluid, known as its head. Losses along a flowing pipeline are lumped into friction, which depends on pipe roughness, excepting slow viscous liquids in the laminar regime.

The profiles of mass flow rate and (usually) standard flow rate are flat along a pipeline operating at a steady state.

A state is *steady* (as opposed to *transient*) if flows of mass (M), momentum $(M \cdot v)$ and energy (E) – physically conserved quantities – are all in perfect balance, unchanging over time. A steady state performs special mass, energy and momentum balances.

Steady state models are easier to implement than dynamic models. All of this can be done via a spreadsheet – albeit quite an involved one. This affords a user of a simulation package the opportunity to do a basic reality check that simulated results agree with manual calculations, and vice versa.* Whenever a design or operations team spends an hour or so setting up a basic Atmos SIM model, they often choose their first scenario to give it some known steady state pressures, and ask the simulator to deduce the *"nominal"* flow rate. Upon seeing a familiar value, they often breathe an audible sigh of relief!

Design tasks solved by steady states

There are a fair number of design tasks that are solved by steady states. What flow can a pipeline transport from A to B: its capacity or *nominal throughput*? How do we ensure that *"the right"* capacity is available through a pipeline – enough, but not too much? It depends on how the entire system will operate.

Having commissioned a pipeline and put in place all kinds of arrangements for it to start operating, it would be a real let down if we then find it incapable of handling the

* This is one check amongst many, to be supplemented by further validation.

throughputs we present it with. *Over-sizing* is bad, not just under-sizing. Over-sizing a pipeline's diameter is not a generous decision if it then risks rendering our flow meters inaccurate! And a pipeline that is too large for its flow rate might force us to select control valves that are comically inefficient.

The operating envelope of a pipeline is the range of flow rates at which it can run. We have already discussed the capacity, and the upper flow bound; what about the lower? Simple liquid products pipelines might be able to operate down close to zero flow, but a very heavy crude pipeline for instance might be dependent on frictional heating to keep the oil warm enough for its viscosity to be low enough to flow when it is pumped. For multiphase pipelines, liquid accumulation is much worse at low flow rates, which also creates a minimum flow limit. And even the simplest liquid pipelines have some sort of lower flow limit imposed by the surge limits on their pumps. The operating envelope might differ with summer and winter conditions, and with the fluid being pipelined: [45]

- Where are our pumps or compressors going to discharge?
- What kind of flow rates are we going to handle?
- What's our delivery pressure?
- How much pressure do we need to feed the system?

To answer these questions, a number of physical parameters are at play. One of many tasks that a pipeline design engineer undertakes is to ask:

> How do we set supply and mid-line pressures to overcome friction, achieving delivery pressure at the design flow rate?

Once confident in the model we have built, we can use steady states for pipe sizing, to decide on diameters, as well as to determine pipe insulation requirements. The costs of construction includes the materials, depending on pipe diameter and wall thickness. Operating costs depend on pressure drop. So these calculations matter. They aren't mere afterthoughts; they are the bread and butter of a process engineering team.

A steady state can be used to describe what pressures and flows are sufficient to meet production targets. This informs decisions about pipe sizes and ratings, to select appropriate equipment and where to locate its stations, and to assess how nominal operation will look in the summer versus in the winter, at peak-forecasted-flow versus least-forecasted-flow, and so forth.

Steady hydraulics are important to those considering the commercial drivers underlying a pipelining enterprise. They are useful for feasibility studies, high-level optioneering, and calibration against historic site data. Calibration against a stable average historic dataset tunes a model's parameters, enabling accurate predictions. For an existing pipeline network, debottlenecking can be important, with studies into expansion alternatives being considered. Plugging different parameters in to simulate many steady states to do, say, a power cost analysis is a basic kind of *optimization*. We discuss that in its own chapter.

In the realm of batch tracking, steady states can tell us what happens if we import crude from a different origin, batching a new product, or changing the order of batches. These are all situations warranting serious examination of how a fluid's differing composition will perform in the pipeline. They are reported as a set of base cases: behavior for an expected composition is compared against *"extreme"* compositions; *sensitivity analysis* for operations.

A steady state tells us what flow rate a given set of *boundary conditions* (BC's) will achieve. This flow rate is the very purpose of the pipeline, and is referred to as its transportation or throughput *"capacity"*.

We also use steady states to select equipment capable of adequately delivering the pressures we need, even at the most challenging ambient temperatures and fluid compositions. We use them to select the best locations for pump or compressor stations. And we use them to estimate each station's overall power requirements, selecting suitable drivers and so on.

All of these decisions are made striving towards an overarching aim: optimal operation and ability to adapt in the future. As such, engineers at every stage of a pipeline project prepare many steady state scenarios to run diverse simulations.

Given some input data, our task might be to determine any of the following pipeline parameters given the others:

- Diameter
- Pressures
- Nominal flow rate
- Location of stations and section length
- Specifications of pumps or compressors

We run through how to perform these calculations by hand, then demonstrate how to quickly carry them out using our simulator using real-life case studies.

Ideal constraints as boundary conditions

Any problem statement, whether tackled manually or computationally, must be *constrained* at the system's boundaries by user-specified conditions. They are met in one of two ways: as a *controlling constraint* or as a *loose constraint*.

If we assign a pressure constraint at every boundary, the flow rate follows from that, and the simulator calculates it for us. If we then attempt to also assign that flow rate as an equality anywhere, the problem is said to be *over-defined* or *over-constrained*.

Instead, at any given moment in time:

- Each control item has a *controlling constraint*, applied as an *equality* *
- Plus it may have further *loose constraints*, applied as *inequalities* †

Atmos SIM never ignores any enabled constraint of either status. It tries to satisfy *all* of them by switching its *controlling* constraint (the equality) where and when needed. In practice a few of the *loose* constraints might happen to be almost exactly met too, but we allow that a loose constraint can be *"more-than-satisfied"*. If a loose constraint says: *"maximum pressure is 20 bar"*, the simulator allows it to be 18 bar (because the inequality $18 \leq 20$ is met). If, and only if, the inequality is about to be violated, say pressure is about exceed 20 bar by going higher than it, do we switch to this constraint as the controlling constraint here. Pressure must then meet it as an equality ($= \ 20$).

Once we have provided enough inputs for our problem statement to be well-defined, a solution is calculated to meet every boundary condition as an exact constraint. This is a unique solution: physics dictates that nothing else could possibly be happening. We must specify enough pressures and flows to calculate all values everywhere (i.e. not under-specified) but never implying a value twice anywhere (i.e. not over-specified). There are pitfalls in cold-start.

Figure 56 – A stretch of pipeline controlled on flow constraints at both ends (flow-flow or F-F)

One pitfall is controlling on flow at both ends of a stretch of pipeline ($\mathrm{F} \rightarrow \mathrm{F}$). In steady state, we know the flow into the line must equal the flow out, so if we specify it at both

* Some call it a *setpoint* or a *mode*. *"Control"* here doesn't refer to logic blocks.

† Linear programmers call this a *slack* constraint.

ends the problem is *over-constrained*. In addition, we haven't specified a pressure anywhere: there's no unique steady solution for pressure; infinite possible pairs of inlet and outlet pressure result in that exact flow rate. So this set of initial conditions is *under-constrained* too: we can't deduce pressure. * We must set an initial pressure at one of the two ends at least, never constraining a section on *"flow-flow"* in a steady state, such as having:

- Flow constraints at both the inlet and outlet of a pipe section $(\mathrm{F} \rightarrow \mathrm{F})$
- Two pump or compressor stations in a row, both on flow control $(\mathrm{F} \rightarrow \mathrm{F})$

The pitfalls of under-specifying include some more subtle situations. Consider a supply (A) controlled on a flow constraint, then some pipe, then a pump (B) on a discharge pressure constraint, then more pipe, then a pump (C) on a suction pressure constraint. This is both under-specified and over-specified:

- Region A to B has an undetermined pressure $(\mathrm{F} \rightarrow ?)$
- Region B to C determines its own flow ($\mathrm{P} \rightarrow \mathrm{P}$ implies an F) but the supply (A) also controls on flow, so overall (A – B – C) flow is over-determined $(\mathrm{F} \neq \mathrm{F})$

It isn't obvious to a user their suggested initial constraints are under- or over-constrained. Atmos SIM auto-switches these setpoints if it detects an issue, and gives the user a message indicating what it did. In summary, to cold-start a simulation by specifying a steady state to initialize, we carefully control on pressure-pressure $(\mathrm{P} \rightarrow \mathrm{P})$ or on pressure-flow $(\mathrm{P} \rightarrow \mathrm{F})$. Some choices of allowed sets of constraints are perhaps better than others. In a gas pipeline, if flow rate is too high, it chokes, which makes it harder to figure out what is going wrong with a configuration. A good place to start is to specify minimum pressures at demand points, compressor suctions, and regulator inlets, and specify maximum flow rate constraints at supplies. That is likely to converge with no choked flow no matter how fouled up a configuration may be.

* In a later chapter, we'll come to solve a transient step. There, we already know the pressure from the preceding step, and so flow-flow control is unique.

5.2 – STEADY HYDRAULICS OF LIQUID PIPELINES

The operating envelope of a liquid pipeline

Flow velocity limits in a liquid pipeline

A liquid pipeline should never flow faster than a certain velocity, dependent on the viscosity of the liquid. Typically this is 1 or 2 m/s. If a pipeline is capable of moving liquid at, say, 4.5 m/s, this might still be within the bounds of the *"speed limit"* beyond which severe erosion occurs, but incurs unacceptable frictional losses. Instead, velocity can be made more reasonable using larger diameters or looped lines, weighing up capital and operating costs of each approach. [45]

Pressure limits in a liquid pipeline

Figure 57 – A longitudinally ruptured pipeline (National Transportation Safety Board NTSB)

Some locations are liable to conditions of over-pressure, which would exceed the *maximum allowable operating pressure*; an MAOP violation. A pipeliner is legally required to consider that their steady state operating conditions don't represent the highest pressure seen by their pipeline. As we will see in a later chapter, transients cause pressures to rise well above the steady state pressure. An unacceptable over-pressure excursion causes the pipeline to burst. We explore the consequences of a release of fluid from a pipeline in the chapter on leaks and ruptures.

Other locations are liable to fall below *lowest allowable operating pressure*; an LAOP violation (under-pressure). The LAOP limits in a liquid pipeline are designed to keep operating conditions safely above the liquid's vapor pressure, so it does not form a vapor bubble in a condition called *slack-line* or *column separation*. If that bubble collapses, extreme conditions can arise, causing shock waves in a pipeline. This collapse can happen to the bubble as it flows along and experiences a change in pressure, a scenario that can destroy a pump impeller by cavitation erosion. Damaged equipment can have a domino effect, causing high fatigue loads on other components. Through fatigue, erosion or corrosion, under-pressure events could rupture a pipeline.

Getting close to violating either limit *at any point along the pipeline* is worth notifying the operator about. Alarms are configured in this industry, like any other, as *"Low-Low"*, *"Low"*, *"High"* and *"High-High"*. The meaning of these depends on the specific pipeline's operating philosophy, and the item of equipment in question. For a tank holding liquid, low-low may refer to a situation in which *"we're about to lose ability to draw from the tank"* and high-high may refer to a situation in which *"the tank is nearly overflowing"*. If events messages, alerts and alarms are raised, they can be escalated, and appropriate corrective action can then be taken in accordance with standard operating procedures.

The hydraulic profile of a liquid pipeline

To examine how a liquid flows along a pipeline, we look at elevation plotted as a profile along its length, along with selected variables: pressure (P) with limits, head (H) with limits, and a type of flow rate or velocity. Some pipeliners might also be interested in temperature (T), in-line and ambient. The resulting charts are collectively known as *hydraulic profiles*, an insightful graphical feature.

Although a real pipeline has varying elevation, diameter and liquid properties, it's insightful to make statements about the hydraulics of a *flat* liquid pipeline:

- In the laminar flow regime, pressure drop is proportional to flow rate
- In a turbulent regime, pressure drop is roughly proportional to flow *squared*

As we saw in the chapter on applying conservation laws in a simulator, we need not build and solve a full hydraulic model to draw a liquid pipeline's hydraulic profile. Even without all that effort, a simulator can calculate pressures and heads throughout a steady liquid pipeline using a *"simple hydraulic profiler"*. Based on pipeline parameters and metered pressure data, the only variables that affect it are the *density*, *diameter* and the presence of *column separation* or *slack flow*. We will discuss this *slack* later.

A pressure profile (P) must not violate either of its allowable limits at any point along the length of a pipeline. Those are straight lines, perhaps more stringent at sensitive regions, called *maximum* and *lowest allowable operating pressure* (MAOP and LAOP). A profile of pressure along a pipeline traversing rough terrain shows us *"pressure drop"* doesn't necessarily correspond to flow direction. A steep descent may cause pressure to *increase* in the direction of flow! For this reason, we look at another quantity as well.

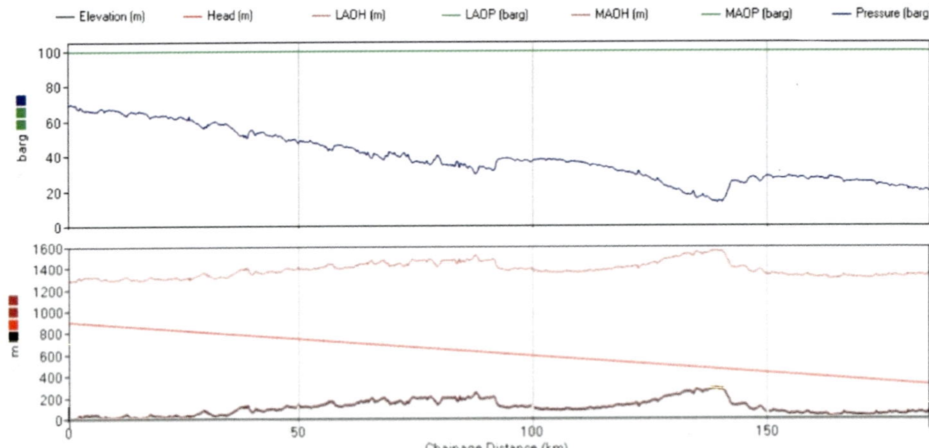

Figure 58 – A liquid hydraulic profile in Atmos SIM. Pressure limits (top and bottom lines in the upper plot) are flat, while the pressure profile shows the effect of elevation. Elevation relief is visible in the head limits (lower plot), but the head profile is flat. Both convey we're within limits

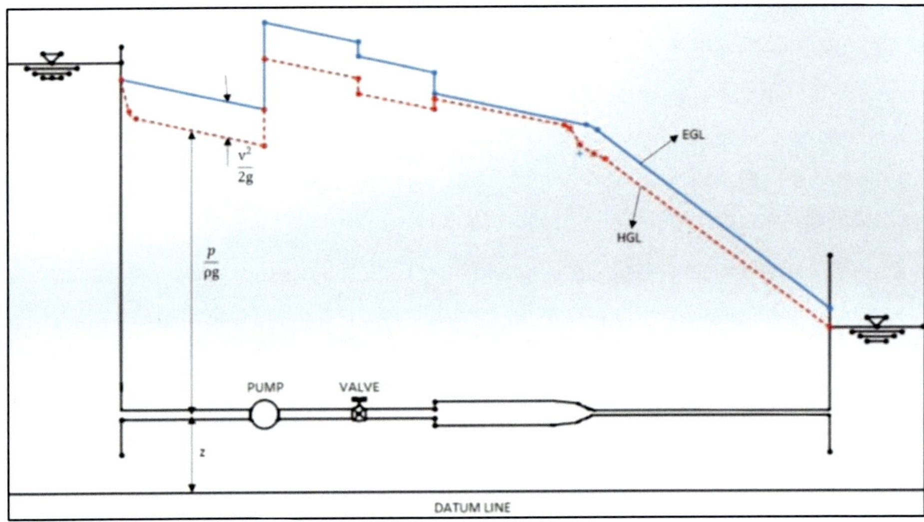

Figure 59 – A flat pipeline between two reservoirs, with a mid-line pump, regulating valve, and changes in diameter. The total head profile (energy grade line EGL) is shown in solid blue. The piezometric head, being the static head plus elevation (hydraulic grade line HGL) is shown in dotted red. Note that both the gradients for the larger diameter section in the middle of the line should be flatter than for the small diameter, we kept the diagram unchanged from its source. Mayankrajput052 / Wikimedia Commons / CC BY-SA 3.0

The *hydraulic head* of a flowing fluid is a concept we explore in the *Primer* appendix. It is the height a vertical column of that fluid would rise if a tube was attached to the pipeline. Head (H) has three components, which sum together to give the total height: pressure head, elevation head (simply the pipe's elevation at that point), and finally a portion due to the kinetic energy of a moving fluid (which is typically small in pipelines):

$$H = \frac{P}{\rho \cdot g} + z + \frac{v^2}{2 \cdot g}$$

The total head indicates the total mechanical energy of the flow. It can be instructive to look at the portion of that not *"tied up"* as kinetic energy, known as *piezometric head*, and consisting of *pressure head* (also called *static head*) plus *elevation head*.* [34] This is equivalent to the head above some *datum*.

A *head* profile shows the frictional losses in a pipeline unobscured by the variations in elevation which clutter up the pressure profile. Instead, the *limits* vary with elevation on a head profile. Equivalent *maximum allowable operating head* and *lowest allowable operating heads* (MAOH and LAOH) have a shape akin to the elevation profile. This shows us the effect of a change in diameter, fluid or flowrate more clearly. Here we see head plotted along a pipeline whose diameter expands midway, at which point the slope of the head plot changes. This would be far more subtle on a profile of pressure.

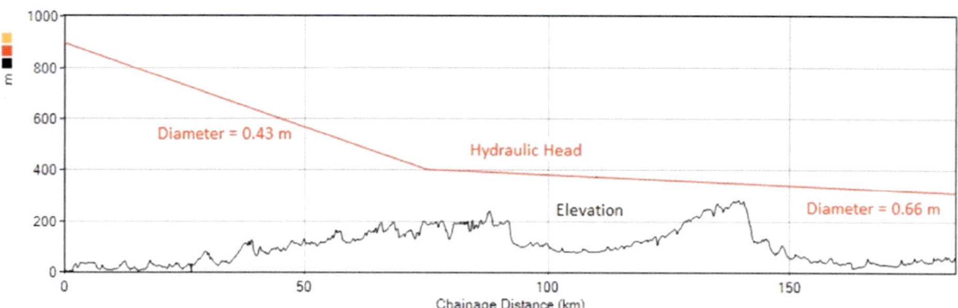

Figure 60 – A profile of *total head* clearly shows us the effect of a change in diameter.

Using the head profile, a liquid pipeline operator might want to determine the most efficient speed at which to operate their pumps, given a set of constraints. Those constraints might include the standard flow rate to be delivered, the pressure at key locations and the flow velocity. Another limit governing how best to operate a pumping station is the *actual flow rate* at a pump's suction, which must be maintained in the operating region of the pump. In pipelines transporting liquid products, actual flow rate

---

* In the water industry, engineers often refer to a profile of total head as the *energy grade line* (EGL). They also talk about a *hydraulic grade line* (HGL), the profile of the piezometric head, which is pressure head plus elevation head.

is typically close to the standard flow rate, but not so close that we can ignore it entirely. When determining whether the pump is operating on the map, the actual flow rate is required. So it can be useful to select which quantities the user is interested in viewing on a hydraulic profile, depending on the task at hand.

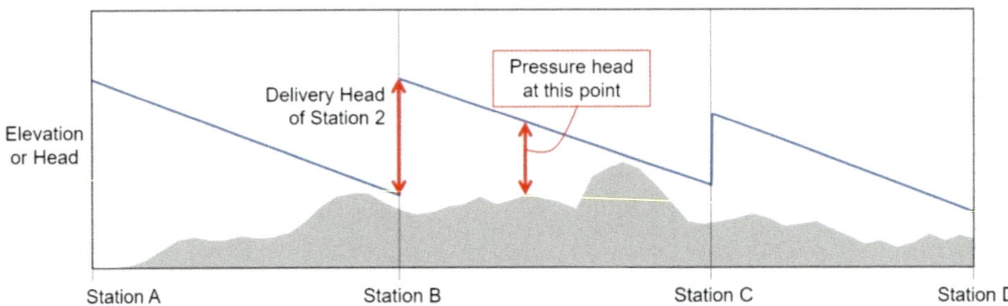

Figure 61 – The hydraulic profile along a liquid line with intermediate pump stations. The head profile (blue) has a gradient associated with a particular flow rate.

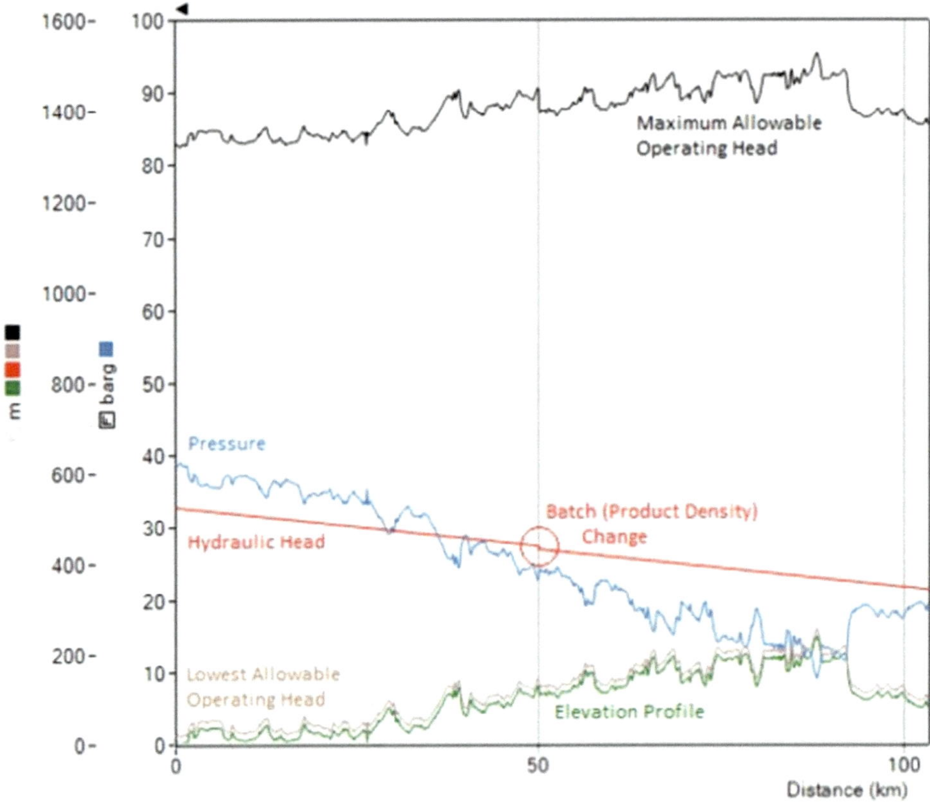

Figure 62 – A hydraulic profile of a liquid pipeline. Pressure (blue) increases against the flow, wherever the elevation descends steeply. In contrast, head (red) uniformly drops throughout. A kink is visible, along with a subtle change in gradient, at a batch interface.

Continuity of steady liquids: the mass balance

In our earlier discussion of conservation laws, we saw how a mass balance expresses the adage: *"what goes in must come out"*. In steady flow, an almost-incompressible liquid is described by a special case of the mass equation. In this scenario, the flowing fluid is conserving actual volumetric flow rate (Q): *

$$Q = A_1 \cdot v_1 = A_2 \cdot v_2$$

This is interesting: it lets us easily relate velocity (v) and cross-sectional area (A), for a steady, incompressible liquid, i.e. one with constant density (ρ).

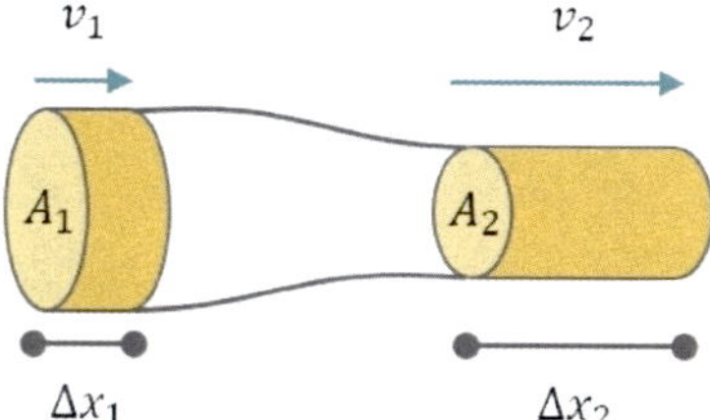

Figure 63 – Continuity dictates a liquid's velocity is faster through a smaller diameter.

This isn't just stating the obvious: in a liquid pipeline, this relation deduces the velocity at changes in diameter. It also has profound implications. Say a piston pushes an incompressible liquid through a small diameter pipe, moving it some distance. Pascal's law says pressure immediately transmits to push on an outlet. If that outlet is a large diameter, it moves more mass a shorter distance.

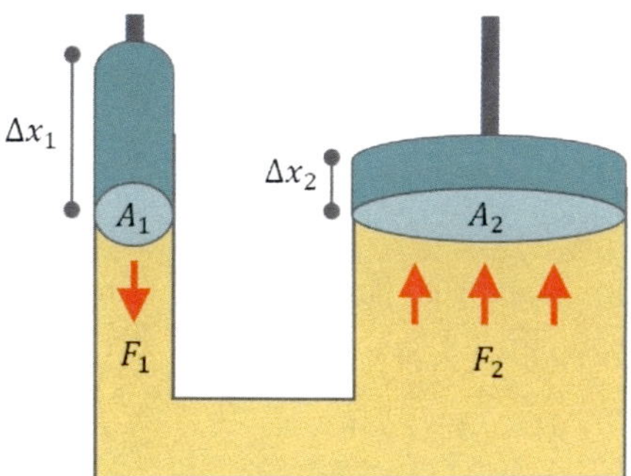

Figure 64 – A hydraulic press is analogous to amplifying a force (F) by means of a lever.

* Some people refer to this as *"the"* continuity equation, which is misleading.

Isothermal steady liquids: a simple energy balance

The three conservation laws we encountered earlier capture the underlying physics. They always hold true, provided our one-dimensional approximation holds up. In this section, we introduce a couple of *special cases*, because those are sometimes useful for pencil-and-paper analysis, to confirm simulator results. They should only be used with great caution, as they are not universally true. The mass and momentum equations combined give us Bernoulli's principle for the special case of an incompressible liquid. The development and uses of Bernoulli's Principle are covered in detail in the *Primer* appendix (see the companion text to this book). We modify this equation to include friction as our pressure-throughput relation for manually calculating the steady hydraulics of liquid pipelines. In a steady state with no temperature effects, the energy equation gives us this same result, but it does *not* if we include heat transfer.

If we look at a flow as a moving *control mass* – a *closed* system – the *First Law of Thermodynamics* says the change in internal energy (du) is the sum of the heat flow into the control mass (q) and the work done on the control mass (w):

$$\mathrm{d}u = q + w$$

It is more intuitive to consider a pipe element as an *open system*, a *control surface* permeable to the flow of fluid, so our pipe is a bounded *control volume*.

Figure 65 – We think of a stationary open system, rather than a moving closed system.

This means that we also need to consider the heat and energy convected into and out of the control volume. In the chapter on conservation laws, we expressed conservation of energy in a form that applies to fluid flow in a pipe. We call this our *energy equation*:

$$\frac{1}{A}\cdot\left(\frac{\partial(e\cdot A)}{\partial t}+\frac{\partial(e\cdot A\cdot v)}{\partial x}+\frac{\partial(P\cdot A\cdot v)}{\partial x}+P\cdot\frac{\partial A}{\partial t}\right)+4\cdot U\cdot\frac{T-T_y}{D}=0$$

We have taken our conserved quantity (y) to be specific energy (e). We define it as:

$$e = u + ½\cdot v^2 + g\cdot z$$

- 1st term: rate of change of energy-density accumulating in our element
- 2nd: difference in energy-density convecting across our element
- 3rd: difference in gravitational potential energy across our element
- 4th: difference due to a change in cross-section across our element
- 5th: heat transfer with the ambient medium in the wider environment

Setting time-derivatives to zero, neglecting temperature effects, and neglecting the work done by pipe expansion, we can write this in terms of mass flow rates (Q^{mass}):

$$M \cdot (q + w) = Q_{\text{out}}^{\text{mass}} \cdot \left(u + \frac{P}{\rho} + \frac{v^2}{2} + g \cdot z\right)_{\text{out}} - Q_{\text{in}}^{\text{mass}} \cdot \left(u + \frac{P}{\rho} + \frac{v^2}{2} + g \cdot z\right)_{\text{in}}$$

British universities call this the *steady flow energy equation* (SFEE). If there is zero heat transfer ($q = 0$), and zero externally-imparted work ($w = 0$), we cannot say that there is no change in the internal energy of the fluid ($\mathrm{d}u = 0$) in general. We see that what is conserved in this special case is a little different to Bernoulli's principle, because aside from one special case (an ideal gas undergoing a process with zero temperature change), the internal energy (u) of a fluid might change and so must be accounted for:

$$\left(u + \frac{P}{\rho} + \frac{v^2}{2} + g \cdot z\right)_{\text{A}} = \left(u + \frac{P}{\rho} + \frac{v^2}{2} + g \cdot z\right)_{\text{B}}$$

In addition to the three terms summing to a fluid's total *"mechanical energy"*, as per Bernoulli's principle, there is also internal energy as part of the sum that remains constant between two points (A and B).

Forces in steady liquids: the momentum balance

We can alternatively derive Bernoulli's principle by integrating the momentum equation from one point to another. The momentum equation can give us units of energy when we do this, because without convection, it essentially says:

$$F = m \cdot a$$

And integrating force (F) with respect to distance ($\mathrm{d}x$) gives us units of energy. It is possible, and in some ways less objectionable, to derive the famous form of Bernoulli's principle by picking up our last, special-case expression for conservation of *momentum* rather than conservation of energy. This special case is for steady incompressible flow:

$$\frac{\partial P}{\partial x} + \rho \cdot g \cdot \sin[\theta] + \frac{\rho \cdot f \cdot v \cdot |v|}{2 \cdot D} = 0$$

The final term applies losses along a pipeline.

Pressure-throughput relation: incompressible flow

We saw how for a steady incompressible liquid, Bernoulli's principle can be derived either from conservation of energy or conservation of momentum. It describes a so-called *"mechanical fluid energy"*, equivalently stated in units of energy per unit volume (*pressure*), energy per unit mass (*specific energy*), or energy per unit weight (*head*).

Pressure P form	$P+\frac{\rho\cdot v^2}{2}+\rho\cdot g\cdot z=\text{constant}$	$\mathrm{Pa}\equiv\mathrm{kg}\cdot\mathrm{m}^{-1}\cdot\mathrm{s}^{-2}$
Specific energy head e or $H_{\mathrm{J/kg}}$ form	$\frac{P}{\rho}+\frac{v^2}{2}+g\cdot z=\text{constant}$	$\mathrm{J}\cdot\mathrm{kg}^{-1}\equiv\mathrm{m}^2\cdot\mathrm{s}^{-2}$
Length-units head H form	$\frac{P}{\rho\cdot g}+\frac{v^2}{2\cdot g}+z=\text{constant}$	m

This principle is only concerned with a flow with no shaft work, heat transfer or friction. Bernoulli's principle, at its core, says that this *"mechanical energy of a fluid"* that is flowing at steady state can manifest as any one of three distinct terms:

- Static pressure / pressure-volume work / static head *
- Dynamic pressure / kinetic energy / dynamic head
- Gravitational potential energy / elevation head

It is *not* an expression that universally holds true. It only applies if and only if:

- There are no local losses (to shocks, sudden bends or fittings)
- Flow is steady and incompressible, and there are no frictional losses
- There is no heat transfer and no externally imparted shaft work

For special subset of scenarios, this expression is useful for *local analysis*. It predicts what will happen across a shock wave, or through a valve or orifice. But can it serve as a pressure-throughput relation in a liquid pipeline? An almost-incompressible liquid has a constant density (ρ) by definition. Flow continuity dictates that its velocity (v) remains constant throughout any pipeline of the same diameter. In fact, the term is small even if there does happen to be a change in diameter. The kinetic energy is almost unchanged along pipelines carrying almost-incompressible liquids.

---

* Some authors prefer to lump in this pressure-volume work (often called PV-work, and being the work done by fluid expansion) together with the shaft work as *"overall work"*.

Therefore it is often reasonable to neglect the change in its kinetic energy term. If we can do so, then when a liquid pipeline changes elevation, that manifests as a *change in pressure*, either a pressure drop or a pressure increase, calculated by rearranging:

$$\frac{\Delta P}{\rho \cdot g} = \Delta z$$

But we can't apply this *"lossless"* Bernoulli's expression along a stretch of pipeline. Overcoming friction to maintain steady flow incurs losses. As we look further along a flowing pipeline (or follow a clump of fluid during its journey through the pipeline), we see its *"mechanical fluid energy"* falling as it goes; it is not conserved along a pipeline. *Head loss* (ΔH) is classified into two types:

$$\Delta\left(\frac{P}{\rho \cdot g} + \frac{v^2}{2 \cdot g} + z\right) = \text{Local Losses} + \text{Frictional Losses}$$

Any head loss that is *not* attributable to friction is due to fittings, sudden bends and other local effects. This is negligible in a pipeline; it only affects stations. *

The kinetic energy term is negligible in a liquid line

The act of smoothly transitioning from a small diameter to a wider section has almost no effect on pressure, but does affect velocity and Reynolds number. Consider two scenarios, one with a constant diameter, another with expansion:

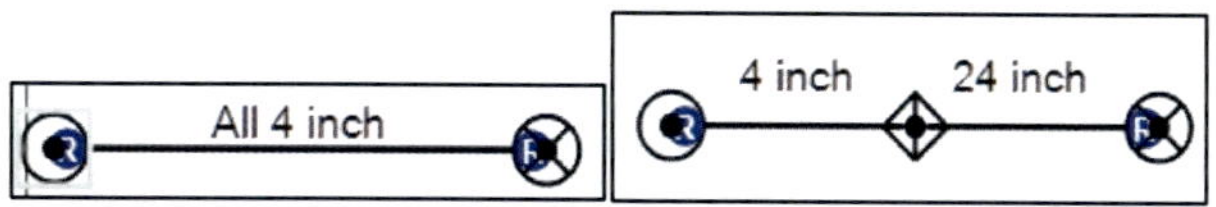

Figure 66 – A short flat 4-inch pipeline (left) and again but transitioning to 24-inch (right)

* If you find this to be a bold claim, it may be reassuring to recall values listed in handbooks like *"CRANE 410"* let us express losses as equivalent lengths of pipe (L/D). The worst bends would only amount to few dozen metres. Pipework on process plant incurs pressure drops across many bends and fittings, whereas a pipeline is big, long and straight. Outside stations, far more frictional loss is associated with the pipe wall and the fluid flow itself than with the fittings.

We fix the temperature to be constant throughout the pipeline, so density is only affected by pressure. Simulating the two scenarios with *water* as the fluid, its pressure was barely affected at all, and its density remained constant:

	All 4-inch		4-inch to 24-inch		
	Near end	Far end	Near end	Far end	
Pressure	2.000	1.935	2.000	**1.968**	atm
Velocity	1.219	1.219	1.219	**0.039**	$\mathrm{m \cdot s^{-1}}$
Reynolds	112 000	112 000	112 000	**20 000**	–

And simulating again with a *crude oil*, its pressure too was barely affected, so its density stayed constant. Only flow velocity and Reynolds number changed:

	All 4-inch		4-inch to 24-inch		
	Near end	Far end	Near end	Far end	
Pressure	2.000	1.922	2.000	**1.961**	atm
Velocity	1.219	1.219	1.219	**0.039**	$\mathrm{m \cdot s^{-1}}$
Reynolds	2 640	2 640	2 640	**470**	–

The effect of a gradual change in diameter (a constriction or expansion) on the pressure of a liquid is trivial because it only affects velocity, which only affects the *kinetic energy* term of Bernoulli's equation. That term is trivial for liquids. For example, if water at atmospheric pressure flows at 2 m/s to descend 10 m, its kinetic energy term (i.e. dynamic head) constitutes only 0.05% of the total energy change due to the change in elevation.

Friction in a liquid pipeline

To calculate frictional losses, we first need to determine the *flow regime* by calculating our Reynolds number. But Reynolds number depends on flow velocity, which we don't know: it is precisely the quantity we want to calculate!

In practice, a pipeliner will know a ball-park estimate of flow rate, so they know if the flow regime is laminar, transitional, smooth-pipe, middling, or rough-pipe turbulent. Having calculated the Reynolds number and guessed the flow regime, we use this as a basis for calculating the friction factor. This is the last parameter needed to solve our pressure-throughput relation.

The effect of friction factor is proportional to velocity-squared, always acting to oppose flow ($-v \cdot |v|$), and is inversely proportional to pipe diameter (D):

$$\frac{\mathrm{d}P_{\text{friction}}}{\mathrm{d}L} = \frac{-\rho \cdot f \cdot v \cdot |v|}{2 \cdot D}$$

If we think of it as an actual volumetric flow rate (Q) per unit area ($\frac{1}{4} \cdot \pi \cdot D^2$), we equivalently write that friction is affected by the fifth power of diameter:

$$\frac{\mathrm{d}P_{\text{friction}}}{\mathrm{d}L} = -\frac{8 \cdot \rho \cdot f \cdot Q \cdot |Q|}{\pi^2 \cdot D^5}$$

Note that these statements are approximate, as friction factor (f) is completely independent of velocity (v) only in the limit of very high Reynolds numbers.

We should be aware that there is a risk not just in undersizing a pipe, but also in oversizing it. Avoiding an excessively low flow rate imposes a more subtle upper limit on diameter than merely the upfront price of building the pipeline.

Say we wish to check whether some diameter is appropriate for achieving a certain flow rate through a certain pipeline. Relating its pressure drop to that flow rate is readily solved as a *capacity scenario* in a pipeline simulator.

By hand, the Colebrook-White correlation obliges us to perform iterations: *Diameter* → affects *flow regime* → affects *friction* → affects *flow rate* → affects *flow regime* (and so forth). Today, even this *"manual calculation"* is less awkward than it was in the past as spreadsheet software lets engineers iterate.

Determining the friction factor

In the laminar flow regime, for an almost-incompressible liquid, the analytical *Hagen-Poiseuille* equation for the pressure loss (ΔP) can be derived from first principles:

$$\frac{\Delta P}{\rho} = 32 \cdot \frac{L}{D} \cdot \frac{\mu \cdot v}{\rho \cdot D}$$

Except for the heaviest crude lines, however, pipeline flows are generally turbulent. In non-laminar regimes, the friction is only known empirically, using correlations. The *Darcy-Weisbach* equation tells us that the frictional head loss (ΔH) is proportional to velocity squared (v^2):

$$\Delta H = f \cdot \frac{L}{D} \cdot \frac{v^2}{2 \cdot g}$$

Quantifying the effect of friction in a completely general statement has proven elusive. Instead, it is characterized by several available correlations. Each one uses a different formula to calculate the same empirical dimensionless friction factor which we feed into our equation for pressure losses.

All *"flow equations"* are based on an assumed expression for a dimensionless correlating function, the *friction factor* (f).* All non-one-dimensional effects are wrapped into it, including the effect of turbulence if present. Typically, the value of f is between 0.01 and 0.1. It is calculated using *Reynold's number* and *relative roughness*. This relationship is described by empirical formulae. Depending on our choice of friction factor equation, roughness might have absolutely no effect at, say, $Re = 40\,000$. This is physically correct!

There are several widely-used friction factor correlations:

- Colebrook-White
- Weymouth
- Panhandle A and Panhandle B
- AGA (American Gas Association)

---

* As it is an arbitrary empirical multiplier, several definitions of friction factor are used. The Darcy friction factor is four times the Fanning friction factor. Either might be seen tabulated or plotted on a Moody chart, without stating which! A handy shibboleth to distinguish the two is to look at the laminar regime's straight line: 16/Re is Fanning's friction factor. 64/Re is Darcy's friction factor.

Each of these friction factor correlations is based on its own experiments, and is tailored to work only for certain fluids within certain conditions. A team exercises engineering judgment to choose one that works for their purposes.

The most widely adopted in the pipeline industry is the Colebrook-White correlation, sometimes referred to as *"the Colebrook equation"*. It covers a broad range, including all flow regimes pipelines tend to operate in, and the transitions between them. A downside of this formula is that friction factor needs to be found *iteratively*, as it is a function of itself:

$$\frac{1}{\sqrt{f}} = -2 \cdot \log_{10}\left[\frac{\varepsilon/D}{3.7} + \frac{2.51}{\mathrm{Re}\cdot\sqrt{f}}\right]$$

Where a pipe's *relative roughness* (ε/D) is its absolute roughness divided by diameter. A larger pipe made by the same manufacturing process has a different diameter (D); despite having the same absolute roughness, its relative roughness will be lower. Reynolds number (Re) characterizes the flow regime.

There are non-iterative approximations to the Colebrook-White equation. One example is the Swamee-Jain equation. Listing them all here would make a *long* list, but if we encounter any of them we must bear in mind that it is only a good approximation within certain conditions. They come in handy for hand calculations, but there is no reason to use an approximation in a simulator: it would needlessly introduce an error. Atmos SIM just iterates the equation to solve it.

It may seem comical for anyone to be using a look-up chart in the 21st century. But the Moody diagram is so useful that engineers still make frequent reference to it. It is a graphical representation of the Colebrook-White correlation alongside the analytical result in the laminar flow regime on the far left. The jump between friction factor in laminar flow and friction factor in turbulent flow is shown as being quite sharp, raising uncomfortable questions. This is known as the *transition region*.

Taking a moment to look at a Moody diagram lets us make a number of insightful observations. Even a minuscule absolute roughness ($\varepsilon < 0.01\ \mu\mathrm{m}$) causes significant frictional head loss. Notice that as relative roughness (ε/D) drops by four orders of magnitude, the rough-pipe turbulent friction factor only decreases from 0.1 to 0.01. A whole lot of this behavior can be seen if we look closely at the Moody diagram, so although we have an algorithm for friction factor, its iterative nature is less appealing to many engineers who have grown accustomed to using this graphical method. Even those who are happy to iterate will look at the chart to make sense of friction.

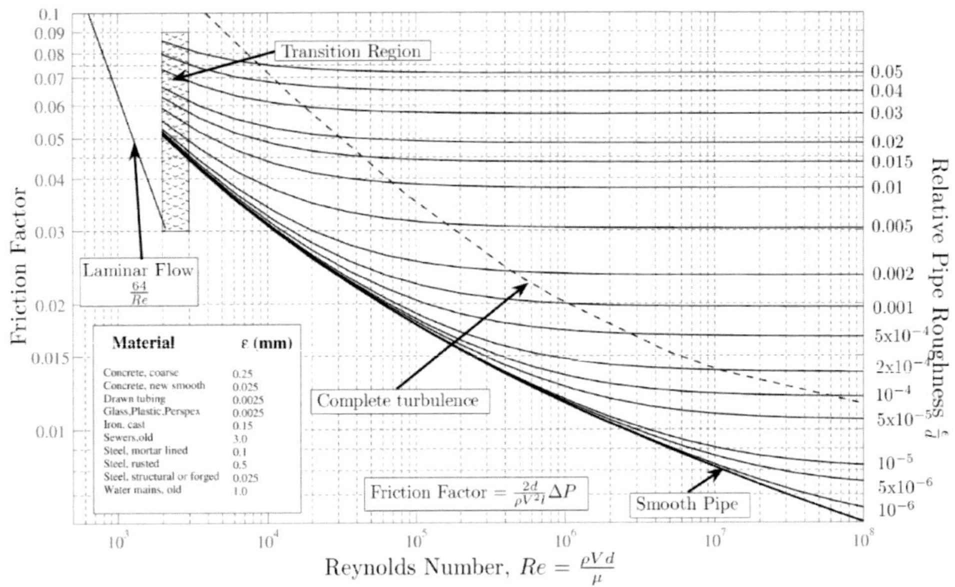

Figure 67 – The Moody diagram giving us the friction factor (f) based on the Reynolds number (Re) of a flowing fluid and the relative roughness of a pipe (ε/D)
S Beck and R Collins at the University of Sheffield / Wikimedia Commons / CC BY-SA 3.0

Maintaining a back-pressure to keep a pipeline tight

Back-pressure is usually maintained at a liquid pipeline's delivery point. Typically, a liquid pipeline starts from a storage tank, often atmospheric, as its supply. Fluid travels through pumps, then along the pipeline, and ends up at a delivery storage tank. Starting at the pipeline's limit at the delivery facility, at atmospheric pressure, calculate the static head from the tank's height. Then, at our desired flow rate, calculate frictional losses. A meter run might lose 0.3 bar, a strainer 0.7 bar, and a tank another 0.7 bar. Summing those together, we end up with 1.7 bar back-pressure, required as a bare minimum to maintain that flow rate into the delivery station. In many pipelines, yet more back-pressure is needed.

This additional back-pressure prevents the pressure at elevation peaks along a pipeline falling below the fluid's vapor pressure. A crude oil or refined liquid flowing reduces in pressure particularly when flowing uphill, potentially causing it to flash into its vapor phase. These are the locations along a typical pipeline which are most susceptible to becoming only partly filled with liquid, a phenomenon known as *column separation* or *slack-line*. One might think that an empty pipeline is, like a bus carrying fresh air, nothing to worry about, albeit pointless. This first impression couldn't be more wrong. Liquid pipeliners care a lot about slack, and some pipelines operate routinely in slack flow, or expect a degree of column separation in shut-in. To simulate slack flow, we must consider the physics taking place more closely. It is the subject of a later chapter.

If a pipeliner wishes to avoid this condition, a tight-line operation controls a pipeline's pressures and flow rates to maintain it in a liquid phase throughout. So it is important that even a *"single-phase"* liquid model can *detect* and *handle* the presence of a slack region, even if it does not fully capture its behavior.

Inside a slack region, pressure is guaranteed to be constant to a high degree of accuracy. But liquid might be flowing forwards or backwards at any instance in time. Luckily, we know that in *steady* conditions, a slack region always flows downhill, so the *head* slope indicates flow direction, even within a slack region.

As a first pass, simulate what the back-pressure at the terminus would look like along the pipeline as a hydraulic gradient – a profile of the effect of friction. Then, consider where the head must go over a high point: if pressure there is below vapor pressure, more back-pressure is needed at the delivery facility. Just a couple of atmospheres of additional pressure (2 barg) is a sufficient margin to keep some pipelines tight. For others traversing treacherous terrain, a back-pressure of 28 barg might be needed! [35]

One way of achieving that back-pressure is by installing smaller diameter piping at the destination tank farm. Commonly, additional back-pressure is provided via a *"back-pressure valve"*. This valve imposes a variable resistance at the far end by adjusting its position to run the line close to optimal across a range of operating conditions, much less wasteful than small piping that always burns off extra back-pressure across the board. It can be abstracted as a minimum pressure setpoint on a demand object within our model. Or, if we wish to also simulate some equipment downstream of this valve, it can be simulated as a model item which Atmos SIM calls a regulator (also known as a control valve), setting it to control on a minimum inlet pressure setpoint.

At certain points in the line, such as any mid-line deliveries to low-pressure systems, it is essential to *reduce* the pressure using regulators. Regulators on liquid pipelines can also be used to control flow rate, to avoid developing excessive pressures, and to ensure flow is directed along its intended route.

5.3 – DRAG REDUCTION AGENTS

Why these additives are used

Drag reducing agents or *drag reduction agents* (DRA) are a class of additives that make hydrocarbon liquids more amenable to pipelining. They all serve to suppress frictional losses in some manner. Exactly *how* depends on which type is used, and often it is not perfectly understood. They are used to increase flow rate for a given pressure drop. Other options for doing this are adding parallel lines (looping) or raising the operating pressure using pumping stations.

These additives can be broadly categorized as polymeric, fibrous or surfactant in nature. The latter two categories are relatively new to the field, and some pipelining enterprises are wary of trying them without data and experience to back up their performance. Polymers are in widespread use. When added in parts-per-million concentrations, these chemicals serve to dampen a flow's turbulence near pipe walls, reducing drag in a liquid line by as much as 80%.

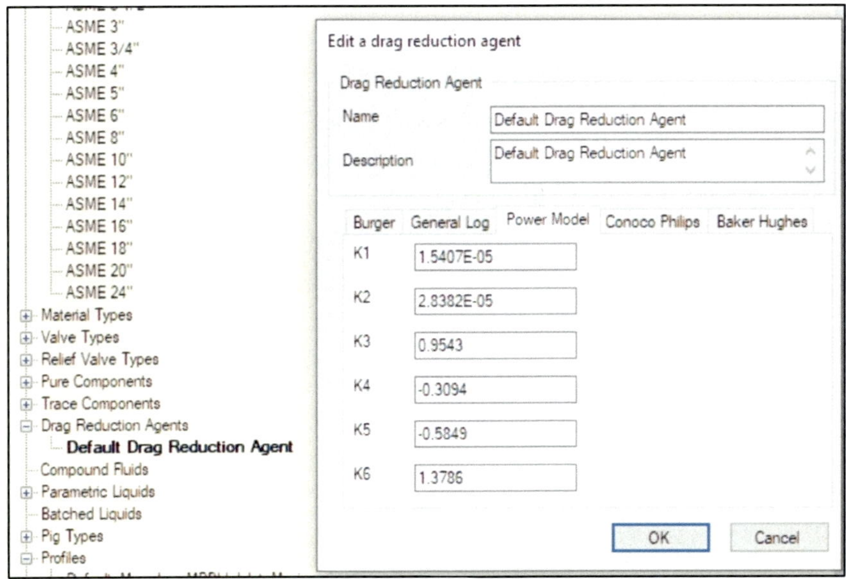

Figure 68 – Assigning parameters to a drag reduction agent (DRA) within Atmos SIM's libraries. A number of possible correlations are available to describe the efficacy of DRA. Similar dialogues let us specify how it degrades on passing through model pipes and pumps.

How these additives are described

What drag reduction agents *are* is both proprietary and not really our business. One guest on the excellent *Pipeliners* podcast called them *"magical unicorn snot"*. A brief look at some of the literature mentions people are using aloe vera, which sounds far more pleasant. What matters to us is not what these additives are made of, but rather what they *do*. The behavior of a drag reduction agent is described by how much it reduces the frictional pressure losses, expressed as an empirically determined percentage called *drag reduction factor* (DRF) or drag reduction effect (DRE). The specifics of how drag reduction agents work, and how they degrade, is quite a specialist topic Ih takes us back to our discussion of flow regimes. A turbulent flow regime reduces the throughput of a liquid pipeline because it increases the incurred drag, which causes part of the flow's energy to be lost to friction.

By drag, we mean that the pipe wall drags on the fluid and slows it down. The essence of what these agents strive to do is to reduce these frictional losses. The smaller vortices slam into the wall roughness in the rough-pipe turbulence flow regime, and that is a very efficient mechanism for transferring momentum from the fluid to the wall, so we want to suppress it. There is just one problem with this explanation: DRA sometimes works in the smooth-pipe turbulent regime as well. As DRA suppresses smaller vortices it is still going to be effective at reduce the transfer of axial momentum throughout the fluid and that will even help in the transitional flow regime.

Polymeric additives consist of very long hydrocarbon molecules, comparable in length to the smaller vortices. For a vortex to form, it must first bend these molecules, so small vortices are suppressed. A larger *"smallest vortex size"* makes for a thicker viscous sublayer. So these additives partially convert our flow from a rough-pipe regime to a smooth-pipe regime, reducing overall drag.

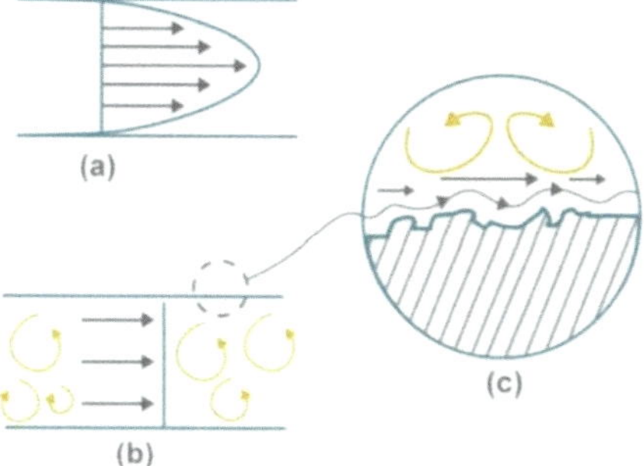

Figure 69 – The velocity profile of a liquid flowing in the turbulent regime (a) is improved by adding drag reduction agent (b). Only larger vortices or eddies can occur anywhere, reducing the turbulence within the buffer region between the laminar sublayer and the turbulent core (c)

Drag reduction agents do exist for gas pipelines working by a different principle to fill irregularities on the rough surfaces of pipe walls. But *"gas DRA"* remains experimental and not in widespread use, so here we discuss only liquid DRA.

It is not pressure change that affects DRA, but it partially degrades as it travels along a line, as its molecules are *"bent"* by all these micro-vortices. Eventually the DRA molecules snap altogether. It degrades as it passes through certain equipment, such as control valves. It severely degrades on passage through a pump or a slack section. This degradation is characterized within a simulator by entering the fraction of DRA lost per unit length in each pipe, and per device for other equipment.

The effectiveness of the drag reduction agent (DRA) is characterized by an empirical *drag reduction effect* (DRE). It appears in the momentum equation as a multiplier on friction factor (f), and is a function of concentration (ppm):

$$f \rightarrow \mathrm{DRE}(\mathrm{ppm}, \ldots) \cdot f$$

Rather than provide a drag reduction effect function, manufacturers now typically give a formula for pressure drop in a pipe, usually underestimating an agent's performance. DRA is not really totally destroyed in a pump, for example. In our experience, each running pump destroys about half of the drag reduction agent passing through it.

The following ad-hoc form for drag reduction effect (DRE) is often reasonable, especially if tuned to match observed behavior:

$$\mathrm{DRE} = \frac{A}{B + \frac{1}{\mathrm{ppm}}}$$

Another way to put it is that a *drag reduction factor* (DRF) acts as a multiplier:

$$\mathrm{DRF} = 1 - \frac{A}{B + \frac{1}{\mathrm{ppm}}}$$

A highly concentrated drag reduction agent (DRA) attains the most effective drag reduction. This is often better than 50% – reducing friction by a half – and today 80% is obtainable in real pipelines. These empirical expressions typically lack sufficient experimental support, especially in transient conditions. This means that this significant effect is poorly characterized, introducing a major error into any simulator's calculations that outweighs all others. It makes online tuning mechanisms particularly important on liquid pipelines that use DRA.

Optimizing how pipelines use drag reduction agents

Enterprises pipelining crude oil and liquid products spend millions on these chemicals, so they want to be absolutely sure they're making the best use of it. Suitably modifying a pipeline's operations can achieve substantial savings, by trading off the energy cost of running pumps against the cost of injecting the right amount of DRA at the right location. Alternatively, adding DRA can increase a pipeline's profitability by increasing the amount of oil it can transport. And some pipelines can only ever operate with DRA.

Tailoring *how* we use DRA fulfils the needs of a pipeline's operation by optimizing its drag reduction effect on flow rate and pressure. This warrants extensive studies into the numerous cases under consideration for any given pipeline. Injection of DRA could do away with the need of some of the pumping equipment altogether, eliminating the considerable costs of pump purchase, pump operation and pump maintenance. [36]

To this end, simulation professionals want more powerful tools at their disposal than merely *capturing* the effect of DRA. Manually setting up, executing and analyzing countless scenarios limits what a team can do with the time available to them. This kind of task is where an optimizer really comes into its own. We discuss optimizers, including those used for studies exploring drag reduction agents, in the final chapter of this book.

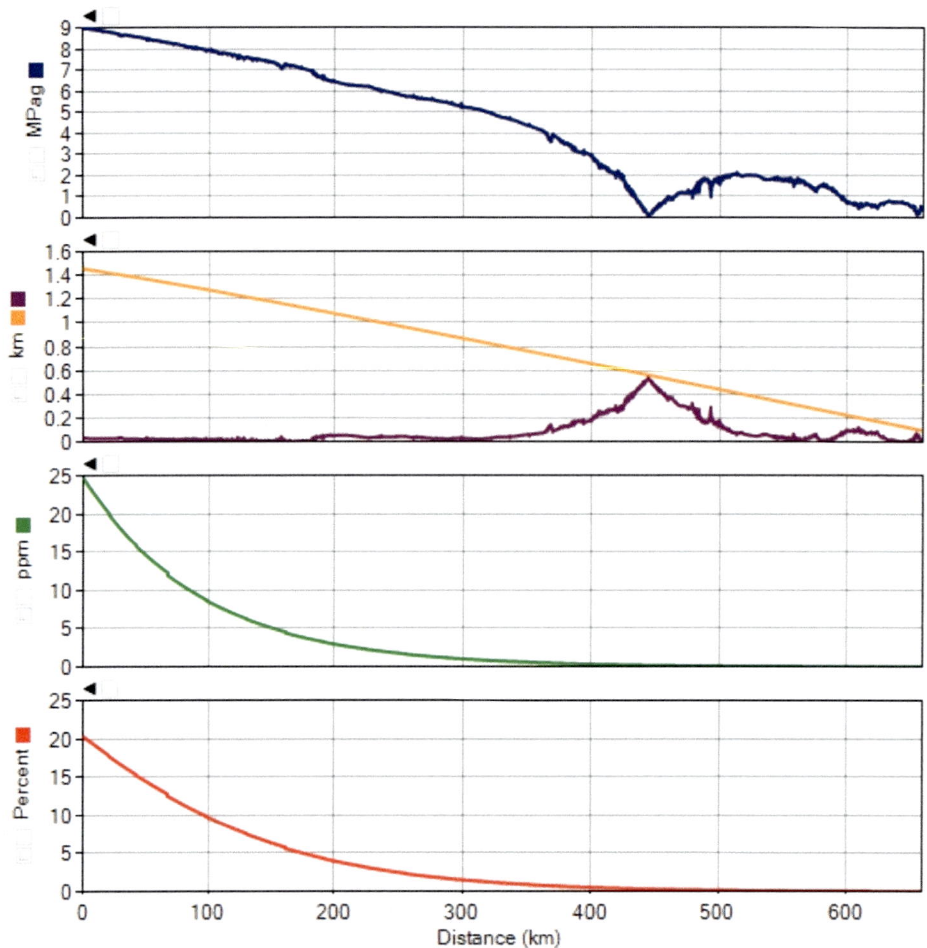

Figure 70 – By the time this liquid product has reached a mountain along its route, the drag reduction agent (DRA) has entirely degraded (ppm) and so has no effect (percent). Another DRA injection point could be placed in the model at a mid-line station to improve the pressure behavior of this liquid, seeing as its head profile touches an elevation peak.

5.4 – STEADY HYDRAULICS OF GAS PIPELINES

Can a gas pipeline ever be considered steady?

Gas pipelines rarely reach steady state. Typically, they are operated in an *imbalanced* condition, so that a section of pipeline is *packing* (accumulating inventory) or *unpacking* (depleting inventory). This process – or the inventory it involves – is sometimes referred to as the *linepack*. Nevertheless, looking at the operating envelope of a gas pipeline that is *"nominally"* steady is a good place to start when analyzing its hydraulics.

Velocity limits in a gas pipeline

The flow velocity always varies along a gas pipeline because even in a single pipe of fixed diameter in steady flow, the gas density changes. The maximum velocity is at the lowest density i.e. lowest pressure, which happens at the far end of a gas pipeline. This is typically 5 to 12 m/s (18 to 43 km/h) in natural gas pipelines. Beyond 15 m/s carbon dioxide corrosion becomes untenable in natural gas pipelines, and faster than 18 m/s the noise becomes unacceptable. A pipeline simulator can alert the user if the fluid velocity approaches these limits at any point along the route of a pipeline.

Pressure limits in a gas pipeline

The *maximum allowable operating pressure* (MAOP) in a gas transmission pipeline is typically well above 100 atmospheres. It could be that a weak stretch at the top of a hill might well be the main factor constraining a pipeline's throughput. A section of pipe could be particularly weak if, for example, a pig got stuck and a pipe was cut to remove it and then welded. But often, just as in a liquid pipeline, it is sections of pipeline whose surroundings warrant special caution that impose the most stringent limits. In a gas pipeline, that caution might be particularly warranted in the vicinity of populated places.

A minimum pressure is often contractually stipulated for a gas delivery. This means the operating philosophy of a gas pipeline incorporates provisions for ensuring a delivery point's pressure never falls below something like 50 barg for example. However, one should be aware that depending on the elevation profile along a pipeline, the pressure may fall well below this elsewhere along the route. This is because the pressure gradient might be in the opposite direction to flow rate in a hilly, high-pressure gas pipeline, for the same reason it can do so in a liquid pipeline. And so, as in a liquid pipeline, *profile alarm* can be set to monitor the entire length of the pipeline, alerting the user if pressure ever falls below the limit at that point. This limit is referred to as the *lowest allowable operating pressure* (LAOP). For transmission lines with high flow rates it is typically 20 to 50 atmospheres. Local distribution networks, too, must maintain a few atmospheres; pressure can't fall to zero, else pilot lights in furnaces would go out.

Both maximum and lowest allowable operating limits (MAOP and LAOP) are set conservatively by applying reasonable margins to stringent *safe operating limits* (SOL). It is excursions beyond that safe limit (not included in the scope of pipeline simulators) which pose a safety hazard, rather than *"merely"* a risk of disrupting routine operations.

Temperature limits in a gas pipeline

A gas pipeline can get so cold that it is at risk of freezing if it is left unchecked through measures like trace heating. The 2021 Texas blackout attests to this; liquid anywhere along a gas pipeline could turn to ice in frigid temperatures, blocking pipes or freezing valves, a phenomenon people call *"freeze-off"*. The coldest temperature along a gas line is an operating limit in a more general sense. Below a certain temperature, a solid phase transition occurs in steel to a brittle phase, which might present a risk of potential fracture in a cold gas pipeline. However, the temperatures needed for the ductile-to-brittle transition are lower than those that occur during normal operation, so this risk is mainly of interest during a blowdown or rupture, where gas temperatures become extremely low. Excessive liquid droplets entrained in a gas pipeline can cause erosion and pipe failure. If the temperature at any point drops below $0°C$, water droplets might freeze! It would be advised to avoid such an operating condition, so we configure a profile alarm to inform us if minimum allowable temperature is approached or violated at any point along the route of our pipeline.

Yet another consequence of gases becoming cooler as they flow is that some natural gas pipeliners find hydrocarbon drop-out unavoidable. This is collected at special stations, tankered away to refineries as natural gas liquids (NGL's). Temperature effects thereby present a risk of corrosion, cracking and hydrate formation if *"free water"* drops out along the pipeline. Water is less soluble at lower pressure and temperature, so as a gas is transported, its water content must be carefully controlled. If impurities such as oxygen, sulfur oxides, nitrogen oxides, or carbon monoxide exceed set levels at certain operating conditions in a gas pipeline, the risk of corrosion is higher still. [15]

In a wet gas pipeline, care must also be taken to avoid formation of hydrates. If pipeline conditions are such that both liquid water and liquid hydrocarbons drop out, gas hydrates can form: solid waxy compounds of water molecules with methane molecules trapped inside that can plug a pipe. Last but not least, the inventory held within a given section is significantly affected by temperature. 10°C changes a gas linepack by something like 3%. There is usually a maximum and minimum allowable linepack in a gas pipeline.

A mass balance for steady gas pipelines

Because a gas is compressible, its density significantly changes as a function of pressure and temperature, which vary along a route. Conservation of mass dictates that as the density of a gas drops along a pipeline (A to B), it must therefore move much faster at the far end than it did at the near end. Consider mass flow rate (Q_{mass}):

$$Q_{\text{mass}} = \rho_{\text{near}} \cdot A_{\text{near}} \cdot v_{\text{near}} = \rho_{\text{far}} \cdot A_{\text{far}} \cdot v_{\text{far}}$$

For stretch with no changes in diameter ($A_{\text{near}} = A_{\text{far}}$), the pressure typically drops along a pipeline without any drastic descents in elevation ($P_{\text{far}} < P_{\text{near}}$), so any gas equation of state will tell us that the density drops too ($\rho_{\text{near}} > \rho_{\text{far}}$). The Joule-Thomson effect causes temperature to become cooler ($T_{\text{far}} < T_{\text{near}}$), causing the density to drop a little less, but this is a comparatively minor effect. Therefore it follows that velocity (v):

$$v_{\text{near}} < v_{\text{far}}$$

Natural gas traverses a flat 100 km constant-diameter pipeline at steady state. It enters at 4.3 m/s, 60 barg, 20°C. At the far end, pressure is 15 barg, and temperature 13°C. How fast is gas flowing at the far end?

As gas density drops from 65 to 15 kg/m^3, it speeds up from 4 to 19 m/s, faster than an urban speed limit (40 miles per hour or 65 km/h), which might become noisy! How fast an acceptable velocity is in a pipeline depends on the gas it carries; some pipeline gases are caused to flow as fast as 30 m/s, but that is not typical.

A flow of gas can be described in terms of an actual volumetric flow rate (Q), as this is an important quantity for ascertaining equipment performance. But as it's not conserved, it's unwieldy for describing gas hydraulics. Instead, gas pipeliners refer to the standard volumetric flow rate (Q_{std}). * For a gas at standard conditions (near atmospheric pressure and room temperature), this is equivalent to a molar flow rate. † No reaction is taking place, so moles are conserved and so standard flow (Q_{std}) is conserved, interconvertible with mass flow via a fluid's standard density (ρ_{std}).

---

* As well as calorific flow rates for compressors and commercial purposes. That conversion factor is only constant if the composition is the same throughout.

† Standard volume happens to almost perfectly conserved when blending two gases, but this is not so universally. Propane + butane exhibit non-ideal shrinkage on mixing.

Considering mass flow rate (Q_{mass}) across a mixing point, incoming streams (1) and (2) blend into an outgoing stream (3):

$$Q_{\text{in}}^{\text{mass}} = Q_{\text{out}}^{\text{mass}} \text{ where } Q_{\text{in}}^{\text{mass}} = \rho_1^{\text{std}} \cdot Q_1^{\text{std}} + \rho_2^{\text{std}} \cdot Q_2^{\text{std}} \text{ and } Q_{\text{out}}^{\text{mass}} = \rho_3^{\text{std}} \cdot Q_3^{\text{std}}$$

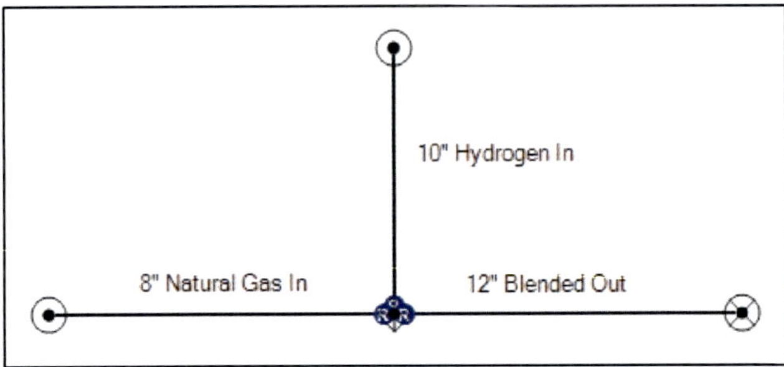

Figure 71 – Blending incoming flows of natural gas with hydrogen in Atmos SIM, reporting the standard densities and standard volumetric flow rates at the inlet and outlet of the junction (though they each remain exactly the same all along each respective leg of the pipeline).

In this way, the standard density (ρ_3^{std}) of a blend is calculated at a junction, as the average of the incoming streams weighted by *standard flow rates (Q^{std})*:

$$\rho_3^{\text{std}} = \frac{\rho_1^{\text{std}} \cdot Q_1^{\text{std}} + \rho_2^{\text{std}} \cdot Q_2^{\text{std}}}{Q_1^{\text{std}} + Q_2^{\text{std}}}$$

In this example: $(0.864 \times 5.9 + 0.085 \times 18.8) \div (5.9 + 18.8) = 0.272 \text{ kg/m}^3$

	Standard density ρ^{std} kg/m^3	**Standard flow** Q^{std} Sm^3/s	**Mass flow** Q^{mass} kg/s
Natural gas in	0.864	5.9	5.132
Hydrogen in	0.085	18.8	1.602
Blended out	0.272	24.7	6.734

Table 4 – Standard flow is conserved blending hydrogen with natural gas at a mixing point

Pressure-throughput relation for compressible flow

Combining the mass and momentum equations for a gas flowing along a pipeline give us a compressible equivalent to Bernoulli's principle, which few seem to refer to, and is occasionally attributed as Ferguson's equation. [38]

A pressure-throughput relation which describes steady flow of a *compressible* fluid is less straightforward than for an incompressible liquid, even if we continue to restrict ourselves to the special case in which our fluid is inviscid.

One relation we encounter is similar to the incompressible Bernoulli equation, with an added dependency on the fluid's *ratio of heat capacities* ($\gamma = c_P/c_V$):

$$\left(\frac{\gamma}{\gamma-1}\right)\cdot\frac{P}{\rho}+\frac{v^2}{2}+g\cdot z=\text{constant}$$

Where does this come from?

To accurately determine the pressure drop of a compressible fluid as it flows through a pipe, we must relate *pressure* (P) to *density* (ρ). This is difficult to do in a hand-calculation. One limiting case is an *adiabatic* flow ($Q = 0$), which is defined as one where no heat enters or leaves the pipe, nor is any generated by friction. If we say that this adiabatic flow is of an ideal gas we then have: [24]

$$\frac{P}{\rho^\gamma}=\text{constant}$$

This is assumed to be true in a short pipe (i.e. a negligible amount of heat added by friction) which happens to be perfectly insulated (allowing no heat transfer). There are local happenings in a pipeline where this comes in handy, but this does not represent what we see in a typical gas pipeline. [24]

The other limiting case is if a flow is isothermal: happening at a constant temperature throughout the pipe. This is equivalent to a steady state with no insulation at all, and ambient temperature being constant everywhere. It is not a terrible approximation for high-pressure gas pipelines, despite them being insulated in reality, at least in terms of their *steady* hydraulics. In this *isothermal* ($T = \text{constant}$) case, for an ideal gas: [24]

$$\frac{P}{\rho}=\text{constant}$$

What happens in reality is that along a gas pipeline, pressures and densities are related in some way in between the two limiting cases, neither adiabatic nor isothermal. ,

As gas velocity increases, so does its kinetic energy. Because total head is conserved, this has consequences. Bernoulli's principle says that if the same fluid flows faster, it has a lower pressure, because more of its *"total mechanical energy"* is *"wrapped up"* within the kinetic energy term. Whereas in a liquid, the kinetic energy term is negligible, in a compressible fluid like a gas, that kinetic energy term can be quite significant, and it changes a lot over long distances.

In gas lines, elevation losses are not so significant unless there is exceptionally hilly terrain in a high pressure section. In a high flow, low pressure natural gas pipeline, the effect of kinetic energy cannot be neglected. The dominant effect for compressors to overcome is *friction*. This friction is substantial, and it is imperative that the expensive compressors employed to boost pressure are capable of maintaining required flow rates in the most challenging scenarios. In our mechanical energy balance for a gas pipeline, we might omit elevation entirely from the pressure-throughput relation.

However, if one is to do this they must be careful to note that in a *high*-pressure transmission pipeline, the effects of elevation can become quite significant. For example, consider this gas pipeline descending from a mountainous region. After each mountain peak until the next valley, the pressure gradient is in the *opposite* direction to the flow. This happens for the same reason as it does in a liquid pipeline: the density of a high pressure gas can become quite substantial, maybe a fifth of the density of a light compressible liquid such as liquefied petroleum gas (LPG). But unless a pipeline is heading substantially downhill all along its route, this is only local.

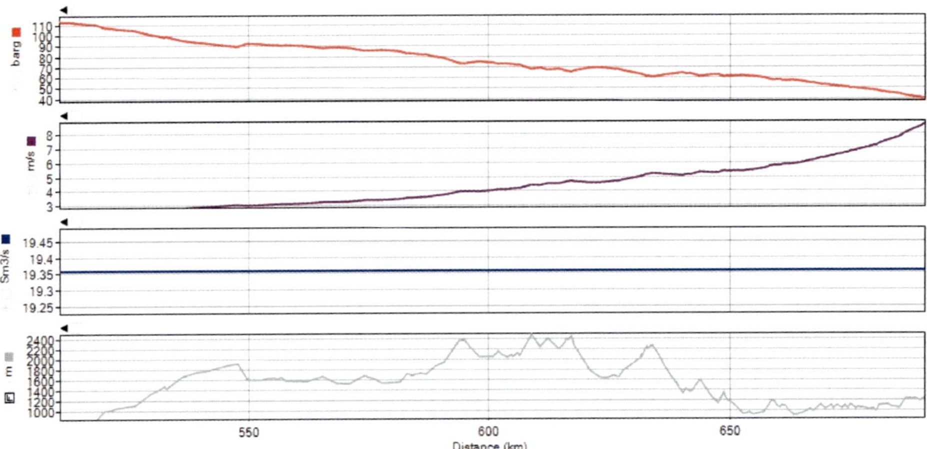

Figure 72 – A high-pressure gas pipeline through mountainous terrain is significantly affected by elevation. This one has several regions (such as the one from 630 km until 640 km distance) along which the pressure gradient is in the opposite direction to the flow.

The differential form of the kinetic energy term might be calculated with reference to the pressure and temperature at standard conditions $(P_{\text{std}}, T_{\text{std}})$ using the concept of *supercompressibility* (Z) which we introduced earlier:

$$v \cdot \mathrm{d}v = -\left(\frac{4 \cdot Q_{\text{std}} \cdot Z \cdot T}{\pi \cdot D^2} \cdot \frac{P_{\text{std}}}{T_{\text{std}}}\right) \cdot \frac{\mathrm{d}P}{P^3}$$

Where actual volumetric flow rate (Q) is related to the velocity (v) of the gas at each point, and so varies along the length of the pipeline.

Substituting for density and the kinetic energy term, assuming average values of supercompressibility and temperature $(Z_{\text{avg}}, T_{\text{avg}})$ over the pipeline's length, and taking account of the fluid's specific gravity (Γ), it is then possible to integrate based on all of those approximations. Some authors use this pressure-throughput relation: [5]

$$P_B^2 - P_A^2 = \text{constant} \cdot \frac{\Gamma \cdot Z_{\text{avg}} \cdot T_{\text{avg}}}{R \cdot g \cdot D^4} \cdot \left(\frac{P_{\text{std}}}{T_{\text{std}}}\right)^2 \cdot Q_{\text{std}}^2 \cdot \left(\frac{24 \cdot f \cdot L}{D} + \ln\left(\frac{P_A}{P_B}\right)\right)$$

The constant depends on the units employed. Again, as for liquids, standard volumetric flow rate (Q_{std}) varies as the two-and-a-half power of diameter (D). Doubling the throughput of a gas pipeline may reduce cost per unit of fluid transmitted by a third. Quadrupling its throughput may make it twice as cost effective. [16] There's no popular name for this expression, so let's call it *"the pressure-squared equation"*. Unlike flow of a liquid, gas flow doesn't depend *linearly* on pressure drop. We instead say *pressure-squared* drop drives *flow-squared*. A gas transmission pipeline operates at very high pressures, so in practice the pressure profile in these pipelines still looks quite linear.

We've seen how to get an expression for pressure drop, and estimate friction factor, for an increment in length. How do we calculate pressure drop if density varies along the line? We can't just multiply by length as we did in a liquid, but are forced to *integrate*. The logarithmic term in the pressure-squared equation accounts for kinetic energy pressure drop. It causes our equation to be *implicit*, meaning it can only be solved iteratively for pressure. In practice, the easy way is to do this kind of computation using an off-the-shelf simulator, and so essentially all we'll need to choose is a suitable knot spacing. At each knot, we calculate density and so forth, and call it a day. But to really understand what this calculation is doing, it helps to perform it manually at least once.

Example: pressure and diameter for a gas pipeline

Working through an example from a reference book demonstrates how tedious even a steady gas line calculation is compared to plugging it into Atmos SIM.

The following is adapted from a textbook by the well-regarded *Wang & Economides.* [5] Gas is gathered at a tie-in point A from processing plants B and C which supply flow rates (in standard volumetric units Q_{std}) of 80 and 50 MMSCFD respectively. It is then transported by pipeline to customers at demand D arriving at a pressure (P) of 500 psi. BA is 1 000 ft long. CA is 800 ft long and 5-inches in diameter. AD is 10 miles long and 10-inches in diameter. This problem assumes temperature (T) is 77°F throughout (i.e. an isothermal scenario), 100% methane, and a relative roughness (ε/D) of 0.001.

Gas pipelines operate in the rough-pipe turbulent regime. In this flow regime, Reynolds number (Re) is so large that its inverse ($1/\mathrm{Re}$) is negligible. Despite Reynolds number varying along a gas pipeline, friction factor (f) can therefore be regarded as the same throughout, only depending on a pipe's wall roughness (ε) and inner diameter (D). For much the same reason, we can treat dynamic viscosity as constant (for methane a value of $\mu = 0.013\ \mathrm{cP}$) as it barely changes and only matters as an order of magnitude. For 100% methane, we can also look up its specific gravity ($\Gamma = \rho_{\mathrm{std}}/\rho_{\mathrm{air}} = 0.554$) and its critical pressure and temperature ($P_{\mathrm{critical}} = 673.6\ \mathrm{psi},\ T_{\mathrm{critical}} = 346.1\ \mathrm{R}$).

We are presented in this textbook example problem with the following questions:

i) *What is the pressure at the tie-in point A?*

ii) *If pressure at B is 1 240 PSI, and we wish for the tie-in point to have no pressure regulators (i.e. incoming pressure from BA equals outgoing pressure to AD), what size diameter should be selected for pipe BA?*

iii) *If diameter of CA is 5-inches, and pressure at C is 1 000 PSI, what would the pressure be at A? To inject stream CA into the mainline AD, by how much does a compressor at outlet of CA need to boost the pressure?*

If we knew all the pipes' diameters, we would have enough information to solve the problem. It would be *"flow-pressure"* ($\mathrm{F} \rightarrow \mathrm{P}$) bounded: A pressure is given (at D) along with flows at all other boundaries (B and C). But the problem does not state one of the diameters (BA), so let's choose one as a first guess and work through the solution.

A *direct manual procedure* to calculate the pressure at the connection point (A) is:

1. Total flow rate into A is $130\ \mathrm{MMSCFD}$
2. Initial guess $\mathrm{Re} = 10\ 000\ 000$ – once pressure is calculated we return to check
3. Supercompressibility is by trial-and-error as it depends on calculated pressure
4. Assume pressure at the tie-in point (A) is $1\ 000\ \mathrm{psi}$ as a first guess
5. Average $P = 0.5 * (1\ 000 + 500) = 750\ \mathrm{psi}$ and $T = 77\ °\mathrm{F}$, charts say $Z = 0.9$
6. Left-hand-side differs from the right-hand-side. So we guess pressure again
7. We repeat the above steps until the left-hand-side equals the right-hand-side
8. Checking $\mathrm{Re} = 11\ 600\ 000$, the Fanning friction factor is $f_f = 0.0049$ indicating that our initial guess for Reynolds number was close enough.

We reach the following answers using this direct manual procedure:

i) Using the direct manual procedure above gives $P_{\mathrm{A}} = 1\ 195\ \mathrm{psi}$ with $Z = 0.9$

ii) With $Z = 0.85$, $\mathrm{Re} = 11\ 000\ 000$, $f_f = 0.0049$, $P_B = 1\ 240\ \mathrm{psi}$, $P_A = 1\ 195\ \mathrm{psi}$, $L = 1\ 000\ \mathrm{ft}$, $Q_{\mathrm{std}} = 80\ 000\ \mathrm{MSCFD}$, we calculate diameter of BA to be 6 inches

iii) With $Z = 0.88$, $\mathrm{Re} = 8\ 800\ 000$, $f_f = 0.0049$, $Q_{\mathrm{std}} = 50\ 000\ \mathrm{MSCFD}$, $L = 800\ \mathrm{ft}$ The pressure-throughput relation calculates $1200 - 960 = 240\ \mathrm{psi}$ is needed.

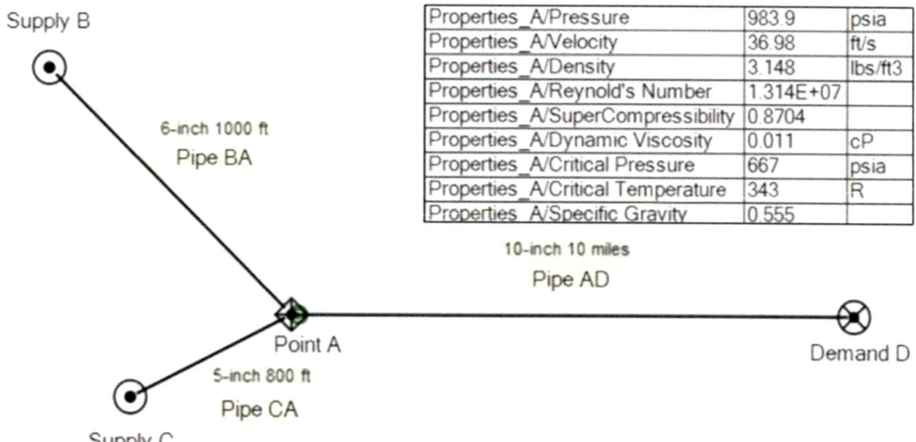

Figure 73 – The solution to this example problem can be reached quickly and easily by trial and error using a pipeline simulator like Atmos SIM. Calculating manually is more tedious.

Parameter	Units	Location			
		Supply B	Supply C	Tie-in A	Demand D
Absolute pressure P	psia	1 018	1 011	984	500
Standard flow rate Q	MMSCFD	80	50	130	130
Supercompressibility Z	None	0.867	0.868	0.870	0.928

Table 5 – The simulated results for this scenario indicate the pressure at A is quite different to the value calculated by a direct manual procedure – we must *discretize* the 10-mile pipeline

The direct manual procedure neglects variation supercompressibility along a pipeline. A *discretized manual procedure* to calculate pressure at the same tie-in point (A) is:

1. Work backwards, taking a short enough distance from demand D (less than 3 000 ft for example) that pressure drop is small (less than 70 psi) so that it is reasonable to assume supercompressibility remains constant ($Z = \text{constant}$)
2. Calculate supercompressibility (Z) at outlet conditions (point D, i.e. 500 psi)
3. Calculate pressure at the segment inlet via the pressure-throughput relation
4. Calculate supercompressibility (Z) at the segment inlet
5. Repeat for the next segment, and so forth, until reaching point A at 10 miles

For further practice questions to get to grips with the pitfalls and tedium of using manual procedures, the textbook from which we adapted this example is worth picking up. [5] In this example we saw calculation can be challenging even in an isothermal scenario.

Friction in gas pipelines

In a flowing gas line, Reynold's number is always rough-pipe turbulent, except when flow is so slow as to be essentially stopped. As gas approaches the limit of stopped flow, the effect of friction becomes negligible anyway, as it scales with velocity-squared. So when a gas line flows so slowly, losses due to friction are orders of magnitude lower than other sources of error: we need not worry about flow regimes in a gas pipeline. Gas flow is *"always turbulent"* for the purpose of calculating friction factor and losses.

How do we calculate the frictional losses incurred along a flowing gas pipeline? The Darcy-Weisbach equation we saw for liquid hydraulics can still be applied. But in a compressible fluid, we are careful because density varies substantially. A rule of thumb is that it remains reasonably accurate so long as the pressure drop it calculates is no more than 10% of inlet pressure (if using inlet or outlet density) or no more than 40% of inlet pressure (if using an average of the inlet and outlet densities). Long pipelines involve higher pressure drops than this, and so we cannot simply plug our numbers into the equation once. There are a number of special-case equations, a few of which appear in the CRANE handbook, including the Weymouth and Panhandle formulas. [24] Both of these equations are surprisingly old (1912 and 1956 respectively) and rather inaccurate, and have been essentially superseded by the Colebrook-White equation.

Calculating friction in gas lines is cumbersome. In fact, hand-calculating friction in a gas line is almost impossible, which is why engineers are not so familiar with such exam questions from their undergraduate degrees! A steady analysis is only just about possible if one negotiates with a spreadsheet. Along the flow direction, pressure head is lost to overcome friction, so pipeline pressure decreases. Nothing surprising yet, the same goes for liquid pipelines. What's peculiar in a compressible fluid, gases included, is that the lower pressure at the far end of a pipeline causes gas to become less dense.

Because continuity dictates that along a pipeline carrying a compressible fluid, the variation in density (ρ) affects its velocity (v) at a constant actual volumetric flow rate, such that the fluid is faster moving at the far end of a pipe, we must be cautious. For a liquid, pressure drop due to friction (or head loss, if we divide through by $\rho \cdot g$) was:

$$\Delta P = f \cdot \frac{L}{D} \cdot \frac{\rho \cdot v^2}{2}$$

To apply this expression for the pressure drop or head loss due to friction in the flow of a *compressible* fluid, we must only apply it to a *very short section*. There is no getting around this even if we conjure up some smart way to integrate velocity and density, because the friction factor (f) is itself affected by velocity. [*] We must split the pipe into slices, small intervals, and applying this equation to each slice in turn, essentially perform a numerical integration.

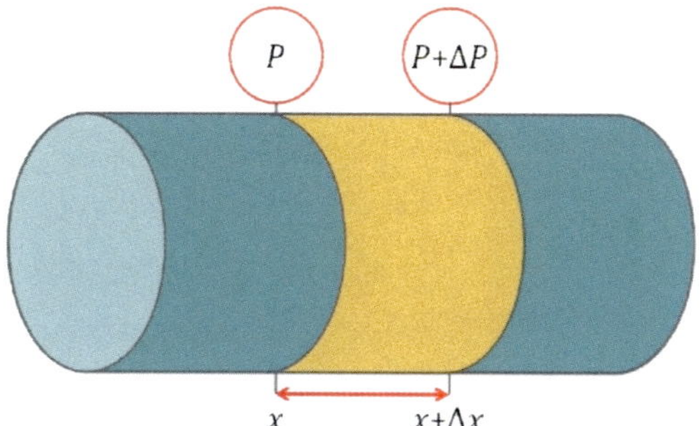

Figure 74 – Pressure varying over a small element of pipe from P at x to $P + \Delta P$ at $x + \Delta x$

As pressure drops across each element, density drops accordingly, and velocity rises as a result. Taking this element and its pressure drop to be infinitesimal, we say both density-drop and velocity-rise are negligible across the element. That is to say: the gas within the element is constant in density and velocity. Bernoulli's principle tells us that total head should be conserved across every element, and the missing portion is the head loss (ΔH):

$$\frac{P}{\rho \cdot g} + \frac{v^2}{2 \cdot g} + z = \frac{P + \mathrm{d}P}{\rho \cdot g} + \frac{(v + \mathrm{d}v)^2}{2 \cdot g} + (z + \mathrm{dz}) + \Delta H$$

Hydraulic profile along a steady gas pipeline

The *"speed limit"* imposed on velocity in a gas pipeline is an order of magnitude greater than in a liquid pipeline. However, a gas pipeline flows at a higher velocity near the end of a pipeline section, because at lower pressure, gas density is substantially lower. This gives us more variables when considering how to increase the throughput of a gas pipeline. We could increase diameter, install more intermediate compressor stations, or operate at higher pressures throughout the pipeline. At a higher pressure, there is a higher density, so the same mass (or standard volume) flows through at a lower velocity. Some gas pipelines operate at pressures of several hundred atmospheres!

Pressure at flowing conditions might be *lower* than at shut-in conditions. One result of our simulation is the *settle-out pressure*, which is the pressure the pipeline reaches if

[*] Gas pipelines all operate in a turbulent flow regime, but don't always operate in the rough-pipe turbulent regime, so friction factor is still affected by velocity.

it were shut in and allowed to settle to hydraulic equilibrium. On inspecting the hydraulic profile of a gas pipeline, i.e. how pressure varies over its length for each steady flow rate, it should be apparent that the settle-out pressure is higher if the line is uniformly at a higher pressure, i.e. at a lower flow rate. It's possible to quickly estimate the inventory held within a pipeline based on settle-out pressure.

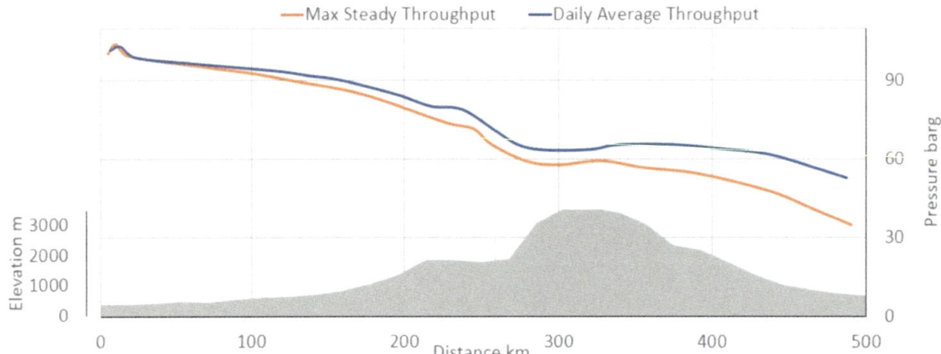

Figure 75 – A textbook example of the pressure profile along a gas pipeline at its average daily flow rate and its maximum steady throughput capacity. This example is unusual; it is *delivery pressure* that changes, as the line is controlled to maintain *supply pressure*.

The estimated settle-out pressure should agree with all same-elevation reporting points following a shut-down.

Effect of diameter on flow in gas pipelines

Pressure drop scales as the inverse fifth power of diameter $(1/D^5)$, just as it does for a liquid. But in a gas pipeline, this means that even a little liquid drop-out causes efficiency to drop drastically. We deal with this by automatically learning the efficiency on a real-time system as a form of online tuning, which we discuss in a later chapter (*Live on Site*). Fully capturing how much liquid is dropping out of a gas pipeline, as water and as condensed hydrocarbons, is a discussion we touch on in the final chapter.

Effect of ambient temperature on flow in gas lines

As the density varies significantly with temperature as well as pressure, we also account for heat transfer between our fluid and its ambient environment. An overall heat transfer coefficient says that all heat transfer is at steady state. Calculating the throughput of the pipeline at warm ambient temperatures, we find it is capable of carrying significantly less flow rate than in winter. A typical gas pipeline has steady *throughput-capacity* scenarios for summer and winter.

Case study: Langeled as a hydrogen pipeline

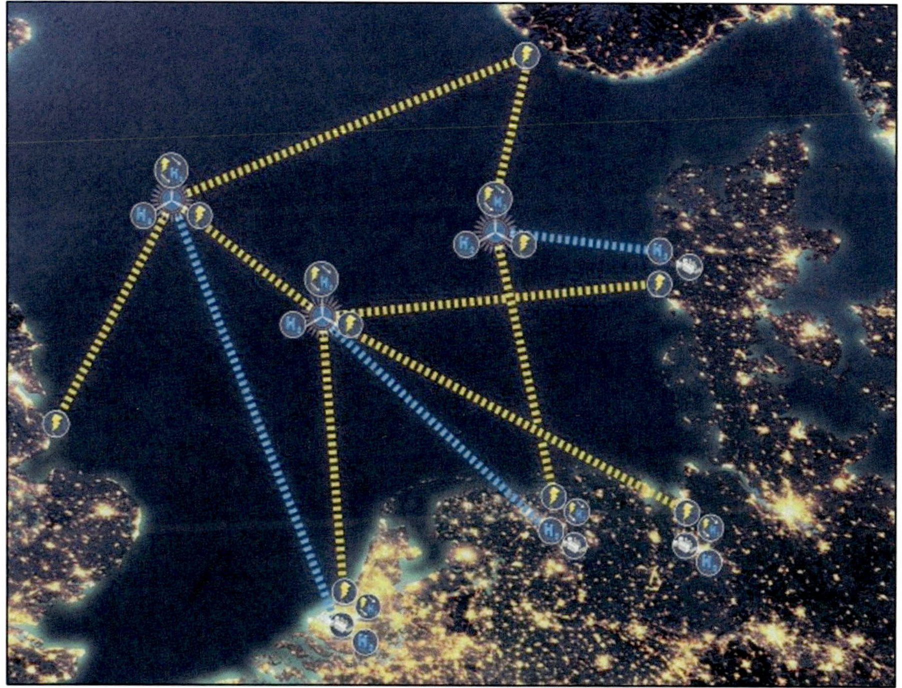

Figure 76 – A network of proposed undersea cables and hydrogen pipelines linking Europe with offshore platforms (with permission from North Sea Wind Power Hub)

Langeled is one of the world's longest subsea pipelines laid deep on the seabed. Gas extracted from beneath the Norwegian sea traverses the first leg of its journey which involves rising up a shear underwater cliff face. It is then processed on land in Norway, and over 70 million cubic meters per day flows a further 1 200 km across the North Sea to reach consumers in the UK. Modelling this pipeline's diameter and route parameters, we can predict the flow rate through it. Many offshore platforms in the heart of the North Sea are existing or planned sites for the world's largest wind turbines. As part of expanding that electrical generation, it is proposed that there be constructed centralized hubs to generate hydrogen, which might be blended into existing gas transmission pipelines, such as Langeled. This saves laying down substantial copper, adds seasonal storage capacity, and helps to decarbonize un-electrifiable sectors.

The first question is: how much energy can this pipeline currently transport? Simulating the original capacity scenario on Atmos SIM, we plot the profile of a portion of this pipeline's route and examine how it looks with natural gas. This scenario (not necessarily the line's true capacity) transmits 35 GW net.

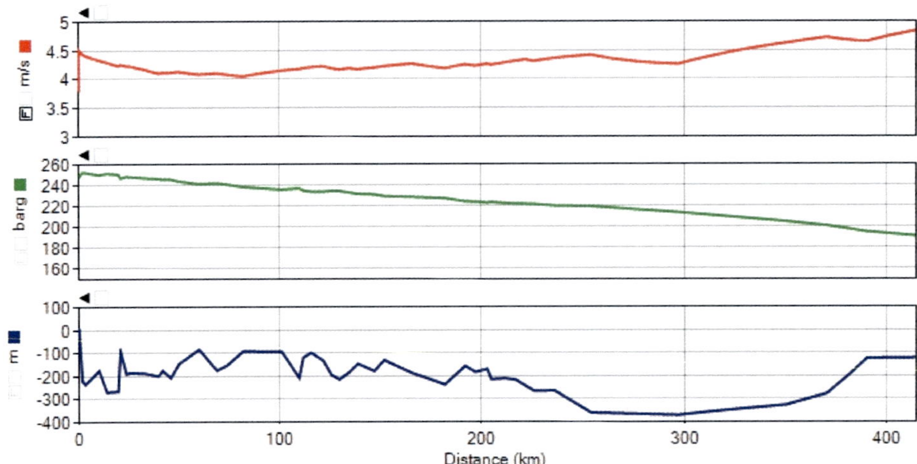

Figure 77 – A hydraulic profile of velocity, pressure and elevation. ***Natural gas*** **in this steady state has a standard flow rate of 974 Sm³/s, delivering 35 GW net calorific energy.**

The next question is: what happens if we operate the pipeline at the exact same pressures at each station, but instead of natural gas, it carries pure hydrogen? The net calorific energy transmitted in that scenario is reduced by around a third to something like 25 GW. The velocities are about four times higher, and the standard volumetric flow rate is around 30 times higher. It appears that hydrogen service would warrant a total re-think of operating philosophy.

The third interesting question becomes: at the same operating conditions, what energy could it transport as a 15 mol% blend of hydrogen in natural gas?

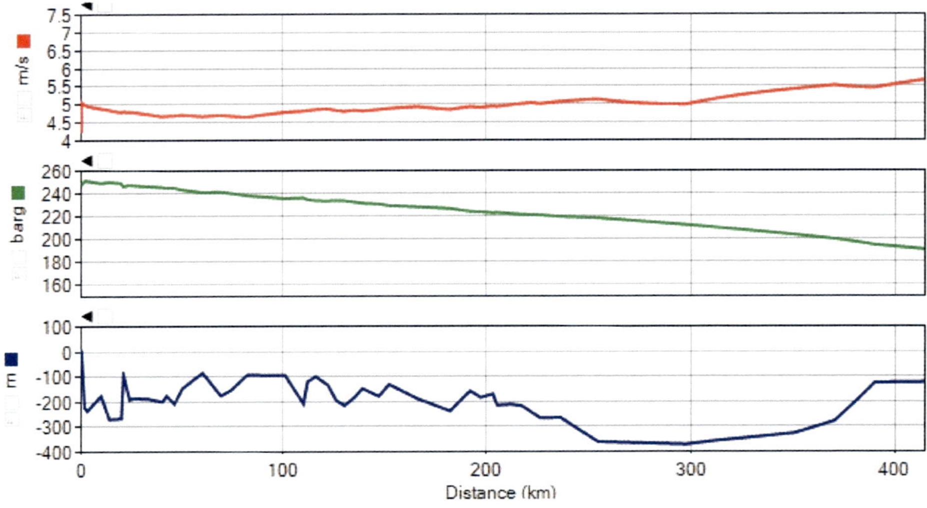

Figure 78 – A hydraulic profile of the same scenario, but with a 15 mol% hydrogen blend in ***natural gas.*** **In this steady state, standard flow rate is 973 Sm³/s, delivering 31 GW net.**

As can be seen on the hydraulic profile, the velocities are far more similar to the original operating conditions. The standard flow rate is almost unchanged, and it carries nearly the same calorific energy. Blending hydrogen into natural gas is not as jarring a change as converting a pipeline to carry pure hydrogen.

Here we focused on the behavior of the pipeline itself. A compressor station needs retrofitting to compress hydrogen to the same pressure as originally. We go into the details of compressors and compression stations in a later chapter.

5.5 – Calibrating the hydraulic model

Scale times and stable calibration datasets

How do we know what flow rate to expect for a given pressure drop? If the line hasn't yet been built, we are unsure, relying on *nominal* or *design* values. Once real data is available from metering on site, a SCADA historian keeps a record. The state of a pipeline is never truly steady: it is transient, changing over time. Pressures and flow rates are transmitted throughout a pipeline at the speed of sound. Changes in composition or inlet temperature propagate at the flow velocity. Changes in ambient temperature along the pipeline – and the effect of fluid temperature on its surroundings – take a very long time indeed before they settle to anything resembling *"steady"*. [26]

The first step in calibrating a model is to identify a stable operation over the pipeline's characteristic period of time, and take an average over that time-scale. A stable historic dataset forms the basis for calibration to align the offline model's parameters, so its results match averaged values from site. Covering this topic properly would end up becoming its own book. In brief, the questions are "*how stable?*" and *"for how long?"*

The answer to the first question is rather difficult in a gas pipeline which is always subject to changing supplies and nominations over time, packing and unpacking being associated with all kinds of operations and not easy to avoid.

The answer to the second question concerns the subject of what we might refer to as *characteristic scale times*. How long it takes for a pipeline to settle depends on what it is we are looking at. One of these is throughput time – how long it takes a particle of fluid to traverse the pipeline. Another is settle-out time, the duration it takes for pressures everywhere to equalize if a pipeline was flowing and is then brought to a shut-in condition by closing block valves. In this way, there are actually a number of different characteristic scale times.

There is even a much slower characteristic settle-out time, possibly weeks, over which the gas within the pipeline returns to thermal equilibrium with its surroundings.

Calibrating liquid flow against pressure drop

Figure 79 – A pipeline's roughness and its inner diameter can be calibrated in Atmos SIM (left) to reflect the condition of a real pipe, like this plastic pipe with mineral build-up (right)
Josefus2003 / Wikimedia Commons / CC-BY 3.0

If we don't calibrate a liquid line for frictional losses, we haven't captured how flow rate and pressure drop are related. Unless the flow happens to be in the rough-pipe turbulence regime (Re > $100\,000$) the effect of roughness is relatively small. In the laminar flow regime, a pipeline's wall roughness affects nothing at all. For this reason, a liquid pipeline is typically calibrated by adjusting *hydraulic efficiency* – a multiplier on frictional losses acting regardless of flow regime – rather than adjusting *wall roughness*.

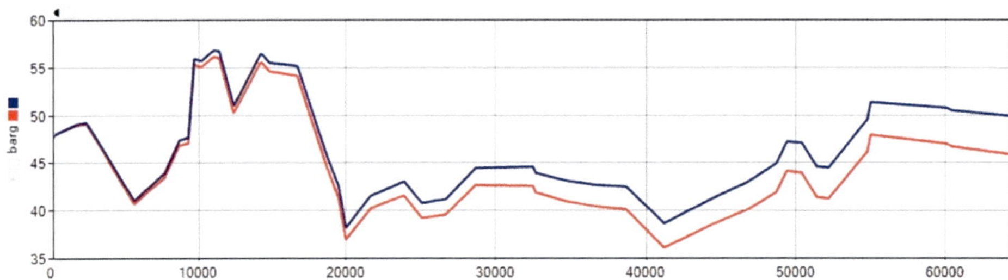

Figure 80 – Adjusting wall roughness at a Reynolds number of 40 000 (transitionally turbulent) affects the pressure profile; results are shown for two scenarios controlled at the same steady flow rate. This difference would be bigger if it were flowing in the rough-pipe turbulent regime.

But *inner diameter* also affects flow rate, so how do we tell which of the two (*diameter* or *efficiency*) needs adjusting? Well: the inner diameter affects the *inventory* as well. If the inner diameter is wrong, a correct flow rate would move a batch or pig through the line at a different velocity than in reality. However, this approach, using linefill via transit time to calculate the inner diameter, is also affected by *temperature* in ways that might be difficult to pick out.

Case study: a liquid pipeline where everything is wrong

If this tuning is happening during commissioning of a new pipeline, the low flow rate might be low for its diameter, making tuning even more challenging. Before calibrating this pipeline, everything is wrong. Let's walk through how to distinguish one parameter from another.

- The fluid properties of the batch are wrong
- The flow meter is not calibrated properly
- The inner diameter is somewhat wrong
- The ground temperature is wrong
- The wall roughness is wrong
- The overall heat transfer coefficient (OHTC) is wrong.

The final issue in the above list would mean that although both in reality and in the model the temperature profile is flat by the end of the pipeline, in reality the temperature profile becomes flat considerably further up the pipeline compared to the model.

To rectify these issues we can look at the following measurements:

- Pressure drop
- Transit time of composition changes as a measure of linepack
- Downstream temperature
- Upstream and downstream speed of sound

The speed of sound is not directly measured, but is implied by the pipeline's transient response via the Joukowsky equation. How do we use these clues to fix the bad model parameters? Debugging and troubleshooting can be an elusive task. There are a few general checks we can perform, as presented in the following seven step process.

Step 1: During shut-in, we look at the measured pressures at different elevations. Assuming the elevation profile is correct, which is usually a safe assumption within a few meters, this should tell us the fluid density. In a liquid pipeline the temperature error won't have much impact on density so we can use this to nail down the fluid density accurately even if everything else is wrong. If we are unsure of pressure meter offsets, we can check many locations. This is easier if the same batch stretches from one meter to the next. We perform this calculation for each of the fluids in the line remaining aware that pressure must be well above vapor pressure at high points for this calculation to work, or else there may have formed an unknown-size vapor pocket at the top of a hill.

Step 2: There are usually flow meters at each inlet and outlet. Do they sum to zero over long periods (say, hours)? If not, one or more of them may be suspicious. If there's slack we must average over longer periods (at least days). Are there ever big jumps at one flow meter that aren't reflected in the sum of all other flow meters? That probably reflects a malfunction, you will need to sort it out before doing detailed calibration.

Step 3: Manually calculate the physical pipe volume, take that inventory and divide it by the flow rate. Then compare it with the observed transit times of batches or pigs. Batch interfaces are better for this check, because pigs slip at a rate that is uncertain. If the transit times don't agree with observation, either the flow meter is suspicious, or the inner diameter is wrong. The inner diameter is unlikely to be *very* wrong, but we can use the result of step 2 to help here: if the outlet flow meter disagrees with the inlet flow meter; and the former is consistent with transit times while the latter is inconsistent, then it is probable the outlet flow meter is correct, and the inner diameter is about right.

Step 4: If we are interested in getting the wavespeed right to perform surge analysis or certain other applications (like leak detection) that benefit from the hydraulic transients being exactly correct, then we need to nail down the bulk modulus. Do observed sharp pressure and flow surges at the same location agree with the Joukowsky equation? When checking this we want to look at the shortest possible time interval while getting good data that isn't fouled up by the limited scan rate. What happens before a reflected wave gets back from the next downstream point that controls pressure or flow? If the pressure surge height disagrees with the Joukowsky hand calculation, adjust the fluid bulk modulus in the input parameters to the Bulk Modulus equation of state so it does. If we pick up the wave arriving at a distant pressure meter on SCADA, we can also deduce wavespeed, using the adiabatic bulk modulus (unless the fluid is isothermal in reality, which is only really possible in an uninsulated exposed pipeline).

Step 5: Now we adjust pipe efficiency or roughness, to get flowing pressure drop right. If pressure drop is very wrong, we should instead try adjusting diameter. A 5% change in inner diameter gives a 25% change in dynamic pressure drop. If the ends are at different elevations, we must account for that effect on pressure drop before deciding how far off we are. If we are pumping a heavy crude oil, we adjust viscosity in addition to (or perhaps instead of) roughness. Crude viscosity changes a lot with temperature, so for crudes we may want to iterate this step with the next two steps (6 and 7), getting the temperatures right. For a products or water pipeline, temperature has minimal impact on frictional losses, so temperature can be sorted out independently (if indeed we bother getting it right at all as it isn't necessary for those fluids in some applications).

Step 6: The downstream-end temperature in a liquid line should be higher than the ground temperature; if you are measuring it to be lower, you probably have the wrong ground temperature. If you are measuring it right at the ground temperature, but your model is showing it significantly higher than ground temperature, you should check where the temperature meter is. Does it sample the fluid? If it's actually measuring the outside surface of the pipe, especially if it's outside the insulation, it may just be basically measuring the ground temperature and be pretty useless for tuning. A measurement taken outside the pipe but inside all the insulation and cladding should be a pretty good estimator of fluid temperature.

Step 7: If step 6 has passed but our modelled downstream temperature is much warmer than the ground temperature, while the measured temperature is only slightly warmer, we should turn up our overall heat transfer coefficient (OHTC), or if using thermal shells we should turn up the thermal conductivity of the shells. When doing detailed testing of this it is better to be aware of what the recent flow rate history of the pipeline was; if it was recently operating at 50% flow, or shut down – recently meaning *"within the last week"* – this can foul up this check. Try to find a period where the pipeline was operating at roughly the same flow for a week.

A final tip is that one should probably start with pipes fairly rough. If we start with a small roughness such as 1 micron, we need to tweak the roughness substantially until we see any effect at all. Whereas if we start at 50 microns and expect to turn it down, it is practical to adjust it in a way that is expected to have a noticeable effect.

Calibrating gas flow against pressure-squared drop

The roughness of a typical gas transmission pipeline can affect its flow rate at a given pressure drop by a significant amount.

What bad thing happens if we don't calibrate our model of a gas pipeline? Well: learnt efficiency handles the effect of roughness in an online scenario simulating in real-time anyway (see the *"Live on site"* chapter), but inventory is a function of inner diameter, which might vary substantially along a wet gas pipeline because of liquid dropout (as discussed in the *"Beyond single phase"* chapter).

Moreover, if we have a stable averaged dataset from the field, we ought to ensure the model is well-aligned for the purposes of conducting offline studies. If we have brought in a model from our live system and intend to use it for offline studies, and it already includes learnt efficiencies, then essentially the live model has performed the calibration for us.

5.6 – Calibration with the Tuning Assistant

As a matter of routine due diligence, then, there are a number of tuning cases to be performed on a pipeline model until we can say it is well-calibrated and ready to go. The common saying goes: *"if at first you don't succeed, try and try again"*. When attempting various guesses of the various parameters at play, repetitive work wastes valuable effort performing manual trial-and-error. The amount of effort involved in this cannot be allowed to become an excuse for unacceptable shortcuts. Luckily, as for any repetitive task, it can be automated with an assisted calibration tool. Atmos SIM's *Tuning Assistant* automatically finds suitable values to calibrate our model. It does this by solving steady state, improving its guess, and solving again:

- *"What roughness achieves this pressure drop at that flow rate?"*
- *"What soil conductivity achieves this temperature at the destination?"*

Figure 81 – Many hydraulic and thermal properties of a pipeline model can be calibrated.

The flow coefficient can be calibrated for a resistance, valve or regulator. For a pipe, Atmos SIM can calibrate the wall roughness, overall efficiency and / or inner diameter. It can also calibrate thermal parameters, including the overall heat transfer coefficient, ground thermal conductivity, burial depth and the velocity of the ambient medium (air or seawater for example). We shall look further at what those are in the next chapter.

This empowers the user to perform one-click tuning of any pipeline property. In fact, it goes even further: if we, say, tune a pipe's roughness, and then go on to tune the surrounding ground's thermal conductivity (as we see in the thermal model chapter), we must go back and tune pipe roughness again! To avoid this hassle, an automated Tuning Assistant lets us calibrate two pipeline properties at once. Once a suitable value has been identified for each tuned parameter, those values can be automatically applied throughout the model.

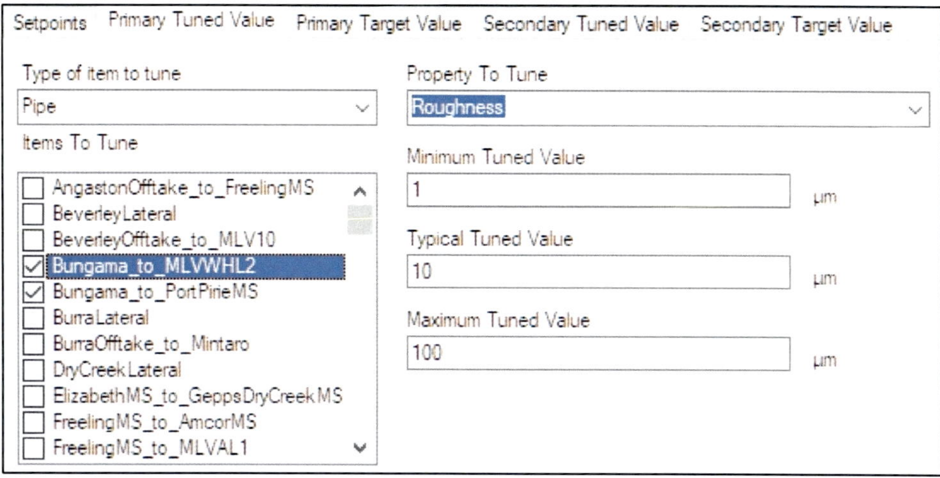

Figure 82 – Instructing the automated Tuning Assistant in Atmos SIM by first specifying which values to adjust as well as the possible range of the adjustable property

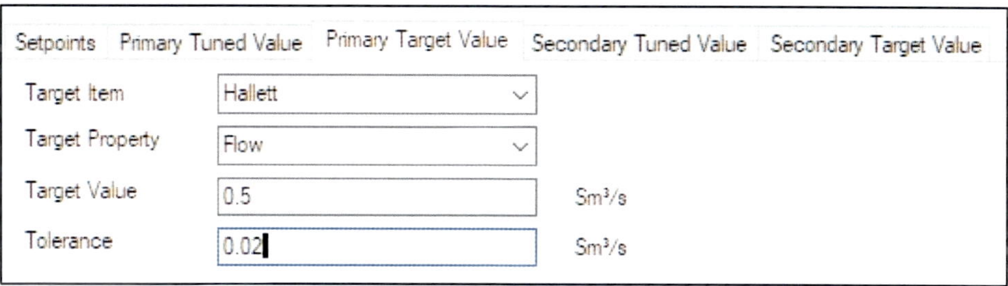

Figure 83 – Instructing the automated Tuning Assistant in Atmos SIM by then specifying what target the primary tuning should seek to achieve within some tolerance

The way the automated Atmos SIM Tuning Assistant works is broadly similar in its principle to a kind of multi-variable optimization.

6 – TEMPERATURE: THE THERMAL MODEL

Figure 84 – A well-insulated pipeline and a geothermal power plant in Iceland

> *"The companies that run [an oil] pipeline suggest that if flows drop too low, the line could be compromised. Others say the industry's numbers are wrong, part of a campaign to open up off-limits areas like a wildlife refuge to drilling... [claiming] the engineering problems are solvable with upgrades and modifications, such as heaters"* [37]

6.1 – WHO CARES ABOUT TEMPERATURE?

Temperature in gas and liquid pipelines

Although most of the density difference along a gas pipeline is due to pressure rather than temperature, perhaps 10% or 20% of the change in density is due to the change in temperature. Temperature might change by 10% in absolute terms from station discharge to station suction, and that change in the absolute temperature of a gas linearly affects its density. This means that in a gas pipeline, an isothermal model simply won't do; a linepack calculated without consideration for differences in temperature along a route would be significantly wrong. Worse still, a flowing gas may be 10% or 20% less dense at the far end of the pipeline than at its inlet. For the same flow rate, velocity is inversely proportional to density. In turn, friction is a function of velocity-squared. This means that ignoring thermal effects by pretending a pipeline is isothermal could cause frictional losses in a gas line to be 40% off!

That gas pipeliners obsessively strive to understand the temperatures in their network should not come as a surprise, given its centrality in any equation of state describing gas densities, and thus inventory in their pipeline. Along with calculated dew point, temperature is also used to estimate the risk of hydrates. But should liquid pipeliners be in the least interested in how warm it is today?

All heat flows are balanced, unchanging over time
Temperature falls due to drop in density as pressure drops (AB)
Temperature rise across compressor (B)
Temperature fall across cooler (C)
Heat transfer to (AB) from the ambient environment
Heat transfer from (CD) to the ambient environment
All temperatures remain constant over time

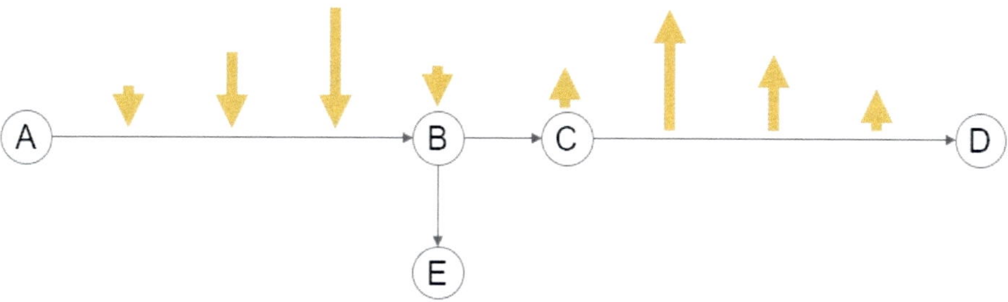

Figure 85 – Heat transfers in hypothetical natural gas pipeline that is flowing at a steady state. In reality, all these temperature changes are practically never truly balanced at a steady state.

Of course, extremes of temperature have an effect on the pipe material itself. If a metal becomes too cold, it undergoes a *brittle transition*, and it must then be dug up and replaced. If a pipe is warmed too suddenly, it weakens and can become prone to fracture. But this isn't the only way temperature affects a pipe. Temperature, along with pressure, causes pipes to expand and contract. We revisit the consequences of this phenomenon in the chapter on fast transients.

We mention this here to point out that even when temperatures don't reach an extreme, pipeliners do see the effects of temperature in their liquids and gases.

Temperature in steady liquid scenarios

Accurately estimating temperatures in a liquid pipeline allows the designer of a pipeline conduct offline studies to select a suitable insulation material for the pipe. Operators and planners use it to assess corrosion rates. But a liquid pipeliner could just feed in the nearest temperature sensor for a ball-park estimate: any swings in temperature are relatively small along the length of a liquid pipeline.

One use case in which closer attention must be paid to the temperature along a liquid pipeline is model-based leak detectors. Those always require careful thermal calibration as part of a broader need for obsessively accurate physics. We discuss model-based leak detectors later, noting for now that this accounts for why many of the best papers presented on thermal models at conferences of the *Pipeline Simulation Interest Group* (PSIG) have focused on liquid lines.

When a fluid like diluted bitumen or heavy crude becomes colder, it becomes more viscous. Eventually, below its *pour point temperature*, it ceases to flow. Crude on a cold day might become too thick to effectively pump through a line. Keeping *"viscous fluids"* amenable to pipelining is a known challenge in cold climates. Such fluids are diluted carefully to control their viscosity by blending, or injecting chemicals known as pour point depressants. The aim is to make the fluid flow easily at all the temperatures along the pipeline. Otherwise, the line temperature would have to be deliberately maintained using in-line heaters or trace-heating. There is essentially an imposed limit on minimum allowable temperature for a pipeline to be operable at the desired flows and pressures.

But what about the countless liquid pipeline models *not* deployed to detect leaks, nor to transport very viscous fluids? In many liquids, the temperature barely affects density enough to impact arrival times (unlike linepack in a gas pipeline). The amount of liquid held within a stretch of pipeline doesn't change by much over time. There is a perception that those pipelining water or certain liquid products need not think particularly closely about temperatures and the thermal model. This perception is not unfounded, but it should be noted that a good thermal model is always important for certain kinds of liquid, and for certain tasks.

If the liquid being pipelined is highly compressible, such as liquefied petroleum gases (LPGs), its density can change by a lot. This directly affects the frictional and gravitational pressure gradients. Neglecting the thermal model when, say, performing offline studies for these compressible liquids might give misleading results, affecting the calculation of important quantities like pipeline throughput.

For incompressible liquids, we can often get everything we need from a purely logistical batch tracker combined with the simplified physics of a simple hydraulic profiler. That approach is used to watch for violations in allowable pressure, and keeping tight lines out of slack. The hydraulics here are simple: the pressure (P) varies along the length of a pipeline (x) as a function of the liquid's density (ρ), the acceleration due to gravity ($g = 9.81\ \mathrm{m} \cdot \mathrm{s}^{-2}$), change in elevation ($z$), flow velocity ($v$), pipe diameter ($D$) and friction factor ($f$):

$$\frac{\partial P}{\partial x} + \rho \cdot g \cdot \sin[\theta] + \frac{\rho \cdot f \cdot v \cdot |v|}{2 \cdot D} = 0$$

And in the case of monitoring a live system in real-time, in this simple model, temperature only affects the *friction factor* (f), and that friction factor can be simply learnt for each section of pipeline bounded by` a pair of pressure meters. Say we wish to extend this simplified hydraulic model to predict the friction factor for planning future operations. Would a *"simple"* thermal model suffice?

It turns out performing even basic tasks requires our simulator's thermal model to be very accurate if the fluid is particularly viscous, even if there is no pour point issue. In a crude oil or a liquid product, frictional losses depend strongly on viscosity, which is *exponential* in temperature. Friction is independent of viscosity in the rough-pipe turbulence flow regime, but liquid lines don't always operate there. So a small error in temperature drastically affects the calculated friction.

Even for a fluid like gasoline, which doesn't have significant viscosity even at cold temperatures, some hydraulic scenarios such as surge studies still require a thermal model to be accurate. The problem with an isothermal surge study is that the isothermal speed of sound for gasoline – speed of sound calculated by an isothermal model – is about 15% off. But the magnitude of a pressure surge caused by e.g. a sudden stoppage of flow is linearly proportional to the speed of sound, so that means that in turn will be 15% off. This means that an isothermal model is really only safe within a transient solver for fluids like water where the isothermal and adiabatic speed of sound are nearly identical.

Temperature is very important if there are vapor bubbles due to column separation, or if the line is slack. A fluid's vapor pressure is a small term multiplied by a term that is exponential in temperature, and in some cases that exponential in temperature overcomes the small term. This means that the pressure at which slack appears might strongly depend on temperature. If our temperature causes absolute vapor pressure to change by a factor of 100 from 0.001 bar to 0.1 bar, that is not significant; it is a mere 0.1 bar off if we get it wrong. But if it causes absolute vapor pressure to change from

0.1 bar to 10 bar, that is a huge difference, and if we get that wrong our model results are potentially going to be useless. Maybe the first 10°C of error causes it to go from 0.001 bar to 0.1 bar, but then the next 10°C of error causes it to jump to 10 bar.

Therefore, to predict the onset of slack, which is a fair expectation even from a single-phase pipeline simulator, we need a good thermal model. To go further and fully model the fluid's behavior during slack flow, we need a high-performance application, along with a truly excellent thermal model. All kinds of physical and thermal effects must be tuned to near perfection to get it right. We revisit that topic in a later chapter.

How can we model temperature?

There are different angles from which we can tackle heat transfer. The simplest cases are an *isothermal* approach, and an *adiabatic* approach. An isothermal approach doesn't bother with a thermal model at all; it fixes a constant temperature, and runs with that. An adiabatic approach simulates a thermal model with zero heat transfer through the pipe walls. Those are the two extremes. What a pipeline does in reality must lie somewhere in between.

In a few specific cases, using an *overall heat transfer coefficient* (OHTC) is acceptable. However, in some senses, it is a step backwards. It is probably no more accurate than an isothermal approach or an adiabatic approach. If we're doing design, and are truly only interested in steady state results, then OHTC is okay and will match the results of a more sophisticated thermal model. But we had better be sure that is all we'll do with the model! One might expect that all transients in a pipeline must settle out in a few hours, and so can be safely ignored. But the temperature of the ground around a pipeline is a thermal transient that might not settle out even after many days.

In the limiting cases – the isothermal approach and the adiabatic approach – we don't need to think about exactly how heat is transferred through the soil around a buried pipe. But for any real scenario (the in-between cases), we do. In many pipelines, a *transient ground thermal model* is indispensable. This approach closely models the temperature of each concentric shell around the pipeline, considering its surroundings. This feature is not offered by every simulator, and it can be computationally intensive, so it is worth spending some time understanding why it is needed.

6.2 – HEAT CAPACITY OF A SOLID OR A FLUID

For any substance, fluids and solids alike, total heat capacity (C) is the energy it takes to raise its temperature by one unit. It is usually expressed per unit mass (M), and so referred to as *specific* heat capacity at constant pressure (c_P) or constant volume (c_v):

$$Q = M \cdot c_P \cdot \Delta T$$

$$\text{or } Q = M \cdot c_v \cdot \Delta T$$

Many liquids, such as water, behave such that their heat capacity's value is almost the same for a process done at constant pressure (specific isobaric heat capacity c_P) as its value at for a process done at constant volume (specific isochoric heat capacity c_V) – i.e. at constant density. For some incompressible liquids, like water, the two specific heat capacities are quite similar in value to one another, but not universally identical. [*]

The ratio of heat capacities can be significantly different from 1 in some liquids and in gases. The easier of the two heat capacities to measure is the isobaric heat capacity. [†] For a gas, it is also relatively easy to measure isochoric heat capacity, because a gas fills any rigid container it is enclosed within. [‡] For a liquid, it is far easier to measure isobaric heat capacity than isochoric heat capacity, because to keep the volume of a liquid truly constant requires expensive equipment given that a liquid does not fill the entirety of a container. Instead, for a liquid, the isochoric heat capacity is found via thermodynamic relations – it is calculated from its isobaric heat capacity, having separately determined the fluid's coefficient of thermal expansion (α) along with its compressibility (κ) which is equivalently expressed as the inverse – bulk modulus (B).

---

[*] Liquid property reference tables often just present *"heat capacity", "specific heat"* or *"specific heat capacity"* without specifying whether it's isobaric (c_P) or isochoric (c_V). In that context it is always isobaric – it is difficult to measure isochoric heat capacity of a liquid, so when needed it is generally computed from the isobaric heat capacity using the relation on the next page. There is a prevalent but mistaken belief that the ratio of heat capacities is nearly 1 for mostly incompressible liquids. While this is true for liquid water at standard temperature and pressure (1.006), it is untrue for most hydrocarbon liquids. Products and crudes have a ratio of 1.2, which means the true speed of sound in those liquids is about 10% off the value which would be obtained from a calculation that omitted this. For accurate hydraulics in pipeline simulation, we must account for it.

† *"Easy"* now that we know how to do it! Determining heat capacity per unit mass of a gas was the subject of the prize of the French Academy of Sciences for the year 1812, motivated by discrepancies in measurements at the time.

‡ Experiments to determine the isochoric heat capacity of a gas (i.e. acting at constant volume) presented difficulties that weren't overcome until the 1880s. Clausis examined the topic establishing the first link between thermodynamics and kinetic theory.

A thermodynamic relation for the difference between these two heat capacities is stated in terms of the coefficient of thermal expansion (α) and isothermal bulk modulus (B_T):

$$c_P - c_v = \alpha^2 \cdot \frac{B_T \cdot T}{\rho}$$

For an ideal gas, the right hand side simplifies to just be the universal gas constant (R):

$$c_P - c_v = R$$

This is approximately true for a real gas, but only at moderate pressures. What about gases at not-at-all moderate pipeline pressures? Or what about liquids? For those, the relation depends on density (ρ), i.e. on the equation of state. One of the advantages of using a modern pipeline simulator such as Atmos SIM is that it calculates all of these interactions for the user.

The ratio of heat capacities (γ) is defined as isobaric divided by isochoric heat capacity:

$$\gamma = \frac{c_P}{c_v}$$

One should note that although this ratio is about 1 for water it can be as much as 1.2 or more for liquid hydrocarbon products. It is significant in all gases, typically taking a value between 1.2 and 1.4. This heat capacity ratio is often referred to as the *adiabatic index* (or the *isentropic index*) for reasons we shall discuss in the chapter on compressor thermodynamics.

At a given temperature, Atmos SIM calculates the specific isobaric heat capacity of a pure gas component by assuming that it changes linearly with temperature and interpolating. This is a reasonable approximation over the range of temperatures encountered in a pipeline simulator. For a mixture, we cannot reasonably make such a simplification. We calculate one heat capacity for the mixture, then use the above thermodynamic relation to calculate the other. We do so based on a rather complicated *rule of mixtures*: in a real gas, the overall heat capacity of a mixture is unfortunately not a simple weighted average of the heat capacity of each constituent, because the presence of other molecular species gives rise to complex interactions.

6.3 – THE FLUID THERMAL MODEL

Temperature profiles in steady flow

The *fluid thermal model* is solved simultaneously with the hydraulic model. In a steady state all heat flows and temperatures are perfectly balanced and *unchanging* over time:

$$\frac{\mathrm{d}T}{\mathrm{d}t} = 0$$

But this certainly doesn't mean temperature is constant throughout the pipeline. Even if we posit a perfectly insulated pipeline, as any fluid flows, the very fact of its flowing will affect its own temperature profile. We can capture the effects of *steady-state* heat transfer with a relatively simple energy balance expression, involving only three terms with neither accumulation nor depletion over time: [38]

$$\frac{\mathrm{d}T}{\mathrm{d}x} \approx \Delta T_{\text{friction}} + \Delta T_{\text{state}} - q_{\text{conduction}}$$

It's not always a strict equality because the specific heat capacity, for instance, may change with pipeline conditions, but it's a very reasonable approximation.

The temperature profile in a flowing pipeline is affected by two heat source terms in addition to conductive heat transfer with the surroundings. One heat source raises the temperature due to friction ($\Delta T_{\text{friction}}$). Another heat source term changes the temperature due to change in thermodynamic state (ΔT_{state}). The relative importance of the two source terms depends on whether the fluid in the pipeline is a liquid or a gas. Accordingly, whether a pipeline operates warmer than or cooler than its ambient environment (once it has reached flowing thermal equilibrium) depends on whether it contains a flowing liquid or a flowing gas. This in turn affects the direction of conductive heat transfer with the environment ($q_{\text{conduction}}$). As a function of temperature, the conductive heat transfer depends on the *overall heat transfer coefficient* (U):

$$q_{\text{conduction}} = U \cdot A \cdot (T - T_{\text{ambient}})$$

This is an ordinary differential equation whose solution is well-defined, provided ambient temperature is reasonably constant throughout our section. It tends towards this steady state *"long-range"* solution as its asymptote:

$$T \to T_{\text{ambient}} + \frac{\Delta T_{\text{friction}} + \Delta T_{\text{state}}}{U \cdot A}$$

The relative importance of the three effects defines a line's temperature profile.

Liquids flow warmer than ambient

Looking along the steady profile of a *liquid* pipeline, it is warmest at pump outlet. Then, as our liquid flows along its route, only two effects are at play:

Heat is typically lost to the ambient environment because it is typically cooler ($Q_{\text{ambient}} > 0$), and friction generates heat as the liquid flows ($\Delta T_{\text{friction}} > 0$). In a typical liquid, those two effects dominate: temperature is not significantly affected by flow-induced changes of the thermodynamic state ($\Delta T_{\text{state}} = 0$). [38]

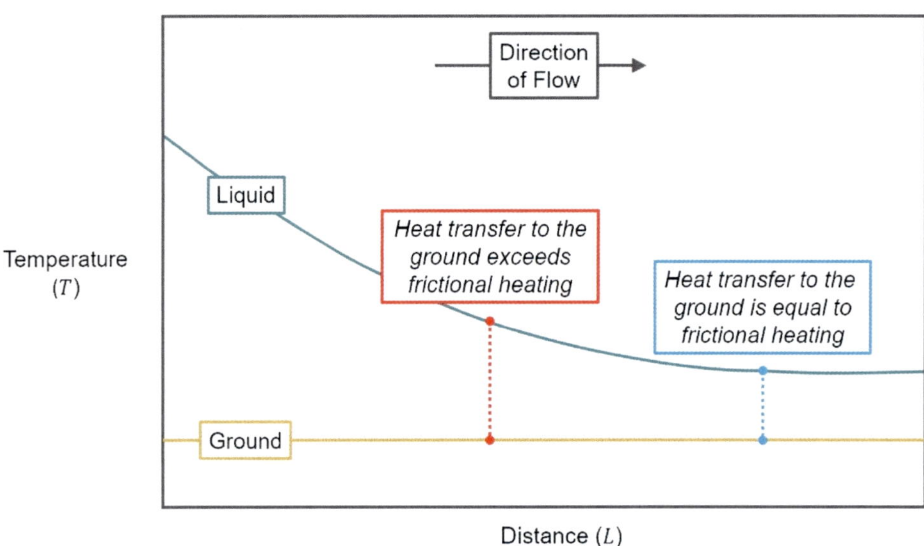

Figure 86 – The temperature profile of a flowing liquid. A liquid flows warmer than ambient. [38]

Liquid lines operate at temperatures warmer than ambient. The liquid is heated by pump inefficiencies, so it enters warm. As it flows along a pipeline, the work done by friction turns into heat warming up the fluid, though this effect is small. The significant head loss along a liquid pipeline translates energy into frictional heating that is usually affected more by pipe diameter than by wall roughness, except the fluid is flowing in the rough-pipe turbulence regime.

At thermal equilibrium, the frictional heating balances the heat lost through the pipe walls, which means the fluid temperature must be higher than the ambient temperature. Left to its own devices, a typical liquid pipeline usually settles to an equilibrium temperature warmer than the ambient environment, typically by 3°C, plausibly by more (7°C is quite normal).

Gas regulators and the Joule-Thomson effect

Connecting a high-pressure gas pipeline to a lower-pressure point (reaching another pipeline or a consumer) is achieved via a pressure-reduction station. In these facilities, pressure is reduced by expanding the gas through a control valve which some people call a regulator or a throttling valve in this application. As the pressure is lower, the density is lower. So this process is referred to as a *fluid expansion*, and is considered a change in thermodynamic state. It brings about a change in temperature: the *Joule-Thomson* effect. In most gases this manifests as a cooling down: a temperature drop ($\Delta T_{\text{state}} < 0$). For example, natural gas typically incurs half a degree of temperature drop ($\Delta T = 0.5°\text{C}$) per atmosphere of pressure drop ($\Delta P = 1\ \text{atm}$), a rule of thumb to ±10%. A notable exception is hydrogen, which warms up on expansion as its Joule-Thomson coefficient (j) has the opposite sign to that of most gases at pipeline conditions.

The expansion of a fluid can be considered to be isenthalpic ($\text{d}h = 0$) if three quantities are negligible compared to the change in enthalpy.* Those quantities are heat transfer to or from the surrounding ground, the change in gravitational potential energy due to elevation, and the change in bulk kinetic energy of forward motion. Gas moving through significant pressure loss over a short distance – for example at a *throttling valve*, or a *shock* – can be considered isenthalpic. Over longer distances, heat flow through the pipe wall becomes significant. But it is worth thinking about how temperature varies along a perfectly insulated pipeline, and then moving on to consider the heat transfer taking place between a fluid and its surrounding environment.

One might expect that during an isenthalpic expansion, a fluid always gets warmer – after all, we are increasing its internal energy (u) by changing the PV-term, which is perceived to represent heat content. But it turns out that for most gases at pipeline conditions (excepting hydrogen and helium), isenthalpic expansion causes a gas to get colder. The molecules are necessarily farther apart. The forces between molecules in most gases are attractive at long distances. Energy must come from *somewhere* to achieve that distancing, for the expansion to overcome the molecules' attraction to one another. That energy comes out of kinetic energy at the molecular level, which is thermal energy. This is why the Joule-Thomson effect lowers the temperature of most gases at pipeline conditions.

---

* Enthalpy is internal energy (u) plus pressure over mole-basis density (P/ρ).

Gases flow cooler than ambient

Looking along the steady profile of a gas pipeline, it enters very hot at the inlet. This is at a compressor station discharge, usually brought down to some sensible (though still hot) temperature using a heat exchanger called an after-cooler.

As gas flows along a pipeline, pressure drops along its route, reducing its density. So the Joule-Thomson effect acts here too, and as we said, for most gases, it is a heat sink term cooling down the temperature ($\Delta T_{\text{state}} < 0$). This is more significant in a gas pipeline than the frictional heating ($\Delta T_{\text{friction}} > 0$).

Beyond a certain distance along the route of a gas pipeline, temperature eventually settles such that this Joule-Thomson sink term balances with the term for conductive heat transfer to the ambient environment through the ground ($Q_{\text{ambient}} > 0$). Exactly how cool depends on that ground heat transfer.

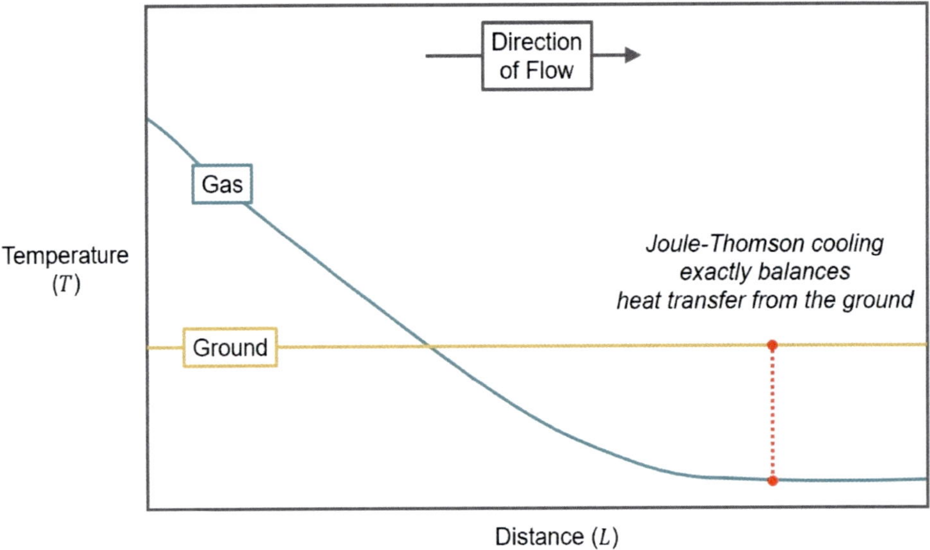

Figure 87 – Temperature along a flowing natural gas pipeline settles cooler than ambient. [38]

As a gas goes *faster* further along a pipe, it becomes *colder*, up to a point. Though both are consequences of the reducing density; neither is caused by the other. Predicting the delivery temperature (T_{out}) accurately is important in a gas line. This is a function of Joule-Thomson coefficient (j in SI-units of K/m), and a constant (a) describing heat transfer coefficient and fluid heat capacity:

$$T_{\text{out}} = \left(T_{\text{in}} - T_{\text{amb}} + \frac{j}{a}\right) \cdot \exp(-a \cdot L) + T_{\text{amb}} - \frac{j}{a}$$

At lower flow rates, pressure changes less, so heat transfer with the environment is more significant and there's less cooling over a given distance. Natural gas flowing at a higher flow rate *"expands"* more, and so experiences more Joule-Thomson cooling, therefore settling at a cooler equilibrium temperature by the time it is received at a faraway station.

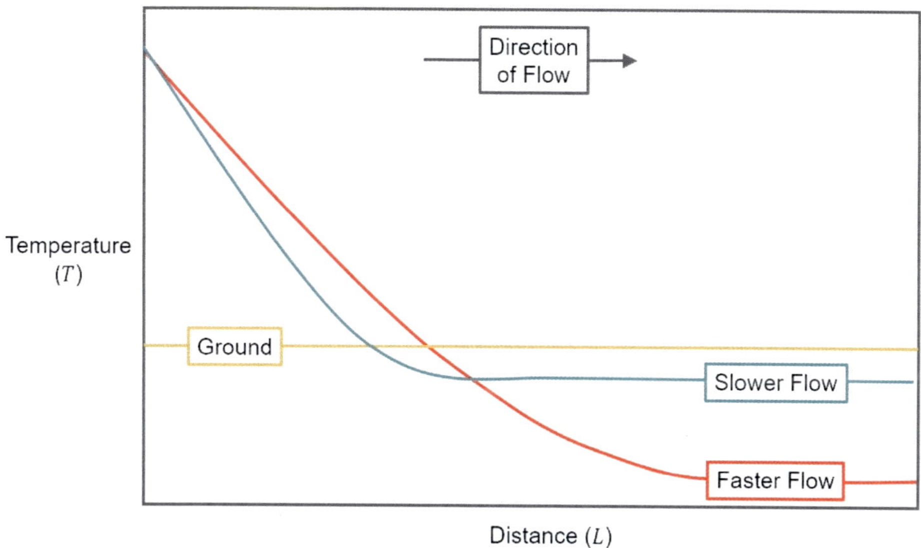

Figure 88 – Faster-flowing natural gas reaches a lower equilibrium temperature due to the Joule-Thomson effect.

Why do we need a transient fluid thermal model?

A fully transient fluid thermal model is extremely computationally-intensive, often responsible for the majority of a simulation's total performance draw. The thermal equations are themselves rather complicated, and the calculated values they require from an equation of state are also nastier. Why is it that a more straightforward approach – a steady fluid thermal model – won't do? Can't we get away with it as just another simplifying assumption? The answer is categorically no.

For a gas pipeline, the reason is easy: flows along a gas pipeline are never steady, because part of it might be packing while another section is unpacking.

For a liquid pipeline, the reason warrants a little more thought.

Temperature in transient liquid scenarios

Figure 89 – A temperature gauge on an insulated pipe, reading in degrees Celsius

When we consider operations and transients happening within a liquid pipeline, the accuracy of temperature calculations holds special importance. To conduct accurate studies of viscous liquids such as crude oils, or compressible liquids such as LPG, using a transient thermal model is vital, including a transient *ground* thermal model. Depending on what we're looking at, neglecting to do this might give misleading results. Big temperature changes at an inlet create big thermal transients along the pipeline. Consider a pipeline receiving crude batches from tanker ships loaded offshore, where every ship is at a wildly different temperature. On that pipeline, performing meaningful studies offline, or leak detection online, requires getting the thermal model exactly right.

Some crude oils and liquid products have a ratio of heat capacities ($\gamma = c_P/c_V$) of around 1.1 or more. Methane, for comparison, has a ratio of heat capacities of 1.3. If we run with the wrong temperatures – an isothermal model, equivalent to a model with excessive heat conduction to the inner layer – the speed of sound (c) differs by a factor of $\sqrt{c_P/c_V}$ compared with the other extreme: an isentropic (i.e. a reversible adiabatic) model with zero heat transfer between the pipeline and the ground.

The gamut between adiabatic and isothermal is controlled by the details of the inner shells in a *transient ground thermal model*. We would potentially end up with as much as 5% error in several results, such as how fast a surge propagates, and the magnitude of a surge in flow terms if we are controlling on pressure at the boundary conditions.

Thermally challenging scenarios

The fluid thermal model can be challenging

The ground thermal model rarely raises issues, so long as we do not strive to model anything too extreme, such as the transient heat transfer across a steel pipe wall in certain scenarios. It is generally fair to assume that a steel pipe is at the same temperature as the fluid inside it, and Atmos SIM always makes this assumption: both in the fluid thermal model and the ground thermal model.

The fluid thermal model, on the other hand, can take years of experience to get right. Here are a few scenarios that any simulator might find troublesome unless it has been designed specifically to handle them elegantly.

Temperature as a boundary condition

One constraint is imposed exactly as an equality at each end of a pipe, as a boundary condition. This can be either pressure or flow. On the basis of this pair of boundary conditions, an offline model or an online model calculates the hydraulics within that pipe.

The temperature profile is not calculated in the same way. Disturbances in pressure or flow affect the pipeline in both the upstream and downstream directions, because momentum can carry energy upstream. In contrast, temperature changes are only carried *downstream* by a flowing fluid; there is no mechanism for carrying heat upstream. Therefore we only need a boundary condition at the upstream end of a pipe to apply the energy equation. *

Instead, temperature is calculated based only on the upstream temperature, along with ambient temperatures along the route of a pipeline, based on the thermal properties of the fluid, the pipe walls and their environment. This presents special challenges in certain conditions which might well arise during a scenario or during operation live on site, without us even realizing.

---

* Atmos SIM's state estimator doesn't deduce the temperature at a given point. It is only possible to do thermal tuning, because temperature changes slowly, and Atmos SIM's state estimator works one time-step at a time. It's possible to wrap thermal state estimation into a Kalman filter based approach, where state estimation, tuning and leak detection/location are all handled by one algorithm.

Inflow or outflow at both ends of a pipe

What does a simulator's thermal model do if fluid flows *into* a pipe from both ends? In Atmos SIM, we address this situation by identifying the knot at which there is inflow from both directions, and we drop its energy equation, replacing it with an additional temperature boundary condition from where fluid in that direction is flowing backwards into the system.

If fluid flows *out* of a pipe at both ends, coping with that situation is trickier. Any solver of partial differential equations will need an extra equation here, but we already used energy conservation equations at every knot. Atmos SIM addresses this by imposing a *synthetic* boundary condition, stipulating that temperatures $T_{\text{up}} = T_{\text{down}}$ at the knot where fluid flows out in both directions.

Stopping or surging flow

If flow is stopping and / or surging back and forth, we don't necessarily know which knot, or even which pipe, is flowing in which direction. A wrong guess can lead to model instability.

One option is to solve the hydraulic part first, then solve the thermal part using those flows. This is called a *leapfrog* thermal model, in contrast to a fully coupled model. It might produce an artefact, an error that doesn't shrink during transients, even if we take shorter time-steps, though this error might be small for liquid models. Atmos SIM does not use a leapfrog thermal model.

Another option is to re-run a step if our guess turned out to be wrong; this is how Atmos SIM handles it.

6.4 – THE GROUND THERMAL MODEL

Coupling the ground and fluid thermal models

The ground thermal model starts at the *outside* of a steel pipe. The pipe wall's thermal mass is treated as if it were part of the fluid in the fluid thermal model, except not convected. The two thermal models' interface is the outer diameter.

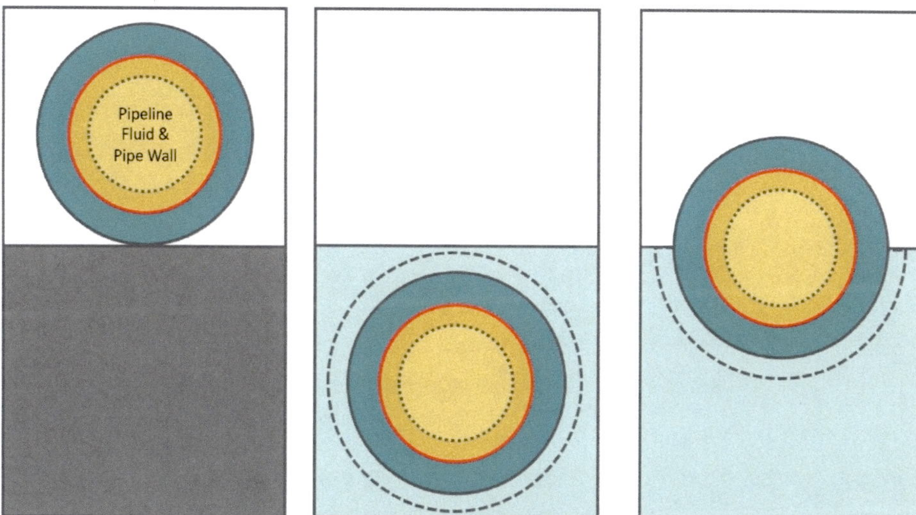

Figure 90 – The pipe wall (gold) is at the same temperature as the fluid in the thermal model. Any thermal layers (teal) beyond the outside of the pipe wall – including cladding and soil – are considered part of the *ground* thermal model, along with convection to the ambient medium.

The reason for this is to allow for long model time-steps while keeping the ground thermal model stable as it simulates the rapid transfer of heat through a metal pipe wall. Conduction through the pipeline fluid, convection between it and the pipe wall, and conduction through the pipe wall, are all assumed to be so fast that the temperature – at any given point along the pipe – is everywhere the same throughout the fluid and pipe wall, and around the outside of the pipe.

The fluid thermal model is coupled to a *ground* thermal model as follows. When the ground thermal model runs, it needs an inner temperature boundary condition at the metal pipe's outer diameter. For this, it uses the preceding step's result from the fluid thermal model. When the fluid thermal model runs, it needs the heat transfer at that knot. It uses the preceding step's result from the ground thermal model.

What's happening around the pipeline?

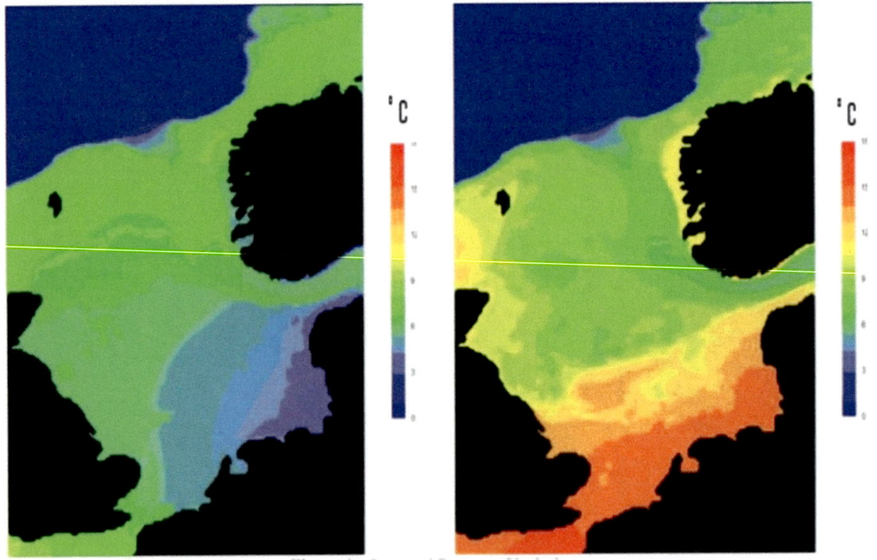

Figure 91 - Ambient temperature maps of the North Sea at different times.

Knowing what the temperature of the ambient medium is in the first place can be a challenge. Cross country pipelines might traverse rugged terrain. As the weather changes along the route, ambient temperature is rarely measured in real-time at more than a handful of locations. Rules of thumb for how ambient temperature varies with elevation (i.e. the air is cooler higher up!) are better than rudimentary interpolation. Some operators purchase daily weather forecasts to simulate their pipelines.

Heat transfer between a pipe and its environment

There is always heat transfer with the environment, because there is always a driving temperature difference between a pipeline fluid and its ambient environment. In a liquid pipeline, friction generates heat everywhere. In a gas pipeline, isenthalpic expansion vs. intermolecular forces consumes heat everywhere. There is also heating in compressors, and a bit of heating in pumps. So every steady pipeline is at some equilibrium where the rate of heat generation/consumption is balanced by heat transfer through the pipe walls. All pipelines exchange heat with their ambient environment to some extent.

If the pipe wall is made of steel, it is assumed to be at the fluid temperature. Heat (q) is transferred radially between a pipe and its ambient environment by two mechanisms happening in series: *thermal conduction* through solids and *thermal convection* with the ambient medium (the ambient air or water).

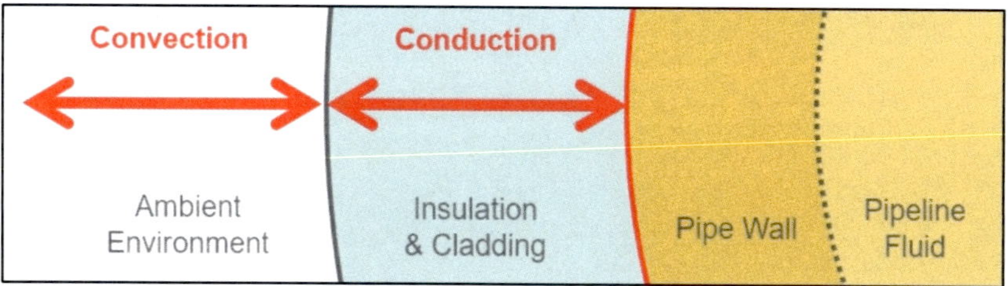

Figure 92 – Radial heat transfer in an unburied pipeline: convection with the ambient environment, and conduction through layers of insulation and cladding.

If a typical pipeline is unburied, the outermost solid layer is the cladding. If it is buried, it is the soil. We treat each solid layer as a hollow cylinder, and call it a *shell*. Heat is transferred across each shell's circumferential surface area (A), geometrically calculated from its inner diameter (D) and the knot length (Δx):

$$A = \pi \cdot D \cdot \Delta x$$

The total heat transferred per unit surface area (A) is the heat flux ($\dot{q}$). This is what we will refer to in our discussion, to avoid getting distracted by geometry.

The overall heat transfer coefficient

Instead of going to the trouble of considering what the ground is made of, could we consider the *overall* lumped effect of everything between the wider ambient environment and the pipeline? Between the temperature of the fluid, which we also assume to be the temperature at a steel pipe's outer diameter (T_{fluid}), and the temperature of the ambient medium (T_{ambient}), we could define a single lumped parameter – an *overall heat transfer coefficient* (U) abbreviated OHTC. This tells us the rate of heat transfer (q) across the circumferential area (A):

$$q = U \cdot A \cdot (T_{\text{fluid}} - T_{\text{ambient}})$$

This approach is straightforward to solve, and is a widely used ground thermal model. All pipeline simulators historically started off with it. But it assumes the soil and cladding are at thermal equilibrium: that heat transfers are all *steady*.

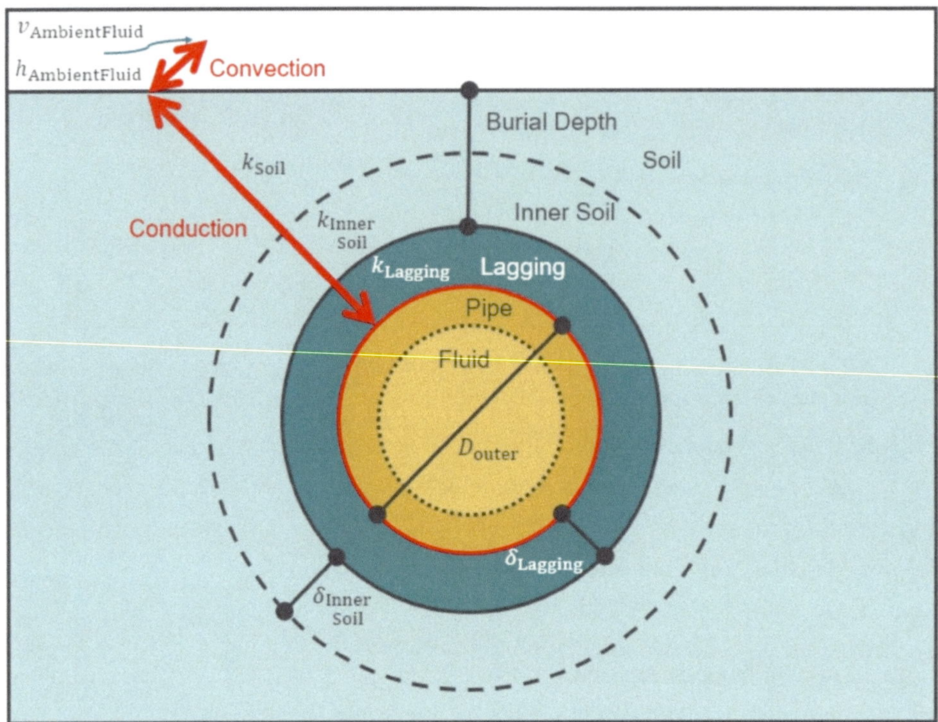

Figure 93 – Almost all pipelines have shells of cladding and soil. In those, conduction (k) rather than convection (h) is the rate-limiting mechanism. Conduction doesn't settle to equilibrium quickly enough to be considered steady state: we need a transient model. [38]

We must be careful not to overlook the limitation of a steady thermal model. It assumes every distinct heat transfer mechanism is unchanging over time, and in perfect equilibrium; a succession of steady states – just like we discussed in the chapter on hydraulics, but here our system is the ground's thermal model.

In a real pipeline, the thermal environment in many cases is never truly steady. This makes an OHTC approach a poor choice in a thermally transient scenario. It gives wildly misleading results if transients happen, be those hydraulic transients in the pipe, or thermal transients in the pipe or in its surroundings.

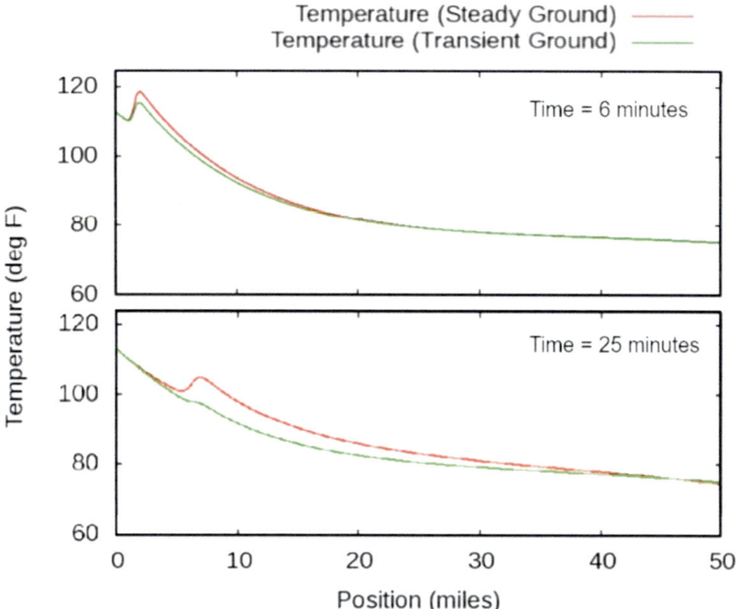

Figure 94 – A steady ground thermal model might settle to the wrong temperature, introducing a persistent error which then propagates to other results along the pipeline.

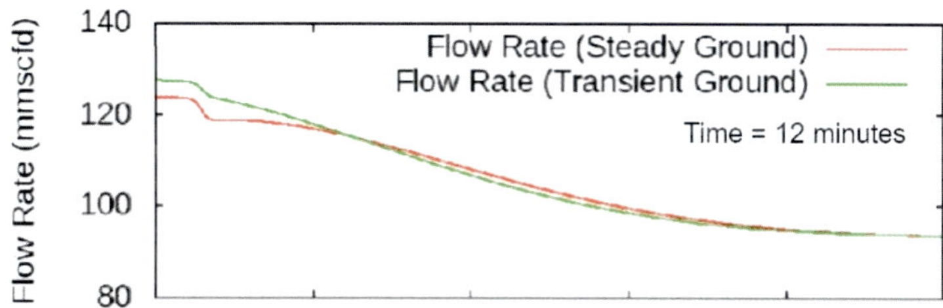

Figure 95 – A consequence of the wrong temperature profile (in the preceding figure) is a wrong flow rate profile. A transient ground thermal model is essential in this case.

We should not be assuming thermal shells are always in steady state; that assumption implies that if a supply temperature changes in our pipeline, the surrounding ground instantly snaps to a new thermal equilibrium at the next model time-step. Such assumptions are so thoroughly false, they often give worse results than if we ignore temperature changes altogether (isothermal)!

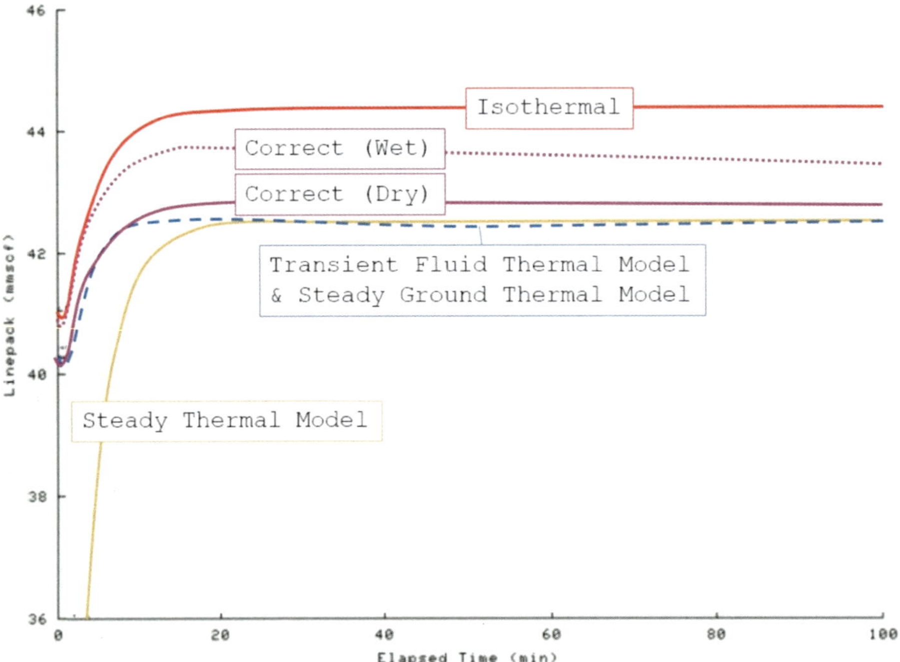

Figure 96 – When this gas pipeline undergoes a transient scenario, the linepack from a steady fluid thermal model is initially worse than unchanging temperature (isothermal)!

Eventually, people realized that although the ambient medium can usually be assumed to always be in thermal equilibrium, this assumption almost never holds true for everything between the environment and the pipe (the *ground*). A *transient* ground thermal model is necessary to capture the transient reality of conduction. We must define *thermal shells* for the cladding, and for the soil. Transient heat transfer mechanisms change over time, and *affect one another*! Each shell of soil or cladding has its own heat capacity, gradually changing in temperature as the pipeline itself heats up and cools down. This coupled effect can be quite significant. Only by adopting a sophisticated *transient* approach to the ground thermal model can we accurately quantify the true heat transfer.

Nonetheless, users may wish to set an OHTC rather than configuring transient thermal shells. For example, some users wish to run their calculations using an OHTC despite recognizing it to be a bad idea, to validate against reference results from other legacy software. So it's still made available in Atmos SIM.

The conductive (k) heat transfer through the thickness (δ) of all the thermal layers and the convective (h) heat transfer with the ambient environment are mechanisms that act in tandem, and are related to the *overall heat transfer coefficient* (U) as follows:

$$U = \left(\frac{\delta}{k} + \frac{1}{h}\right)^{-1}$$

This sweeping abstraction lumps in the combined effects of several distinct mechanisms beyond a pipe's diameter: *conduction* through any insulating layers (such as cladding, and the soil if the pipe is buried), and *convection* with the pipeline's *ambient environment;* water if subsea, or air. Let's look at these more closely.

If the pipeline is buried, heat is conducted through the soil. If it is offshore, the true burial depth might well be uncertain. It might lie directly on the seabed, surrounded by circulating seawater, dominated by convection. Convection governs how seawater or air circulates around our pipe. If it sits above ground exposed to air, convection dominates along with radiative heat transfer.

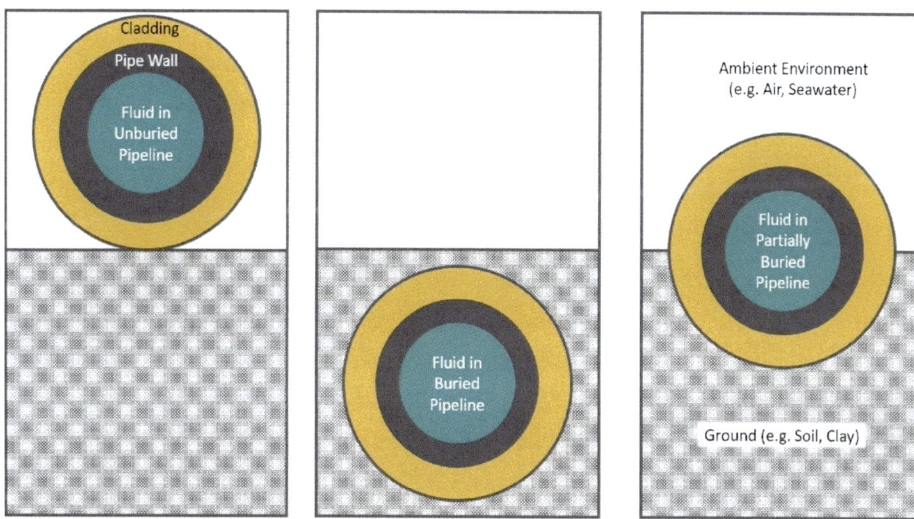

Figure 97 – A pipeline could be laid unburied (left) or buried underground (middle). An undersea pipeline's exact degree of burial on the seabed changes over time (right).

Convective heat transfer with the ambient medium

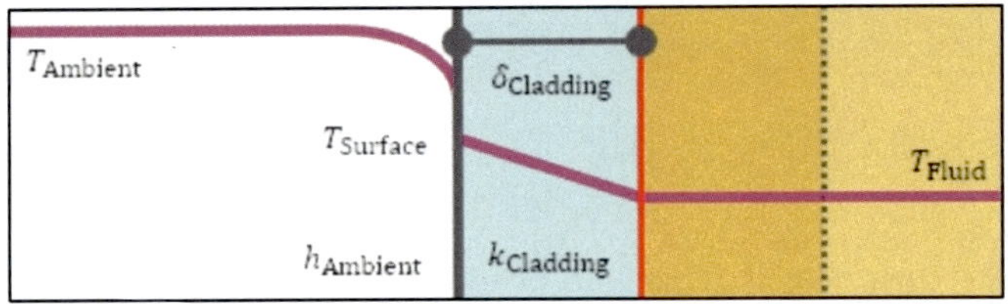

Figure 98 – The radial temperature profile across an unburied pipeline.

Convection transfers heat between the last solid surface and the ambient medium. It does not offer much thermal resistance compared to conduction through solid layers (cladding and soil). As such, it is only in one special case that convection is the *rate-limiting mechanism*. It limits the rate of heat transfer to/from a pipeline which is completely exposed: that is to say, a pipe that is unburied and not covered by any cladding whatsoever. Thankfully, such a *completely* exposed pipeline is not the norm. It is difficult to model such a pipeline accurately: it is affected when a wind blows, or a cloud blocks sunlight.

In any pipeline that is buried or has cladding, convection *quickly settles to an equilibrium*, far quicker than conduction does. Because it reaches its *steady* rate of heat transfer so quickly, it is reasonable to treat convection as a *succession-of-steady-states*, represented by a heat transfer coefficient. This gives us good time-averaged behavior over the timescales we are simulating.

Thermal convection [*] does *not* give rise to a linear radial temperature profile. To fully capture it would entail a three-dimensional model. Instead we simplify the ambient medium's heat transfer with the outer shell's surface by saying the medium's thermal resistance acts in a film near the surface. There isn't one in reality, but it is a reasonable abstraction. On this basis, we empirically define a convective heat transfer coefficient which is often called the *film coefficient* (h), via Newton's Law of Convection: [†]

$$\dot{q}_{\text{convection}} = h \cdot (T_{\text{surface}} - T_{\text{ambient}})$$

Experimental results record correlations to calculate this film coefficient as a function of the ambient medium's thermal properties and the ambient velocity. Thermal properties are themselves temperature dependent, so we lump the heat transfer between the solid surface and the bulk of the ambient medium. Properties are evaluated at the *film temperature* (T_{film}), taken as the average of the surface

---

[*] Some recent authors seem to prefer using the term *advection* in this context.

[†] Often referred to as *Newton's law of cooling*, but it is equally true for warming.

temperature (T_{surface}) and the ambient temperature (T_{ambient}). If there is even the slightest seawater current or wind speed, the velocity of the ambient medium (v_{ambient}) means *forced* convection takes place. This is a much more efficient heat loss mechanism, dominating the *natural* convection of a stationary medium ($v_{\text{ambient}} = 0$). When calculating, we choose the greater of the two film coefficients (for forced convection versus for natural convection).

In reality if there is a slow wind or current, what's going on is complex, in between forced and natural, but it is not usually vital that we get that exactly right, because heat flow well outside of the pipe wall, past any burial and insulation, has minimal effect on the minute-to-minute or even hour-to-hour behavior of the temperatures within the pipe.

Ambient environment and mechanism		**Film coefficient** $\mathbf{W \cdot m^{-2} \cdot K^{-1}}$
Air	Free / natural convection	2 to 25
Air	Forced convection	10 to 500
Water	Free / natural convection	50 to 3 000
Water	Forced convection	50 to 10 000

Table 6 – Typical film coefficients for forced and natural convective heat transfer

Figure 99 – Some pipelines are not buried because of permafrost; whether the wind is blowing might then significantly affect the heat transfer between a pipeline fluid and its surroundings

Conductive heat transfer through solid shells

Across a solid shell, such as cladding or soil, the temperature profile in the radial direction is linear. The gradient of temperature in each shell is proportional to its *thermal conductivity* (k), by Fourier's law of conduction, the steady-state rate of heat flux is:

$$\dot{q}_{\text{conduction}} = \frac{k}{\delta} \cdot (T_{\text{fluid}} - T_{\text{surface}})$$

Layers of soil shallower than 20 meters undergo seasonal heat transfer. Layers shallower than 0.1 meters undergo diurnal (day / night) cycles. A transient ground thermal model is a true accurate picture of heat transfer by conduction. The transient equation that applies here is the *heat equation*:

$$\frac{\partial T}{\partial t} = \text{constant} \cdot \frac{\partial^2 T}{\partial x^2}$$

It says that the rate at which temperature changes over time depends on the second derivative of temperature at that point with respect to space. *

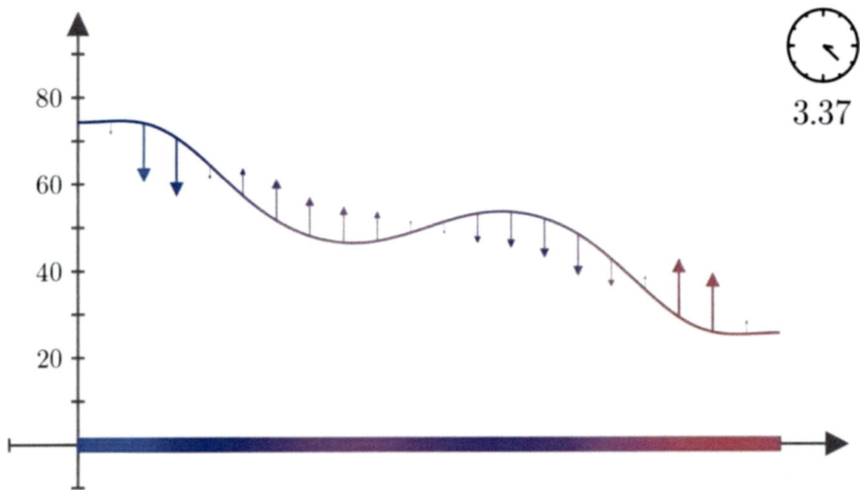

Figure 100 – The heat equation says that temperature at a point varies over time at a rate proportional to how different from its neighbors it happens to be at the moment. A point warmer than its neighbors cools down (Credit: 3Blue1Brown on YouTube)

We apply the heat equation to a series of individually defined shells. Each shell's thermal parameters are taken to be constant over time. This approach verifiably gives us reliable predictions in transient scenarios. Happily, it happens not to be much more

* When Fourier sought a solution to this equation, he invented Fourier series.

intensive than the simpler *steady* approach for the ground thermal model; it is the transient *fluid* thermal model that represents a significant performance hit.

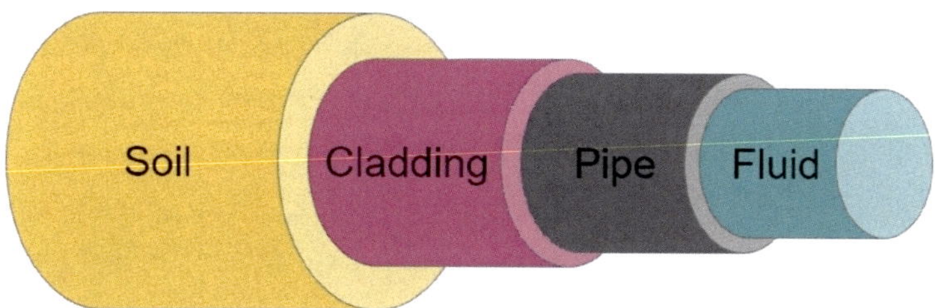

Figure 101 – The fluid is surrounded by concentric *shells* in a transient ground thermal model. The last of those is a soil shell if the pipe is buried. More shells can be defined if soil properties are different in close proximity to the pipe, or if there are several layers of cladding / insulation. Usually, a steel pipe is held at fluid temperature given its high conductivity.

Shell material	Thermal conductivity $k\ \mathrm{W \cdot m^{-1} \cdot K^{-1}}$	Specific heat capacity $c\ \mathrm{J \cdot kg^{-1} \cdot K^{-1}}$	Density $\rho\ \mathrm{kg \cdot m^{-3}}$	
			Minimum	*Maximum*
Steel	20	500	7 750	8 050
Polyethylene	0.5	1 550	910	970
Dry sand	0.6	800	1 300	1 800
Wet sand	2	1 700	1 700	2 200
Dry clay	0.3	1 100	1 300	1 600
Wet clay	0.7	2 300	2 000	2 500
Dry loam	0.2	1 100	1 300	1 700
Wet loam	1.1	2 100	2 000	2 200

Table 7 – Typical thermal conductivities and heat capacities for ground heat transfer shells.

Beneath the ambient medium of air or water, soils come in many varieties. We see mud, stone with gaps filled with air, stone with gaps filled with water, and the whole range of the clay-loam-sand triangle. Water and rock content varies, as does ice content in some climates. If we know how wet or dry a soil is, whether it is large sand, small sand, clay or loam, and its porosity, we can make a fair estimate of its thermal conductivity, heat capacity and density.

Over the operational lifetime of the pipeline, the properties of the ground vary as it undergoes all kinds of thermal changes. A hot pipe desiccates the soil around it; a buried oil or liquid product pipeline is essentially surrounded by drier soil than we might expect. Similarly, a cold pipe draws in moisture towards nearby soil layers, so a buried natural gas pipeline is surrounded by wetter soil than we might expect. In some cases, then, the soil's heat capacity is *itself* a function of temperature! However, it is not typical to go so far in including those levels of detail when configuring a pipeline for simulation. We should bear in mind, however, that the *"granularity"* of a thermal model should not be neglected; if soil properties are very wrong by a big factor (say, threefold), we run into problems!

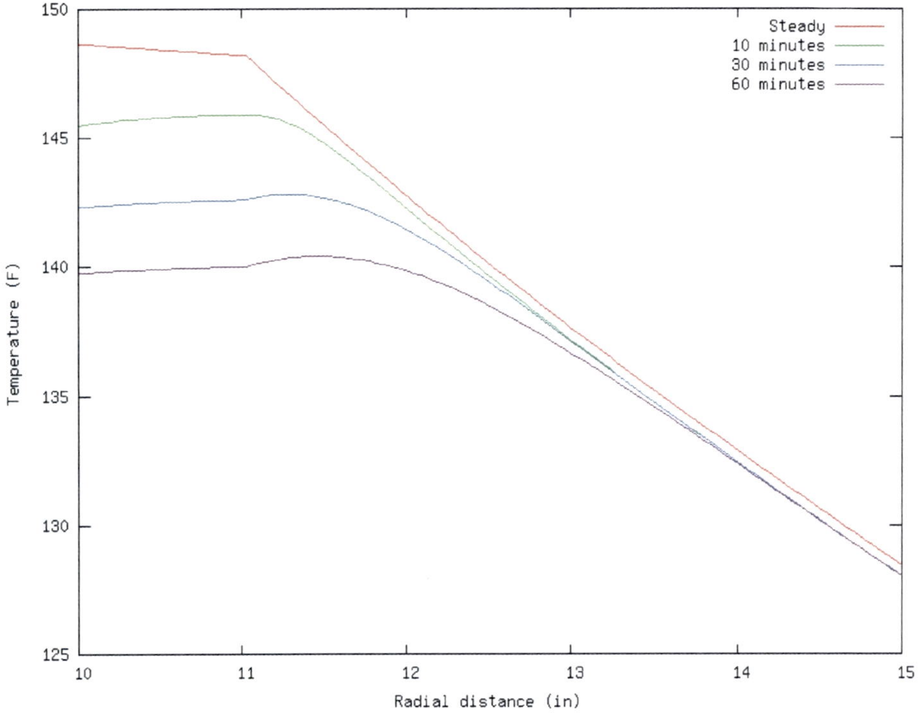

Figure 102 – The radial temperature profile at various instances in time during a transient scenario, as calculated by the dynamic ground thermal model. This diagram simulates a scenario with extremely high thermal conductivity, because it has a major effect 12 inches out after an hour. This is much faster than anything we would normally see. We would ordinarily expect changes 1 or 2 inches out after an hour.

6.5 – CALIBRATING THE THERMAL MODEL

Thermal properties are uncertain

A pillar of building and maintaining trust in a pipeline simulator is reliably making accurate predictions. Only then can it support a user's diverse needs.

Usually, nobody knows the ground's thermal properties particularly well. Many heat exchange parameters, notably the soil's conductivity, and the velocity of the ambient seawater or air, vary significantly along a pipeline. By estimating these parameters, we introduce uncertainties, so it is good practice to perform a *sensitivity analysis*, and suitably qualify our predictions. They don't need to be perfect, only good enough for our purposes.

These thermal properties might also vary over time with the weather. Even where transients are not so drastic as to change over the course of a scenario, slow changes do take place, making it difficult to keep thermal parameters right over many months. This motivates developers of pipeline simulation software to look into how best to learn the thermal parameters. The long-term rate of heat transfer is effectively time-averaged, so average rainfall might be more important than day-to-day weather. The fact that soil moisture is inevitably unknown to some degree suggests that automatic tuning of soil thermal conductivity in the region immediately around a pipeline might be worthwhile.

Outlet temperature is not enough

The simplest question we can ask about a simulator is: if we don't tell it about a mid-line meter, can it guess exactly what that meter says?

Conditions throughout *any* pipeline, gas and liquid alike, are subject to limits. No routine operation may, at any point along the pipeline, exceed that point's maximum (MAOP) or lowest allowable operating pressure (LAOP), for example. Stricter requirements stipulates that no pressure excursion may under any circumstances violate the *safe* operating limits (SOL). The terrain along a route makes it almost certain that a potential violation will be at an unmetered point.

Therefore, if a simulator is telling us that doing something differently would save us wasting energy in some way, such as compressing less vigorously or operating at a lower pressure to avoid letting fluid pressure down through a back-pressure regulator, we should first be comfortable trusting its calculations at every unmetered point. Otherwise, what the simulator predicts would be immaterial!

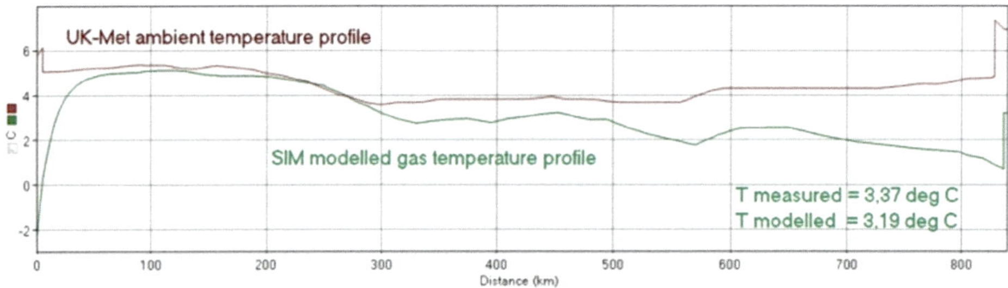

Figure 103 – Ambient and pipeline gas temperature profiles. This model appears to be well-calibrated as simulated and measured temperatures at the downstream end agree. But are we sure the thermal model always gives us the right temperature *along* the line?

We want to spend enough effort on getting the thermal model right, but not too much. Even so, the thermal model is important. Not only must our thermal model achieve an accurate *outlet* temperature: for our model and simulator to make accurate arrival time estimates, and simulate other physical results accurately, we must ensure simulated temperatures are always accurate throughout the full length of a pipeline's entire route.

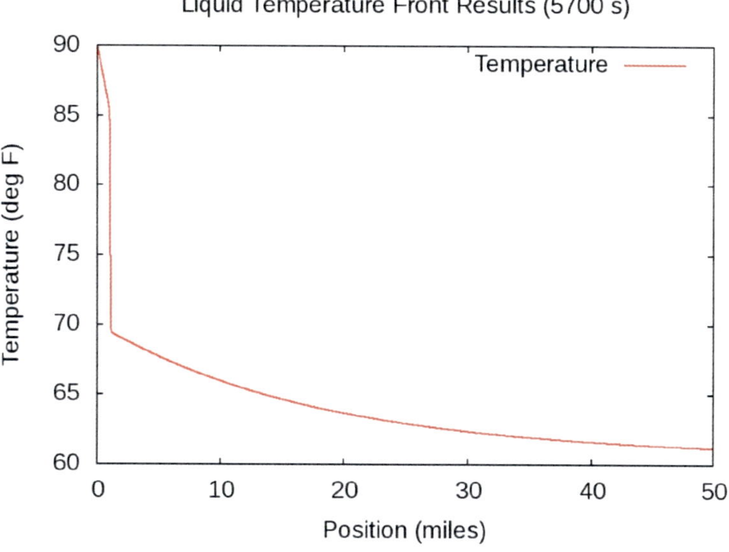

Figure 104 – A profile evolves over time as a liquid's temperature front proceeds.

There are substantial uncertainties in several thermal parameters. Therefore, it's reasonable to adjust these parameters to calibrate a gas pipeline. *

One uncertainty is *burial depth*. If no information is available on how deeply a pipeline is buried, it is reasonable to assume about a meter. The exact value doesn't make much difference to transient thermal effects in the ground, as they are all mediated by the soil closest to the pipeline. However, a wrong burial depth gives us the wrong long-term *equilibrium* rate of heat transfer. That long-term equilibrium rate of heat transfer is also wrong if we are wrong about other thermal properties. These parameters are practically unknowable to a sufficient accuracy: a soil's thermal conductivity varies depending on such matters as whether it has rained recently.

This has ramifications beyond the conditions along the pipeline being wrong. We are not too concerned about matching downstream temperature, as it is only really a measure of environmental properties in the last short stretch of pipeline, whereas *transit time* is a measure of environmental properties along the whole route.

* Adopting the same strategy to calibrate the thermal model for a compressible liquid pipeline is more challenging, as it also hinges on making predictions (like estimating arrival time) based on *batch operations*. In principle it is possible.

Inventory affects transit time

We saw in our discussion of the mass equation that although manual hand calculations based on conservation of species are valid, tracking composition in this way in a pipeline simulator gives rise to too much numerical dispersion. Instead, let's consider a composition front or a pig moving through a pipeline.

The pressure setpoints at the ends and compressor setpoints in the middle affect the velocity profile. For an incompressible liquid, inventory is only a function of those setpoints. For a compressible fluid or a gas, inventory is a function of those setpoints, the thermal situation, and the history of both.

An important simulated result is *transit time* of batches or composition fronts. In an online system it lets operators know when they must open and close valves, for instance. We therefore need these transit times to be accurate, for an accurate estimated time of arrival time (ETA), It depends on *inventory*:

$$\text{Transit Time} = \frac{\text{Inventory}}{\text{Averaged Flow Rate}}$$

This equation holds true for conserved quantities (mass or standard volume).

Transit time in a gas pipeline

Let's consider the case of a gas line: the transit time for composition changes is the total linepack ($\mathrm{Sm^3}$) divided by the average standard flow rate (Q_{std} in $\mathrm{Sm^3/s}$). Total linepack is a result of the simulation and depends on the pressure profile, which we know quite accurately, and the temperature profile, which we don't know accurately. In a gas pipeline the temperature profile along a pipeline – and its history – affects its density profile, and therefore the inventory contained in its volume (what gas pipeliners refer to as the *linepack*). This means that the accuracy of a simulated inventory depends on simulated heat transfer, which is sensitive to changes in the ground thermal parameters such as the thermal conductivity of the soil. Inventory, in turn, affects the arrival time. A standard volume of gas moving at some flow rate must pass through the entire inventory to reach its destination. The more the inventory, the longer the transit time at a given standard flow rate. Transit time is *directly proportional* to inventory. *

---

* Although most pipeliners rarely look at velocity profile, it can be instructive to do so to gain a new perspective. The same standard flow rate is equivalent to a slower velocity profile if there is more inventory (linepack) within a pipeline.

Therefore transit time is sensitive to all of the input parameters to the thermal model. Some of those are quite uncertain, notably a soil's thermal conductivity. We must therefore tune the temperature profile, or some thermal properties that affect the temperature profile, in order to achieve accurate transit times.

The first challenge is measuring transit time. A gas pipeline's transit time can be deduced using gas chromatographs (GC's). One of these composition meters is usually installed at the start and end of a gas pipeline. If there are identifiable composition changes * moving through the system fairly regularly, then we can get a series of estimates of transit times by matching up composition changes entering the line with composition changes exiting the line. In doing so, we expect there won't be much change in the shape of the composition front over even a long pipeline, because turbulent mixing is slow.

Transit time in a liquid pipeline

The arrival of a composition front at the downstream end is indicated by a liquid batch detector (density, opacity, etc.). However, it is uncommon to tune a liquid pipeline's arrival time in the same way as in a gas pipeline, because a liquid's linepack (inventory) doesn't change by much, so we can't usually tune transit times on liquid lines by much.

If a liquid pipeline is way off the mark in terms of its transit time, that either reflects a configuration error, or miscalibration of the flow meters. If there is more than one flow meter and they all agree, that makes a configuration error much more likely. An exception is if there's regularly slack happening, in which case the transit times give us an estimate of the slack volume.

A notable exception is if the pipeline fluid is a very compressible liquid, such as liquefied petroleum gas (LPG), or if a batch of such a liquid is present. Transit time in such a pipeline is affected by temperature like in gas pipelines, but less so on a percentage basis. LPGs are heated in pumps more than incompressible liquids, which also creates a driving temperature gradient.

On a viscous liquid pipeline, the pressure drop is actually a good measure of the temperature profile. Considering what this implies, we should realize that having adjusted a thermal parameter to calibrate our temperature profile, we must always go back and adjust hydraulic parameters to re-calibrate pressure drop against flow rate.

---

* Some gas pipeliners refer to an identifiable gas parcel as a *composition peak*.

Which thermal parameter to tune

In the steady states chapter, we calibrated a pipeline's *hydraulics*. We matched pressure drop to flow rate by adjusting several hydraulic parameters, such as the roughness and inner diameter of pipes, and resistance of items. In the vast majority of pipelines, heat transfer is important, so it too should be calibrated. A good procedure for calibrating thermal parameters is a more involved affair.

If a section of pipeline is unburied, and we're sure about the thermal properties of its cladding, we tune its overall heat transfer coefficient. If the pipe is buried, we could choose any one of several following thermal parameters, including ambient temperature (T_{ambient}), or inner shell thermal conductivity (λ).

Tuning any of these can reproduce a measured transit time (in a gas pipeline) by altering the linepack. Here we shall tune the soil's thermal conductivity rather than the ground temperature. But it's not clear which of the two is more uncertain. Depending on burial depth, we might expect either of the two to change more than the other over time. Which we choose as our tuning parameter is somewhat arbitrary. Choosing soil thermal conductivity as our thermal tuning parameter does have downsides. If fluid temperature is very close to ambient temperature, adjusting it has little effect. Also, we can only turn it down as far as zero, and beyond a certain point turning it up ceases to have any further effect because the fluid is then at the ambient temperature. Tuning *ground temperature* gives us a dial we can turn arbitrarily far in either direction.* But an advantage of using conductivity is that the effects of tuning it are felt much faster than the effects of tuning ambient temperature.

The ground's thermal conductivity and heat capacity have a lot more effect on the short-term gas temperature than ambient temperature does. Let's simulate two identical steady-state 0.91 m (36-inch) diameter pipelines 100 km in length, as a test scenario. We fix their conditions to be identical and constant throughout, at a standard flow rate of 60 MSm^3/d, and an inlet pressure of 100 barg. Inlet temperature is 40°C and ambient temperature is 5°C everywhere. One of them is buried at a depth of 2 meters in a dry soil of thermal conductivity $1\ \mathrm{W \cdot m^{-1} \cdot K^{-1}}$. The other is buried at an equal depth under a wet soil of thermal conductivity $3\ \mathrm{W \cdot m^{-1} \cdot K^{-1}}$. This gives rise to a difference in inventory of 3%, thereby noticeably affecting the transit time at this flow rate.

* This is analogous to choosing our hydraulic tuning parameter to be efficiency rather than roughness; it is an adjustable knob that doesn't saturate.

Therefore, in seeking to improve a model's inventory - and thus its arrival time estimates – we can consider a soil's thermal conductivity to be a valid choice of heat transfer parameter to calibrate buried and partly-buried pipe sections.

Legacy pipeline simulators did not pay close enough attention to heat transfer. An old model might have simply shifted the entire ambient temperature profile up or down to make a modelled outlet temperature match a metered value. But typically only the ambient temperature profile along the last few kilometers influences metered outlet temperature. Such a non-physical fudge-factor had adverse ramifications on inventory, and estimated arrival times of composition fronts, peaks and pigs. Today's state of the art pipeline simulators take a smarter *physical* approach. Getting temperatures right gets the densities right, and lets us accurately track composition. If a model over- or under-estimates arrival time, we must revisit the thermal parameters of fluids, pipes and soils.

We calibrate heat transfer parameters in an offline model to affect inventory, which in turn can match modelled arrival time and outlet temperature against a historic dataset. A soil's thermal conductivity strongly affects heat transfer. To get the temperature profile right, we must tune thermal conductivities to match measured downstream temperatures, as well as the arrival times of a pig or composition front. This tells us the whole temperature profile is right.

This brings up a more general question: if the arrival times are correct, why should we worry about getting the temperatures right everywhere? It turns out getting the model physics right is extremely important if we are trying to detect and locate leaks, and somewhat important if we are performing surge studies. This is because fluid temperature has a moderate effect on the speed of sound, with a 6°C temperature error producing about a 1% error in speed of sound; and that in turn affects the transient response of the pipeline by about 1%, as we described in the earlier section on the Joukowsky equation. 1% isn't much, but small errors in transient response can degrade leak location significantly.

Tuning soil conductivity can also act as a bucket compensating for unknowns in the wider thermal environment.

What is the target transit time

Having established that downstream temperature is less important than transit time, because the former is affected by ambient conditions in the final stretch of pipeline, whereas the latter is affected by temperatures everywhere, how do we go about finding the true transit time for us to set it as a target?

It is rare that the temperature profile on an operating pipeline actually reaches an equilibrium. We just spent a solid chapter of this book building a transient fluid and ground thermal model to cope with this! So how can we find a stable dataset we use as a basis for thermal calibration? The answer is we average over the relevant characteristic scale time. This depends on the pipeline's length, inventory and flow rate. And this is before considering that we must account for seasonal variations; we ought to average those out over a year!

For example, take a 100 km gas pipeline with an average flow velocity of 3.5 m/s. Its transit time would be around 8 hours. If a composition meter had a sample interval of 15 minutes each, this gives ±3.1% as a best achievable deviation in estimated time of arrival (ETA) and a worst-case deviation of ±6.2% between two composition meters. The same sample intervals on a 400 km pipeline flowing at the same average velocity would involve a best achievable ETA deviation of ±0.78% and a worst-case of ±1.6%. The soil's thermal conductivity is then tuned to achieve our pipeline's target accuracy.

To determine the true arrival time, we track a gas parcel or liquid batch. Then we consider the ratio of true and simulated transit time as our error factor:

$$\text{Transit Time} = \text{Arrival Time at Delivery} - \text{Arrival Time at Supply}$$

$$\text{Transit Time Error Factor} = \frac{\text{Simulated Transit Time}}{\text{Actual Transit Time}}$$

If operating conditions were identical to the controlled reference case used for the roughness tuning for the duration of the gas batch transit time, then the tuning of arrival time is simple. However in reality this is never the case, so we are forced to assume linearity between the controlled reference case and the transient pipeline behavior for the duration of the gas batch transit time. This linearity assumption enables the *Transit Time Error Factor* calculation to be directly applied to the offline controlled reference case. We assume that the transit time error factor doesn't change over our timescale:

$$\text{Target Inventory} = \frac{\text{Controlled Reference Inventory}}{\text{Transit Time Error Factor}}$$

The calculated target inventory can now be used to tune arrival time.

Sampling the error in arrival time typically spans several weeks, given the nature of a long pipeline's transit time. A good way to accelerate tuning is to feed a previously logged historical dataset to Atmos SIM as a replay run.

A historic state should be used as the starting point for replaying historic data for the purposes of this sort of tuning. This historic state completely records the condition of a pipeline simulation as it was at a precise moment in time. It captures all simulated data across the pipeline network, including the operating conditions – pressures, flows and temperatures – as well as the fluid's compositional profiles throughout every route.

This dataset should include a high stable flow rate, and a low stable flow rate.

To determine the discrepancy between simulated and actual transit time of a gas parcel from source to sink, we need to find a suitable *arrival-time sampling period*. This should be conducted on as many composition changes as possible to reduce any anomalies within the data and to account for seasonal variations, where the temperatures are observed for the warmest and coldest months. This discrepancy time sampling records three moments. Firstly, the instance when a composition change enters from upstream (in both the real operation, and in the simulator). Secondly, the instance it is detected leaving downstream. And thirdly, the instance it is simulated to leave downstream.

The transit time error factor can then be computed with the aforementioned equations.

Tuning thermal properties

Manual calibration

The procedure for thermal tuning is similar to that for hydraulic tuning. To perform an offline calibration, we take an average transit time as our target. A manual approach is to update soil thermal conductivity, simulate, then compare simulated inventory against target inventory as calculated above. If we are outside our target tolerance, we update soil conductivity and repeat.

Tuning hydraulic capacity and ETA of gas batches involves properties which directly affect one another. This means that we must tune roughness, then arrival time, then go back and tune roughness again, and so forth – an iterative approach which is tiresome if done manually.

Typical manual tuning steps are:

1. Calibrate hydraulic parameter: maximum flow rate versus pressure drop
2. Sample the transit time error from the operating data
3. Calibrate heat transfer / inventory tuning
4. Calibrate soil conductivity to achieve accurate ETA
5. Offline roughness tuning using the Tuning Assistant (re-adjust point 1)
6. Deploy tuned configuration online
7. Online transit time error sampling (point 2)

Then we iterate: repeat the same steps until the target transit time error is met.

Assisted calibration

The Tuning Assistant automates these iterative steps. It can calibrate thermal parameters, simultaneously recalibrating hydraulic parameters accordingly. A number of pipes are selected along with the property to tune. This allows pipes which, for example, are unburied to be left as is when tuning soil thermal conductivity. A tuning range for the selected property and a target are specified, along with a tolerance. The entire procedure for tuning inventory followed by roughness is automated into a single step in the Tuning Assistant. We repeat this procedure until *transit time error* is within tolerance.

The Tuning Assistant has been designed to automate offline simulations for controlled reference cases to achieve the correct hydraulic behavior. The controlled reference case refers to the state of the physical pipeline, where the instrument values for the pressures and flows are within a steady state condition. The Tuning Assistant automates an iterative approach for tuning multiple pipeline parameters. It can calibrate hydraulic parameters (such as roughness or resistance) so that a pressure drop matches a flow rate, while simultaneously calibrating thermal parameters (such as soil conductivity) so ambient temperatures and fluid outlet temperature match measured values, and the transit time is accurate enough.

The tuned thermal and hydraulic parameters can then update the model configuration with one click to accurately predict inventories and arrival times.

Automatic learning

To learn thermal parameters in real-time, we take ongoing estimates of transit time as a target. This learning continues throughout the course of an ongoing simulation, gradually adjusting a thermal parameter to tune it automatically.

On each model step, we figure out the direction of the discrepancy in transit time at the moment, and nudge the property being tuned a tiny bit in the direction that corrects it. The magnitude of the update to ground temperature (if we are tuning that) is proportional to the magnitude of the transit time error, but must be kept small – the complete transit time error should only be corrected over a number of model steps many times as long as the actual transit time. Ideally we want it to be ten times longer, although if that much data isn't available a shorter interval is possible – never less than three times longer.

7 – PUMPING AND COMPRESSION

Figure 105 – A pumping station along a water pipeline

"Confusion about entropy always increases"

7.1 – MOTIVATION

Pumps are used for liquids and compressors are used for gases. To those who are new to the field, this may seem like a trivial distinction, but it's important not to conflate terms because they are quite different to one another. Though both operate on broadly similar principles, they are not at all synonyms; a gas pipeline operator may (rightly) exhibit signs of confusion or doubt our expertise if we talk about their pumps! Simulating them presents special challenges, and not every application of a pipeline simulator needs to model them closely. In a real-time model deployed only for leak detection, instruments are required to be good enough that they can isolate each station interior, so in that application simulation results don't model pumps or compressors and, more importantly, don't depend at all on a model of pumps or compressors being accurate.

As the modelling of this equipment is arguably rather difficult, and is not a topic that authors in our industry have devoted much attention to, the reader should rest assured that it is worth our attention. Implementing pump and compressor models in a pipeline simulator is essential in many applications, especially if we must focus on accuracy.

In a real-time simulation live on site, knowing what pumps or compressors are doing helps operators troubleshoot what's going on. It shows the operating point on a convenient pump or compressor map. It calculates power – if power is measured to be way in excess of what is expected, something might be wrong with the equipment. If multiple units are operating, but each of them isn't separately instrumented, a pumping or compression station model adds value by figuring out how the flow and head (shaft work) are split amongst them. These models have also come in useful when deploying a real-time simulator on unusual pipelines whose instrumentation is peculiar, lacking or unreliable. It might simply be that the discharge temperature is not metered, or there might be challenging stations where the operator is not even sure what is running!

And in offline design studies, design engineers consider potential changes in pipeline operations or future upgrades. They're particularly interested in how pumps and compressors perform in many scenarios. Selecting and operating them in an optimal manner offers substantial long-term savings. In an optimizer, simulators identify good ways to operate a pump or compressor. Even a simple single-station optimization can save power, or deduce what to do during scheduled maintenance. We'll discuss optimizers in a later chapter.

The point to appreciate here is that it can be worth paying close attention to pumps and compressors in a pipeline simulator for numerous applications, because these complex technologies matter to many users in a wide range of use cases. Their performance calculations are complicated by several little appreciated facts. The first of these is that what a pump or compressor does will always depend on the pipeline system it is placed in. The performance of the equipment is intertwined with the pipeline itself, which – it can be argued – is probably the largest process people have ever operated in history.

7.2 – ABSTRACTING OUR SYSTEM'S HYDRAULICS

Pipeline resistance curves and operating envelope

It's said Soviet engineers wary of being blamed for under-design chose to oversize a pump and a pipeline's diameter. Control valves are only linear within a certain range of positions. To avoid being too near the closed position, engineers were forced to specify a comically tiny control valve, wasting tremendous energy in routine operation! [48]

We discussed in the chapter on steady hydraulics that getting a fluid through a pipeline at a certain flow rate requires a differential head to overcome the resistance imposed by elevation rise and fluid friction along the route. A consequence of a pipeline's resistance to flow being itself a function of flow rate is that a pump or compressor cannot be associated with achieving a single pressure in all circumstances, or with discharging a set flow rate by itself. How a pump or compressor performs depends on its application – how it is deployed within a station, and the pipeline system it serves. The relationship between head required and flow rate achieved is quadratic. Pipeline resistance is represented by a *"system curve"* on a head-versus-flow plot. System curves are compared for different diameters or regulator positions (seen as different gradients), or for different amounts of static lift (seen as different head-axis intersects).

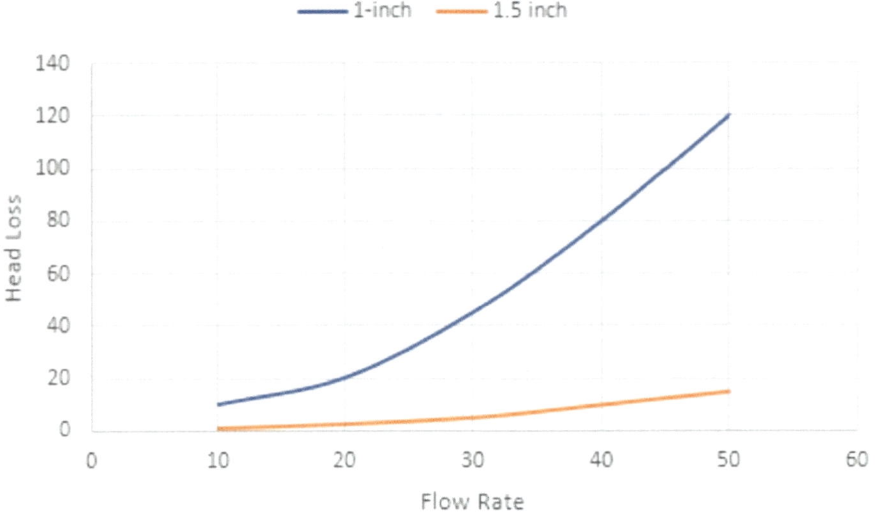

Figure 106 – The system curve of a short pipeline if it has a 1-inch diameter (top) or a 1.5-inch diameter (bottom). The former incurs more head loss to deliver a given flow rate.

Because the properties of the fluid in a pipeline may vary, and in some cases the pipeline itself may also vary depending on the route a fluid is taking for example, almost every pipeline has not just one system curve, but a whole family of system curves. So a set of system curves can be drawn to compare the resistance to flow through the same pipeline in its various configurations. [58] The most difficult curve will be for the most viscous fluid through the most restrictive pipeline route on the coldest day. The most permissive curve will be for the opposite. These two limiting curves are seen as bounding an *"operating envelope"*, whose limits are minimum and maximum flow rate, along with minimum and maximum pressure.

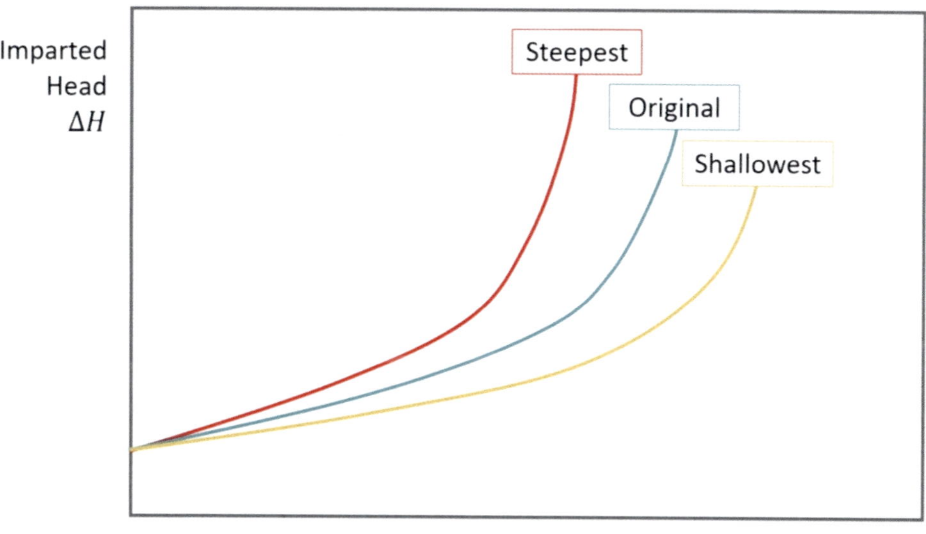

Figure 107 – A nominal system curve, compared to the steepest and the shallowest curve. In this case, only dynamic head changes, so all three intersect the head axis at identical static lift.

An alternative way of looking at this is to say that for each fluid being transported, in the case of a batched liquid pipeline for example, the pipeline system would have a different operating envelope if it were entirely filled with that fluid. Once a pipeline system reaches its natural throughput flow capacity there is a law of diminishing returns: one can only practically force so much flow rate through a given diameter of pipeline and a given set of resistance items (like regulators). For each pipeline configuration – pipe size, valve position, etc. – there is some practical range of flow rates, with a trade-off between capital outlay and operating expense. This makes system curves useful in the design stages and for operations planning, especially in conjunction with pump or compressor curves as we shall see shortly. They help quantify the operational costs and upfront costs with due consideration for how a pipeline will respond to a station.

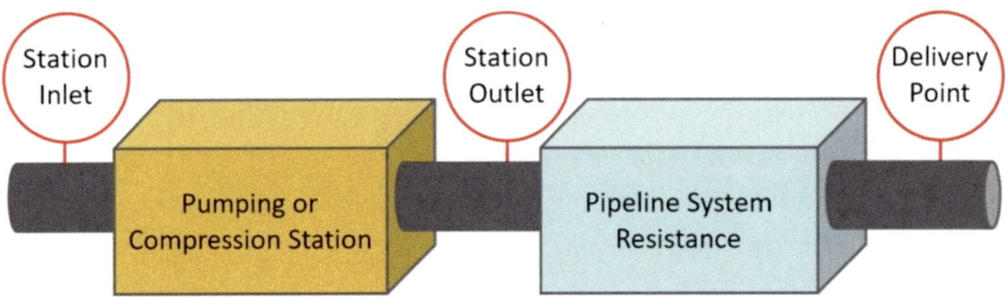

Figure 108 – Performance of a pump or compressor station depends on the pipeline it serves.

Discharge head, static lift and losses

The overall difference between the levels in a liquid supply tank and discharge tank is referred to by different names.[*] We shall call it the *static lift* (ΔH_{static}):

$$\Delta H_{\text{static}} = z_{\text{delivery}} - z_{\text{supply}}$$

We don't think much about static lift in gas pipelines because although part of the pressure at a given point is indeed due to the weight of the gas above it, this is a trivial amount of head relative to the operating pressure of a gas transmission pipeline. In liquid pipelines, static lift is more important, but even then the static lift is not usually the only obstacle to getting the fluid flowing.

Consider a scenario transferring a liquid from a higher supply tank to a lower delivery tank. A pump must still impart enough head to overcome *friction* and *fittings*, despite not having to raise liquid above its original supply elevation by the time it reaches its delivery point. The *discharge head* (ΔH_{dis}) must provide additional head for these *losses* (ΔH_{losses}) associated with friction and fittings:

$$\Delta H_{\text{dis}} = \left(z_{\text{delivery}} - z_{\text{pump}}\right) + \Delta H_{\text{losses}}$$

In most liquid pipelines, the system resistance is mostly due to frictional losses which are far more significant than the static lift. It is a pump's duty to overcome the system resistance, whatever it may be at the resulting flow rate. But a pump can only perform its duty if the *suction head* available is sufficient for its needs.

[*] One of these names is *total head*, which can be mistaken for another quantity.

Suction head

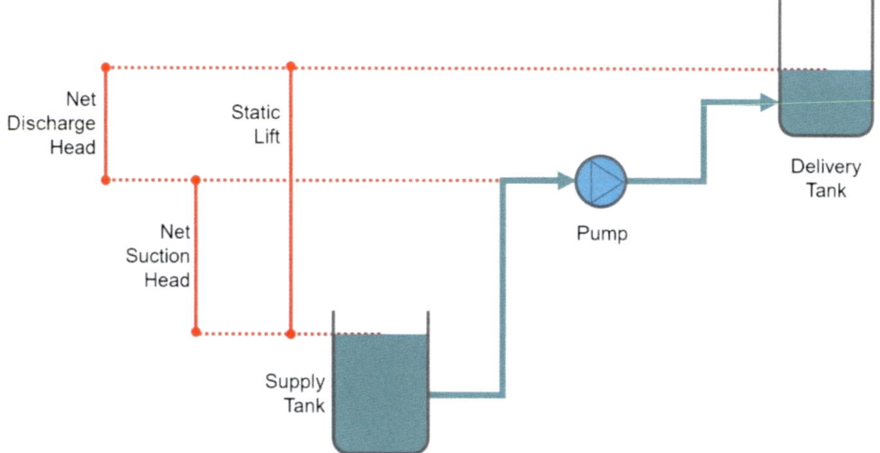

Figure 109 – This pump is located above the liquid level within a supply tank. The resulting *negative* suction head can only be overcome by creating a partial vacuum. This arrangement won't work on a typical pipeline: a normal pump won't be able to pull any liquid from the tank.

Unlike gases – which fill their container – liquids always have a level. Even in a pipeline that is completely full, we can think of the pressure as equivalent to the level liquid would reach in a thin vertical stopped tube attached to the pipe. This level is called the head. We say that at the suction side of a pump there is a *suction head*. Depending on the atmospheric pressure (P_{atm}), the density of the liquid (ρ) and the elevations of the liquid level in the supply tank (z_{supply}) and of the pump (z_{pump}), this might end up being a *negative suction head* (ΔH_{suc}). That is to say, it is possible that $\Delta H_{\text{suc}} < 0$:

$$\Delta H_{\text{suc}} = \frac{P_{\text{atm}}}{\rho \cdot g} + z_{\text{supply}} - z_{\text{pump}}$$

This is the age-old problem of having to lift water, brine or whatever liquid from a lower level of elevation than that of the pump itself. One solution popular in the modern water industry is selecting what is known as a *"can pump"*, which is specially designed to be suitable for placement in less accessible spaces, and it is positioned at a lower level than the main pumps. A more common solution is to select and install a special pump which creates a partial vacuum to *suck* liquid upwards into it.

However, most pumps require that the suction side has a net positive head available. In fact, they require the suction head available to always exceed a set stipulated value.

Net positive suction head and damaging cavitation

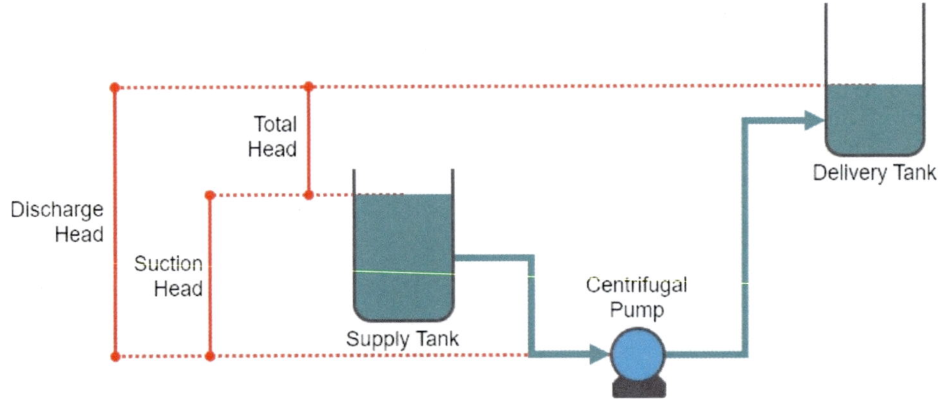

Figure 110 – This centrifugal pump transfers liquid from a higher-level supply tank. The resulting net positive suction head available (NPSHa) must account for losses to friction and fittings in suction-side piping, entering the pump still meeting the net positive suction required (NPSHr).

In addition to the atmospheric pressure (P_{atm}) and the density (ρ) of the liquid, we must account for vapor pressure (P_{vap}), losses due to fittings ($\Delta H_{\mathrm{fittings}}$) and friction in suction pipes ($\Delta H_{\mathrm{friction}}$). There is technically a term to account for the incoming velocity at the suction side (v) but in practice this is very small. This overall result is known as the *net positive suction head available* (NPSHa):

$$\mathrm{NPSH_a} = \frac{P_{\mathrm{atm}}}{\rho \cdot g} + z_{\mathrm{supply}} - z_{\mathrm{pump}} - \Delta H_{\mathrm{fittings}} - \Delta H_{\mathrm{friction}} - \frac{P_{\mathrm{vap}}}{\rho \cdot g} - \frac{v^2}{2 \cdot g}$$

The density (ρ) may be affected both by the composition and the temperature. A volatile liquid's challenging vapor pressure (P_{vap}) warrants special attention.

Even if NPSHa is calculated to be positive, it sometimes might not be enough for the pump in question. A pump's internal components move very fast, so the pressure somewhere inside the equipment might still fall below the fluid's vapor pressure, causing damaging cavitation* to occur at that location. This happens when some liquid boils into vapor bubbles, which then damage solid surfaces as they collapse.

Although people casually say cavitation is to be avoided, strictly speaking only excessive cavitation is damaging. To avoid damaging cavitation, as a property of their pump (on its *pump map*), manufacturers stipulate a *net positive suction head required* (NPSHr) based on experiments for each actual volumetric flow rate (Q). NPSHr is often

* Pump cavitation is also widely referred to in industry as *choking*, though it has nothing whatsoever to do with the choked flow that happens in gas pipelines.

stated as NPSH3 where the *"3"* refers to a 3% derating in delivered head compared to what would otherwise be expected. Another way of thinking about this is that the manufacturer is saying any cavitation that causes less than a 3% derating in delivered head isn't considered damaging. [40]

The NPSHr value stated on the map is based on a liquid with effectively zero vapor pressure, usually water. A liquid with significant vapor pressure, such as crude oil or its products, therefore needs additional suction pressure on top of the pressure equivalent to the stated requirement. Some people lump that into NPSHr, others consider it to be part of the NPSHa and compare that against the original stated NPSHr. However people look at it, they must always account for vapor pressure when ensuring their NPSHa satisfies the NPSHr.

The requirement for net positive suction head is more stringent at higher flow rates, because as a liquid moves faster the pressure drop between the pump suction and the point inside the pump of minimum pressure is greater. To increase a flow rate through a pipeline using the same pump, it is likely we will need to increase the net positive suction head available (NPSHa). This could be achieved by supplying from higher tank levels i.e. more static head at the origin. Or it could be achieved by reducing the losses incurred in suction pipes.

In practice, it is usual to be conservative and apply a further allowance as a generous margin of error on top of the required value. This margin is determined by whatever is recommended by a given company's guidelines. It is intended to compensate for uncertainties and for dissolved gases like air. A rule of thumb in the water industry is 1.5 meters, and in wastewater 2.2 meters.

All pumps, including both centrifugal and reciprocating pumps, present their own net positive suction head requirements. An exception is small *"booster pumps"*, specially designed to *"boost"* pressure so flow enters the suction side of subsequent main pumps with enough net positive suction head available.

7.3 – PUMP OR COMPRESSOR PERFORMANCE

A pump or compressor steady flow energy equation

As pipeline fluids must be intensely pressurized, generating pressure to drive flow and control sophisticated operations are part of daily routine operations. A good pump or compressor accommodates a degree of flexibility throughout a range of operating conditions and its behavior is expected to serve us according to known rules.

An ideal pump or ideal compressor immediately delivers a set amount of useful energy without restriction, doing exactly as it is told by the user. All this energy is perfectly added in a useful manner, appearing as pressure in the fluid. In the spirit of great models ignoring whatever they can get away with, this simple ideal abstraction is convenient in a project's early design stages, or for certain scenarios where it is adequate or preferable to directly add energy to the fluid.

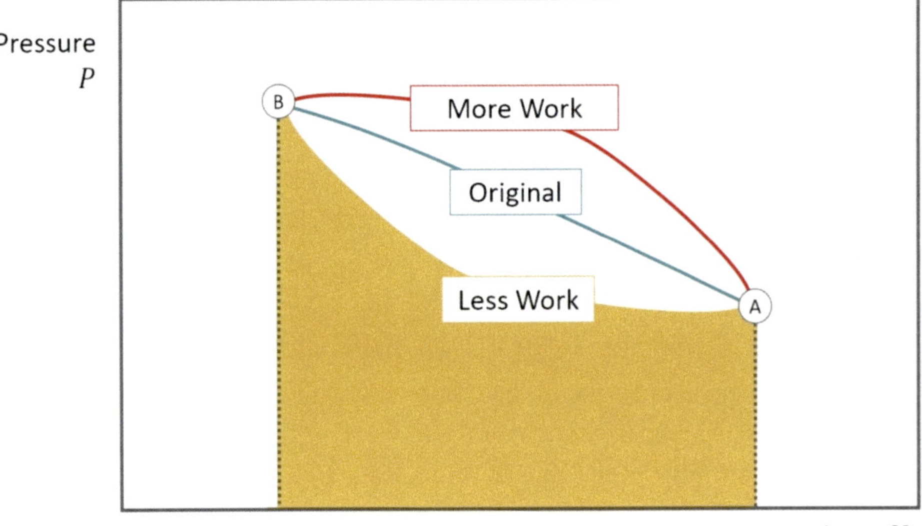

Figure 111 – A pressure-volume diagram of a compressor; the change in enthalpy – or any other thermodynamic property – is the same for all paths, whereas the work done is path-dependent.

Any pump or compressor affects a fluid from the outside world in two ways: transferring heat (q) and doing work on it (w). Both of these are impossible to calculate directly. This is because they are path-dependent quantities; knowing a fluid starts at state A and ends up at state B is not enough, since there are many different paths going from A to B, each incurring different heat and work. The way out of this conundrum is to calculate heat and work indirectly. A thermodynamic property is a *state function*, meaning that it is purely a function of the thermodynamic variables describing the current state of the system (such as pressure and temperature). It characterizes a process based only on where it starts and ends, without considering points in between. This kind of variable is useful to work with because it changes by the same amount regardless of the path taken.

Familiar variables like pressure, temperature and speed are measurable. To relate them to one another, we are obliged to deal with a host of less tangible notions; *enthalpy*, *entropy* and *efficiency*. Many people use these, but few find them intuitive. For engineers, what matters is they're useful. Without needing to delve into their true meanings, these quantities are intertwined according to known relations, and so can be treated by the great majority of us as mere intermediate steps.

$$h\ ,\ s\ ,\ \eta$$

Enthalpy (h), entropy (s) and efficiency (η) are quantities that help represent the unique behavior of a pump or compressor.

We want to capture the performance of a pump or compressor across its entire *operating envelope* or *map*, the full range of pressures and flow rates it can handle. A few key parameters fully describe how a fluid is pressurized by equipment. Any of these form the basis of our problem statement, depending on what we know. We usually know the *suction conditions* including pressure (P_{suc}) and temperature (T_{suc}). We usually know either the suction-side actual volumetric flow rate (Q_{suc}) or the discharge pressure (P_{dis}), and we need to calculate the other. We must also calculate two further unknowns: firstly, the total rate of shaft work done by the pump or compressor, i.e. the shaft power ($\dot{w}$), and secondly, the temperature rise (ΔT).

When we analyzed a length of pipe, we kept in mind that it is a *flowing* process. So instead of a *closed system* containing a set mass within an impermeable boundary, we considered an *open system* with fluid free to enter and leave a control volume fixed in space. In a pipe, fluid was pushing its own way through. Now let's consider our control volume to be a pump or compressor enticing fluid through and pressurizing it, so parcels of fluid depart at a higher pressure.

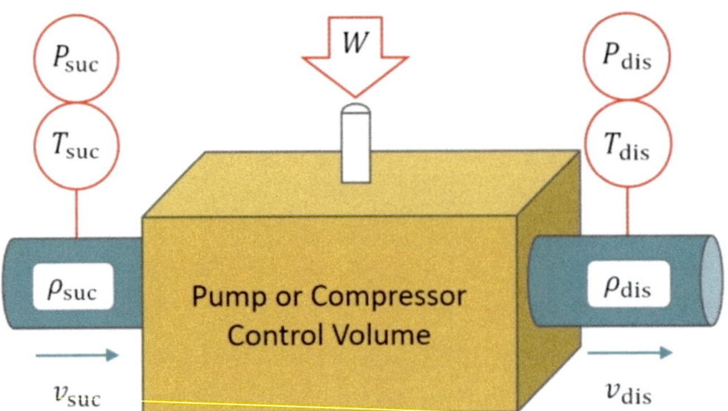

Figure 112 – Parcels of fluid flow into and out of a pump or compressor. The equipment is considered as a *control volume*, an open system bounded by a control surface. Work (W) is imparted via a shaft, to take the fluid from suction to discharge conditions. Heat flow through the walls of the box is negligible, because the fluid is so effective at carrying away excess heat.

We can ignore anything that happens inside a pump or compressor beyond steady state - the equipment's volume is so small, it can't build up or lose any significant amount of energy on the scale of the pipeline as a whole. The angular momentum in the impeller of a spinning compressor is actually significant – simulators attempted to model this in a transient fashion to improve estimates of behavior during start-up and spin-down. However, it is difficult to get an accurate picture of how the impeller couples to the fluid, so Atmos SIM uses a purely steady model of pumps and compressors.

A steady flowing energy balance expresses the First Law of Thermodynamics in a convenient form, which is known by various guises. Some people refer to it by a coinage, calling it the *steady flow energy equation* (SFEE) or similar:

$$q + w = \mathrm{d}u + \mathrm{d}(P \cdot \rho^{-1}) + \mathrm{d}(\rho \cdot g \cdot z) + \mathrm{d}(½ \cdot \rho \cdot v^2)$$

This equation captures the full effect of heat (q) and mechanical work (w) on a flowing control volume's energy content by examining the change in internal energy (u), flow work ($P \cdot \rho^{-1}$), * potential energy, and kinetic energy terms. In a pump or compressor, we can say that elevation and fluid velocity are about the same at suction as they are at discharge, slashing the potential ($\mathrm{d}(\rho \cdot g \cdot z)$) and kinetic energy ($\mathrm{d}(½ \cdot \rho \cdot v^2)$) terms. Pumping or compression happens so quickly that there is no time for significant heat transfer (q) during the short period a fluid spends in the pump or the

* Often called *"PV-work"* ($P \cdot V$) as instead of density people equivalently denote it as the pressure multiplied by the specific volume or, on a mole-basis, by the molar volume.

compressor. Assuming this to be true, we can stop carrying that term around too, leaving us with:

$$w = \mathrm{d}u + P \cdot \mathrm{d}\rho^{-1} + \rho^{-1} \cdot \mathrm{d}P$$

The right hand side is the definition of enthalpy change ($h = u + P \cdot \rho^{-1}$). The work done by any pump or compressor increases the fluid's enthalpy, suction to discharge. Our first task is to calculate the shaft power for a given flow rate of fluid. This depends on the ratio of suction to discharge pressure, which we call the *pressure ratio* or the *compression ratio*.

Power, work and head

Calculating a pump or compressor's shaft power lets us select an appropriate driver or predict what power an existing driver will draw in a given scenario. We do this by first determining the ideal hydraulic power required: what we must impart as a minimum to raise the flowing fluid's pressure. Both shaft power and hydraulic power are legitimately powers; their SI-units are Watts.*

Doing shaft work on a fluid can be stated interchangeably in several ways. For a pump pressurizing an incompressible liquid the imparted head (ΔH) is:

$$\Delta H = \frac{\Delta P}{\rho \cdot g}$$

The height-units head (H, SI-units m) is used for liquids, where it has a physical significance. No matter which liquid is being processed, the same pump can raise a given flow rate of it to the same height. Gases do not have a surface but rather fill their container. So people talk about compressors' mass-specific energy (SI: $\mathrm{J} \cdot \mathrm{kg}^{-1}$), but still call it *"head"* because it is just height-units head multiplied by acceleration due to gravity (g). The above equation, for instance, is for a pump, but we can state it as follows with specific energy units head (which we shall denote $\Delta H_{\mathrm{J/kg}}$ rather than Δe):

$$\Delta H_{\mathrm{J/kg}} = \frac{\Delta P}{\rho}$$

The two definitions of head are equivalent and nominally interchangeable, but in practice stating head in energy-units would confuse a liquid pipeliner, and stating it in height-units would confuse a gas pipeliner. As far as we can tell, this is an unwritten and unspoken convention. We mention it here just as a heads up.

---

* Imperial units mask this; even SI-unit authors say horsepower for shaft power.

Temperature rise

There is inevitably a temperature rise due to pumping or compression. In an incompressible liquid, whose pump is not expected to warm up the liquid at all in ideal conditions, this is entirely due to inefficiency:

$$T_2 = T_1 + \Delta T_{\text{inefficiencies}}$$

In a compressible liquid or in a gas, temperature rises significantly even in a perfectly efficient and reversible (i.e. *isentropic*) process of compression. By definition such a process may not exchange heat with its environment (i.e. it is *adiabatic*). If it seems odd that the act of changing a fluid's density even in a frictionless perfectly efficient manner raises its temperature, see the footnote. *

If a compressor operates perfectly efficiently, the gas follows an *isentropic* path through it and reaches the discharge pressure at the *isentropic discharge temperature* (T_{2s}) that is already very different to the suction temperature (T_1). This is impossible in a real system. In reality, inefficiencies become low-grade heat appearing as an additional temperature rise (ΔT_η) higher than that ideal:

$$T_2 = T_{2s} + \Delta T_\eta$$

The overall temperature rise is neglected in most texts about pumps handling incompressible liquids. In compressors, there is a very substantial temperature rise. In pumps handling compressible liquids it is also important. For one, a rule of thumb states that a propane pump, for example, must not see a temperature rise of 8°C or so. Secondly, temperature affects density, which is used to calculate the power drawn by the pump or compressor, and to calculate how head imparted to the fluid manifests as pressure. Furthermore, calculating discharge temperature accurately in our model is vital to ensure suitable heat exchangers are available if needed, and to simulate how fluid flows downstream of the equipment if there is no temperature meter at discharge.

Temperature obliges us to dig into thermodynamics. This is rather complicated, so we first discuss incompressible liquid pumps and then gas compressors. Compressible liquids have aspects in common with each of the two categories.

---

* This is one major reason aircraft store jet fuel in their wings rather than their fuselage. As the wing moves fast it compresses the air ahead of it, causing what the airline industry calls *total air temperature* (TAT) to be significantly warmer than the ambient *static air temperature* (SAT), and allowing flight at higher altitudes. Even if atmospheric temperature falls below the fuel's freezing point, the air meeting the wing is at a much warmer temperature than it originally was, due to the temperature rise of compression.

Enthalpy and entropy

As we follow a flowing parcel of fluid, it carries two forms of energy along with it: internal energy, plus the pressure-volume work done to it (which some may casually refer to as *"making room for a fluid"*). Note that the second term is not really a conserved energy in the usual sense, despite having specific energy units, but treating it as such is a convenient shorthand that lets us wrap the work done by or on the environment into our equations. The sum of these two quantities is defined as the enthalpy:*

$$h = u + P \cdot \rho^{-1}$$

The calculations make use of changes in enthalpy, so defining a zero point is arbitrary, so long as we are working to a consistent set of *reference conditions*. Accordingly, what matters is the differential enthalpy ($\mathrm{d}h$):

$$\mathrm{d}h = \mathrm{d}u + P \cdot \mathrm{d}\rho^{-1} + \rho^{-1} \cdot \mathrm{d}P$$

As enthalpy is defined completely in terms of state functions, so it is itself a state function. This makes our result useful: We've related path-dependent work done on the fluid (w) to the fluid's path-independent change in enthalpy ($\mathrm{d}h$).

$$w = \mathrm{d}h$$

This all seems esoteric so far. How do we make any use of this information? Well: although we can measure neither internal energy nor enthalpy directly, we are able to calculate them from measurable quantities via *thermodynamic relations*, equations known to always hold true for any process. A change in internal energy ($\mathrm{d}u$) is a function of a change in entropy (s) and density (ρ) as follows:

$$\mathrm{d}u = T \cdot \mathrm{d}s - P \cdot \mathrm{d}\rho^{-1}$$

Entropy in this context is a measure of energy degradation. An increase of entropy indicates that energy has been converted from valuable, concentrated forms – say, a canister of compressed air – to worthless, dissipated forms – like lukewarmth or vibration. For our purposes, we need not internalize how physically meaningful entropy can be. It's an intermediate variable; to us, what matters is that it remains constant throughout any reversible adiabatic process.

---

* We could choose any thermodynamic state function and arrive at the same answers. Every state function fully expresses the state in terms of two valid variables. We can, say, given pressure and temperature, find everything else on that basis.

Back-substituting our thermodynamic relation for internal energy change ($\mathrm{d}u$), a term cancels, so the enthalpy (h) is a function of two variables: entropy (s) and pressure (P):

$$\mathrm{d}h = T \cdot \mathrm{d}s + \rho^{-1} \cdot \mathrm{d}P$$

This has an extraordinary physical subtlety; during the process, all useful work raises pressure (P), and all wasted work goes into entropy (s). The same cannot be said for temperature, density or any other combination of variables.

Efficiency

Entropy can be regarded as a sort of measure of uselessness. The Second Law of Thermodynamics says it always inevitably increases whenever something happens: even if entropy decreases locally within a system, it can only do so if entropy has increased by a greater amount elsewhere in the universe. This implies the heat death of the universe, and got some philosophers quite upset.

The ideal head rise (ΔH in length units or in specific energy units) exactly corresponds with the enthalpy added to a fluid (Δh) by *"perfectly efficient"* equipment ($\eta = 100\%$). The external power draw ($\dot{w}$) of any real equipment is higher than this ideal hydraulic power because of inefficiency. * A real pump or compressor's *efficiency* (η) is the ratio of the useful ideal enthalpy rise ($h_{2s} - h_1$) to the achieved enthalpy rise ($h_2 - h_1$):

$$\eta_s = \frac{h_{2s} - h_1}{h_2 - h_1}$$

Where the subscript (s) refers to a state at the same discharge pressure reached by a perfectly efficient process that wastes no work. This is referred to an *isentropic* process throughout which the entropy (s) stays constant. This way of defining efficiency is used by some compressor manufacturers but it is inconvenient because even if two series compressors have the same isentropic efficiency, the isentropic efficiency of the whole process will be different. To get around this most compressor maps instead give a *polytropic efficiency* (η_p) defined as the ratio of useful enthalpy change to total enthalpy change that would need to be applied at each tiny step across the pressure-entropy map to get from suction conditions to discharge conditions. It can be seen as equivalent to the isentropic efficiency of an infinite series of infinitesimal identical compressors.

$$\eta_p = \frac{\rho^{-1} \cdot \mathrm{d}P}{\rho^{-1} \cdot \mathrm{d}P + T \cdot \mathrm{d}s}$$

---

* As external work done is transmitted from a driver to the pump or compressor via a shaft, it is referred to as *shaft work*, and its rate (external power) as the *shaft power*.

Each of these two definitions of efficiency has a notional head associated with it, the isentropic head (also called the adiabatic head) and the polytropic head. The shaft work done per unit mass by a compressor equals the adiabatic head divided by the adiabatic efficiency, and also equals the polytropic head divided by the polytropic efficiency. While shaft work is a real physical quantity, these two types of efficiency and head are essentially useful intermediate steps in our calculations.

Analyzing a pump or compressor's performance relies on finding the ideal enthalpy change, then correcting for the efficiency to deduce the real enthalpy change, and thus the conditions. For many fluids, no enthalpy tables are available. We rely on what can be referred to as *performance calculations*. Those are based on values retrieved from *performance maps*: the way that a manufacturer tells us how their equipment performs.

7.4 – Types of pump and compressor

There are a host of ingenious designs for pumping liquids and compressing gases. Screw pumps are used for very viscous liquids, for example. But they are one of a handful of exceptions to the rule. Almost all pipeline pumps and compressors work by one of two fundamental principles: either *centrifugal* or *reciprocating*. Once datasheets are available, a thorough characterization of the equipment's performance is possible. We can then replace the black-box ideal equipment, and model the characteristic behavior of each unique centrifugal pump, reciprocating pump, centrifugal compressor or reciprocating compressor.

Reciprocating pumps and compressors are *direct-acting*. They work on the principle of *positive displacement* (PD), directly moving a piston to push on the fluid. This means they impart high pressure ratios per stage to a nearly constant actual flow rate at a given speed. A reciprocating pump or compressor can only process a relatively low flow rate, so people often arrange them in parallel within stations to achieve a combined pump or compressor capacity capable of handling a pipeline's total throughput.

Most widely deployed pumps and compressors on large pipelines today are *centrifugal*. These work by a dynamic principle – they don't push a fluid directly, but rather coerce it to flow from a smaller impeller at the suction side to a larger volute. This increase in cross-sectional area slows down the velocity raising the pressure when it's flung out of the top at the discharge side. A variable-speed centrifugal pump or compressor can impart a limited range of head to a wide range of flow rates. They are therefore used quite often in both series and parallel combinations within a station, to impart a greater combined head to a given flow rate and to accommodate larger flow rates respectively.

The *head-flow curve* is different at each shaft rotational speed because the equipment achieves higher pressures at lower flow rates in a manner unique to each speed. And there is an *efficiency-flow curve* at each speed, wasting more power when operating at certain combinations of pressures and flow rates. [*] These curves and their limits, along with a couple of others we shall discuss in detail later, are collectively referred to as a map, a performance curve, or a rating curve.

It is sensible to design a pipeline with a degree of resilience, remaining capable of delivering some specified flow rate even if a given pump or compressor, or station, is unavailable. This makes a graph of head versus flow especially useful if one wishes to consider possible combinations of available pumps or compressors. Just as a pipeline's system curve has an operating envelope, so the pump or compressor curves might be available in a number of distinct configurations.

All the intricacies of the centrifugal pump or compressor geometry are captured in the manufacturer's performance map. For each speed at which a wheel can spin, a centrifugal pump or compressor imparts *head* (ΔH). This is referred to in length units for a liquid pump, or in specific energy units for a gas compressor. The head it imparts depends on the actual volumetric flow rate at suction conditions (Q_{suc}). We refer to this relationship as the *head-flow curve*. Efficiency (η) is also stated in some appropriate form, relating the delivered hydraulic power to the requisite shaft power a driver inputs.

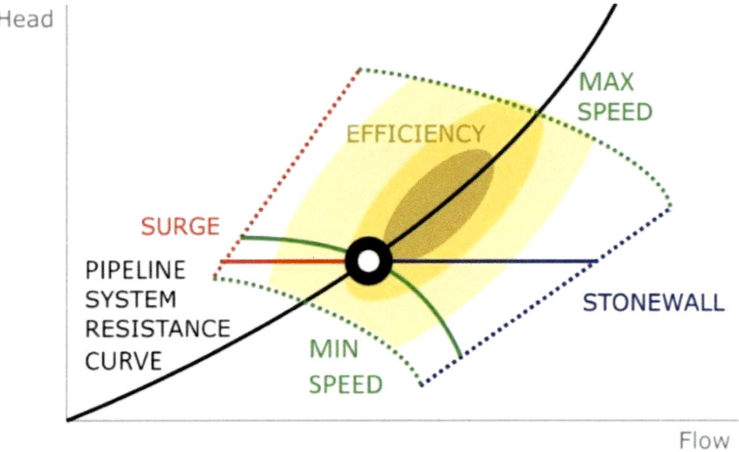

Figure 113 – A schematic of how a centrifugal pump or compressor map relates flow rate to head. The pipeline system curve (black) is superimposed onto this map. The operating point is where this system curve intersects with the centrifugal equipment's operating speed (green).

---

[*] Some refer to the head-flow and efficiency-flow curves the *head characteristic* and the *efficiency characteristic*, and collectively as the *characteristic curves*.

A centrifugal pump or compressor map can come in various guises, and this is worth pausing to appreciate. The only general map relates *head* (ΔH) to *actual volumetric flow rate* at suction conditions (Q_{suc}). Many maps state the pressure ratio ($P_{\text{dis}}/P_{\text{suc}}$) instead of the head, or they state standard flow ($Q_{\text{suc}}^{\text{std}}$) or mass flow ($Q_{\text{suc}}^{\text{mass}}$) instead of actual flow at suction (Q_{suc}). The flow rate a map presents is hidden under the guise of an innocuous term like *"capacity"*, which reveals itself once we look at its units. A non-general map applies only to some implied set of reference conditions and some implied reference fluid. To use it for anything else, it needs correcting. Correcting a non-general map to see how it will work in our own application can be a surprisingly finicky task.

If we're given a map that isn't in units of head vs. actual volumetric flow rate at suction conditions, then we must convert it into those units. A manufacturer doesn't provide this *"generalized form"* of a map not because they're unhelpful, but because of the practicalities of the rig they use to test their equipment when characterizing its performance. If it is a pump, they probably used water, and measured the pressure at discharge. We convert pressure ratio into head by correcting for water's density to make the map generally usable for other fluids.

Another point to note here is that because for an almost-incompressible liquid the actual volumetric flow rate (Q) is very close to standard flow (Q^{std}) – within a percent or two – it's tempting to take them as equivalent. But this is frankly an approximation we don't need to make. A pipeline simulator strives to be more accurate than to dismissively wave away a 1% error as an approximation! And if a centrifugal pump is operating near the rightwards end of its map, a 1% flow error can manifest in our results as a much larger error in the calculated head.

Finally, centrifugal compressor maps may give the isentropic head and isentropic efficiency (often called the adiabatic head and adiabatic efficiency), or alternatively the polytropic head and polytropic efficiency, as we discussed in the previous section. For centrifugal pumps pressurizing nearly-incompressible liquids there is no meaningful difference between the two forms of efficiency.

7.5 – AFFINITY LAWS

There are usually *"bends"* in a centrifugal pump or compressor's head-flow curve that are challengingly non-linear. One popular way of representing these curves analytically is to assume the equipment obeys a set of *affinity laws*, which are usually referred to as the *fan laws* (in the case of a compressor), or the *pump laws* (in the case of a pump). These are approximations that apply to a centrifugal pump or compressor operating near its design speed at low pressure ratios. The idea is to make all speeds' head-flow curves equivalent by scaling the flow rate by speed, and the head by speed-squared:

- Volumetric flow rate varies directly as speed $Q \propto \omega$
- Head imparted varies as speed-squared $\Delta H \propto \omega^2$
- Power drawn varies as speed-cubed $\dot{w} \propto \omega^3$

These affinity laws are widely used to predict a new operating condition based on an old operating condition. But they cannot tell us what the new efficiency will be.

It is worth mentioning, here, that a pump handling a high-viscosity liquid might have a special correction applied to its map to account for that.

7.6 – PUMPS IN LIQUID PIPELINES

Pump performance

Pumping an incompressible liquid

The density of an almost incompressible liquid only varies slightly. The density at a set of pressure (P) and temperature (T) operating conditions is calculated from standard density (ρ_{std}) at standard conditions (P_{std}, T_{std}) via the bulk modulus equation of state:

$$\rho = \rho_{\text{std}} \cdot \exp\left(-\alpha \cdot (T - T_{\text{std}}) + \frac{P - P_{\text{std}}}{B_T}\right)$$

There are two parameters in the exponential. The dependence on temperature is governed by the coefficient of thermal expansion (α), and the dependence on pressure is governed by the isothermal bulk modulus (B_T).

We said that the work done (w) on a fluid equals its change in enthalpy (h):

$$w = \Delta h$$

The minimum work (w_{min}) required to pressurize a fluid is the work done in an ideal reversible ($\eta = 100\%$) adiabatic ($q = 0$) process, a combination equivalent to saying the process is *isentropic* ($\text{d}s = 0$). Why is this? Consider the definition of enthalpy:

$$h = u + P \cdot \rho^{-1}$$

A change in this becomes:

$$\text{d}h = \text{d}u + P \cdot \text{d}\rho^{-1} + \rho^{-1} \cdot \text{d}P$$

The first term the change in the fluid's internal energy, which if we consider the *closed system* of following a clump of fluid around can be substituted out of our way by the thermodynamic relation:

$$\text{d}u = T \cdot \text{d}s - P \cdot \text{d}\rho^{-1}$$

The term involving change in density ($\text{d}\rho^{-1}$) cancels, leaving us with:

$$\text{d}h = T \cdot \text{d}s + \rho^{-1} \cdot \text{d}P$$

Assuming the process is isentropic ($\text{d}s = 0$) means only one term remains. The head imparted (ΔH) in an isentropic pump to take an incompressible liquid from suction to discharge pressure is therefore:

$$\Delta H = \frac{P_{\text{dis}} - P_{\text{suc}}}{\rho}$$

This corresponds to a lower discharge enthalpy (h_{dis}) than we would expect to see when using a real – i.e. inefficient – pump. To *"correct"* the isothermal shaft power, an efficiency is then applied from the map. The temperature rise of incompressible liquid through a pump is entirely due to inefficiency ($\eta < 100\%$):

$$\Delta T = \frac{1}{c_V} \cdot \frac{P_{\text{dis}} - P_{\text{suc}}}{\rho_{\text{ref}}} \cdot \left(\frac{1}{\eta} - 1\right)$$

Assuming a pump is isothermal is saying that this temperature rise is zero:

$$\Delta T = 0 \text{ i.e. } T_{\text{suc}} = T_{\text{dis}}$$

The incompressible heating equation here doesn't come from thermodynamics but rather from inefficiency. The power lost to inefficiency in a pump turns into heat, and that heat has to go *somewhere*. Of all the places that heat can go, *"being carried away with the liquid"* is an immensely more effective channel of heat transfer than any of the other possibilities. So to a good approximation, we can say that the heat due to inefficiency is all carried away with the liquid.

Pumping a compressible liquid

For very compressible liquids, such as liquefied petroleum gas (LPG), we use special equations of state to calculate the change in density (ρ) with pressure and temperature. The head imparted by a pump in this case is calculated much in the same way as for a compressor.

Even in a perfectly efficient pump (one that is reversible and adiabatic, i.e. isentropic), a compressible liquid has an isentropic temperature rise, such that the isentropic discharge temperature is higher than the suction temperature:

$$T_{\text{dis}} > T_{\text{suc}}$$

This is because the fluid's increase in density is always associated with a rise in its temperature. This isentropic path can be calculated using a fluid properties calculator such as those provided by research institutes in the United States of America (NIST) or in Europe (GERG). Then we can add the heat generated by inefficiency to that value, in exactly the same way as we did with a pump moving incompressible liquid.

Atmos SIM calculates the isentropic power including the effect of temperature rise in a compressible liquid.

A real pressurization process is not isentropic, especially for a compressible fluid. This is widely appreciated for compressors. Compressor manufacturers don't provide an isentropic efficiency (also called adiabatic efficiency) on their maps, but rather a *"polytropic efficiency"*. We shall discuss the rationale for this when we come to talk about compressors – both specify the same behavior. Because pumps handling compressible liquids are not isentropic, we approach them in a similar manner in Atmos SIM, calculating the power and discharge temperature via integration in the same manner as we do for a gas compressor.

Pump and compressor control strategies

Let's briefly clarify a few terms for the lines surrounding a pump or compressor.

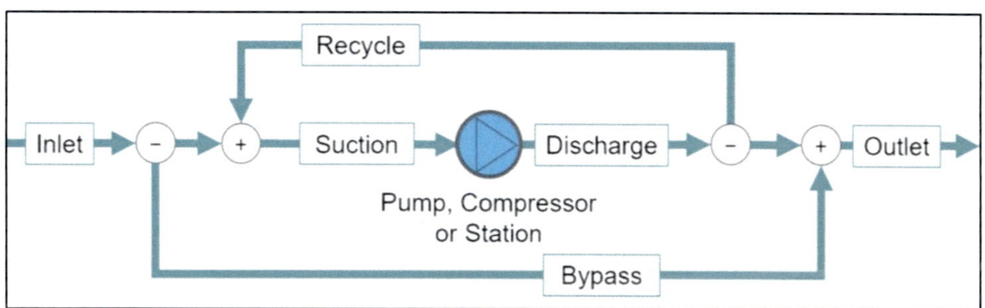

Figure 114 – The inlet, bypass, suction, discharge, recycle and outlet lines around a pump, compressor or station.

There are only a handful of dials at our disposal that we can directly adjust to get the unit to do what we want. We may adjust the speed at which the equipment is operating: *"speed control"*. Or we could adjust the position of a regulator / control valve. By carefully feeding it signals from meters via a controller, the position of a regulator can be adjusted to maintain a pressure, a flow rate of any variety, or a temperature, at that point or elsewhere in the system. The variable to be controlled, and the feedforward or feedback logic used to control it, is a big topic we can only touch on, which we do later. But whatever we are *trying* to control, we end up actually adjusting the speed of the pump or the compressor, or the position of a regulator / control valve.*

* In practice, a pressure regulator might differ to a flow control valve in internal construction, but both are abstracted as the same model item in pipeline simulators, so in our industry we say regulator or control valve interchangeably.

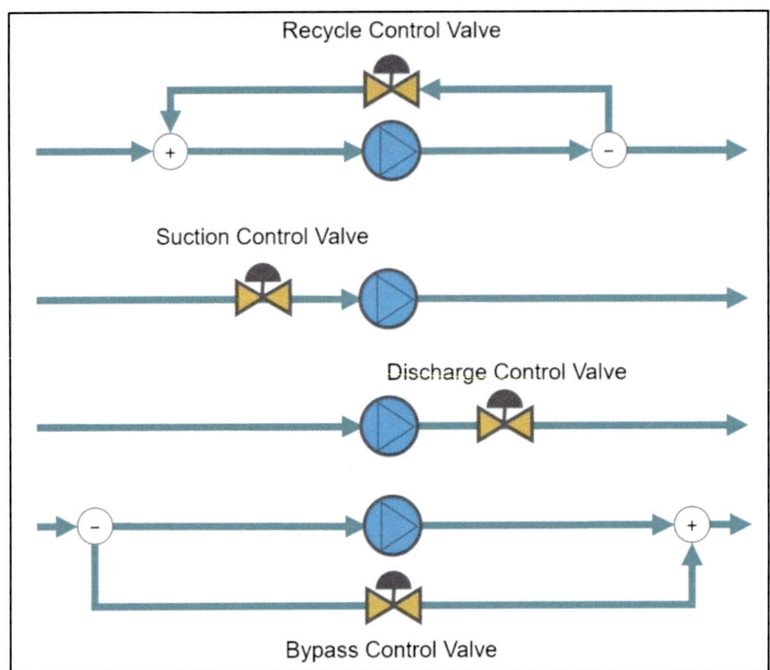

Figure 115 – Recycle control, suction control, discharge control and bypass control strategies around a pump, compressor or station. Not all of these necessarily make sense in all cases.

A control valve could be placed on the discharge line (*"discharge throttling"*), the suction line (*"suction throttling"*), or the recycle line (*"recycle control"*). Which of these approaches makes sense or if will work at all in a pipeline depends on the type of equipment: a reciprocating pump, a centrifugal pump, a reciprocating compressor or a centrifugal compressor. Each has its own peculiarities. Moreover, a pump or compressor often isn't acting alone but in tandem with others in a station, adding further possibilities like controlling a combined recycle for several pumps together (or several compressors together).

Reciprocating pumps

There are many designs of positive displacement (PD) pumps. Three of those are reciprocating pumps: piston pumps, plunger pumps and diaphragm pumps. The essential actions of the liquid-transferring parts of all three of these are the same. [41] Other positive displacement pumps exist, falling under other subcategories. [*] It is conventional, at this juncture, to present diagrams of them, but that doesn't actually concern process-level modelling: we leave it to the remit of mechanical engineers.

The flow from a simple reciprocating pump is uneven over time. This is bad in many applications. Pulsation through meters affects measurement accuracy. Vibration through pipes and equipment could damage them by fatigue failure. To smooth out a reciprocating pump's flow rate, several measures are taken. One of these is using a double-acting design rather than a single-acting one, so that a suction stroke is simultaneously a discharge stroke at the other end. Another is using more than one cylinder (simplex), typically two (duplex) or three cylinders (triplex). [57] A pipeline simulator assumes this pulsation to be negligible, and vibration to be out of scope. Accounting for a reciprocating pump being single- or double-acting and simplex, duplex or triplex, is easy arithmetic. The performance calculations and control strategies are the same regardless.

A reciprocating pump efficiently imparts a wide range of heads to a narrow range of flow rates. It is usually deployed for achieving high pressure at low or moderate flow rates. It almost enforces a flow setpoint: if we shut a valve on its discharge line (a bad idea), it continues trying to push the same flow rate of liquid through anyway, raising the pressure until something gives way and bursts open (*"preferably a relief valve"*!). [43]

In a pure flow setpoint, the flow rate is determined only by the pump's speed. But in reality there are two further affects at play in a reciprocating pump. Firstly, some liquid slips back around the piston – and this is by design, as it keeps the piston lubricated. Carefully checking references for reciprocating pumps tells us they can be assumed to have around 3% slip due to leakage through valves, and that this is independent of flow rate and head, so it shifts the entire map slightly to the left. Secondly, reciprocating pumps move a little less flow at the highest developed heads, because of the slight compressibility of the liquid. [42] A third effect, giving the line a slope, only happens in

---

[*] Rotary lobe pumps, for instance, are occasionally seen in some applications. We mention this because reciprocating pumps are strictly a subset of positive displacement pumps, but because they are far more common than other types (except centrifugal) many people say *"PD pump"* without specifying if they mean reciprocating or rotary.

rotary pumps, and is described as a slight counter-clockwise rotation in the literature, but it does not concern reciprocating pumps. [43] A reciprocating pump's head-flow curve [*] is near-vertical at each reciprocating pump speed. The slight reduction in flow at higher imparted head is marked in red on the diagram, as a leftward bend at the top.

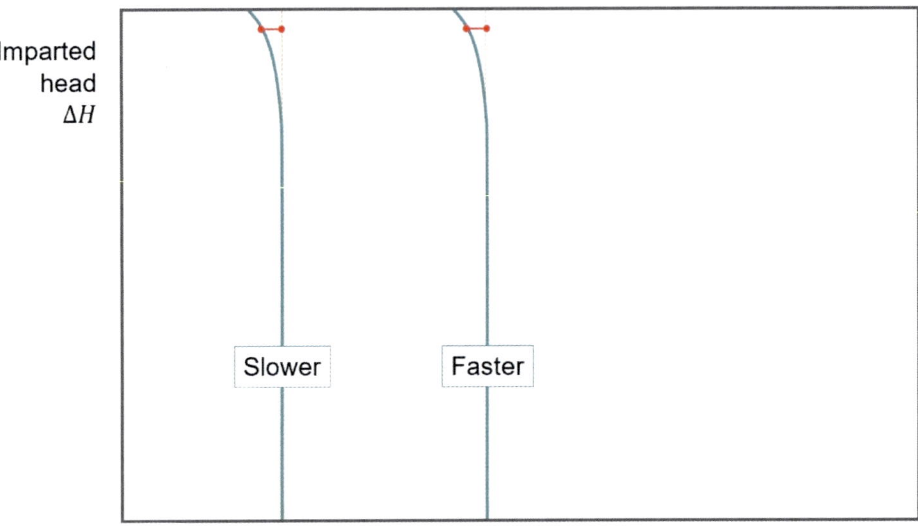

Figure 116 – The head-flow curve for a reciprocating pump at a given speed is almost vertical except for a slight flow reduction at higher imparted heads caused by the compression effect. The effect of leakage (called *slip* or *slippage*) is not visible as it moves the entire line leftward.

The first popular control strategy for reciprocating pumps is adjusting its speed. If it reciprocates slower, it will thereby deliver less flow rate. How else can we control flow?

If we place a control valve on its discharge line, this rotates the system curve counter-clockwise. Because the head-flow curve is vertical, the reciprocating pump delivers whatever power is needed to push the same flow rate through anyway, generating more pressure for that control valve to burn off. Controlling a reciprocating pump flow rate by using a discharge control valve will not work. It only worsens the wear rate and shortens the life of the pump equipment. [43]

Suction throttling has the same effect as discharge throttling, namely rotating the pipeline system resistance curve counter-clockwise, so that doesn't work either. It just places an unnecessary restriction on the suction side, particularly undesirable given a

---

[*] The intuitive choice engineers are familiar with is head on the vertical axis and flow on the horizontal. Inexplicably, reciprocating pump maps are not usually drawn that way around. Nonetheless, we choose the former schema (and are not the only authors to raise an eyebrow at the latter convention).

reciprocating pump typically presents a more stringent net positive suction head required (NPSHr) than does a typical centrifugal pump.

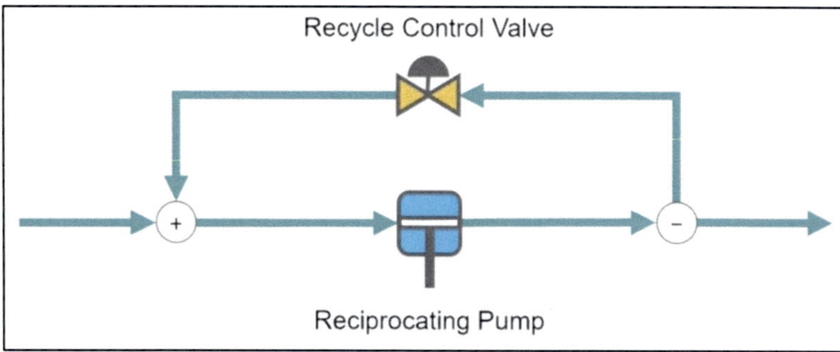

Figure 117 – A reciprocating pump with a recycle control valve.

This leaves the only viable control strategy for a reciprocating pump – aside from speed control – as recycle control. Placing a control valve on the recycle line leaves flow rate through the pump unchanged, but the pressure imparted to the process is lowered. The flow rate delivered inlet-to-outlet is lower than the flow rate through the pump by the amount flowing through the recycle line.

This recycle control scheme can be equivalently viewed from two perspectives. We could say overall effect of the pump-and-recycle system has changed (akin to a change in the pump head-flow curve). Or we could say that varying the recycle flow affects the pipeline system resistance curve as seen by the pump. For this reason, on the head – flow diagram, some people prefer to draw the pump curve differently, while others prefer to draw the system curve differently.

Centrifugal Pumps

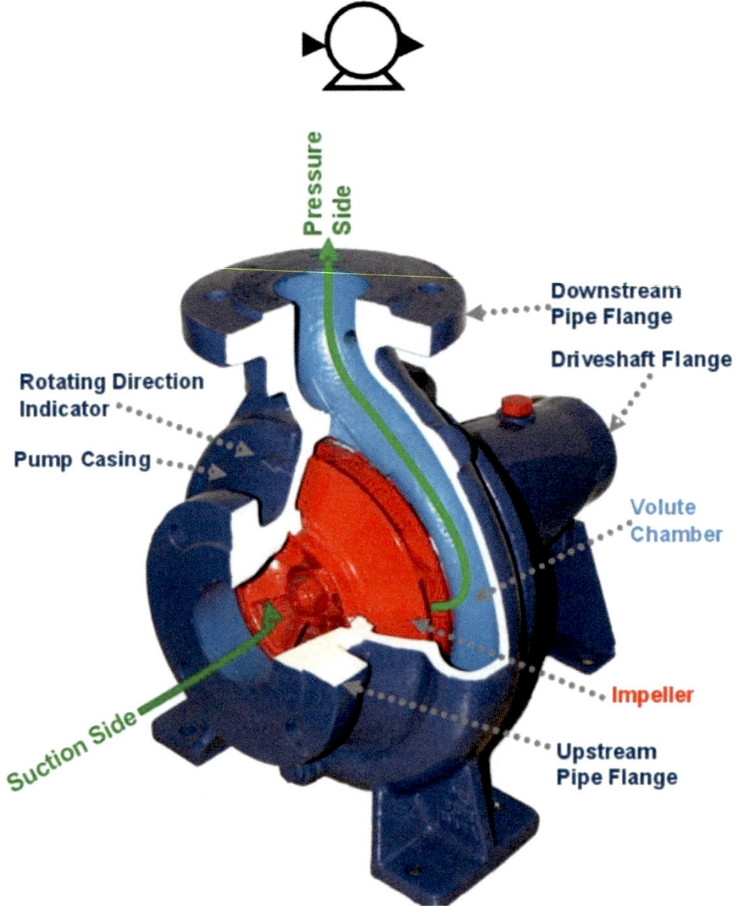

Figure 118 – The distinctive spiral form of a centrifugal pump or compressor is called a *volute*.[*] We abstract overall behavior from the suction side to the discharge side, without having to consider the inner workings of the equipment or its blade geometry, etc.

Maps, operating point and limits

The performance map of a centrifugal pump is general if it relates suction-side *actual volumetric flow rate* (Q_{suc} in SI units of m^3/s) to several parameters at a given *rotational speed* (ω in RPM). [†] These always include the imparted head (ΔH in SI units of m) and

[*] Those familiar with other languages may know the volute as a *snail-like spiral*.

[†] We say *"suction-side"* because while the mass flow going into a pump is the same as the mass flow coming out of it, this is not true for the actual volumetric flow. A real liquid

pump efficiency (η in %) at each volumetric flow rate. Instead of drawing an efficiency curve per speed, we could draw efficiency contours showing *efficiency regions* superimposed on the head-flow curves.

For a pump, another parameter is essential, so it's also stated on a map: the net positive suction head required (NPSHr in SI-units of m) at the impeller.

A map might further helpfully state shaft power (in SI-units of kW). This is intended to aid with selecting a suitably rated driver, and it is stated for the fluid with which the equipment was tested. Pumps are usually tested with water, so if we are pumping a different liquid, the pump draws a different amount of power. This serves as a check for our calculations. It also illustrates how at an excessively low flow rate, there is a region of the map that is so inefficient that power draw *increases* despite the power delivered to the fluid decreasing. And how beyond a highest flow rate the imparted head drops off causing the power draw to drop off too.

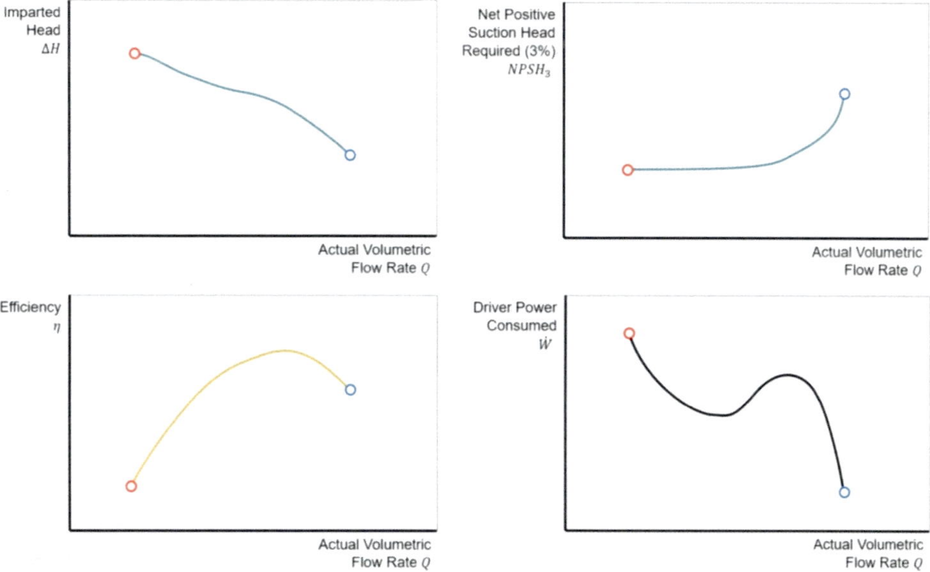

Figure 119 – A pump performance map is presented with at least three sets of data, often four. Usually they are superimposed on the same chart, here they are split out for clarity. One is the imparted head for each speed (head-flow curves). Another is net positive suction head required at the impeller (NPSHr). A third is the efficiency for each speed (alternately presented as efficiency contours). And finally, some pump maps also present the power drawn at a flow rate.

---

is slightly compressed by the higher pressure, which means discharge-side actual volumetric flow is slightly less than at the suction-side.

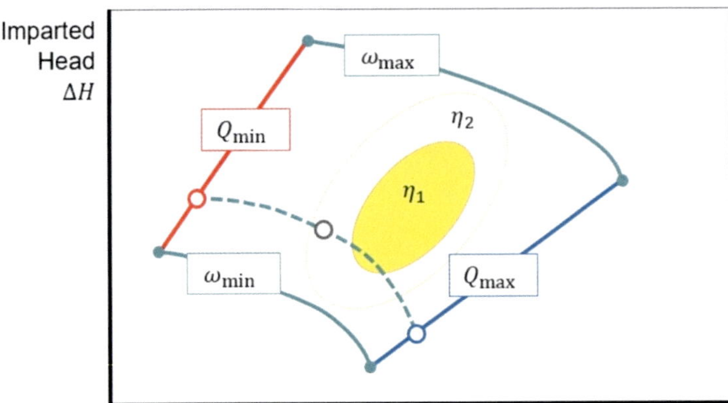

Figure 120 – An operating point in the second efficiency contour on a centrifugal pump map.

An operating point displays the conditions at which a pump or compressor operates on its performance map. It might present the efficiency as a series of contours.

A centrifugal pump's valid operating region is a contained region on its map. At the edge of its map, a pump approaches these operating limits:

1. Pump maximum speed ($\omega_{\max}$) when a driver can't ramp up its power further
2. Pump minimum actual volumetric flow rate ($Q_{\min}$), either because of:
    a. *Cavitation* when too much net positive suction head is required
    b. *Temperature rise* when the pump is operating too inefficiently
3. Pump minimum speed ($\omega_{\min}$)
4. Pump cavitation i.e. a maximum actual volumetric flow rate ($Q_{\max}$)

Atmos SIM can tell if a pump stays within the bounds of its map as composition and operating conditions change.

Minimum suction pressure and pump cavitation

We mentioned earlier that a pump manufacturer stipulates that a net positive suction head required (NPSHr) must be satisfied to avoid damaging cavitation. In basic analysis, many pipeliners simply impose a conservative suction pressure limit for pipeline simulation. In reality, this is a minimum *head* limit, one that varies according to the manufacturer's map. It is preferable to take a more sophisticated approach, comparing the suction-side pipeline system and the fluid parameters at the operating conditions against the NPSHr curve. Perhaps many pipeliners don't take the time to do this because typically heads in a pipeline are so large that NPSH is not the constraining factor in operation.

Minimum flow and pump surge control

Some centrifugal pumps can go to zero flow. At this operating condition, a pump isn't imparting enough energy to the liquid to overcome the pipeline system's resistance, perhaps due to a lack of power from the driver. But it is still imparting head to the fluid – known as *shut-off head*. Where does this energy go? It makes the pump become progressively hotter over time, as it is wasted entirely as heat.

> Atmos SIM can simulate pump shut-off head, but because it doesn't simulate heat loss from the pump itself to the air in the enclosure, it doesn't tell us how fast a pump overheats over time.

In contrast, the vast majority of main pumps used on pipelines cannot be operated below a certain flow rate. The actual volumetric flow rate at the suction side (Q_{suc}) must exceed the lowest allowable actual flow (Q_{min}) for the present rotational speed (ω) it is operating at ($Q_{\mathrm{suc}} > Q_{\mathrm{min}}$ at ω). Below this limit, the pump is prone to a condition known as *pump surge.* Flow oscillates violently between the forward and backward directions with dire consequences on the pump itself – severe vibration, temperature spikes, and rapid changes in axial thrust. Pipeliners must avoid it or else risk damage to equipment.

> Atmos SIM shows us that even if a pump can tolerate surges, the wider process might be susceptible to catastrophic failure.

Even if a robust pump and its valves, meters and fittings withstand occasional, brief surging, the wider process it is used within (a pipeline) may be susceptible to failure if surging arises. Avoiding pump surges is a prime concern for designers and operators.

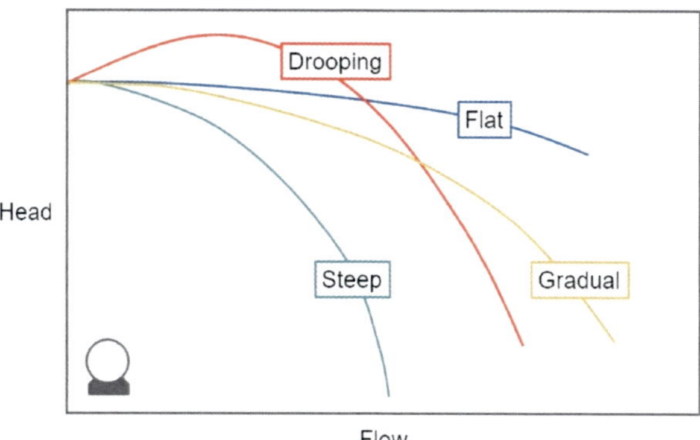

Figure 121 – A centrifugal pump's head-flow curves can be classified as flat, gradual or steep. Some centrifugal pumps have head-flow curves which exhibit *drooping* at low flow rates.

Many real pumps' head-flow curves have an upward-sloping portion at low flow rates, known as *drooping*. This means that for a given head, there are two different flow rates supported by the pump, one on the left side of the peak head and another on its right side. If flow rate becomes low enough to enter this two-solution region, the pump might oscillate between the two different flow rates. This is a physical phenomenon: a flow instability as flow approaches its minimum, which for some pumps is a minimum flow rate as per the pump surge limit, and for others that limit is at zero-flow i.e. *"shut-off"*.

Measures are taken to avoid pump surge in normal operation. One conventional approach is pumping to excess discharge pressure, then throttling across a control valve to reduce it down again. This works because operating at higher head at the same flow pushes the pump to higher speed, which moves the operating point to the left on the head-flow curve at that speed. And that means it's closer to the maximum head, which the pump curve can only produce at one flow, so there is no surging.

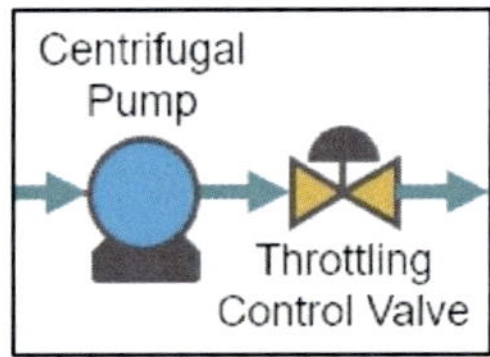

Figure 122 – Pumping to excess discharge pressure then throttling across a control valve.

Alternatively, a recycle path increases the flow rate through the pump, sending a proportion of discharge flow back to the suction side. This actually works in exactly the opposite manner of the previous fix; it pushes the operating point to the *right* on the pump map, and moves it out of the two-solution region. The recycle valve can be controlled to adjust its position based on either the flow rate or the pressure at the discharge side of the pump.

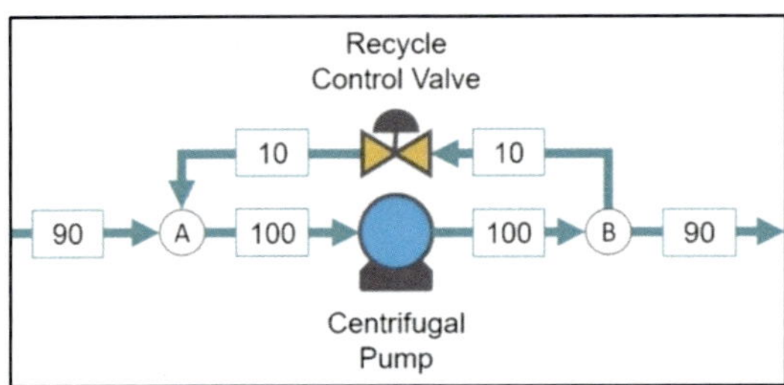

Figure 123 – Below some minimum flow limit, a possible control strategy is to send a portion of the discharge flow back to suction via a recycle line. [44]

To avoid wasted energy, one might consider adopting an alternative strategy, in tandem or in lieu. A device adjusting the driver's speed to meet the required discharge pressure could be used. Or pump sparing could be used, placing multiple pumps in parallel chosen to meet all throughput flows. The trade-off is that both of these options involve capital investment, and introduce complexity.

If a surge condition arises in exceptional circumstances, measures are taken for surge relief, such as a relief valve or a surge tank. Reverse flow is particularly dangerous at a pump's discharge side, as it could cause the rotor to spin backwards. A fast-closing and tight-sealing shut-off valve is typically installed to prevent this from happening.

Atmos SIM can implement surge relief strategies via control logic, to protect a pump.

Inlet Standard Volume Flow	0.02143	Sm3/s
Recycled Standard Flow	0	Sm3/s
Total Standard Suction Flow	0.02143	Sm3/s
Outlet Standard Volume Flow	0.02143	Sm3/s
Surge Fraction	∞	Percent
Inlet Actual Volume Flow	0.02151	m3/s
Surge Actual Flow At Current Speed	0	m3/s
Total Actual Suction Flow	0.02151	m3/s
Outlet Actual Volume Flow	0.02151	m3/s

Figure 124 – Atmos SIM shows how some pumps can run down to zero flow without surging.

Atmos SIM does not let a pump operate in a surge condition. If a pump map does have the positive slope region, it automatically hides that and assume an almost-flat negative slope there. But it is aware this is a surge region and takes appropriate corrective action or issues a warning if a pump operates there. *

---

* An Atmos SIM optimizer, which we discuss in a later chapter, always avoids operating any pump in its surge region.

Maximum flow and pump cavitation

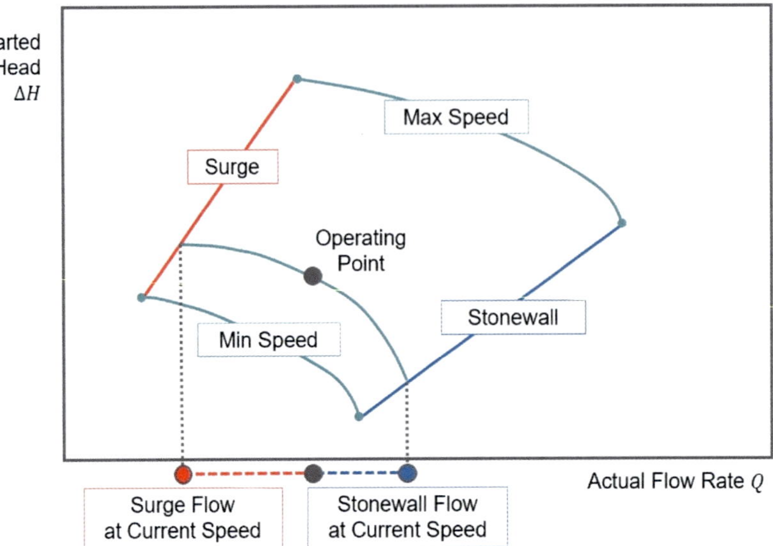

Figure 125 – Surge flow and cavitation flow for a given centrifugal pump's operating speed. The surge margin and cavitation margin tell us how far these are from the current operating point.

When a pump is operating past the maximum flow rate specified on its map, it is in *cavitation*. This is very bad for a pump. Atmos SIM doesn't try to simulate this cavitation, it just treats a pump's head-flow curve at a given rotational speed as dropping off sharply to zero head at flow rates beyond the rightmost edge of the pump map. One might expect we could just extend the pump curve out as a straight line past the manufacturer curve, but actually we can't extrapolate like that. Head drops off to zero very quickly, immediately to the right of the end of the manufacturer-supplied curve.

Atmos SIM assumes a linear drop to zero head in the region from 100% to 105% of the maximum flow rate on the map. In this operating condition, pressure ratio falls right down to become 1:1 – i.e. no pressure is added by the pump from suction to discharge.

Pumping stations for liquid pipelines

Pumps are selected and arranged very carefully, to ensure that by working together in various configurations they're suitable for a pipeline's system curve.

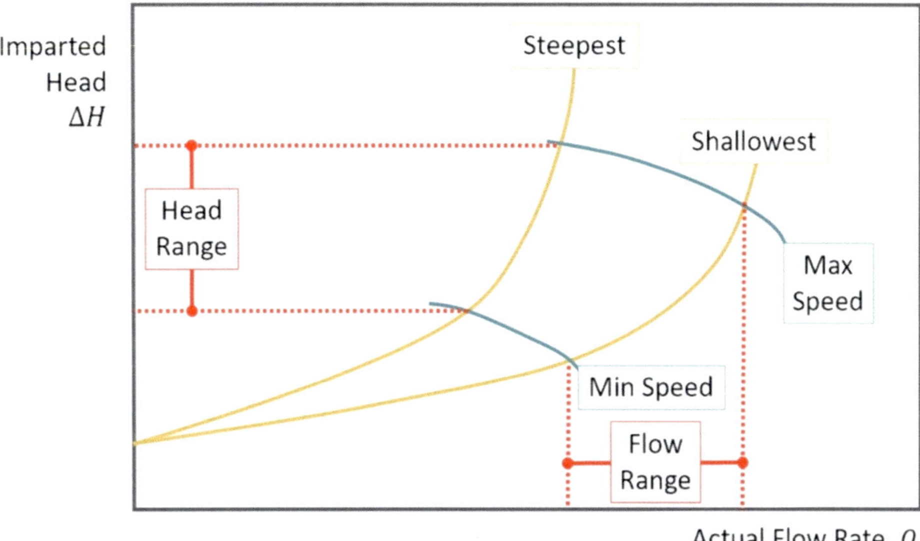

Figure 126 – A system curve (yellow) superimposed on a centrifugal pump map (teal) shows the operating ranges of imparted head and actual flow rate through the pipeline (red).

A system with substantial frictional losses compared to the static lift has a steep system curve. Placing pumps in series makes a station well-suited to these high-friction loss pipelines. Though it is unusual (and surprising), it may be that pumps in series share a shaft. Those can be coupled in the model using logic blocks linked to control them so that they always operate at the same speed.

Conversely, in a low-friction system, adding a second pump in series would push both pumps *away* from their best efficiency point (BEP). In such systems, arranging pumps in parallel works better. A *bank* models parallel pumps operating together.

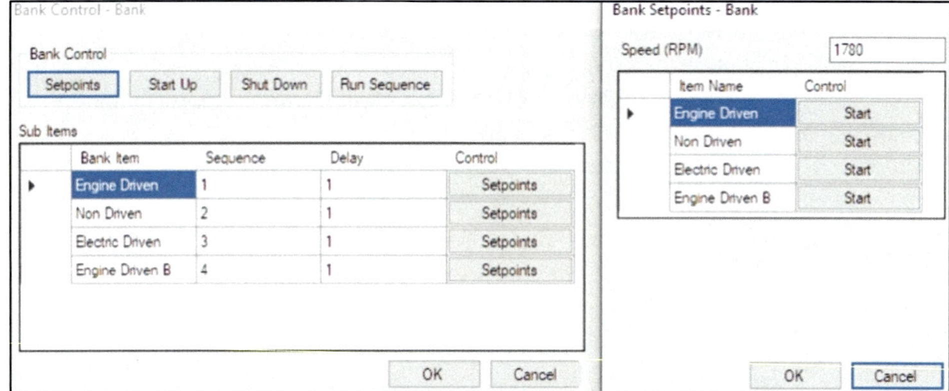

Figure 127 – Pump bank user interfaces in Atmos SIM.

In a station with multiple pumps operating in tandem – in parallel, or in series, or both – a single combined recycle line may be provided to serve all the units.

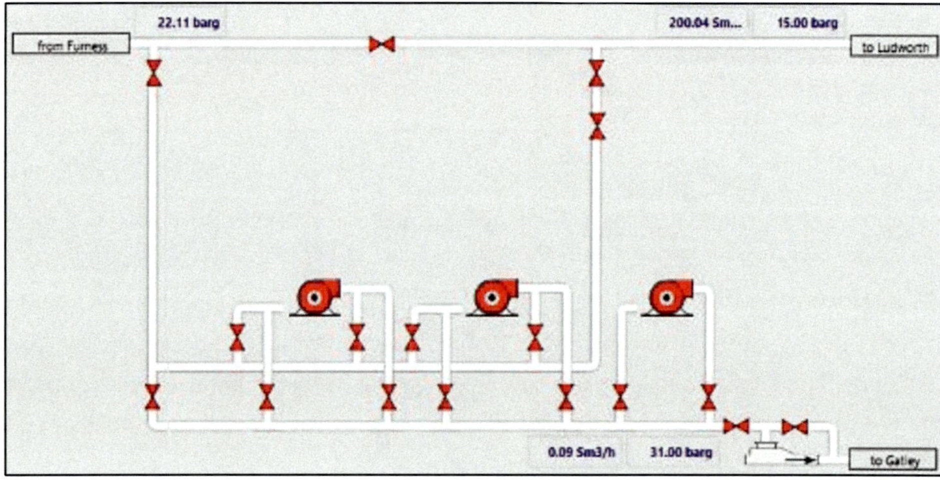

Figure 128 – A pumping station as shown on the Atmos HMI.

An *inlet station* (also called an *injecting station* or *origination station*) typically includes a booster pump followed by several main pumps placed in series. A booster pump might be a small, single-stage, fixed-speed centrifugal pump. On some pipelines, a vertical can pump might be used for this same purpose. What matters is that it requires low net positive suction head (NPSHr). This inherently means that a booster pump is capable of providing little head itself. Its purpose is merely for generating enough head at the station's inlet to meet the suction head requirements (NPSHr) of the subsequent mainline pumps. [58]

A station typically has a throttling control valve at its outlet, whose position is adjusted to regulate the station discharge pressure. Mid-line stations are then spaced around

65-80 km apart, [45] [19] typically consisting of fixed-speed main pumps in series with a throttle, or zero or more fixed-speed main pumps in series with a variable-speed main pump. Another set of pumps is often placed in parallel as spare.

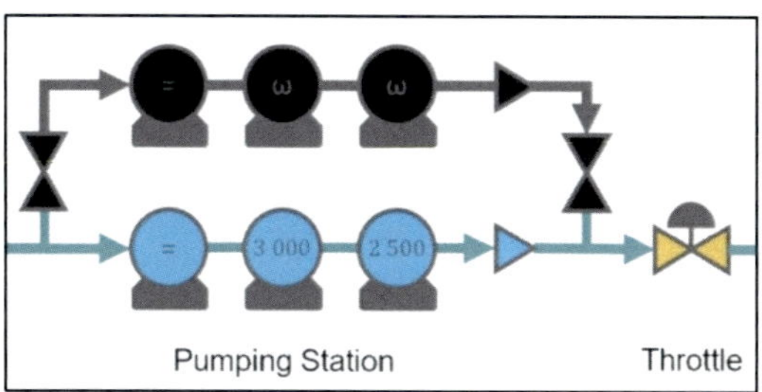

Figure 129 – A booster pump followed by two main pumps in series. A parallel set is spare.

For each pump, Atmos SIM presents the current and recent operating points as a trace on an operating point chart. The operating point is at the intersection of the system curve with the pump's head curve at the current operating speed.

It can be useful to consider the aggregate effects of a pumping station as a sort of *"station curve"*. Doing so would require negotiating the peculiarities of every individual pump's operating envelope, their efficiency contours, their net positive suction head required, and the shaft power deliverable by their drivers.

7.7 – Compressors in gas pipelines

Compressor performance

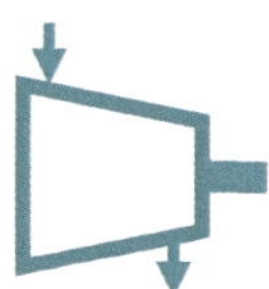

Isothermal, isentropic or polytropic?

A compressible fluid changes temperature as it is pressurized, because its density changes. This means more work is done to compress a mass of gas than to pump the same mass of incompressible liquid to the same pressure. A pump simply has *"an efficiency"* which is defined as the ratio of useful work to shaft work. However, for a gas compressor or for a pump that is handling compressible liquids, we get different results depending on which of three paths we choose:

- *Isothermal* ($P \cdot \rho^{-1} = \text{constant}$) – a reversible process with the exact amount of heat transfer to maintain suction temperature at discharge.
- *Isentropic* ($P \cdot \rho^{-\gamma} = \text{constant}$) – a perfectly reversible process with zero heat transfer such that the temperature rises from suction to discharge. An isentropic process going from a real compressor's suction conditions to its discharge pressure always under-estimates the real discharge temperature. It can be viewed as consisting of a series of small steps each of 100% efficiency.
- *Polytropic* ($P \cdot \rho^{-n} = \text{constant}$) – an irreversible process with heat transfer, as well as a temperature rise. A polytropic process consists of a series of small steps each with a constant efficiency less than 100%; the polytropic efficiency. If the compressor's suction and discharge pressure and temperature are known, we can compute the appropriate polytropic efficiency to match them.

Path	Constant	Work	Heat	Exponent
Isothermal	Temperature	Reversible	Diathermal	1
Isentropic	Entropy	Reversible	Adiabatic	$\gamma = c_P/c_V$
Polytropic	Efficiency	Irreversible	Adiabatic	n

Table 8 – The definitions of isothermal, isentropic and polytropic processes

Manufacturers describe how a real compressor differs from *"perfect"* isentropic behavior with an *"efficiency"*, which tells us how much power is wasted raising the entropy instead of the pressure. There are two different common ways of defining efficiency, an *isentropic efficiency* (η_s) which is the ratio of isentropic work to actual work to achieve the discharge pressure, and *polytropic efficiency* (η_p) which is the ratio of isentropic incremental work to actual work that we would apply at each incremental change in conditions from suction to discharge state to end up with the correct total work. These two efficiencies are generally in the same ball-park, but they aren't identical because the gas properties change as the gas is compressed. Either efficiency can describe the exact same real-world compressor behavior, it just needs to be coupled with the appropriate calculation for that particular flavor of efficiency.

Polytropic efficiency (η_p) has a convenient property lacking in isentropic efficiency (η_s). It is defined such that if we place several units (e.g. compressors) in series, each with the same polytropic efficiency individually, the combined overall polytropic efficiency of the whole process is still the same value as the per-unit polytropic efficiency.

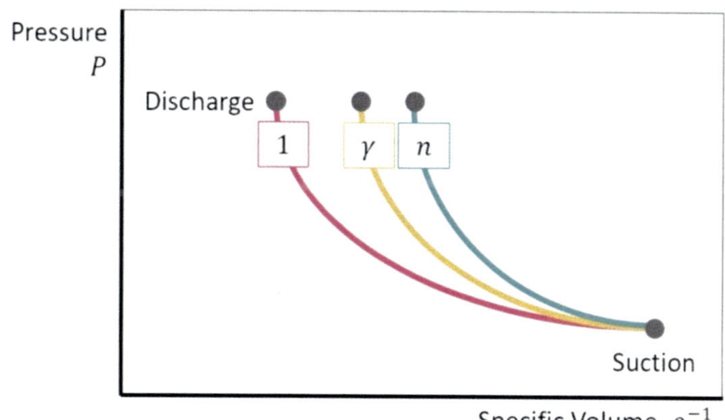

Figure 130 – The curves of an isothermal (1), an isentropic ($\gamma = c_P/c_V$) and a polytropic (n) compression on a pressure-volume diagram (adapted from the GPSA Engineering Data Book).

In the next sections we shall consider these three possible paths to calculate:

- the isothermal compression work
- the isentropic compression work and isentropic temperature rise
- the polytropic compression work and polytropic temperature rise

Isothermal compression

Here we must be careful. A major assumption often goes unstated in reference works on pumps and compressors. Work in thermodynamics is always defined:

$$w = \int P \cdot d\rho^{-1}$$

An isothermal compressor delivers maximum useful hydraulic work to a fluid. Textbooks assume an isothermal path because it is easier than real estimates. But isothermal compression rarely occurs in pipeline applications, so we only cover it here because it is helpful in explaining some aspects of compression.

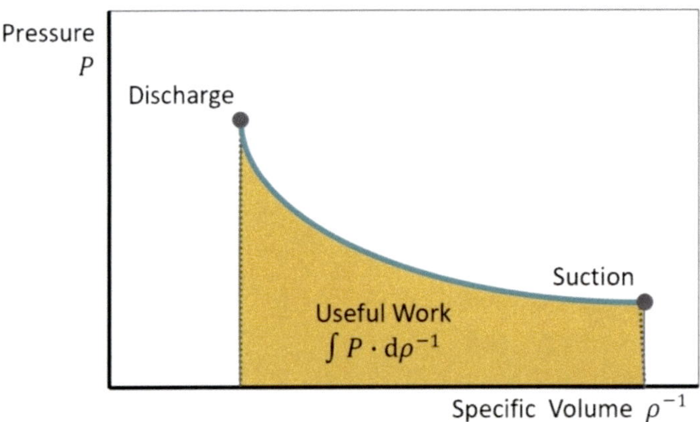

Figure 131 – The useful hydraulic work delivered to a fluid lies under the pressure-volume curve.

The isothermal path gives an interesting insight too. Because temperature stays the same from suction to discharge ($T_{\text{suc}} = T_{\text{dis}}$), for an ideal gas we can say that the product of pressure (P) times density (ρ) remains constant:

$$P \cdot \rho^{-1} = \text{constant}$$

The value of this constant depends on temperature, so we can draw isotherms. If we assume this is happening to an ideal gas, we express *"PV=NRT"* in the following more convenient form, using the mass-specific density (ρ) so that we can more clearly see that the work is dependent on the fluid's molar mass (m_r):

$$P \cdot \rho^{-1} = \frac{R \cdot T}{m_r}$$

In plain terms, a compression at a higher temperature will require more work.

We get this reassuring formula for isothermal compression work of ideal gas:

$$w_T = \frac{R \cdot T}{m_r} \cdot \ln\left[\frac{P_{\text{dis}}}{P_{\text{suc}}}\right]$$

This is only really useful as a quick check (and a lower bound). Real work is much greater than this. An arrangement approaching the isothermal limit is equivalent to infinite stages, each imparting identical compression ratio, with complete inter-cooling incurring no pressure drop. This is impossible in practice, as no reasonable equipment can transfer heat away from an industrial compressor fast enough to achieve it.

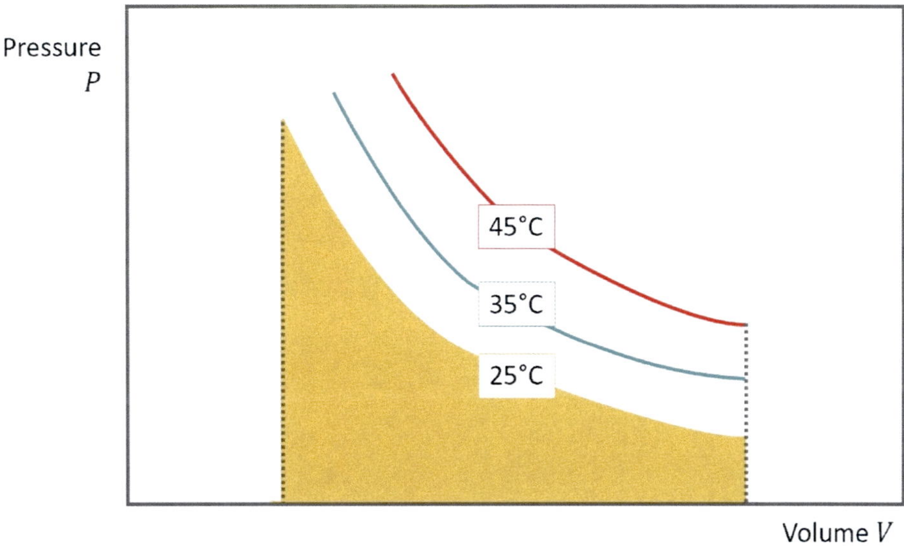

Figure 132 – Isotherms on a pressure-volume diagram. Compressing hot gas takes more work.

As a gas undergoes significant compression, it becomes extremely hot. It is vital we determine quite how hot it will be, both to calculate the density properly, and to ensure available heat exchangers can cool the discharge temperature to an acceptable level in operation. As we strive to describe gas compression, it becomes apparent that this discharge temperature too presents a challenge.

Isentropic compression

Assuming a process is *reversible* and *adiabatic* is equivalent to saying that it is *isentropic*. This assumption also under-estimates compressor work, albeit less drastically than the foregoing isothermal assumption. In an isentropic process, the pressure (P) and density (ρ) at suction and discharge conditions are related as follows by the *ratio of heat capacities* ($\gamma = c_P/c_V$), a ratio which in this context is also known as the *isentropic index* or *isentropic exponent*:

$$P_{\text{suc}} \cdot \rho_{\text{suc}}^{-\gamma} = P_{\text{dis}} \cdot \rho_{\text{dis}}^{-\gamma}$$

$$\rho_{\text{dis}} = \rho_{\text{suc}} \cdot \left(\frac{P_{\text{dis}}}{P_{\text{suc}}}\right)^{\frac{1}{\gamma}}$$

A special case of work is work done in a reversible process. This special case of work occurs with no pressure differential. We say it happens so gradually it is reversible: over an infinitesimal step. This *reversible work* can be expressed:

$$q - w = \int T \cdot ds + \int \rho^{-1} \cdot dP$$

We integrate because the density (ρ) varies with pressure and temperature as related by an equation of state. We note that this is the mole-specific density. If the process is also adiabatic (i.e. negligible in heat transfer) we can say that it is *isentropic* (i.e. reversible adiabatic). Isentropic compression work (w_s) is:

$$w_s = - \int_{\text{isentropic}} \rho^{-1} \cdot \mathrm{d}P$$

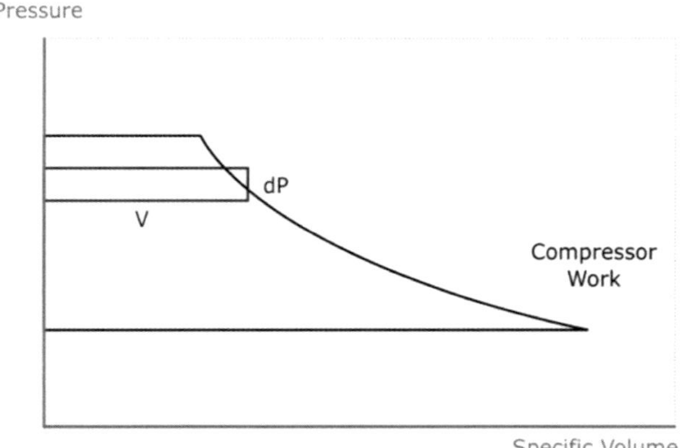

Figure 133 – The work done in an isentropic process lies behind the pressure – volume curve. Its area can be estimated numerically by splitting it into discrete pressure intervals (rectangle).

Numerous authors casually refer to this quantity ($\rho^{-1} \cdot \mathrm{d}P$) as an unqualified expression for general work, which is quite misleading. It only holds true in the special case where all work done goes into the useful part of enthalpy. This underlying assumption makes the isentropic path an idealization, like the isothermal ideal. It is greater than the work input via an isothermal compressor, but still less than the shaft work we provide in any real irreversible compressor.

Pausing to consider this equation, the simplest way to save energy is trivial: reduce the pressure differential ($\mathrm{d}P$). Must we generate all this pressure, or could we avoid some of the pipeline's losses, or improve its control strategy? In the early days of pipeline optimization, software optimizers were able to provide significant cost savings just by coming up with operating plans that involved throttling away less of the generated head. Today, presumably a trade-off between capital outlay and operating expense was already resolved at an earlier phase, when the pipeline hydraulics were examined to establish the *system curve*. The problem at hand here is how best to achieve the required compression ratio (CR), not to question it!

The only other measure left, then, is to increase the density (ρ). Incoming gas at the suction side can be made denser by *pre-cooling*, or *inter-cooling* could be placed between compression stages. Most common of all is *after-cooling*, but this does not affect a compressor's performance. It just reduces the discharge temperature to acceptable levels, and also reduces losses in the downstream pipe somewhat, since the density there is increased.

Although some heat is inevitably lost to the surroundings, the heat transferred to the fluid passing through a real compressor is far more significant, given the substantial throughput of fast moving gas. So contrary to first appearances, the adiabatic assumption is perhaps not too unreasonable. Indeed textbooks mention that people routinely use this relation for reciprocating compressors. However, the isentropic relation doesn't give us a truly accurate picture of the behavior of a real compressor, so the resulting state must either be directly corrected with an isentropic efficiency, or another approach must be used. [46]

It is the *irreversible* aspect of an isentropic assumption that doesn't really hold true, rendering hydraulic power calculated based on an isentropic compression inaccurate. An *isentropic efficiency* (η_s) corrects for this. [*] The problem is when we come to apply this efficiency in practice, we find we don't know what it is! The supercompressibility (Z) decreases as a real, non-ideal gas progresses through successive stages towards discharge, so a mass flow rate (Q^{mass}) of fluid manifests as a smaller actual volumetric flow rate (Q) in the later stages of a compressor, reducing the head delivered. As pressure climbs, it makes the gas denser, which helps a little, but the power reduction far exceeds this effect. In short, across any multi-stage compressor, the isentropic efficiency is far worse than would be implied by applying the per-stage efficiency to the overall process. In effect, there are many different *isentropic efficiencies*. [47] [49]

There is progressively less actual volumetric flow rate as a gas passes through successive compressor stages.

Needless to say, it is a lot of trouble to calculate an efficiency for every stage of a multi-stage compressor. Even if we do that, we find that it gives a poor estimate of *discharge temperature*. Say a compressor is perfectly efficient (reversible and adiabatic, so

[*] Commonly referred to as an *adiabatic efficiency*.

isentropic). If we compress methane by, say, a compression ratio of 1.5 from a suction temperature of 25°C, the discharge temperature is a whopping 54°C. A real compressor with isentropic efficiency of say 80% has a discharge temperature that's higher still, at 62°C. We see that most of the temperature rise is already present in an ideal compressor. This is very different to almost-incompressible liquids like water, crude oils, or batch products, where isentropic compression produces zero temperature rise. For this reason, gas compression necessarily involves thinking about cooling the discharge gas. Pumping compressible liquids is similar.

In the earlier formulas we stated for isentropic processes, the ratio of heat capacities ($\gamma = c_P/c_V$) at a given gas composition was taken to be constant and used as the isentropic index. For non-ideal gases and for substantial compression ratios, this can be a bad assumption. The isochoric heat capacity (c_v) becomes infinite at the critical point, so this is particularly an issue with any fluids operating in that region. Work and temperature rise each follow separate exponential laws, so some define one isentropic index to track temperature (γ_T) and another to track volume i.e. work (γ_V). [47 51]

For an ideal gas undergoing an isentropic process, the equation of state is:

$$\rho = \left(\frac{R \cdot T}{m_r \cdot P}\right)^{-1/\gamma}$$

An isentropic compressor taking this ideal gas from suction to discharge pressure imparts the following energy-per-unit-mass, i.e. head ($\Delta H_{\mathrm{J/kg}}$) *per compression stage*:

$$\Delta H_{\mathrm{J/kg}} = \frac{R \cdot T_{\mathrm{suc}}}{m_r} \cdot \frac{\gamma_V}{\gamma_V - 1} \cdot \left(\left(\frac{P_{\mathrm{dis}}}{P_{\mathrm{suc}}}\right)^{\frac{\gamma_V - 1}{\gamma_V}} - 1\right)$$

For a multi-stage compressor, this calculation must be done multiple times. That is to say: we can't just apply it once to the overall multi-stage compressor.

The isentropic temperature rise for this ideal gas, per stage, is calculated using the isentropic efficiency (η_s):

$$\Delta T = \frac{T_{\mathrm{suc}}}{\eta_s} \cdot \left(\left(\frac{P_{\mathrm{dis}}}{P_{\mathrm{suc}}}\right)^{\frac{\gamma_T - 1}{\gamma_T}} - 1\right)$$

For a real gas undergoing an isentropic process, the equation of state is:

$$\rho^{-\gamma} = \frac{R \cdot T \cdot Z}{m_r \cdot P}$$

An isentropic compressor going from suction to discharge pressure imparts the following isentropic head per compression stage to a real gas ($\Delta H_{\mathrm{J/kg}}$):

$$\Delta H_{\mathrm{J/kg}} = Z_{\mathrm{avg}} \cdot R \cdot T_{\mathrm{suc}} \cdot \frac{\gamma_V}{\gamma_V - 1} \cdot \left(\left(\frac{P_{\mathrm{dis}}}{P_{\mathrm{suc}}} \right)^{\frac{\gamma_V - 1}{\gamma_V}} - 1 \right)$$

This must be calculated separately for each stage of a multi-stage compressor.

The isentropic discharge temperature for this real gas, per stage, is:

$$T_{\mathrm{dis}} = T_{\mathrm{suc}} + \frac{Z_{\mathrm{suc}}}{Z_{\mathrm{dis}}} \cdot \frac{T_{\mathrm{suc}}}{\eta_s} \cdot \left(\left(\frac{P_{\mathrm{dis}}}{P_{\mathrm{suc}}} \right)^{\frac{\gamma_T - 1}{\gamma_T}} - 1 \right)$$

Isentropic efficiency is an insightful calculated value but is not practically available for us to use it. This makes isentropic assumption unwieldy to deal with, so for centrifugal compressors make a subtly different assumption – a polytropic path.

Polytropic compression

We instead pick a path with a different, more useful property. If we split this path into any number of steps, each has the same *polytropic efficiency*. Over each infinitesimal step ($\mathrm{d}P, \mathrm{d}T$) the polytropic efficiency is the same through the entire process, in all stages. For several stages in series, we are allowed to sum the head from each stage, or alternatively consider the head to be across one overall unit, applying the same efficiency in both approaches to get the same result. It is polytropic head and efficiency which are provided to us by manufacturers, in the form of performance maps. Although the polytropic approach is the best assumption in practice, it presents an inconvenience: a polytropic path conserves no thermodynamic variables. It merely follows some arbitrary *polytropic index* (n) as the exponent:

$$P_{\mathrm{suc}} \cdot \rho_{\mathrm{suc}}^{-n} = P_{\mathrm{dis}} \cdot \rho_{\mathrm{dis}}^{-n}$$

The expression for a polytropic process is identical to the isentropic expression in all respects but this exponent.

$$\rho_{\mathrm{dis}} = \rho_{\mathrm{suc}} \cdot \left(\frac{P_{\mathrm{dis}}}{P_{\mathrm{suc}}} \right)^{1/n}$$

Unlike the isentropic index (γ), the polytropic index (n) is <u>not</u> the ratio of heat capacities ($n \neq c_P / c_V$). It is empirical, and it varies with pressure, temperature and composition.

Given a polytropic efficiency (η_{p}) at a set of conditions, the polytropic index (n) can be calculated from isentropic index (γ) as follows:

$$\frac{n-1}{n}=\frac{\gamma-1}{\eta_{\mathrm{p}}\cdot\gamma}$$

The polytropic method still doesn't give an accurate discharge temperature for non-ideal gases. A typical gas pipeline compressor spans such a large range of temperatures and pressures from suction to discharge that the ideal gas approximation doesn't really hold: the heat capacities (c_P and c_V), and supercompressibility (Z) are not constant. So again, we distinguish between one polytropic index tracking temperature rise (n_T), and another tracking volumetric – i.e. work – calculations (n_V). [47] To calculate this during every iteration at every step in space and time would be too resource-intensive, so Atmos SIM's compressor temperature is calculated via a unique, innovative approach.

By substituting in the ideal gas equation of state we find that – in terms of pressure and temperature – the polytropic path for an ideal gas obeys:

$$P^{1-n}\cdot T^{n}=\mathrm{constant}$$

For a compressor, the polytropic head ($\Delta H_{\mathrm{J/kg}}$) imparted to take this ideal gas from suction pressure to discharge pressure is:

$$\Delta H_{\mathrm{J/kg}}=\frac{P_{\mathrm{suc}}}{\rho_{\mathrm{suc}}}\cdot\frac{n_V}{n_V-1}\cdot\left(\left(\frac{P_{\mathrm{dis}}}{P_{\mathrm{suc}}}\right)^{\frac{n_V-1}{n_V}}-1\right)$$

And the polytropic discharge temperature for compressing this ideal gas is:

$$T_{\mathrm{dis}}=T_{\mathrm{suc}}\cdot\left(\frac{P_{\mathrm{dis}}}{P_{\mathrm{suc}}}\right)^{\frac{n_T-1}{n_T}}$$

For a real gas, the equation of state includes a compressibility factor (Z) which varies along the polytropic path, so in terms of pressure and temperature we now have:

$$P^{1-n}\cdot(Z\cdot T)^{n}=\mathrm{constant}$$

The polytropic head for compressing this real gas is:

$$\Delta H_{\mathrm{J/kg}}=\frac{R\cdot T_{\mathrm{suc}}}{m_r}\cdot\frac{n}{n-1}\cdot\left(\left(\frac{P_{\mathrm{dis}}}{P_{\mathrm{suc}}}\right)^{\frac{n-1}{n}}-1\right)$$

And the polytropic discharge temperature for compressing this real gas is:

$$T_{\text{dis}} = T_{\text{suc}} + \frac{Z_{\text{suc}}}{Z_{\text{dis}}} \cdot \frac{T_{\text{suc}}}{\eta_{\text{s}}} \cdot \left(\left(\frac{P_{\text{dis}}}{P_{\text{suc}}} \right)^{\frac{n-1}{n}} - 1 \right)$$

But this is still an estimate, based on compressibility at suction and discharge conditions. We could instead perform a numerical integration, calculating the density using some suitable equation of state (such as Peng-Robinson). By that method, the polytropic head for compressing this real gas is found more accurately, and the polytropic discharge temperature is more accurate too:

$$\left.\frac{\mathrm{d}T}{\mathrm{d}P}\right|_{\text{path}} = \frac{1}{c_P} \cdot \left(\frac{1}{\eta_{\text{P}}} - 1 \right) \cdot \rho^{-1} + T \cdot \left.\frac{\mathrm{d}\rho^{-1}}{\mathrm{d}T}\right|_P$$

Which we integrate numerically over the path. As we alluded to earlier, this more accurate form is necessary if the fluid gets anywhere near its critical point. For dry natural gas, which is mostly composed of methane and is guaranteed to be far from its critical point at pipeline conditions, the algebraic approximations are good enough. For something like carbon dioxide, it is necessary to integrate.

Calculate the real non-ideal temperature rise across a compressor. Compare simulated power and temperature assuming ideal gas equation against via Peng-Robinson.

It is notable that in performing all of the above analysis and calculations, all we needed to think about was a generic compressor abstracted as a black box. Next we shall take a closer look at the types of compressors installed along gas pipelines: reciprocating compressors, and centrifugal compressors.

Reciprocating compressors

A reciprocating compressor, also known as a piston compressor, or as a *"recip"*, works on the principle of positive displacement (PD). It was the most common type of compressor installed on pipelines in the past, so it is often present within compression stations, especially along local distribution pipelines or older trunk pipelines.

Within a certain range of low flow rates, either centrifugal or reciprocating compressors can be chosen for service. At their core, reciprocating compressors are cheaper to buy and install, far more efficient, and simpler to operate than their centrifugal counterparts. It is practical considerations, pulsation and reliability, that have seen them fall out of favor in recent decades. Centrifugal compressors dominate today; reciprocating compressors in pipeline applications are chiefly found as a legacy from a bygone era. Nonetheless, this legacy still needs modelling. Moreover, they continue to be chosen for temporary installations, and wherever cheaper capital expenditure is preferred over ongoing cost-effectiveness to operate and maintain. There are also niche applications where reciprocating compressors remain the best choice, such as for hydrogen. [48]

A reciprocating compressor can efficiently impart a wide range of heads to a narrow range of flow rates. At a given operating speed it can only handle a set actual volumetric flow rate (Q), and only the head ($\Delta H_{\mathrm{J/kg}}$) varies. A speed setpoint for this device is in many situations almost equivalent to a flow setpoint. But if we attempt to push flow against too much head, a little flow leaks backwards, so it won't quite maintain the requested flow. Instead it allows a bit less flow. This is why a reciprocating compressor's flow depends on head, albeit weakly.

It is worth pausing to note that because a gas is compressible, a given volume (V) at a higher pressure (P) as would be achieved at higher imparted head ($\Delta H_{\mathrm{J/kg}}$) is equivalent to more standard volume (V_{std}), so standard flow varies.

For a reciprocating compressor, spin-up duration, spin-down duration and minimum speed are set directly. Operating speed, a few hundred or thousand revolutions per minute (RPM), affects mechanical efficiency (η_m):

$$\eta_m = \frac{\text{delivered power}}{\text{shaft power}}$$

Atmos SIM expects the user to input a reciprocating compressor's performance as a *displaced* volume and a *clearance* volume. We calculate these from equipment datasheets provided by the manufacturer. *Displaced volume* is typically 2 to 8 liters, displaced by a piston as it traverses the length of its stroke. A single-acting cylinder only has valves at the *head-end*, so its displaced volume ($V_{\text{displaced}}$) is stroke length (L) multiplied by the cylinder's circular cross-sectional area, calculated from its diameter, called its *bore* (D_{piston}):

$$V_{\text{displaced}} = \frac{\pi}{4} \cdot D_{\text{piston}}^2 \cdot L$$

A *double-acting* cylinder also has valves at the *crank-end*, so a full stroke additionally displaces volume at that end, whose cross-section excludes the rod's diameter (D_{rod}):

$$V_{\text{displaced}} = \frac{\pi}{4} \cdot \left(D_{\text{piston}}^2 + (D_{\text{piston}}^2 - D_{\text{rod}}^2)\right) \cdot L$$

A stroke does not completely fill this space; a *clearance ratio* (Λ_C) is left between the cylinder head and the piston, typically 15% of displaced volume:

$$\Lambda_C = \frac{V_{\text{clearance}}}{V_{\text{displaced}}}$$

This clearance might be adjustable by opening or sealing *pockets*, modifying the reciprocating compressor's performance by adjusting its flow capacity in a series of *load steps* to suit the operation. *Clearance* ($V_{\text{clearance}}$) is not the only factor reducing the volume delivered by a reciprocating compressor. There is also some inevitable *leakage* of gas moving backwards (V_{leakage}). Therefore:

$$V_{\text{delivered}} = V_{\text{displaced}} - V_{\text{clearance}} - V_{\text{leakage}}$$

The reduction in delivered volume caused by these effects is described by its ratio to displaced volume. We then introduce a definition for a ratio known as *volumetric efficiency* (Λ_V): *

$$\Lambda_V = \frac{V_{\text{delivered}}}{V_{\text{displaced}}}$$

The physical correlation we use for volumetric efficiency (Λ_V) is:

$$\Lambda_V = 100\% - 5\% - \frac{V_{\text{clearance}}}{V_{\text{displaced}}} \cdot \left(\left(\frac{P_{\text{dis}}}{P_{\text{suc}}} \right)^{\frac{\gamma - 1}{\gamma}} - 1 \right)$$

* We represent this as a capital lambda to be clear it's unlike other efficiencies. To be clear, a reciprocating compressor's volumetric efficiency (Λ) is nothing to do with its energy efficiency, as captured by the *isentropic efficiency* (η_s), nor with the pump or compressor's mechanical efficiency (η_{m}). It is not about energy loss.

Where the ratio of discharge to suction pressure ($P_{\text{dis}}/P_{\text{suc}}$) is the compression ratio. In a double-acting reciprocating compressor, there are two volumetric efficiencies, one for the *head-end* (Λ_{HE}), and another for the *crank-end* (Λ_{CE}).

There can be up to a dozen cylinders in a single frame. It is quite common to have several cylinders per stage, known as *throws* in this context. A three-stage, six-throw compressor would have two sets of three cylinders working in parallel. Each set is a stage, with a common suction and discharge. This means we can directly multiply the effect of each throw by the number of throws (N) working in parallel for a given stage.

In Atmos SIM, for throws we use banks. To use stages at the same speed, we link them via control diagrams.

From the manufacturer's data sheets, a pipeline simulator expects the following inputs to model a reciprocating compressor:

- Whether the pistons are single-acting or double-acting
- Number of stages, and number of throws per stage
- Cylinder bore per cylinder – this may differ in each stage
- Stroke length per cylinder – this may differ in each stage
- Volumetric efficiency, or a pair of volumetric efficiencies

There are also additional components, such as unloader valves, which aren't necessarily pertinent to pipeline simulation. A mechanical efficiency (η_{m}) is a user-input constant, along with the actual volumetric flow rate (Q), polytropic index (n), a real gas correction based on the supercompressibility of the gas (Z), the suction and discharge pressures ($P_{\text{suc}}, P_{\text{dis}}$), and the reciprocating compressor's power ($\dot{W}$):

$$\dot{W} = P_{\text{suc}} \cdot Q \cdot \frac{\gamma}{\gamma - 1} \cdot \frac{Z_{\text{suc}} + Z_{\text{dis}}}{2 \cdot Z_{\text{suc}}} \cdot \frac{1}{\eta_{\text{m}}} \cdot \left(\left(\frac{P_{\text{dis}}}{P_{\text{suc}}} \right)^{\frac{\gamma-1}{\gamma}} - 1 \right)$$

Reciprocating compressors range from 35 kW to many MW. The isentropic index (γ) is used here, equal to the ratio of heat capacities (c_P/c_V). So the volumetric efficiency (Λ_V) is effectively wrapped up inside actual volumetric flow rate (Q).

The hydraulic power delivered by a reciprocating compressor is less than that delivered by the compression stroke in isolation. This is because what matters is the overall net work done by a cycle of operations happening in succession.

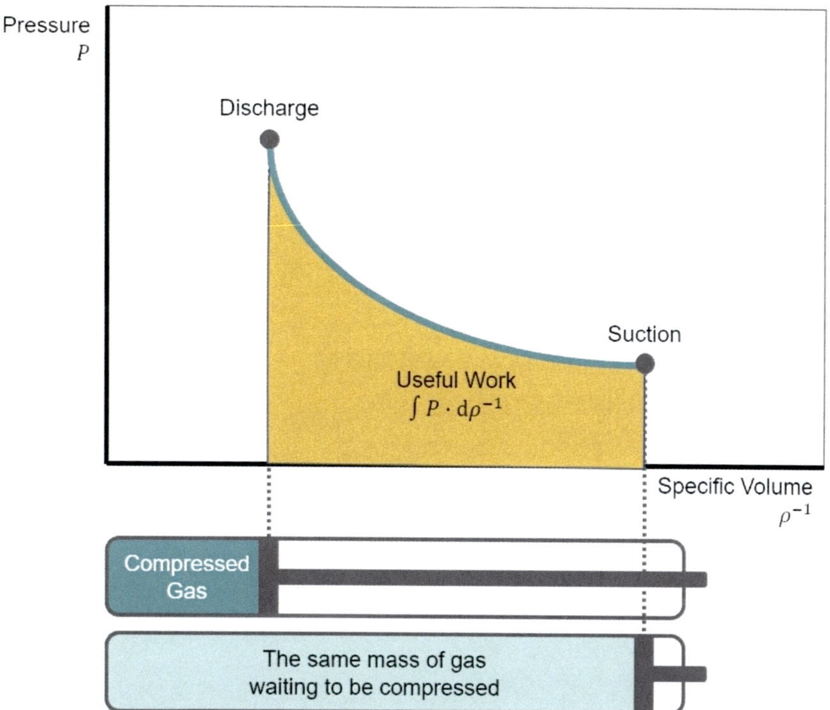

Figure 134 – A pressure-volume diagram of work done ($P \cdot \mathrm{d}\rho^{-1}$) by the compression stroke of a reciprocating piston compressor. This shows the compression stroke in isolation.

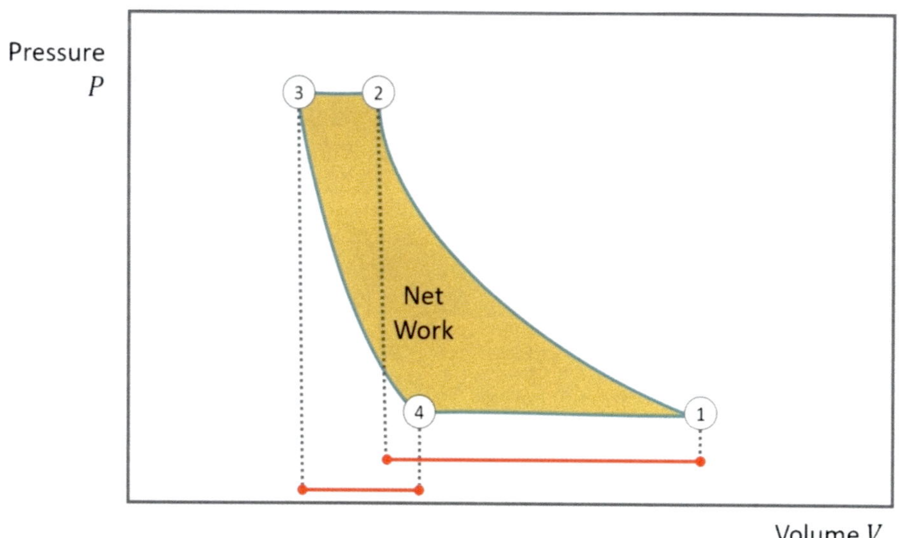

Figure 135 – The net work delivered to the fluid by a reciprocating compressor's overall cycle. This figure represents a single-stage positive displacement compression cycle with clearance. If there were no clearance, the shaded area would extend horizontally to the pressure axis.

Centrifugal compressors

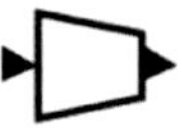

Rotation Speed	17.5	kRPM
Inlet Pressure	18.99	barg
Outlet Pressure	51.23	barg
Inlet Temperature	20	°C
Outlet Temperature	91.69	°C
Driver Input Power	1.265	MW
Driver Output Power	0.9734	MW
Shaft Power	0.9442	MW
Hydraulic Power	0.7186	MW
Adiabatic Efficiency	76.1	%
Polytropic Efficiency	78.79	%
Polytropic Head	120	kJ/kg
Inlet Mass Flow	5.184	kg/s

Figure 136 – A centrifugal compressor's symbol in the Atmos SIM schematic (top) and simulated results: speed, head, efficiency, suction and discharge pressure and temperature, and flow rate.

Centrifugal compressor maps and operating points

A variable-speed centrifugal compressor typically operates within a pipeline in one of three modes. Either it delivers a set flow rate by varying the discharge pressure, or it maintains a set discharge pressure by varying the flow rate, or it just runs as fast it can.

Atmos SIM models a compressor operating to maintain a set flow rate or to maintain a set discharge pressure

Unlike centrifugal pumps for liquids, centrifugal compressors don't have a minimum suction head requirement – the fluid is already a gas so there is no danger of cavitation. A single-stage can deliver low compression ratios up to flow rates of $70\ \mathrm{Sm}^3/s$. The compression ratio across a stage depends on gas properties; it is typically around 1.4 for natural gas whereas that same stage produces a pressure ratio of 1.6 for air. [49] [50] Multi-stage compressors are deployed to achieve the substantial compression ratios needed to transport high gas flow rates through long-distance transmission pipelines. [51]

A manufacturer presents their compressor's capabilities as a set of head-flow and efficiency-flow curves on a performance map. The map doesn't reach zero flow. There are strict operating limits at the flat and steep parts of the curve, representing minimum flow (surging) and maximum flow (choking, also called stonewall). The centrifugal compressor at a given speed has some flexibility within the bounds of this operating range of flow rates. Its speed is controlled by monitoring surge margin and choke margin to keep it operating efficiently.

Atmos SIM displays this information as curves for the energy-units head ($\Delta H_{\mathrm{J/kg}}$) delivered to an actual volumetric flow rate (Q), with superimposed contours telling us the polytropic efficiency (η_p) at any given point on that head-flow map. The two ways of presenting the information are equivalent, as shown on the next page.

During start-up and shut-down, the compressor operates in an inefficient region of its map, so gas temperature is liable to suddenly spike.

If a compressor map states a power draw at a given flow rate, it is for the test rig they used to characterize it and is probably based on air. It should not be expected that it will be the same for the pipeline gas in service: a centrifugal compressor's hydraulic power depends strongly on the molar mass of the gas.

Gas compressors on a pipeline are typically coupled with a gas turbine as their driver. They are known to perform slightly differently depending on ambient conditions, so manufacturers may provide a compressor map for summer and another for winter temperatures.

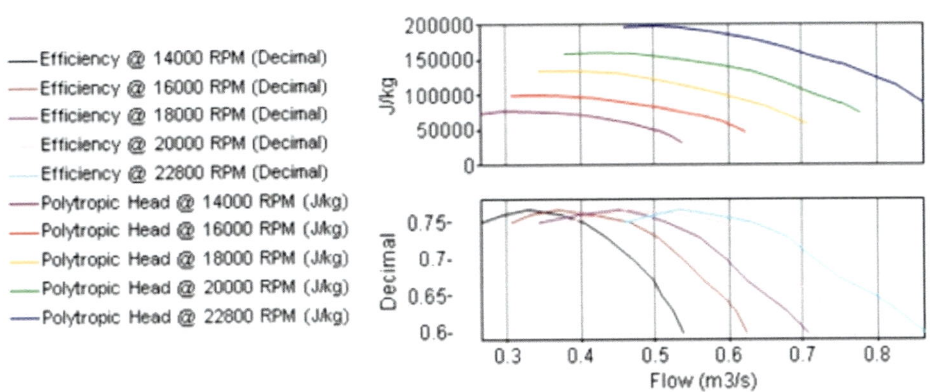

Figure 137 – A separate plot of head-flow and efficiency-flow curves for each compressor speed

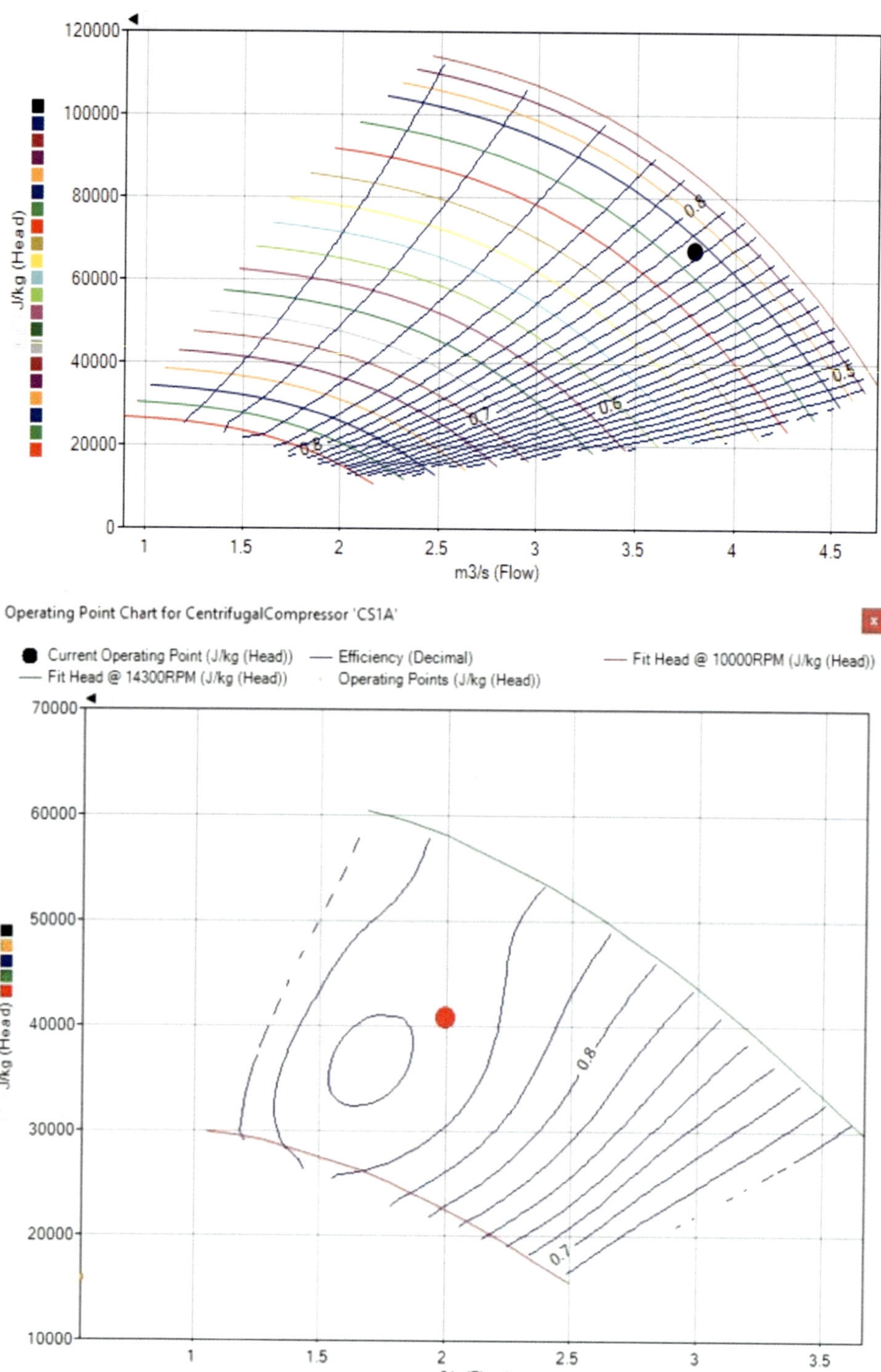

Figure 138 – An Atmos SIM compressor map displays the operating point on two equivalent plots. Efficiencies are equivalently represented either as curves (top), or as contours (bottom).

Minimum flow, stall and compressor surge control

Centrifugal compressors are subject to various forms of flow instability at low flow rates. Although we talk about *"surge"*, these can be classified as several distinct phenomena:

- *Stall*, where boundary layers separate from the moving parts [52]
- *Mild surge*, where flow oscillates back and forth
- *Deep surge*, where flow reverses suddenly

However, pipeline models do not usually need to distinguish between the mechanism or mode of instability – they just need to track whether the unit is operating in the unstable region. Once the compressor has entered unstable flow, it's unlikely that any simulation will be able to produce an exactly accurate picture of its behavior anyway.

So a pipeline simulator represents this all as a single limit – a *minimum flow constraint* – at which the actual volumetric flow rate at the suction side of a centrifugal compressor is insufficient for its current speed. In this part of the map, the head-flow curve becomes nearly horizontal, representing the maximum pressure ratio possible at that given speed, operating *"on the flat part of its curve"*. A compressor is at risk of inadvertently being brought into this condition if its drive speed is suddenly reduced, particularly if this causes discharge pressure to drop faster than the pipeline downstream. Attention is given to setting up a control system to avoid it, especially in scenarios like spin-down.

Atmos SIM models a badly operated compressor going into surge, and how it can recover from it.

Inlet Standard Volume Flow	6	Sm3/s
Outlet Standard Volume Flow	6	Sm3/s
Recycled Standard Flow	0.9316	Sm3/s
Total Standard Suction Flow	6.932	Sm3/s

Inlet Actual Volume Flow	0.2871	m3/s
Outlet Actual Volume Flow	0.1354	m3/s
Surge Actual Flow At Current Speed	0.3316	m3/s
Total Actual Suction Flow	0.3316	m3/s

Figure 139 – Atmos SIM automatically models internal per-unit recycling in each compressor, keeping the surge fraction from dropping below 100%.

How do we maintain enough surge margin without disrupting normal routine operations? A *"surge line"* can be calculated based on a straight line regression of the compressor speed versus minimum actual flow rate ($y = A \cdot x - B$). A further *"surge control line"* (SCL) might apply a 10% multiplier above that line, for a typical compressor's performance map. Further lines can be defined in parallel. The following is a typical centrifugal compressor's anti-surge control.

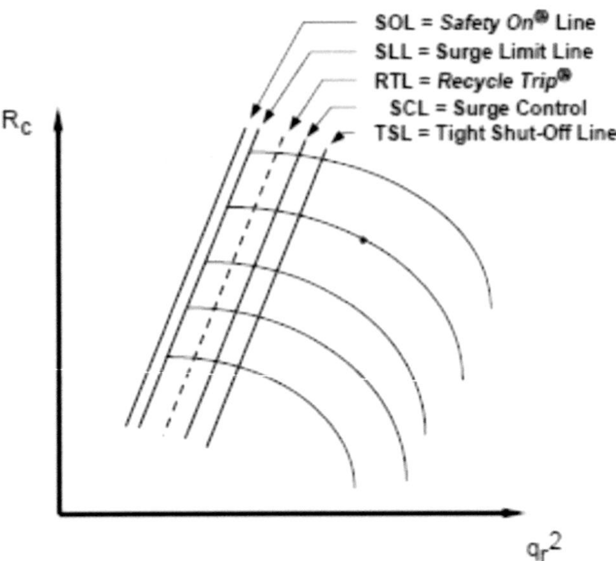

Figure 140 – An anti-surge controller's limits on a compressor map (from a vendor)

- Proportional-Integral response: when the operating point moves left of *surge control line* (SCL), open the anti-surge valve to increase recycle.
- Open-loop response: when the operating point moves left of what this vendor calls a *recycle trip line* (RTL), ratchet the valve open in an immediate step.
- When the operating point moves left of what this vendor calls a *safety-on line* (SOL), the surge control line is moved rightwards. This reduces the possibility of future surge cycles, and counts the number of surge cycles this compressor experienced since last reset.
- Beyond a defined *tight shut-off line* (TSL), the actuator control signal is reduced to zero.

The anti-surge controller also includes high discharge pressure limiting control, and low suction pressure limiting control. If the individual train's suction or discharge pressure exceed a predefined value, the anti-surge valve is opened to prevent that pressure excursion from proceeding any further.

When the actual volumetric flow rate falls too low, an anti-surge control system can implement a number of measures.

The most common approach is *recycling* some of the flow rate from the discharge side back to suction side. In this context, recycling is also known as *"spillback"*. The centrifugal compressor model item in Atmos SIM includes a convenient internal recycle. This built-in surge control moves the operating point to the right, increasing suction volumetric flow at the same head until it's past the surge line – the actual surge control valve is abstracted away.

It can be useful for some studies to disable this feature, and rely instead on a *"regulator"* model item controlling a recycle via a logical scheme. A typical system includes one anti-surge controller per compressor train. Its primary objective is to maintain each compressor's operating point a safe distance to the right of its performance map's surge limit line. It operates automatically, with next to no operator intervention. It decides how to approach compressor surge based on a pre-configured surge line, analyzing the input signals from pressures and temperatures at suction and discharge, and flow rate at suction. It uses this information to modulate an *anti-surge valve* (ASV). When the operating point deviates negatively from the surge line the valve tends to open, and vice versa (if deviating positively, the valve tends towards a closed position).

In a station with multiple compressors operating in parallel, a single combined recycle line might be provided to serve all the units. A pertinent question that arises when recycling is how warm or cool a recycled discharge-side gas is when it mixes with incoming flow rate on the suction side. It might be a hot recycle, or a cold recycle starting downstream of a compressor station's outlet cooler. Recycled flow is usually cooled first, especially if there is a significant pressure ratio, in order to avoid a high discharge temperature. Otherwise, a recycled flow of fluid that is too hot can actually exacerbate surge rather than abating it.

An alternative or complementary measure is *throttling upstream* to reduce suction pressure, lower the density of the gas, and thereby increase the actual volumetric flow rate which corresponds to the same standard flow rate. There are other use cases requiring recycle without suction throttling.

Both of these ways of avoiding compressor surge – recycling, and/or suction throttling – are widely adopted, long proven to be simple, effective and reliable. A downside of recycling and throttling alike is that both these measures waste energy. Surge is such a concern that many systems are set up conservatively, continually throttling and recycling far in excess of what is reasonably needed. It has been found that reviewing surge characteristics and control algorithms can deliver energy savings by allowing safe operation closer to the surge limit.

The most efficient method of surge control takes an entirely different approach: deploying a variable speed drive adjusted by a controller, with the operating point adjusted in real-time. This comprises an outlay of capital investment and introduces complexity, so it is a project that entails substantial offline studies.

Maximum flow, minimum pressure ratio, and stonewall

Stonewall in compressors, also called *choking* or *overload*, occurs when there is excessive flow through for a given speed. It has not received as much attention as compressor surge, despite many manufacturers having identified stonewall as a region to be avoided in operation. [53] It takes place for one of two reasons.

In low-speed equipment, if the flow rate is excessive, losses increase until the pressure ratio drops to 1:1. Referring to it as choking is arguably misleading. This can be set as the basis for a maximum flow constraint in a pipeline simulator.

In high speed equipment, as flow increases, the velocity of the gas approaches the speed of sound (c) at the fastest point moving along a compressor stage. This is truly *choking* in the sense that gas physically cannot move any faster – the highest possible flow is reached because the gas cannot accelerate further. A *choke limit* or *sonic limit* can be imposed on a compressor map to prohibit its operation below some minimum pressure ratio. A pipeline simulator can directly impose this constraint. In practice it might require the installation of additional throttle valves, or selecting more complicated recycle and anti-surge valves. [53]

On a pipeline, stonewall tends to happen when the head required (Δe) is so low that a compressor runs excessively quickly. This is rarely encountered, as most pipelines are designed to minimize diameter, and so require substantial head at high flow rates. If it does occur, operating at maximum flow ($\dot{V}_{\max}$) in this way doesn't necessarily turn out to be detrimental, so long as it is only for brief periods. In fact, because little power is consumed in stonewall, compressors are commonly factory-tested whilst operating in this condition.

A consequence of the fan-law approach is that curves at different speeds can't hit choke at different *flow-over-speed*, so Atmos SIM takes the choke *flow-over-speed* to be the maximum from any of the different speed curves.

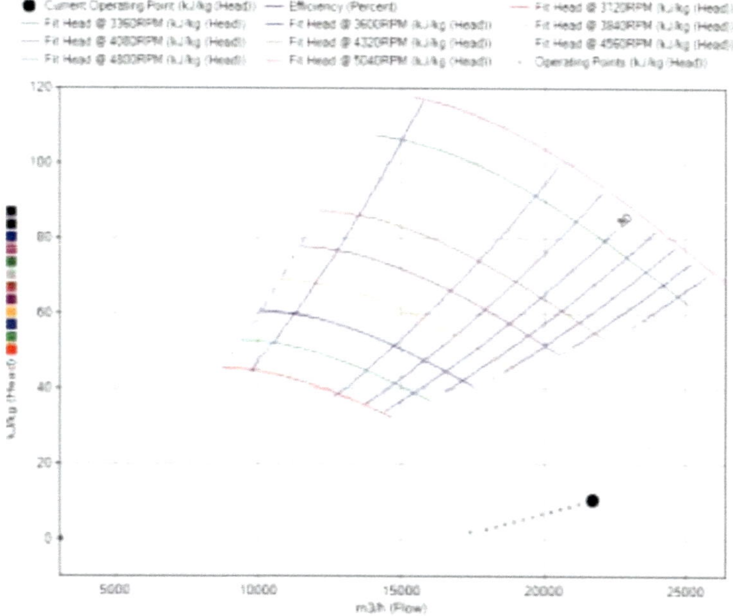

Figure 141 – A compressor whose choking flow-over-speed is 5.2 m³/hr/rpm. Here we see a particularly egregious violation of that limit, with an operating point being far beyond practical.

---

3 120 rpm corresponds to a choked flow of 16 340 m³/h and a +5% flow-at-zero-head of 17 200. At 5 050 rpm, it corresponds to a choke flow-at-zero-head of 27 800, so might develop a little head and be a bit below the maximum.

It is possible to represent the choked flow limit in a more sophisticated way by fitting a choked flow line on the right-hand-side of the flow-versus-speed performance map, in a similar way to the surge line fitted on the left-hand-side. As part of a compressor's control philosophy, we might need to raise an alarm if any individual compressor is in choked conditions for longer than some period, typically 15 minutes and again if it persists for longer than 2 hours.

Warmer or lighter gases are more compressible

It takes more power to compress a warmer, lighter gas of lower molar mass. A lighter hydrocarbon has a larger supercompressibility factor (Z).

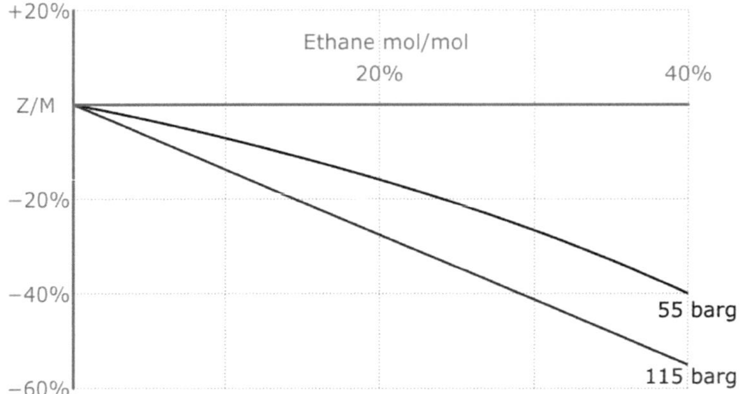

Figure 142 – How supercompressibility for a given molar mass (Z/m_r) varies with composition (ethane content) at different pressures. [54]

As suction pressure (P_{suc}) varies, speed correction (ω) maintains flow stability. The operating temperature (T), the fluid's supercompressibility (Z) and its ratio of heat capacities (γ) affect the *operational envelope*. Most significant, though, is molar mass (m_r); a compound lighter by as little as 5% can leave some compressors struggling to cope. [47 54]

The reason for this is visible on the map. When a compressor rotating at the same speed processes a lighter gas, it attempts to maintain discharge pressure, imparting more head and less flow rate, and risks going into surge. Conversely, the same speed with a heavier gas generates less head and more flow, and risks going into stonewall. As such, attention is devoted to close control, promptly adjusting discharge pressure in response to composition.

> Compressing the same mass flow of a gas with a lighter composition – i.e. lower molar mass – draws more power to achieve the same pressure ratio. This might violate a limit unless the pressure ratio or the flow rate is adjusted.

For multistage gas compressors, a heavier or lighter gas than the rated molar mass makes the operating envelope much narrower than would be expected from the design case. Within a certain range, off-design conditions can be handled, but this flexibility reduces as one strays further from rated values. In extreme circumstances, the impact can be so severe that the first stage can be at stonewall whilst the last stage is at surge; a condition in which the operational envelope is zero. Eventually it's worth re-rating the compressor: changing its components to suit the operation. [54]

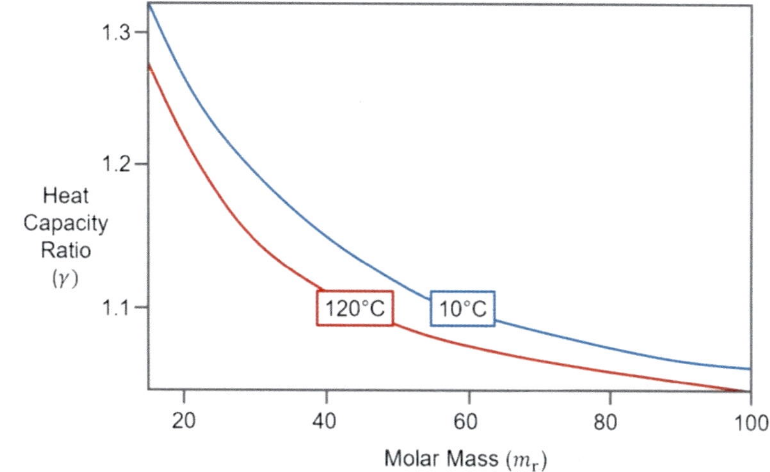

Figure 143 – Ratio of heat capacities (γ) against molar mass (m_r) for various temperatures (T). Adapted from the Gas Processors Suppliers Association (GPSA).

Usually, variations in these parameters are handled by carefully adapting a centrifugal compressor's operation. Hydrogen, however, presents a case so challenging, with its extremely low molar mass and density, that reciprocating compressors are preferred over centrifugal compressors. It becomes impractical to provide enough compression stages; as wheels are added, it becomes increasingly difficult to keep a rotor balanced.

Compression stations for gas pipelines

Long-distance gas transmission pipelines have mid-line compression stations repressurize gas to maintain its flow, overcoming friction and terrain. These are spaced every 60 to 110 km, and operate year-round, stopping only for incidents, routine overhaul or testing the emergency shut-down system. [16] Compressors, like pumps, are selected to suit a pipeline's system curve. A design point overcoming the system resistance is chosen near the *best efficiency point* (BEP). Atmos SIM tracks a compressor's operating point as it moves around its map as conditions change over time, shown as a historic tail.

Reciprocating compressors were originally the most common type, especially on smaller pipelines. They can be placed in series, but are usually placed in parallel so the operator can add together their flow rates. [58] In recent years, many of these reciprocating compressors have been replaced by centrifugal compressors, often powered by gas engines as the driver. Although centrifugal units are less flexible, they are better suited for delivering higher flow rates. [88]

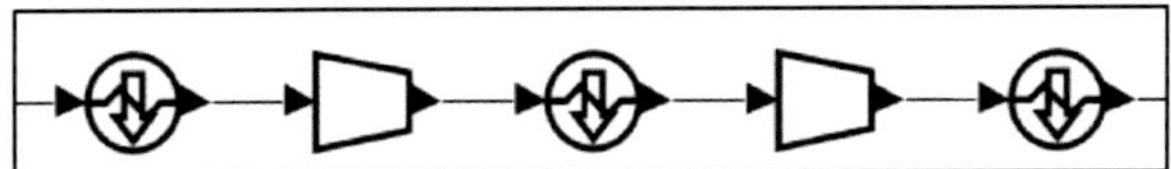

Figure 144 – Two compressors in series, with pre-cooling, inter-cooling and after-cooling. Cooler duty sizes equipment chosen as a heat exchanger, affecting overall compression.

If several compressor stages are needed in series to achieve a compression ratio, although theoretically possible to use any combination of compression ratios with that overall sum, it's most efficient to employ *identical stages*. The theoretical minimum power consumption is for an arrangement approaching isothermal: an infinite number of identical stages, with lossless, complete cooling in between. It practically becomes essential to provide *inter-coolers* as discharge temperature would otherwise exceed $150°\text{C}$. This happens whenever the pressure ratio exceeds $3:1$. These inter-coolers are external heat exchangers rather than integral units within a compressor casing, and each inevitably introduces its own pressure losses, of around 0.5 to 1.0 bar.

It can be advantageous to install a *pre-cooler* ahead of the first compressor stage, as head is proportional to absolute gas temperature at each impeller for the same standard flow rate and compression ratio. [49 55] Following the final stage of compression, the Joule-Thomson effect might provide natural gas with something like a megawatt of after-cooling equivalent *"for free"*, but we often need to transfer many further megawatts of heat using dedicated *after-coolers* so as to meet allowable temperature limits at the final-stage discharge outlet.

The final compression stage in series is most susceptible to surge; its actual volumetric flow rate is lowest.

In a station with multiple compressors operating together, variations in actual volumetric flow rate, molar mass or suction temperature might cause *stage mismatching*, impacting the overall head, efficiency and operating range. [47]

A station might have a different overall surge limit than the individual compressors within it.

For compressors in parallel to be stable, each unit's wheel must be capable of generating the same head. Otherwise, one unit might go into surge whilst attempting to come online as it can't attain the other units' head. Or one unit might, whilst coming online, generate too much head pushing another unit into surge. There are no further restrictions for compressors in parallel, on actual volumetric flow rates, nor on the generated heads. So long as the parallel units' heads are compatible, it is a tried and tested arrangement. [54]

A compression station might be equipped with a combined hot recycle, before the station's outlet cooler, or a cold recycle placed after a station's outlet cooler.

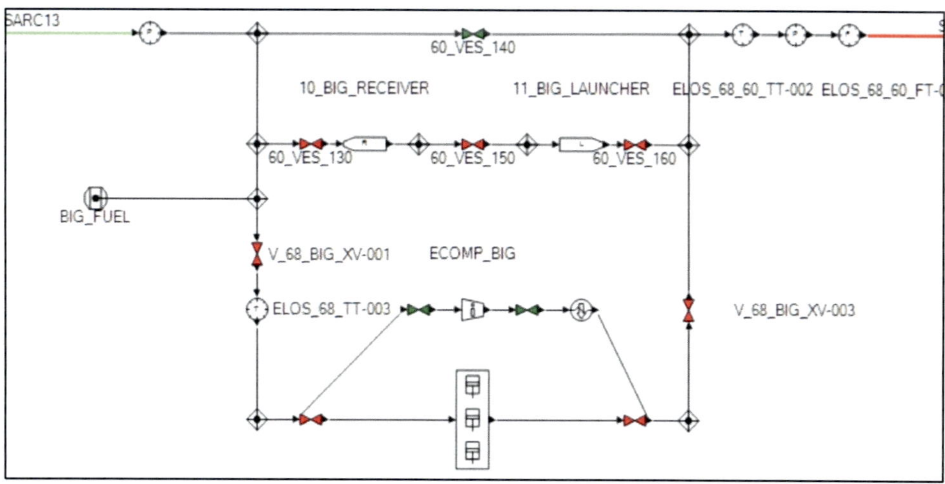

Figure 145 – A bank of parallel reciprocating compressors and an ideal compressor in a compressor station represented within Atmos SIM.

7.8 – DRIVERS: MOTORS, ENGINES AND TURBINES

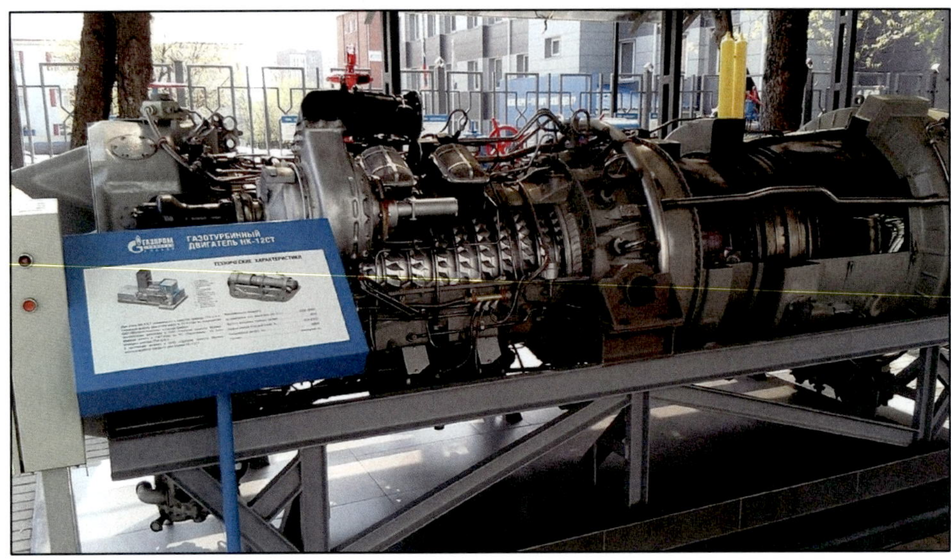

Figure 146 – A gas turbine as used at gas compressor stations in Russia

Prime movers

The performance of pumps and compressors influences a pipeline's operating philosophy. Once suitable centrifugal or reciprocating units are selected, deployed in a series arrangement, or in parallel as a bank, we are comfortable they can impart enough energy to achieve required pressures and flow rates. But what is driving them? The answer is a kind of equipment known as *prime movers*. Deployed in this application they are referred to as *drivers*. Every pump or compressor has a driver, a prime mover consuming fuel or electricity to generate the power running the equipment. These drivers may be fueled by offtake of the fluid being transported. Their overall behavior is non-trivial: If a pump or compressor is fouling up, its driver might actually run faster!

A driver's map, mechanical efficiency and pump or compressor performance translate into four powers. The power input to the driver in the form of a fuel's combustion value or electricity, then power generated by the driver, then shaft power arriving to the pump or compressor, and finally hydraulic power delivered to the fluid. A pipeline simulator can help estimate, assess and optimize capital costs, operating costs and emissions.

Electric motors

If we try to run too fast, we could trip the driver! An electric motor has an electric current rating. Attempting to run above that limit might cause it to trip *"on high amps"*. This has been known to happen inadvertently when an operator changes the composition of a fluid entering a pump or compressor. Atmos SIM relies on rated power and efficiency, as well as a tabulated efficiency map relating the fraction of rated power to an efficiency. Electric motors have become more popular on liquid pipelines recently, while gas pipelines rely more on fueling a prime mover – usually a gas turbine – with an off-take from the gas pipeline to be used for combustion as fuel gas.,

Gas engines and gas turbines

The first prime movers for compressors were reciprocating engines fueled by natural gas (known as gas engines). These are still available and indeed used in certain applications; gas engines operate at lower temperatures than gas turbines, and can be more efficient at generating power. But since the 1970's gas turbines are preferred on pipelines, viewed as being more reliable. Gas is diverted from the pipeline to power these gas-fired engines or turbines via a *fuel offtake*. [16]

A reciprocating engine burns fuel in pistons, but at our level of abstraction we do not think about the famous *"suck, squeeze, bang, blow"* cycle. An engine map lets us treat it as a black box, consuming fuel gas to generate motive power transferred through a rotating shaft. This also means that any pressure oscillations created by the real reciprocating engine – which are certainly there, but hopefully small – are ignored in the simulation. Atmos SIM relies on rated power, fuel factor (volume of fuel consumed to generate power, m^3/W), and a tabulated efficiency map relating each fraction of the rated power to efficiency.

Driver Input Power	1.265	MW
Driver Output Power	0.9734	MW
Hydraulic Power	0.7186	MW
Shaft Power	0.9442	MW

Figure 147 – Driver input power, driver output power, hydraulic power and shaft power as displayed on part of a results grid within the Atmos SIM GUI.

A turbine extracts work from a stream of fluid. Gas turbines, unlike steam turbines, run on combustion products. A proportion of transported natural gas can be taken off a gas pipeline and fed to fuel the gas turbine. A turbine's thermodynamics are remarkably similar to a compressor – the same governing equations apply, just in reverse. A turbine model relies on two user inputs.

A *fuel factor* (SI-units m^3/W) quantifies the energy available as combustion value from a given volume of fluid. And a *heat rate table* relates three dimensionless parameters per row: fraction of rated power, fraction of rated speed, and heat rate. *

Heat rate is the inverse of efficiency (η^{-1}). Based on rated power and operating speed, it is the input power consumed to generate output power to the shaft, specified by vendors in units of $\mathrm{kJ/kWh}$ rather than a fraction or a percentage.

At a given time, a driver operates at a shaft rotational speed (ω_{out}) to provide output power ($\dot{w}_{\mathrm{out}}$). A turbine map of fractions of rated power, fraction of rated speed and heat rates can be considered to be a function that looks like this:

$$\text{Heat Rate} = \text{Heat Rate Map}\left(\frac{\omega_{\mathrm{out}}}{\omega_{\mathrm{rated}}}, \frac{\dot{w}_{\mathrm{out}}}{\dot{w}_{\mathrm{rated}}}\right) \times \text{Rated Heat Rate}$$

To do better than a simple linear interpolation between entries in the table, a third-order polynomial is fitted to the heat rate map. The evaluated heat rate then determines the energy input required by a given turbine as the driver. The combustion value of fuel gas calculates a flow rate to be taken off the pipeline.

In Atmos SIM, the compressor and gas turbine are assumed to be spinning at the same speed i.e. via a directly coupled shaft. If there is transmission gearing (which is unusual) then the manufacturer's turbine maps need their speeds adjusted to be the associated compressor speeds.

A turbine or engine behaves differently depending on the ambient pressure of the outside air as well, which can become important at high altitude locations.

Variable speed drive and variable frequency drive

A reciprocating engine fueled by diesel can be fed more or less fuel so it runs faster or slower, much like pressing one's foot on a car's accelerator. In technical terms it is considered to be a *variable speed drive* (VSD).

An important development for pump and compressor stations along pipelines has been the availability of a cost-effective, reliable, electrically-powered device known as a *variable frequency drive* (VFD). In plain terms, these are big electric motors that spin faster or slower at will. They have become standard drivers on most new or retrofitted pumps in pipeline applications.

---

* These are *rated values* stated at reference design conditions, and may need to be corrected for ambient conditions.

8 – OPERATIONS: CHANGE IN MOTION

8.1 – WHAT CAUSES CHANGE OVER TIME?

Having discussed a pipeline at steady state, and assumed that the internal performance of a pump or compressor can be regarded as a steady state, we now want to consider how to start them up – one of many *pipeline operations*.

Every quantity has an accompanying *gradient*, its rate of change over distance, which is important in its own right. Everywhere we look in the world, gradients across space cause change over time. Looking along the one dimension (length) of a river, we see its elevation profile. The associated gradient causes events to unfold: liquid flows down its slopes. By the same token, in a pipeline, gradients in elevation as well as in other quantities like pressure cause fluid to flow (or not), and heat to be transferred (or not), following the conservation laws we saw earlier.

At the very heart of the numerical methods making up a simulation engine is *calculus*. It is the language formalizing those rates of change to tell us their net result whenever they happen to be imbalanced: accumulation or depletion. It might be that the amount of linepack within a gas pipeline is accumulating or depleting. It might be that the pressure at a point in a liquid pipeline is building up at one moment, then depressurizing the next. Many scenarios happening in a pipeline are not at a steady state, but change over time. They are *transient*.

While studying how heat flows, Fourier stumbled on a problem in the 1800's. Solving it led him to invent a new corner of mathematics. The *Heat Equation* states that at any point, the temperature (T) changes over time (t) at a rate proportional to temperature's second derivative with respect to distance (x):

$$\frac{\partial T}{\partial t} = \text{constant} \cdot \frac{\partial^2 T}{\partial x^2}$$

We saw this earlier, in the chapter on pipeline thermal models. If we consider a temperature profile as values on a discrete mesh, the curvature at some mesh point is a measure of how much a temperature at that point differs from the average of its neighbors on either side. If the temperature profile is linear in space, its curvature is zero, so there is zero change in temperature over time, at least due to conduction. In contrast, if a certain point is a local peak or trough (compared to surrounding points), then its curvature is at a local maximum value, so the temperature at that point changes the most over time.

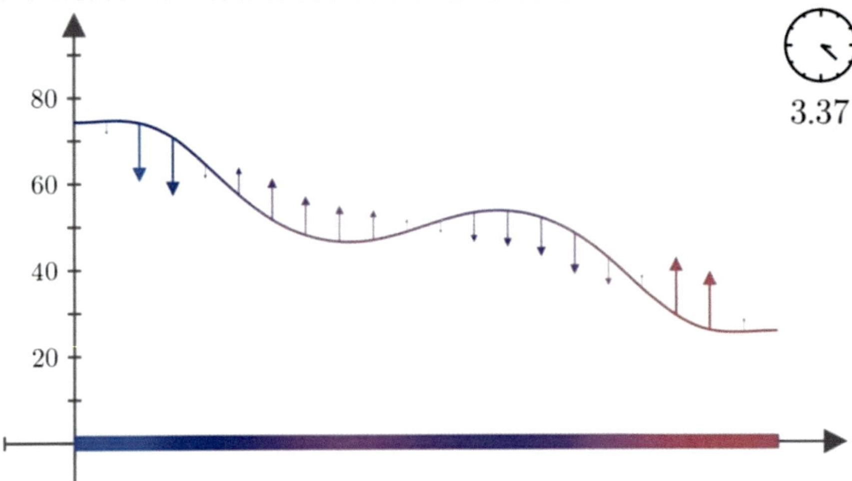

Figure 148 – The heat equation says that temperature at a point varies over time at a rate proportional to how different it happens to be compared to its neighbors at that moment. A point warmer than its neighbors cools down over time. (Credit: 3Blue1Brown on YouTube)

With this in mind, let us cautiously concede that the core ideas underpinning partial differential equations (PDE's) aren't quite as intimidating as they might first appear. Just as we saw how transient thermal effects can be simulated, where they need to be simulated, this chapter turns our attention at when we need to simulate the transient hydraulics that arise during pipeline operations.

To introduce a perturbation in a simulator we change the boundary conditions over time. Any perturbation creates transients in flow. In the absence of further perturbations, the system reaches a new steady state after some finite period of time. Sometimes transients persist briefly, other times they are prolonged.

8.2 – How are pipelines controlled?

Control on paper: narratives and procedures

"Control philosophy" is less stoic than it sounds. It's a definitive narrative of how designers expect a pipeline to operate. Pipeliners must remain in control throughout every operation, and always ensure what happens next is safe. This entails anticipating changes and being prepared with the most appropriate response. A control philosophy is part and parcel of the design of any industrial facility, including pipelines, and helps establish detailed operating procedures. It concerns the technical aspects of an operation. On a liquid line for example, it may say to *"keep suction pressure at station C above 300 psi to avoid slack"*.

As with many engineering applications, the insight we gain by examining a big question (*"How do we control an operation?"*) teaches us more than we had set out to seek. Operating any process obliges us to consider natural principles as well as how humans work with them. Any team tasked with establishing how an industrial operation ought to be executed is burdened with a responsibility for tasks that are not purely technical in scope. There are flaws in psychology; a human-in-the-loop is often the weakest link. Procedures are always liable to evolve with the circumstances. Unless *management of change* (MoC) is good at all levels, there is a proven risk that the designers' original intent will be lost.

> *"Many incidents could have been prevented by better design, and many by better operations. Good operations sometimes compensate for poor design, and vice versa, but that is not something on which we should rely."* [7]

In plainer terms: the way a pipeline is operated on day one is almost certainly not going to be how it will always be operated in the future. Pipeliners often need to come to terms with a new operating procedure, understanding how it can be expected to play out in practice, and dealing with its consequences appropriately. This is one of the many applications in which a pipeline simulator adds value to a pipeline enterprise.

Control in practice: tasks in a pipeline control room

Many pipelines have scores of inlets and delivery points and carry fluids of varying composition. Determining whether it is possible to transfer a quantity of fluid from A to B at some instance in time is not a trivial task. [57] It is even less obvious how best to execute that operation efficiently and without causing disruption. A pipeline simulator helps answer bespoke questions by mimicking the unique operations of each pipeline and presenting the results conveniently.

Experts broadly classify day-to-day pipelining tasks whose responsibility lies with operations teams into four primary workflows, conducted in this order: [58]

- Nominations – can we commit to transporting some fluid from A to B?
- Scheduling – how can our pipeline deliver all the agreed nominations?
- Controlling – what should we do now, and what will happen next?
- Reporting – what have we done?

On a routine basis, at some agreed interval (typically days for liquids, hours for gases), shippers *"nominate"* (i.e. inform a pipeline's nominations team) what fluids they want transporting. Some pipelines rely on longer term contractual obligations, others do not. There is always a due process for agreeing quantities, qualities, origins, destinations and tariffs, [58] intended to assure fairness and integrity. In liquids it is a case of paying more in exchange for reliability: shippers over-book throughput capacity they do not use (*"air barrels"*), to ensure their full deliveries are met if a pipeline curtails a percentage of nominations for any reason, such as unforeseen emergency repairs. [59]

A scheduling team (*"schedulers"*) is responsible for analyzing the nominations. They build schedules to ensure nominations can arrive at the right place at the right time. Gas pipelines respond to frequent demand changes by managing their linepack to keep sufficient quantities of gas available at enough pressure. Already, we see how pipeline simulation can assist these teams with their tasks. Doing nomination and scheduling for a gas or liquid pipeline raises a set of typical questions that can be treated as a multi-variable optimization. A later chapter of this book discusses pipeline optimizers.

Indeed a pipeline simulator is often the *only* way to answer certain questions in a timely fashion. For example, a gas pipeline classed as public infrastructure (a *"common carrier"*) might be obliged to satisfy a regulator that the price they have quoted for a specific nomination is fair. Or a liquid pipeline might have to *"allocate"* or *"pro-rate"* its available throughput capacity if shippers happen to ask for more nominations than can actually be transported that month. [58]

Moreover, for a liquid or gas pipeline alike, quite often a scheduling team must repeatedly adjust their published schedules. This presents a need to automate how these schedules are generated, or else put up with lots of tedious re-work. Eventually, one final schedule is passed on to the control room for execution.

Historically, dispatchers in a control room called up operators in the field to tell them what to do: *"swing this valve"*, or *"start that pump"*, etc. As electronically actuated valves and pumps were introduced along with SCADA systems to run everything, the operator in the field was no longer needed: the roles merged. Today both their duties are carried out by people in a central control room. Some still call these professionals operators or dispatchers, but a more precise term is probably controllers. This team's tasks constitute most of the effort in a control room, and occupy most of its personnel. In their routine duties they're expected to move as much fluid as possible, as efficiently as possible, to meet all of the agreed nominations as published in the final schedule. To do this, they make decisions quickly enough to be effective, but cautiously enough not to be hasty. Their actions are performed in order: *"open this valve"*, *"stop that pump"* then *"adjust this pressure to rise by 10 bar"*. [58]

The team working in a control room is also responsible for appropriately responding to certain situations that are <u>not</u> routine. Some of these situation are quite normal, despite being infrequently encountered. Others, known as *abnormal conditions*, might be fine for now, but cannot be allowed to persist or else eventually such a condition might evolve and become an emergency. When the team or the automated control system deems a situation to be an emergency, an emergency shut-down is carried out (a procedure important enough for a popular acronym: ESD).

"The job of a [pipeline] controller is a bit like an air traffic controller and a pilot [rolled] into one" [58]

This decision-making by control room personnel requires excellent spatial awareness about what is actually happening now (via SCADA and HMI which we discuss in the chapter about simulation live on site), as well as what can be expected to happen next (which we discuss in the chapter on look-aheads). These professionals must have an intimate familiarity with pipeline hydraulics, excellent team-working and a host of other skills that only come through years of thorough training and experience. To this end, pipeline simulators help them with training, assessment and general support so they understand what they did in the past, what they are doing now and what they might do in the future.

Operating a pipeline is a complex task. If we take some action here, what will be the consequence many miles away in a few hours? If some event happens at an unmetered mid-line point, what do we expect to see at the nearest upstream and downstream metering points? If we see a trend over time that differs from what we expect, can we deduce what a probable cause might be? These are teams that expect the unexpected. What if an incoming receipt is interrupted at an origination station because of an issue with an upstream supply point? What if an outgoing delivery is interrupted at a demand point because of an issue with a downstream destination facility? What if several of the pumps or compressors suddenly become unavailable at a mid-line station?

Pipeline operations are captured in a simulator by setting up and running what the industry calls *scenarios*: simulations that commence from some initial state (such as a steady state) and then let events play out over a period of time. An offline scenario is instructed to follow user-set constraints at controlling items.

Control in simulators: ideal constraint setpoints

Initial constraints

We said earlier that an item can be assigned multiple constraints. It satisfies all of them as inequalities (*loose constraints*), and satisfies the most restrictive one as an exact equality (*the controlling constraint*). That means when a new constraint appears, an item might switch the constraint it's being controlled on.

Whenever a simulation commences, if a historic saved state is not available, we must *cold-start*, commencing a transient scenario from an initial steady state. A solver can only progress further steps in time subsequent to that initial steady state. If the initial state is unphysical or unreasonable, it might wreak all kinds of havoc when we try to do anything else after it. To find the initial state, we need an *initial guess* about which constraint is controlling each item. So on every model-control item, the user selects an initial constraint. This is the first combination a simulator tries to reconcile when it seeks the initial steady state. If it finds some of the loose constraints are violated, it will then automatically try other controlling constraints, as we shall discuss in the next section.

Cold-starting a pipeline's hydraulics is much easier to do accurately than cold-starting batches or composition tracking. Errors are inevitably introduced here, lasting a short while until they go away on their own fairly quickly. If the hydraulic transients are slightly wrong, they settle out within minutes in a liquid pipeline, or within hours in a gas pipeline. But if the batch line-up or the fluid composition is slightly wrong, it can take several days for the error to resolve itself: the correct composition or batches takes time to fill the pipeline from the inlet to the outlet.

Whatever we do, for a simulation to recover from an unphysical condition, it needs to be prepared to undo an attempt which went awry. The initial state affects the subsequent simulation, so an experienced user is careful about which constraints are enabled on a model item, and which is initially controlling.

We mentioned before that, during the initial steady state, it's impossible to constrain the pipeline on flow at both ends. Once the transient model is running, this becomes possible for limited periods. We can set the control valve at each end of a stretch of pipeline to control on flow ($\mathrm{F} \rightarrow \mathrm{F}$), and Atmos SIM simulates it. In this scenario, zero flows are stopped, because pressure is remembered from previous steps. Equal flows are steady. Unequal inflow and outflow are an *unsteady* state: i.e. packing, unpacking, filling or draining. For example, a gas pipeline scenario might initialize on a typical pressure for step-zero, then switch all its supplies and demands to known flow patterns. Eventually, unless flow setpoints are exactly balanced, the pipeline packs or unpacks until it switches to pressure control somewhere. This happens so quickly in a liquid pipeline that a user would never generally try to operate that on flow-flow control at all.

Switching between constraints (mode-switching)

A model control item can be told to abide by a set of valid *enabled constraints*. Several can be optionally selected at once. At any given instance in time, only one of that set acts as the *controlling constraint*. Every item operates <u>at</u> the controlling constraint as an equality, without violating the others as inequalities (loose constraints). On each step, each item automatically switches to the most restrictive of its enabled limits as its controlling constraint. Every control mode for every item is handled generically, playing by the same rules as the others. Say a regulator on a gas pipeline is constrained on a flow signal from a custody metering package to meet a nomination, and is additionally subject to minimum inlet pressure and maximum outlet pressure overrides. If pressure falls, as outgoing gas is drawn or compressors stop, an override kicks in. It switches from a constraint of maximum flow ($F \leq \cdots$) to minimum inlet pressure ($P_{\text{in}} \geq \cdots$), so the regulator closes (adjusting its position) such that the pressure at regulator inlet doesn't fall below a limit. This is a switch from maximum flow to minimum pressure. Similarly, if outlet pressure builds up due to valves closing downstream, another override kicks in gradually closing the regulator to never exceed maximum outlet pressure ($P_{\text{out}} \leq \cdots$).

Constraint	Supply	Demand	Regulator	Pump or Compressor
Maximum flow	$F \leq \cdots$	$F \leq \cdots$	$F \leq \cdots$	$F \leq \cdots$
Minimum pressure		$P \geq \cdots$	$P_{\text{in}} \geq \cdots$	$P_{\text{suc}} \geq \cdots$
Maximum pressure	$P \leq \cdots$		$P_{\text{out}} \leq \cdots$	$P_{\text{dis}} \leq \cdots$
Delta pressure			$\Delta P \leq \cdots$	
Minimum pressure ratio				$\frac{P_{\text{dis}}}{P_{\text{suc}}} \leq \cdots$
Minimum temperature		$T \geq \cdots$		
Check	$F \geq 0$	$F \geq 0$		
Maximum speed				$\omega \leq \cdots$
Maximum head				$\Delta H \leq \cdots$
Maximum power				$\dot{W} \leq \cdots$
Maximum position			$X \leq \cdots$	

Table 9 – Constraints for simulating supplies, demands, regulators, pumps and compressors

Almost all inequality constraints in Atmos SIM and on a real pipeline reduce flow rate. That is how it's possible to satisfy them all at once: we get flow rate low enough that the most restrictive constraint is exactly satisfied. The main exception to this is a special type of limit: pump or compressor trips. We shall discuss trip limits in the next section.

Switching part-way into a step and cut-back steps

What do we do if a constraint is violated *during* a time-step? There are two possible options: either we delay switching to the new constraint until the next step, or we cut the ongoing step short, even if this means going shorter than the user-specified shortest allowable time-step. After this switch, the flow rate is going to fall, so the maximum flow limit is still satisfied in the model, going from being treated as an equality constraint to become an inequality constraint. Cut-back steps give us nicer simulated results in several ways. Waiting would otherwise violate constraint setpoints by amounts proportional to step length. If the step length is very long, such as those used in a look-ahead scenario, allowing a violation causes problems. A violation might cause an unphysical state, where some pressure or flow has changed far beyond what is possible. For example, we might end up with pressures of thousands of atmospheres!

Another challenging case is a compressor tripping with reverse flow, or with sub-unity compression ratio ($P_{\text{dis}}/P_{\text{suc}} < 1$). We must have already solved a full step to have even realized that this is happening! Instead of cutting that back, we would have to enshrine it into the official results, switch the constraint at the end of the step, and carry on. On the next step, that compressor has tripped; perhaps it's in bypass enforcing its minimum compression ratio (i.e. 1). So at step start the compression ratio is 0.8, at step end it's enforced to be 1. That's all fine – but when we look at our results, we see occasional one-step violations of setpoints, possibly followed by a slight surge, depending on how violently we bring the model back into compliance on the next step. To get rid of those defects, we'd need to wipe rogue pressures (etc.) from our results.

Atmos SIM does cut-back steps so as to avoid all these artifacts: we never violate any constraint setpoints, and eliminate the need for interpolation. This cut-back approach presents its own challenges which must be overcome. Some of the logical blocks available in the control logic expect a fixed time-step, so we need to ensure they get it.

Another example is as follows. Consider the profile of standard flow rate along a batched liquid pipeline. There are hydraulic transients bouncing around that cause the flow rate at a point to change over time. However, there are also discontinuities at the batch interfaces where the standard flow rate changes as a step moving along with the batch. If we take the step-start and step-end flow rate profiles along a pipe and interpolate linearly to the cut-back point, it is going to be quite wrong around that batch interface! So what we must do instead is:

1. Use the batches and the standard flow (Q_{std}) to compute the actual flow (Q) at step-start and step-end *
2. Linearly interpolate the actual flow (Q) at the partial step
3. Revert the batch positions to their step-start values
4. Re-run the composition tracker to advance batches for the partial step
5. Recompute the partial-step-end standard flows (Q_{std}) from the batch positions and interpolate actual flows (Q)

There are many such complex features in Atmos SIM to allow cut-back steps.

---

* This works because although there is a discontinuity in standard volumetric flow rate across a batch interface, the actual volumetric flow rate stays the same here.

Trip limits on pumps and compressors

In addition to constraints, pumps and compressors are also subject to *trip conditions*. Centrifugal pumps and compressors can follow their setpoint into an operating region that could damage the unit. In the event that this occurs, Atmos SIM will cause the unit to rapidly shut down, an event known as a *trip*. The typical conditions that can trip a unit are many and varied, and are listed in Table 18 and Table 16. Many trip conditions have the opposite *"sense"* of the normal control modes. For instance, a variable speed unit can be controlled on maximum speed and can trip on minimum speed. This is because the normal control limits prevent the unit from exceeding maximum speed, so no trip is needed there. One might ask: why don't we also include a minimum speed limit in the normal control? This is actually impossible, as it can lead the model to situations where there is no solution that satisfies all of the constraints as inequalities. Indeed, this isn't just a quirk of the model, the real pipeline would also behave like this – some of the controllers would fail to enforce their setpoints.

Centrifugal pumps and compressors without built-in recycling can also be set up to trip if they go too far into surge, which might cause damaging oscillations.

An ideal pump or compressor has only one trip limit: reverse flow, like all pumps and compressors. And this trip limit is only enforced if no check valve is enabled because a simulator won't allow reverse flow through a check valve.

If – and only if – no other trip limits are enabled (the user doesn't specify when it should trip), and recycling is disabled, then a centrifugal pump or compressor will always trip when minimum speed is violated, a typical minimum speed for a centrifugal compressor being 3 000 RPM. If no other trip limits have been explicitly specified, and no minimum speed is specified, then Atmos SIM will use the minimum speed indicated by the head map. Otherwise, if the head becomes so low as to trigger minimum speed, that limit is enforced by recycling. A recycle flow rate is calculated to maintain minimum speed, to keep it above the centrifugal pump or compressor's surge line.

8.3 – TRANSIENTS VERSUS STEADY STATES

It can be unclear what pipeline simulation people mean exactly when they say *"dynamics"* or *"transients"*. We'll stick with using the term *transient* – treating it as a noun as well as an adjective – because that is how pipeliners usually talk about changes over time. The intended scope of the term encompasses the full realm of perturbations to a system causing any chain of events. Broadly speaking, changes in a pipeline can be thought of in two further subcategories, *gradual changes* or *rapid changes*. In this chapter, we look at gradual changes, which we call *slow transients*. In the next chapter, we focus on rapid changes, which we call *fast transients*. It's always worth specifying which of the two broad timescales we are referring to.

A pipeline must be controlled to protect its pipes and equipment. A control system is responsible for several tasks. It adjusts pressures and other variables such as valve position to deliver a new throughput. It starts and stops pumps or compressors, or adjusts their speed (loading and unloading). It raises alarms whenever any limits on any operating condition are exceeded at any point, and in certain scenarios it initiates an emergency shut-down (ESD) sequence. It schedules upcoming fluid shipments, expressed as *nominations* of gas deliveries or *batch line-ups* of liquids. It monitors the performance of machinery and pipes for wear, aiding with predictive maintenance. It keeps all pressure surges within acceptable levels, or else alleviates any violations in a predictable manner. All of these scenarios involve big changes over time.

A pipeline model is initially built and validated based on steady states. Having satisfied ourselves that the basis of our model is sound, we then configure operating scenarios to change over time. In Atmos SIM, the same underlying model serves as a starting point for these transient scenarios. Pipeliners use it to configure surprisingly complex operations, exploring the consequences of events their pipeline might encounter, and possible ways of configuring their control system and setting in place the appropriate operating procedures.

It's worth pausing for a moment to ask whether transients are needed at all. Almost everything a pipeline does varies over time, but in spite of this, a lot of people don't use transient pipeline simulators! For certain applications, it is enough to merely use a *succession of steady states* (SSS). In that approach, every step in time is taken to be at a steady state, though it is different to the next. So who needs to go to the trouble and expense of a transient simulation?

Successive steady states are quick solving, and suitable for some operational scenarios. They're good for sizing equipment such as pumps, for expansion studies, and for anticipating how different batches flow along a liquid pipeline. A low-pressure gas distribution network can also be treated as a succession of steady states, because it doesn't hold gas in particularly large quantities as linepack in the pipeline, so it's fair to ignore the transient packing and unpacking over time. In that application, the focus is load forecasting, and for that task there's no pressing need to simulate the full transients in the pipeline.

So what questions can we answer only with a transient simulation? If we ask: *"what if so-and-so happens?"*, and we see an important knock-on effect that is not a steady state, successive steady states won't give us the right answer. A real transient is categorically nothing at all like a succession of steady states. A succession of steady states will assume that a transient happens *too quickly* – after all, a steady state is one in which transients have already settled. But in a real transient, unsteady effects persist and continue to evolve over time.

In certain scenarios, successive steady states give us the correct answer for a question we ask, but they don't let us ask other important questions. In the past, high-pressure gas transmission pipelines used to arrange nominations weeks in advance, and only agree to transport a quantity of gas from A to B if they're capable of doing so at steady state. Today, the lead time is shorter, with gas transmission pipelines operating more flexibly for just-in-time delivery and nomination. A gas nomination changes over time, so accounting for that change with better granularity (1 hour being more granular than 1 day) allows a transient simulator to inform us if a pipeline's throughput capacity is sufficient. This lets a gas transmission pipeline make full use of all its available capacity.

In certain scenarios, the limitations of successive steady states prevent us from using them at all beyond their scope of application. For example, the maximum stress a pipe experiences is caused by hydraulic transients which appear only briefly, a phenomenon which successive steady states cannot capture. Steady state thermal models can also be very wrong; fluids take hours or days until they reach an equilibrium temperature, and the ground takes days or weeks to reach equilibrium.

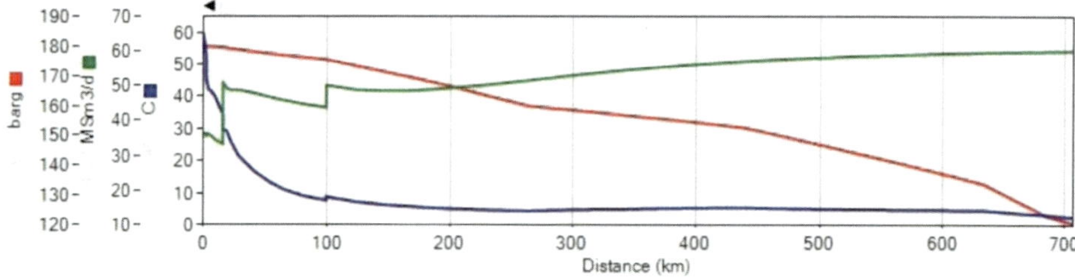

Figure 149 – Profiles of pressure (red), flow (green) and temperature (blue) along a gas pipeline

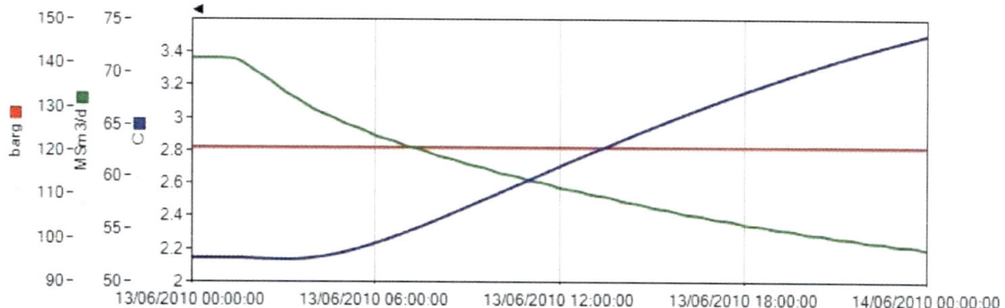

Figure 150 – Trends of pressure (red), flow (green) and temperature (blue) at the line's outlet

Transient simulations are therefore needed for all kinds of important tasks. The user-input parameters to the model are exactly the same: all the hard work is done in the physics engine. A transient simulator implements time-varying terms in its conservation laws, making it a true heavyweight, far more versatile than successive steady states. A fully transient hydraulic and thermal model unleashes computing power on predictive problems, taking us well beyond the realm of glorified spreadsheets. By modelling a pipeline's transients along with its control strategies, they simulate operating scenarios that anticipate the consequences of taking an action or introducing an upset to the pipelining process, whose physics are completely missed by successive steady states:

- Start-up
- Cool-down
- Emergency shut-down (ESD)
- Production planning forecasts
- Availability studies
- Design verification
- Equipment sizing
- Developing control strategies
- Pig planning (offline) and pig tracking (online)
- Operator training
- Supply and delivery switching and variations
- Purging and loading
- Leaks and ruptures
- Loss of pumping or compression stations
- Surge analysis, including water hammer in liquid pipelines
- Linepack management, especially in gas pipelines

Pipeline task	Adequate simulation approach
Liquid batch scheduling	Logistical (no physical simulation)
Pipeline sizing	Steady state
Expansion studies	Steady state
Equipment selection	Steady state
Liquid tank filling or draining	Successive steady states
Liquid batch or pig arrival time	Successive steady states
Low-pressure gas distribution	Successive steady states
Effect of ground temperature	Transient scenario (days)
Linepack in gas transmission	Transient scenario (hours)
Compressor loading / unloading	Transient scenario (minutes)
Emergency shut-down	Transient scenario (seconds)
Surge analysis studies	Transient scenario (split seconds)

Table 10 – Adequate simulation approaches for some common pipelining tasks

8.4 – DYNAMIC OPERATIONS ON LIQUID LINES

What is a transient?

There are pipeline operations which involve change over time but are well represented as successive steady states. There are also operations which cannot be represented as successive steady states but still do not involve fast transients like those used in surge analysis. The language pipeline simulation people use might cause a little confusion here because any change over time might be referred to as "dynamic simulation" or a "transient scenario". For this reason we shall run through a couple of pipeline operations which are quite routine and offer none of the excitement of the fast transients we'll meet later.

Tanks equalizing

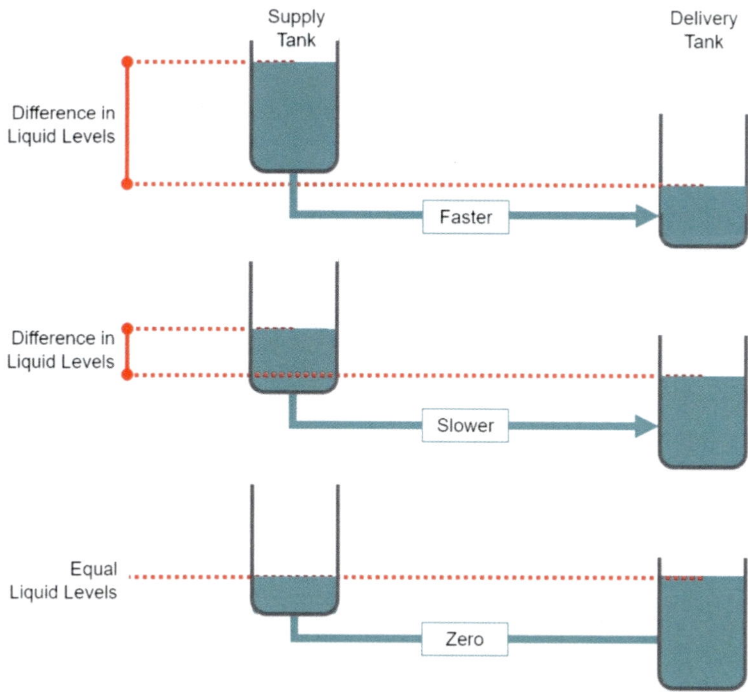

Figure 151 – Liquid levels in atmospheric tanks equalizing by flowing through a pipeline

A liquid pipeline (10", 1 km long) is used to empty a tank (A) into another (B). Both tanks are 10 meters in diameter. Over time, the pressure at the inlet from the supply tank (A) decreases as its water level falls (from 5 meters initially), and the pressure at the outlet into the delivery tank (B) increases as its level rises (from 3 meters initially). How long does it take to complete this operation?

As an atmospheric tank is filled and drained, it exerts a corresponding head. This means that the pressures at the pipeline's supply and demand points change over time, which in turn affects the flow rate (Q) – via a coupled effect.

Simulation step size (seconds)	Predicted duration of operation (minutes)
5	99
10	99
30	100 ± 0.5
60	101 ± 1
120	104 ± 2
300	110 ± 5

Table 11 – The predicted duration of an operation depends on the step size we simulate it with

We have defined *"finished"* as being almost stopped, having the velocity of a tortoise ($v = 0.1 \text{ m/s}$). It turns out this operation takes 99 minutes to complete. We find this answer correctly to within a reasonable tolerance by running a transient scenario with a time-step interval of 60 seconds (1 minute) or closer.

The thing is, any changes in pressure transmit very quickly in a liquid, because the speed of sound in a liquid is something like 1 000 m/s. Moreover, a change in pressure does not cause a meaningful change in the density of a liquid; it is almost incompressible, as we discussed in the shut-in chapter, so the inventory (mass or standard volume of fluid) held within the pipeline's volume is constant.

We note this to say that at any given point in time during this changing scenario, the pipeline is at a steady state. That is to say: if part-way through this scenario, the supply tank had an inlet flow maintaining its level, and the delivery tank had an outlet flow maintaining its level, then the pipeline would simply carry on operating indefinitely, exactly as it does at that instance in time. In that sense, this operating scenario can be simulated as a *succession of steady states* (SSS). If we pause the transient scenario at any point, and apply the same boundary conditions (in this case: tank water levels translated to pressures) by inputting them as the constraints of a fresh offline scenario, it will simulate a steady state identical to the original flow.

We won't list 100 rows here, but will compare some steady states to snapshots from the original transient scenario to demonstrate the validity of this approach.

Time-stamp (minutes)	Level A (m)	Level B (m)	Flow rate (m^3/s)
0	5.00	2.00	0.431
20	4.42	2.58	0.033
40	3.99	3.01	0.023

Table 12 – Each snapshot from this scenario is equivalent to its own steady state

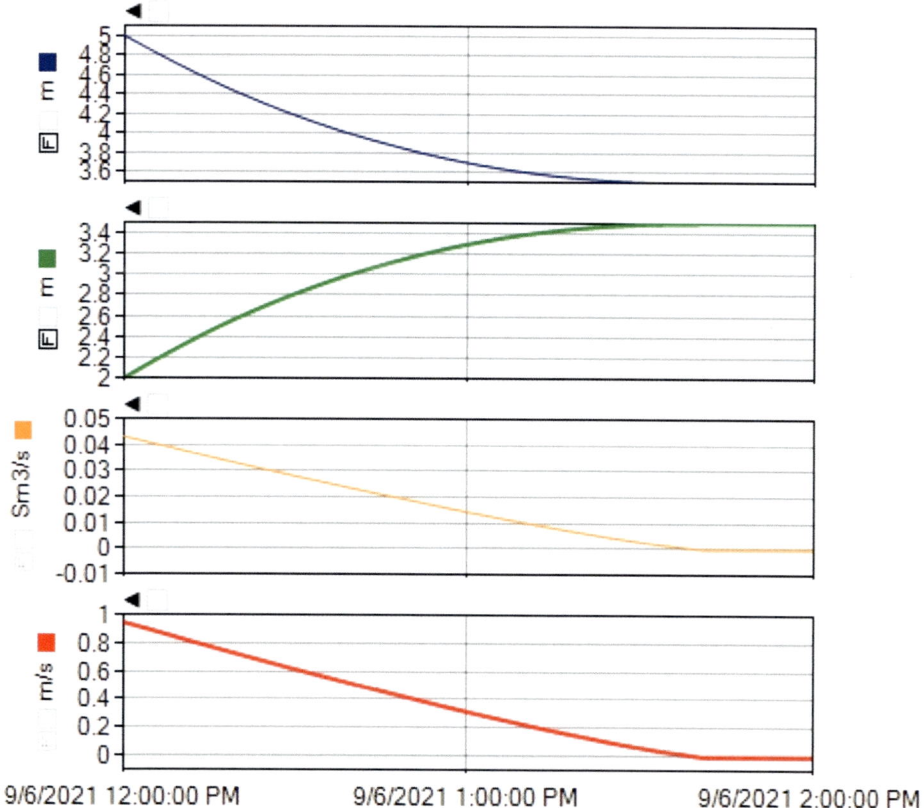

Figure 152 – A transient simulation of the same tank equalization scenario gives these trends for inlet and outlet tank levels, flow rate and velocity respectively

Batching and liquid composition

Numerous scenarios take place when planning a batched operation, including:

- The effect of drag reduction agent (DRA)
- A mid-line off-take controlled on pressure at the end of a short branch
- Tracking the sources of an arriving batch by tracing its origins through mixing and splitting points
- Injecting a compressible batch such as propane into a flat section, to observe how differently inflow and outflow behave as it moves through the pipeline compared to an almost-incompressible batch

Liquid pipelines deliver batches in the right sequence, and contend with the transit time of each batch. Some liquid product batches are interchangeable (*"fungible"*), whereas others are stipulated for delivery as the same molecules. Crude oil pipelines use the concept of batch sequencing and fungibility as well, albeit in a less formal manner some may refer to as the *"common stream"* – which we can summarize as follows. The principle is to avoid a low-grade liquid contaminating an adjacent high-grade liquid, and ensuring a special batch like compressible butane is buffered by suitable neighbors. [58]

Heavy Sour	Light Sour	Light Sweet	Butane	Light Sweet	Light Sour	Heavy Sour

Table 13 – This crude oil batch sequencing cycle places butane beside compatible liquids. [58]

A set of non-physical scenarios are set up to explore the various permutations of sending batches through a pipeline in different orders.

Jet Fuel	Diesel	Jet Fuel	Low-Octane Gasoline	Jet Fuel	Low-Octane Gasoline	High-Octane Gasoline

Table 14 – This refined products sequence places suitable batches upstream and downstream of each batch. The pipeline flows left to right. [58]

When one or several of these scenarios warrant exploring further, the simulator can be instructed to predict the physics of batch movements along the pipeline. Between two adjacent batches, an interface will form within which the liquid is a blend of gradually varying composition.

Figure 153 – Between two adjacent batches, an *interface* might form in which there is a blend

Placing a type of pig known as a sphere between batches can help reduce this, but it doesn't eliminate the interface region altogether. Besides, pigs come with their own challenges, and so aren't actually used to separate most batches on pipelines. This raises questions about what best to do about this interface liquid, whose volume can be substantial.

Do we cut it half-way through, so half of it goes into the upstream batch, and half into the downstream batch? That might be unacceptable if one of these batches must comply with exacting standards, jet fuel being an important example.

Do we cut before it (at its head), or after it (at its tail)? If the interface has formed between high-grade gasoline and low-grade gasoline, the pipeliner might choose to declare contaminated fluid to be low-grade gasoline now.

But sometimes, preserving the quality of one batch is to the detriment of another. In some cases, this interface is cut at both its head and its tail, and diverted into a dedicated tank to be dealt with without affecting the adjacent batches – perhaps blended later or marketed separately. As soon as any quantity of this interface is removed from the pipeline it is known as *"transmix"* – though the naming convention varies, and some companies treat these as synonyms. [58]

Apart from changes in composition over time as batches move along the route and interface mixture forms between them, the actual simulation of batch movements is quite straightforward. Provided all the batches can be treated as being almost incompressible, then at any given instance in time the hydraulics of the operation can be simulated as a succession of steady states. [*]

This approach gives us the wrong answer if one of the batches is compressible, but many pipelines never or rarely transport a batch of something like propane, so the teams operating multi-product pipelines and crude oil pipelines are quite used to the incompressible way of scheduling their batches and planning how they will move

[*] The motion of the batches can be handled separately from the hydraulic model; that must be treated in a non-steady-state manner, of course.

through the pipeline. Incompressible liquids flow perfectly balanced: a change in flow rate anywhere is immediately seen everywhere. For this reason, they are a bit like pushing solids along. The logistical task is analogous to rail wagons forming a trainset, which is why the charts following batches along a pipeline are referred to by liquid pipeliners as railroad charts.

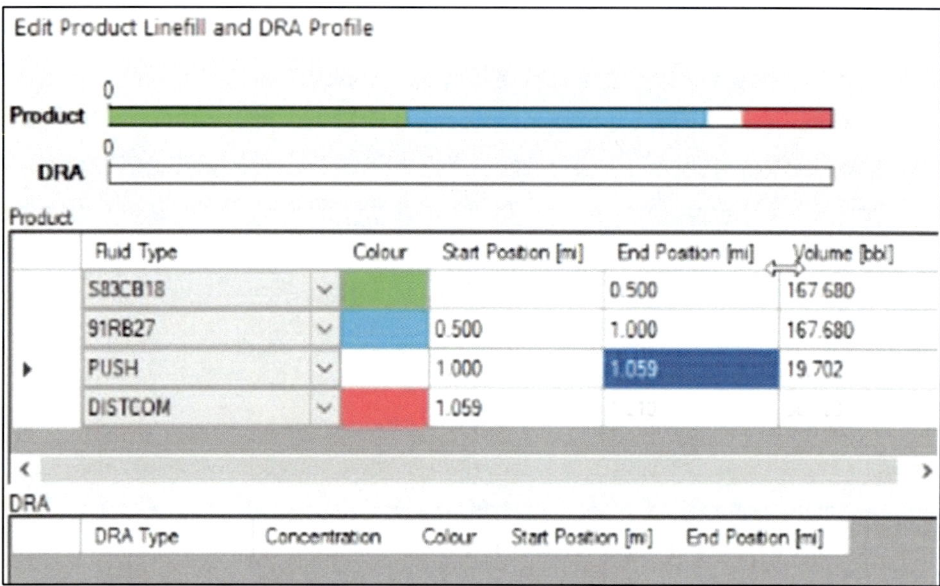

Figure 154 – The series of batches within a pipeline at the start of a scenario can be specified in Atmos SIM as an initial linefill. It is possible to do the same for drag reduction agent (DRA).

As well as keeping track of what is moving along the pipeline, there is a list of batches waiting to be injected. The consecutive batches in series follow an injection schedule. Say all existing linefill consists of light crude, and then we start pumping heavy crude in from the supply. If supply and demand pressures stay the same, how does flow rate vary over time as the heavier batch moves along the line? How long would it take until the line is full of heavy crude?

Batch List

Batch fluid	Volume (bbl)
SYNJET	100
DIESEL	50
93ULP	80

Add Batch

Figure 155 – Atmos SIM allows a queue of batches to be defined in advance by their volume as a *"batch list"*. They are injected in that order from a supply point within an offline scenario.

Changing from one product to another, or a mixed slate, might require a redesign of pumps and equipment, or even the pipeline itself. On a gathering line, flow rates can be quite variable. Moreover 200 MBBL/D of gasoline is quite different to diesel; diesel is denser, more viscous, and less volatile. A set of products to be transported is put together. Studies into the implications of potential future expansions to the pipeline inform enlightened decision-making at the early stages of a design or modification. [45]

Pump start-up

Starting up a pump can be a complex operation. It must follow set procedures. Pipeliners refer to it interchangeably as start-up or spin-up.

A pump is said to be *poorly primed* if vapor cavities have formed as fluid came to rest, such as what might happen if the pump has suddenly tripped. Once pumping resumes, their collapse causes transients. Modelling this is complex, a field of study in its own right (multi-phase flow). Typical hydraulic simulation shows us where the onset of this condition is so that we can avoid it, rather than attempting to simulate its chaotic effects.

Water hammer is likely to occur in a pipeline when a poorly primed pump starts up against air.

An operator might wish to avoid starting up a pump in a liquid pipeline which has gone *slack* along its route. However, this might not be an option in practice. Lots of pipelines operate tight, but if they are left shut down for more than an hour or two, slack could develop. The operator can't just not start the line. Yet if there's slack present, they can't get rid of it until they start the line! Nonetheless, operators must be cautious and aware, as leaks could occur during the start-up. A pipeline simulator could incorporate a slack model, as we discuss in a later chapter.

Dead-heading is another bad idea. This term refers to starting a pump facing into a closed valve so no fluid can flow. A pump should not be available to start until the operator first ensures all relevant valves are open. Either all the valves should be open from source to discharge, or valves should be open to allow recycling of the complete flow through the pump for the initial start-up period.

A pump in Atmos SIM can be configured via custom logic as available to start-up only if certain valves are open.

If the pump is going to supplement another pump already operating in parallel, starting it up might involve *recycling* for a while until it reaches a suitable speed to match the head being produced by the online pump(s). If the new pump is brought online at too low a speed, it will get reverse flow which is bad for the pump – or a check valve will close in which case the pump will be dead-heading, which is also bad. Only after reaching a matching speed is it ready for valves to be switched so it commences pumping its portion of the pipeline's throughput. Pumping with 100% recycle is like a hairdryer in an insulated box, in that the pump is going to overheat unless the operation is properly planned.

A control valve at the discharge side of a pump can be controlled to gradually open during spin-up

Starting a pump forces open the check valve at its discharge-side, accelerating fluid and propagating a pressure rise. A chain of fast transient events involving hydraulic surging could occur along a pipeline if a pump start-up is conducted too suddenly.

Pump operating constraints

We cannot directly use a single-speed reciprocating pump to control on pressure. It is better to think of it as an *imperfect* flow setpoint. Variable-speed reciprocating pumps can be set up to control on essentially the same setpoints as centrifugal pumps.

Constraint	**Pump**		
	Ideal	Reciprocating	Centrifugal
Maximum flow	□	□	□
Minimum suction pressure	□	□	□
Maximum discharge pressure	□	□	□
Maximum hydraulic power	□	□	□
Maximum driver power	□	□	□
Maximum head	□	□	□
Maximum speed	-	□	□

Table 15 – Operating constraints on variable-speed reciprocating pumps and variable-speed centrifugal pumps. Ideal pumps are implicitly variable-speed. A tickbox (□) is a constraint whose selection is optional.

What causes a pump to trip?

A fixed-speed pump protects itself from excessive cavitation by tripping on minimum suction pressure ($P_{\text{suc}} \geq \cdots$). The intention is to avoid suction conditions falling close to the net positive suction head required (NPSHr).

Strictly speaking, one might expect this constraint to be captured in terms of head rather than pressure, because head translates to pressure depending on a liquid's density, and a margin must be added for the liquid's vapor pressure and dissolved gases. But in practice, an operator uses a generous margin suitable for their batched products. Provided their liquids are almost incompressible, the variation is typically within a narrow range, so people consider it reasonable to say that a liquid boils at a given pressure. A notable exception is compressible liquids such as propane or LPG.

Reciprocating pumps trip on a subset of the trip conditions for centrifugal pumps. If no trip conditions are defined by the user then Atmos SIM trips them on minimum speed, reverse flow, or suction pressure less than 1 bar. The user can define custom minimum suction pressure and maximum discharge pressure trips; these would mostly be used with fixed-speed reciprocating pumps.

Trip condition	Pump		
	Ideal	Reciprocating	Centrifugal
Inferred minimum speed	-	□	If no others
Minimum speed	-	□	□
Minimum suction pressure	-	□	□
Maximum discharge pressure	-	□	□
Max. discharge temperature	-	□	□
Minimum surge fraction	-	-	□
Reverse flow	Always, if there is no check valve		

Table 16 – A pipeline simulator's trip conditions for ideal, reciprocating and centrifugal pumps

Case study: compressible batches over mountains

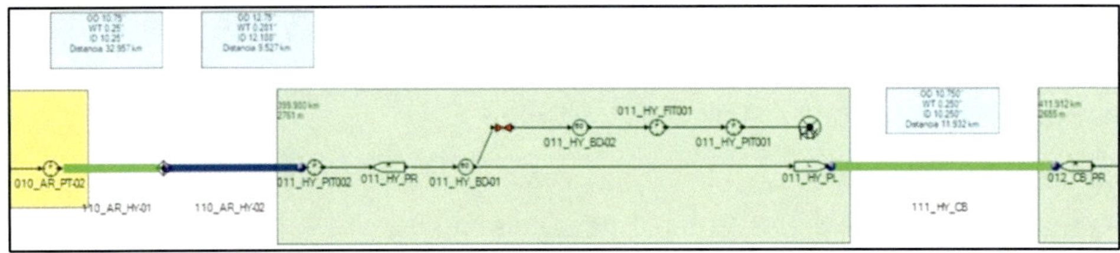

Figure 156 – A station along the route of the pipeline, as deployed on site to predict the arrival times and monitor pressures throughout the system as compressible liquid batches are flowing.

This liquid transportation network covers a country in Latin America. It has 16 pumping stations with a total installed power of 40 thousand horsepower and a length of 3 thousand kilometers of pipelines. The pipeline transports around 25 000 barrels per day of reconstituted crude oil (RECON) for export through a terminal in a neighboring country. It also supplies a remote region with batches of liquefied petroleum gas (LPG) and condensate. To do so, it pumps compressible batches across mountainous terrain. In recent years, the pipeline has been prone to sabotage, disrupting major refineries.

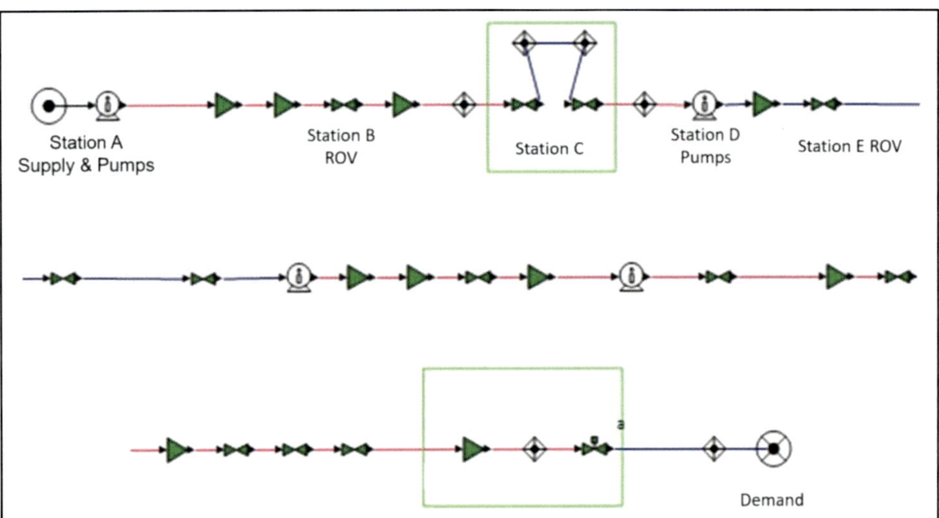

Figure 157 – This pipeline carries batches of crude and liquefied petroleum gas (LPG) across mountain peaks to an export terminal.

The presence of compressible batches in this pipeline leaves it prone to column separation at many points. Many sections are routinely operated in slack flow conditions, so batch tracking on this kind of pipeline is challenging, because it is difficult to predict actual volumes of in-line batches, and where the gaps are.

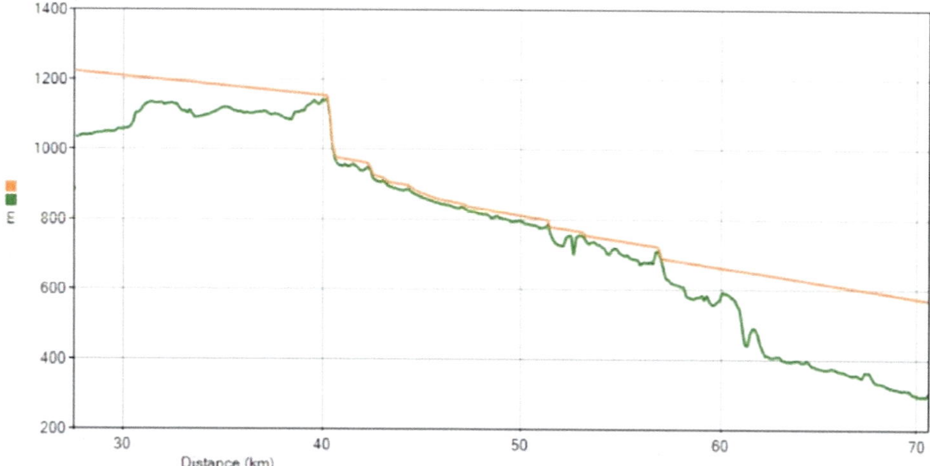

Figure 158 – The head profile along this pipeline shows where the pressure has fallen below the vapor pressure of the fluid batch, so slack flow occurs starting at the top of the first big mountain peak and again at several smaller peaks downstream.

To make accurate predictions of batch and pig arrival times, corrections must be continually made automatically and manually, based on batch detectors at stations along the pipeline's route.

Slow, viscous crude and laminar heat transfer

Using a simple thermal model for viscosity has its shortcomings. In this example we have a run-in with the one-dimensional assumption. Consider a heavy crude line in laminar flow in cold soil. No turbulent vortices are transporting heat radially (across the flow direction). The fluid in the outer lamina loses the initial heat faster than the fluid in the inner lamina: it's in better thermal contact with the environment, and it's moving slower. Over time, the outer fluid becomes more viscous, and becomes denser. Cold, viscous oil on the bottom of the pipe gets colder and more viscous, and eventually stops. The effective pipe diameter is reduced and the process repeats. Eventually the flow stops entirely.

This is one of very few situations the authors encountered where a pipeline simulator's one-dimensional (1-D) approximation breaks down in a manner that can't be accurately compensated for by the usual methods.* In every other common scenario, there is something we can do via tuning, or via adjustments to friction factor correlations, etc. to keep the problem one-dimensional.

---

* The only other such scenario is perhaps very high speed transients in a surge along a liquid pipeline, where there might be some non-one-dimensional flow.

8.5 – Dynamic operations on gas lines

Gas transients are gradual

Gas operations are never truly in steady state: the pipeline is always packing or unpacking, its inventory varying over time. But transients aren't simple! Incorporating the effects of significant compressibility and dynamics acting over time complicates the expressions describing a fluid beyond what is reasonably possible to calculate in a quick spreadsheet. It is only a transient simulation that lets us properly ensure that the planning and pricing of nominations make the best effective use of available pipeline transport capacity. These transient scenarios also allow us to anticipate the progress of a composition front or a pig during its journey along the route of a gas pipeline.

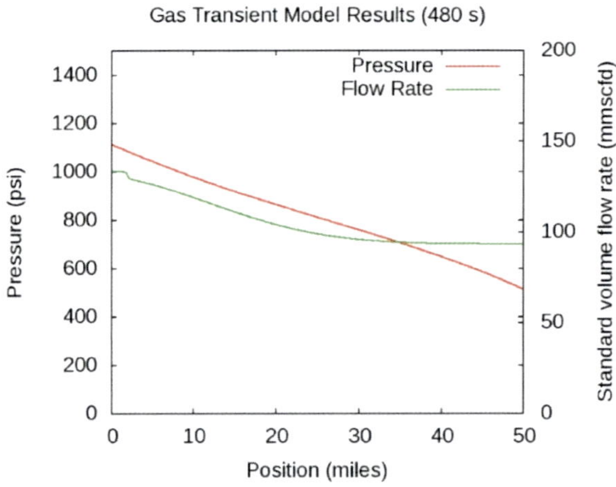

Figure 159 – A snapshot of a transient within a gas line. This profile evolves over time.

A gas pipeline operator needs plenty of time to take actions, as the pipeline is very slow to respond. As gas pipelines rarely reach steady state; even if we are operating in a so-called *"free-flow"* condition through a station, the packing and unpacking always leaves each section in some kind of flow imbalance. This makes predicting upcoming behavior well in advance imperative in operation. A hydraulic transient may cause a gas to compress, giving rise to a compression wave.

The transient behavior of a gas line is all about inventory. An operator can build up linepack by either increasing incoming flow rates or by decreasing outgoing flow rates. Gas pipeliners define the acceptable ranges of inventories and flow rates in each section of the pipeline network during all of its modes of operation. Transient scenarios truly capture the conditions throughout a pipeline as it undergoes these operations. Is there too little inventory during the course of nominations, violating the lowest allowable pressure at a delivery? In practice this might mean the delivery was curtailed to avoid that happening. Or is there too much inventory during any operation, violating maximum allowable pressure at a point along the route?

Compressor start-up or ramp-up

During a *free-flow* condition through a compressor station, its discharge and suction manifolds are at the same pressure. A real life start-up procedure isn't an instant *"turn this on"* instruction like one can do in a model. In reality, the compressor's outlet valve is closed, and a compressor recycles over something like 15 minutes. To start up the first compressor, a suction throttling valve can ensure the compressor train's net pressure gain remains minimal. This facilitates starting multiple compression trains prior to *loading*, such that two trains can be effectively brought online in parallel, whilst avoiding straying into the choked flow region of any compressor's performance map. To ramp-up from free-flow, a master controller discharge pressure setpoint is at the current free-flow pressure prior to start-up. This might be something like 80 barg. After the second compressor comes on and compressors are loaded, this setpoint is gradually ramped up over a few minutes to 10 bar higher than that.

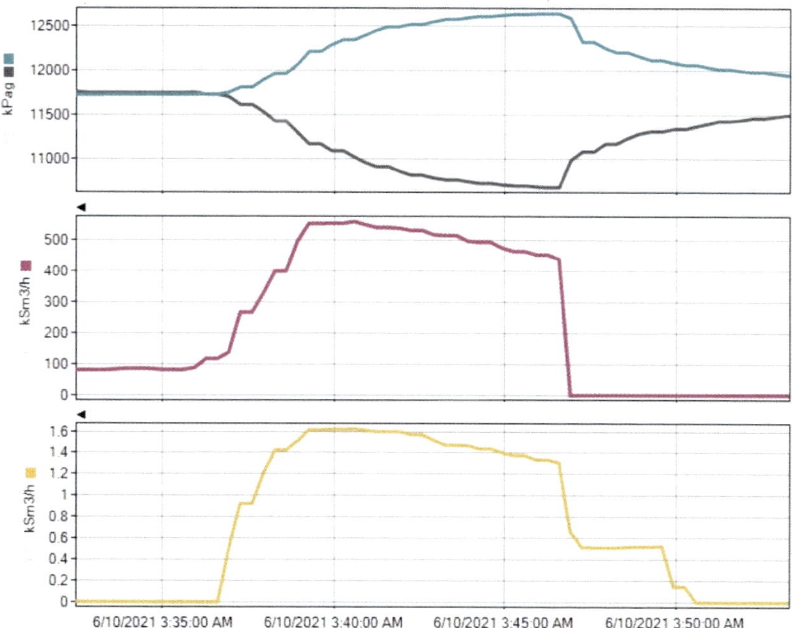

Figure 160 – A compressor spins up, operates for a few minutes, then spins down. During spin-up, suction pressure falls (top – grey) while the discharge pressure rises (top – teal), and vice versa during spin-down. The gas flow rate through the compressor (middle) is achieved by consuming a commensurate flow rate via a fuel off-take (bottom). During spin-down, the mainline flow rate drops to zero while fuel gas doesn't, because the compressor is recycling.

Operating constraints on compressors

A centrifugal compressor can switch constraints to enforce limits on speed, power, suction and discharge pressure, flow rate and discharge temperature. A reciprocating compressor basically enforces a flow setpoint. It displaces the same volume per stroke regardless of pressure, so placing a throttling control valve on the discharge side won't reduce the standard flow rate through it. Instead, a control valve is often placed on the suction side so that the gas entering the compressor is at lower pressure, [58] the result being that the same actual volume caused to flow per stroke corresponds to a smaller standard volume. In Atmos SIM, an internal recycle within the model item lets a reciprocating compressor control on an idealized suction pressure constraint.

Constraint	**Compressor**		
	Ideal	Reciprocating	Centrifugal
Maximum flow	□	□	□
Minimum suction pressure	□	□	□
Maximum discharge pressure	□	□	□
Maximum pressure ratio	□	□	□
Maximum hydraulic power	□	□	□
Maximum driver power	□	□	□
Maximum head	-	□	-
Maximum speed	-	□	□

Table 17 – Operating constraints on ideal, reciprocating and centrifugal compressors. A tickbox (□) is a constraint whose selection is optional.

Compressor turn-down with a suction throttle valve

A compressor's suction-side throttle valve is used during some transient and abnormal operating conditions, such as start-up, or unusually high suction pressure. Normally it is fully open. It can be controlled under *"split-range"* (when a single controller is employed to control two elements) from the compressor train's load-share controller, accommodating the characteristics of the compressor and its driver, including minimum speed and loading requirements.

When a pipeline is operated in *turn-down* (i.e. lower than nominal flow rates) its pressure profile flattens. The head requirement at a compressor station falls, so each compressor slows down accordingly, until it reaches its minimum speed limit. If throughput falls any further, and no suction throttle valve were provided, then compressor stations would no longer be capable of achieving their controlled parameters (like maximum pressure in the discharge manifold).

For this reason, the inlet piping to each compressor train is equipped with a carefully controlled suction throttling valve, to control the throughput after a compressor has slowed to its minimum governor speed. This lets us deliver the lowest pipeline throughputs within the required operating envelope: during low flow rate conditions, the compressor is operating at a low speed, so we tend this valve to closed, reducing the compressor's suction pressure, which in turn increases the actual suction flow rate and keeps the compressor out of surge.

What causes a compressor to trip?

A compressor trips on maximum discharge temperature ($T_{\text{dis}} \leq \cdots$). Otherwise, in some conditions, the discharged gas would become hot enough to cause damage. People informally say they don't want their equipment melting! Some centrifugal compressors on older lines have no dynamic speed control; they are just *"on"*, running at a single fixed speed. A fixed-speed unit cannot control on maximum pressure, or indeed anything – it just runs at its fixed speed. So it protects the downstream pipe from over-pressure by tripping on a maximum discharge pressure. This is mechanically easier to implement than installing an advanced control system with a variable speed drive. Reciprocating compressors trip on maximum discharge pressure and maximum discharge temperature, not on minimum speed, and these are both user-defined limits.

Trip condition	Compressor		
	Ideal	Reciprocating	Centrifugal
Inferred minimum speed	-	-	If no others
Minimum speed	-	-	□
Minimum suction pressure	-	□	□
Maximum discharge pressure	-	□	□
Max. discharge temperature	-	□	□
Minimum surge fraction	-	-	□
Reverse flow	Always, if there is no check valve		

Table 18 – A simulator's trip conditions for ideal, reciprocating and centrifugal compressors

Measuring linepack and scheduling nominations

A gas pipeline acts as a storage facility in its own right. By using packing and unpacking, operators can meet the fluctuating needs of their consumers while maintaining a steady supply, or vice versa. In some networks, both supplies and demands fluctuate. The hydraulics of a pipeline mean that pressure and inventory are intertwined. *Packing* and *unpacking* cause gas pipelines to be in a transient state at all times. This situation can only really be calculated by simulation. It can involve complex effects: for example, even if a pipeline is not nominally *"bi-directional"*, reverse flow can take place in parts of the network during certain operating modes, even whilst supplies are flowing into the pipeline and demands are flowing out! As a section packs or unpacks, gas compresses or decompresses, and so pressure rises or falls over time.

Gas pipeliners describe nominations with some variation on this terminology:

Nomination	Quantity of gas energy content, to deliver within a timescale
Consumer type	Daily pattern expressing nominations in proportion to a total
Peak daily	Maximum daily quantity that we can nominate (energy units)
Peak hourly	Maximum hourly quantity that can be nominated
Linepack	Inventory, the process of packing and unpacking to exploit it
Throughput capacity	Available flow rate at network bottleneck
Packing capacity	Available inventory for packing within MAOP limit
Unpacking capacity	Available inventory for unpacking within LAOP limit

The demand points of various gas consumers typically fluctuate over time according to predictable patterns repeated daily, weekly, and seasonally. A customizable pattern is represented by a *consumer type* in a simulator's type libraries. Most gas networks cannot expect suppliers to feed gas into their system from supply points as fast as their consumers' peak usage, and moreover, even if that were possible, it would take a while for increased supplies to travel through the pipeline until they reach the demand points. That means that in order to keep the customer demands satisfied, the line must be packed in advance. If the operator knows how the pipeline will respond to these fluctuations, value can be provided by accommodating them.

By operating the pipeline to take advantage of *linepack*, using inventory as a buffer store, the operator can smooth out variations in supply and demand. On a *day*'s timescale, it might build up linepack overnight, and then unpack to meet the needs of a cold-weather morning. Such a demand is said to be *peaky* – meaning that it fluctuates significantly over time, occasionally rising to unusually high values before returning to the baseline. On a *week*'s timescale, a pipeline might restore its linepack over the weekend, using that to satisfy peaky demands (predictable and unpredictable) as they arise during the work-week.

Every pipeline has a unique set of customers, so what constitutes a pipeline's peak usage is not a subject one can talk about in general terms. Factories and industrial users might intentionally choose to alter the way they operate to take advantage of variable gas prices, just as they do with electrical prices.

"Operating on linepack", also called *drafting*, might allow a pipeline to meet contractual obligations even during a total supply failure. Whilst waiting for a supply to be reinstated, how long can a pipeline continue to meet its customers required delivery rates? This is called the *survival time*, a figure which evolves and has immense implications for commercial agreements, likely affecting our customers' own fallback contingency plans. Equivalently, it is worth considering how long a pipeline can continue accepting incoming gas supplies if a demand suddenly stops; that is another sort of survival time.

Most gas pipeline operators historically performed their nomination planning assuming steady states. This resulted in under-utilized pipeline capacity: a half-hour or fifteen-minute nomination doesn't bring the entirety of a long gas pipeline to steady state.

> Successive steady states misrepresent transients in gas linepack. A fully transient simulation captures this effect.

The reason steady states are misleading here is that they misrepresent the problem at hand. Consider the dilemma of choosing any steady *"design flow"*. Applying steady state simulation with daily average flows as a basis could lead to serious *under-design* of the system: it misses the peak flow condition. At the other extreme, using peak hourly flows as a basis leads to *over-design*: it suggests the pipeline must handle this flow rate whilst in equilibrium. What we ought to ask, is whether the pipeline can handle a short-lived peak flow rate whilst it undergoes transient conditions. Simply put: we are allowed to use linepack to meet the peak conditions, so long as we restore that linepack afterwards. [54]

A transient simulation has us reach conclusions with all kinds of subtle differences to a sequence of steady states. To meet the required compressor discharge pressure, where would it be most efficient to place extra loops? Steady states suggest that longer loops would be best placed downstream of a compressor, as the compressor discharge is hot, and the extra length removes that heat effectively. But a proper transient simulation reaches an opposite recommendation: positioning the loop upstream of the compressor station is more efficient, because of the buffering effect of linepack on transient flows. The wrong conclusion can be reached if one doesn't perform a transient analysis, particularly where demand is cyclic. The buffering effect of linepack might mean that the receipt flow is approximately stable (e.g. 309-355 kscm/hr) while delivery flow is fluctuating (e.g. 181-470 kscm/hr). [54]

Variable pricing of gas

Determining the price of gas by manual calculation is a tricky business. Trading takes place on a gas pipeline's offers portal, which must assure transparent mechanisms for allocating the available throughput capacity. Pricing capacity involves negotiating some non-trivial matters, such as what happens if a customer curtails their operation if the gas price becomes too expensive for part of the day. Even in its pure form: the technical calculation of throughput capacity is rather involved. Simulation turns the gas pricing puzzle into a sort of transient optimization. We go about tackling the matter as follows:

What is this gas pipeline's throughput capacity
from its bottleneck to the selected demand points:
A) if a particular demand asks for higher pressure?
B) if a particular demand asks for more energy content?
C) if a particular demand becomes peakier?

The rated capacity referred to in such a question is that which is consistently available at all times to selected locations. Which delivery points are selected for this purpose affects the figure: for example, a high pressure delivery point won't have as much flow rate available to it as an adjacent point at low pressure. Every demand has its own calculation for *utilization factor*: the ratio of its maximum demand to the pipeline throughput's rated capacity. Homing in on the remaining flow rate capacity of the pipeline, we quantify the *opportunity cost* of satisfying one customer's requirement, by describing how it limits the pipeline's operational flexibility to accommodate other needs.

If there are X compressor configurations and Y demand possibilities then the number of scenarios considered in answering this question is $X \times Y$. We define a scenario for every viable compressor configuration. Every demand is assigned a minimum delivery pressure, and consumer type, where consumer type is a pattern describing how a typical day looks as a percentage of the total. A peaky demand places strain on a pipeline, and cannot be treated as a flat profile. It is fair to charge more for peaky demands, as the pipeline must sacrifice flexibility elsewhere to accommodate *how* the same quantity of total energy is delivered.

Consumer type is a pre-set pattern of how a nomination is typically spread out across a recurring daily, weekly, monthly or annual interval.

By adjusting the daily total energy delivered at Demand A, we can ask what the capacity (i.e. potential increase in flow rate, defined in energy units) would be at a network's most-limiting point? By iterating until zero capacity remains, whilst satisfying Demand A's minimum delivery pressure. Then, do the same by adjusting Demand B, and so forth.

This provides evidence demonstrating that the pipeline operating company's variable pricing model, uniquely tailored for each individual customer and revised routinely (typically on an annual basis) to account for evolving supplies and demands across the network, is justifiably fair to all parties including other customers. It helps satisfy the regulator that the pipeline operating company is not price-gouging, but rather is pricing their service to reflect the true cost of *sustainably* providing energy content in the manner required by the customer.

Variables that affect the price of delivering gas through a pipeline include *gas quality*: combustion properties and energy content depending on the gas composition, and so pricing can be expressed per energy unit, and some components must be within limits. The price also depends on *delivery pressure*: saving on pressure booster equipment which would otherwise be required for gas turbines or gas engines to be capable of operating during low-pressure periods. Moreover, price depends on *demand pattern*; a peaky usage with spikes at an inconvenient time of day represents a heavier burden than a flat demand profile. Some consumers describe their usage via quarter-hourly data, others half-hourly or hourly intervals. Opting for a finer granularity might even provide some savings in itself!

If peaks are unpredictable – for example a gas-fired *"peaker plant"* which only generates power in response to the electrical grid – they don't follow a set nomination cycle every day. A *no-notice delivery service* allows receipt of gas on demand without making prior nominations, and without paying penalties for daily balancing and scheduling. Should the pipeline operator still set aside that capacity for them just in case? What is the value of remaining *on stand-by* in this way? It can sometimes be reasonable to stipulate that maximum hourly quantities representing this peak do not correspond in this way, as it is highly unlikely that they occur concurrently, and standing by would render several percent of the pipeline's capacity unavailable.

A further question is how to price events that can be accommodated once, but then require the network to take some period of time to recover its linepack, all whilst maintaining routine operations. Such events, if run repeatedly through the simulator, would be demonstrably unsustainable: they would eventually disrupt other customers if they occurred in continuous succession.

Similarly, a customer might wish to utilize the pipeline's inventory for their own purposes, by supplying some quantity of gas inventory, then drawing it back later. Provisions to provide storage using the linepack warrants separate consideration to calculate a fair price. Essentially, the answer is always case-by-case; it depends! How much? For how long? And where?

Case study: compressors in pipeline capacity cases

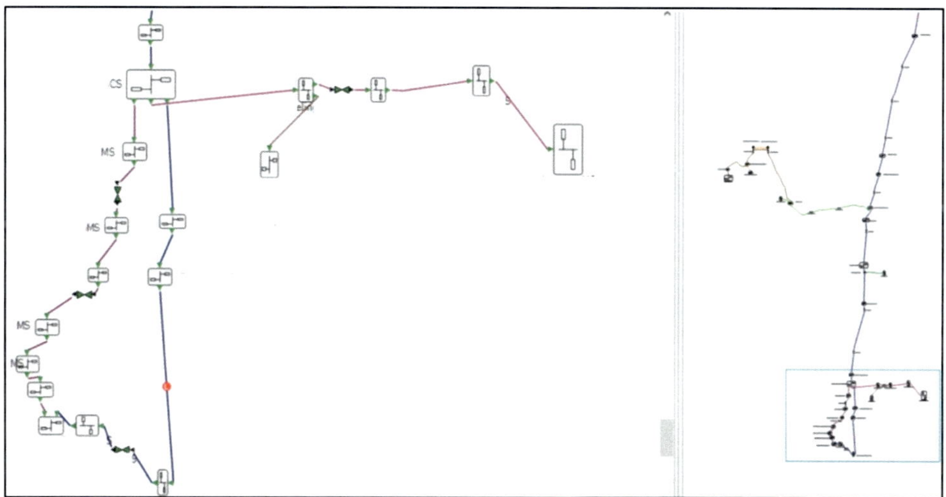

Figure 161 – The pipeline system across the Australian outback includes dozens of stations.

For more than 50 years this pipeline system has transported natural gas from conventional fields, heading south for 780 km from a processing plant across the outback, to supply users in power generation, industrial and domestic applications. The network comprises two major lateral pipelines, one to the regional population centers, and another to industrial users and other pipelines further afield. In total, the network spans 1 180 km. Since the completion of a link pipeline in recent years, it can also transport gas sourced from conventional and coal seam gas fields in a neighboring state. Additional gas receipts are also possible via interconnection with other pipelines.

To operate the pipeline, seven gas turbine driven compressor stations are located about 100 km apart. After-coolers are provided at compressor stations to reduce the risk of stress fracture or over-temperature of the pipeline coating. Both centrifugal and reciprocating compressors are used across the network. The maximum allowable operating pressure (MAOP) of approximately 70 bar has been progressively upgraded to boost capacity and increase security of gas supplies. This included a 40 km loop line near a power station, where pressure limiting is installed to protect the mainline. The compressor stations also underwent major upgrades to their compression equipment, replacing the original packages with modern units, and increasing the driving capacity from 4 200 to 5 300 horsepower. This raised the indicative daily winter capacity of the pipeline by over 10%, from 350 TJ/d to 390 TJ/d. Station control systems were upgraded from antiquated electro–mechanical systems to a modern PLC. The pipeline was enhanced in 2000 by the construction of further loops and upgrading the remaining equipment to higher horsepower specifications. This raised the pipeline's indicative daily winter capacity to 420 TJ/d, delivering 20% more energy than it could originally.

Simulation engineers used Atmos SIM to conduct a number of offline studies. The complex loops, stations and compressor arrangements were configured in the model, as well as the thermal properties of the pipeline's burial to allow versatility of use, so a wide variety of scenarios can be examined accurately. The same model used for real-time simulation was used for offline studies, allowing online saved states to be restored to conduct these studies, and historic meter data to be replayed and compared against offline predictions. One study ran operational capacity modelling scenarios at various compressor configurations, turning each compressor on or off at different stations, and observing the impact of each outage on downstream pressure and flow rate.

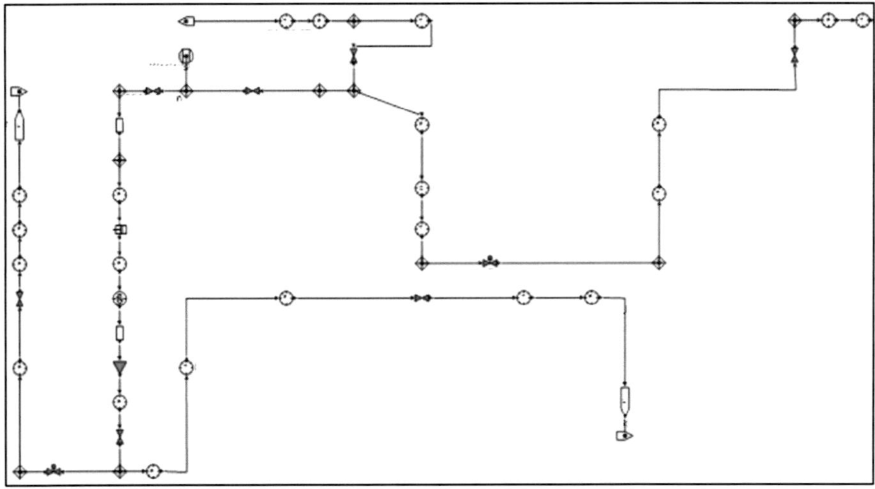

Figure 162 – A compressor station (CS) along the pipeline, with four inlets and outlets.

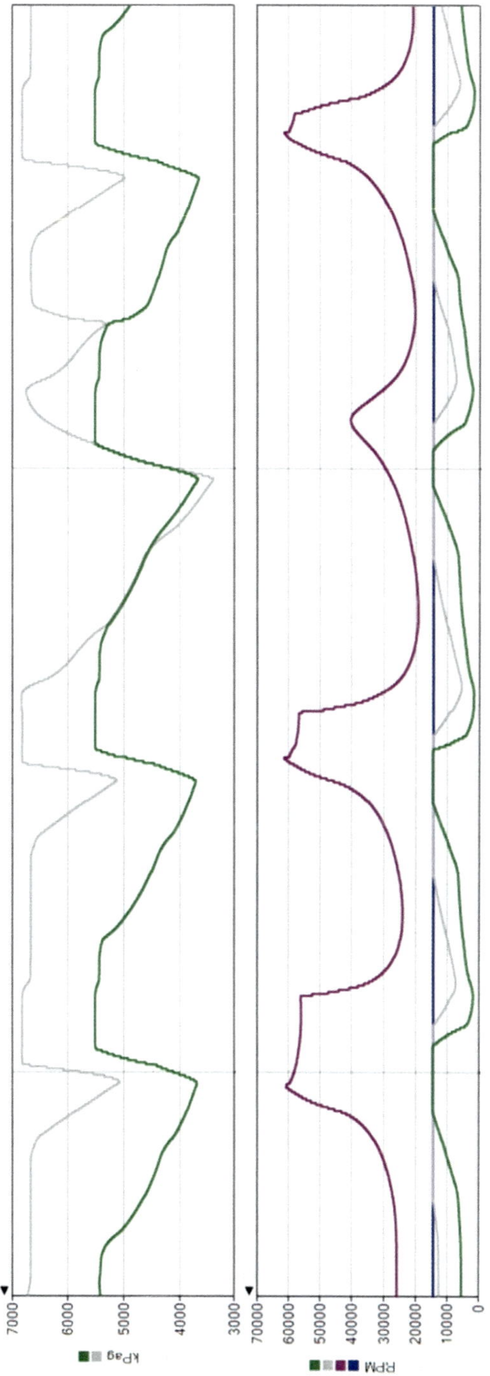

Figure 163 – A trend from a hypothetical offline scenario examining how pressures at two delivery points (top) are affected by how compressors are operated at various speeds (bottom).

8.6 – Pigging operations

Figure 164 – A pig launcher in Germany

An overview of pigging

A pig is a solid device inserted into a pipeline. It is used to inspect or clean as part of a maintenance regime. So called either due to their squeaking as they move through a pipeline, or a popular backronym of a *"pipeline insertion gauge"*, among other versions. There are many kinds of pig, some simply being spheres or dumbbell-shaped solids, others being sophisticated instrumented devices for in-line inspection known as *"intelligent pigs"*. *

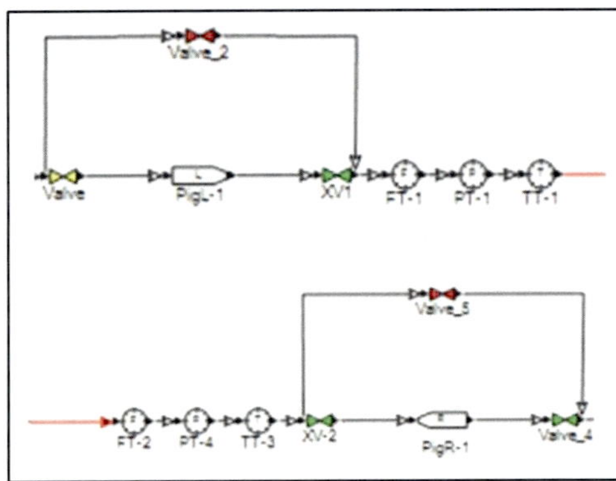

Figure 165 – A simple pig launch station and pig receipt station.

A *pig launcher* allows the insertion of a pig. It travels along a pig route, possibly with others in succession, circumventing any valves which aren't full-bore, and station

* The term *leak-detecting pig* has become ambiguous: either being a device inserted in the pipe, or an actual *sus domesticus* that sniffs out hydrocarbons – both are used.

equipment. Pumps and compressors don't fare well handling what is essentially a heavy solid lump. Along the route, there might be *pig detectors* also called *pig switches*. At mid-line stations, there might be *pig parkers*, facilities to receive and hold a pig until it is launched onwards. A pig catcher, also called a *pig receiver* is accompanied by associated equipment at a receiving station. At the end of the route, valve arrangements might need switching in anticipation of an incoming pig. This makes simulation valuable for providing an estimated time of arrival (ETA). Offline scenarios can plan this in advance, and online scenarios monitor it in a real-time live simulation.

It is quick to build an offline scenario sending a pig along the route of a pipeline, during the course of varying gas nominations or liquid batches. Before pipeline simulators, those responsible for planning these operations struggled with spreadsheets to anticipate a pig's progress. They are delighted when a transient simulator predicts arrival time after a two-day journey within minutes.

Figure 166 – Several pigs could be present in a pipeline at once, moving at different velocities.

It is possible for specially designed pigs to cope with significant changes in diameter. Although changes in pipeline diameter through a pigging route are uncommon, they do exist. When a pig travels through a pipeline there is a limit to how tight a bend it can cope with. Typical minimum radii that a pig can navigate are between five and fifteen pipe diameters. A pig may be specially articulated to pass tighter radii. Occasionally, a pig might get stuck. This is very bad news. In the worst case, the pipeline will have to be uncovered, a cut made into it to free a stuck pig. Needless to say, this is expensive and damaging to the affected pipeline, and a weld inside a pipe may never be clean afterwards. It might be so weak that this stretch of pipeline can never be pigged again!

A pig receipt needs to be carefully planned. At the very least it will be a heavy lump of metal reaching the end of the pipeline that needs to be caught. And in most cases it will likely be preceded with a volume of muck that needs special handling. It is not unheard of for an arriving pig to be pushing out several hundred cubic meters of material ahead of it.

In a real-time model, pig signals indicate a pig has been launched. There may also be pig-received signals, or pig-passed signals at certain points. Pigs are tracked to determine when they are expected to arrive at a delivery point, so preparations can be made to catch the pigs. Current pig location and estimated arrival time to the receiver or any important point are calculated. To verify the position of the pig, indicators called pig detectors or pig switches can be installed along a pipeline. By magnetic or other means, the position of the pig is updated by a pig switch via a *"passed"* signal. These are generally placed at block valves and as an early warning indicator, say 1 km from the receiver. The tracker then relocates the pig's position. This validation can also be used for tuning, making corrections based on what has been observed before.

How do we improve a pig's estimated velocity? A pig generally travels at, or just less than, the speed of the fluid. This difference is described by a *slippage* factor, from which a simulator infers the pig's velocity based on the flow velocity. In general, a normal pig's slippage factor should be less than 10%, but a damaged pig might be bypassing at 30%. It could even become a *lost pig*! Pigging is carefully planned. A pig's slippage depends on a flow rate at its current position, which often varies over time. A separate set of slippage factors can also be defined for pigs moving in the reverse direction. The slippage factor of a pig type and the slippage factor of a pipe are automatically learnt by Atmos SIM during operation, updating values gradually based on a forgetting factor.

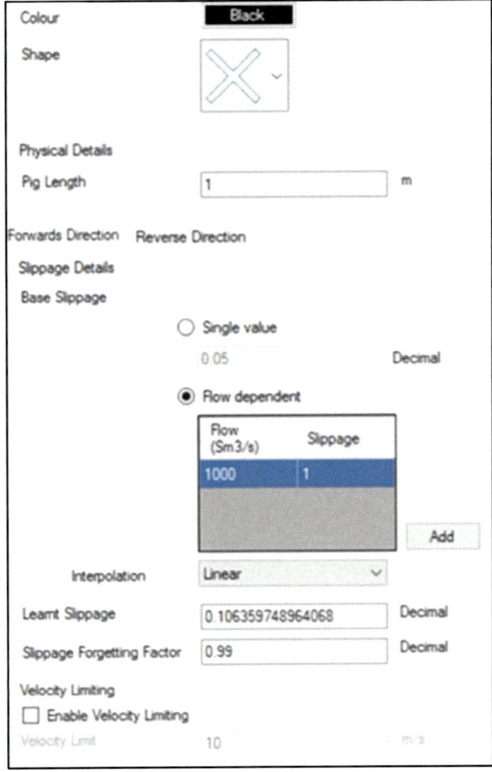

Figure 167 – A dialogue configuring the properties of a pig in Atmos SIM – flow dependent *"base"* slippage in the forward and reverse directions, interpolation method for the slippage, a forgetting factor for learnt slippage, and velocity limiting.

Pigging a liquid pipeline

Physically, a liquid pig may be a simple dumbbell shape, an inflated container or a long device with integrated instrumentation. If there are batches in laminar flow, or if the pipeline is ever stopped with a heavy batch uphill of a light batch, there is a lot of mixing if there's no separator pig. Spheres generally separate liquid batches, for instance, diesel from petrol. Several spheres are injected together for better separation. Scrapers clean by clearing the inside of a pipe wall, removing any deposits. Intelligent inspection pigs check the structural integrity of the pipes.

Sending a pig through a liquid line controlled on flow rate can look a lot like sending a heavy batch through it; the pressure at the origin traces out the elevation profile over time. This is because pressure drop for a given flow rate depends on density, and density changes over time as batches move along a pipeline. Pigs can perform many maintenance functions as they move along with the flow. As a pig traverses long distances, it is important to have accurate estimates of arrival times, so that operators know when to divert flow and remove the pig. A pig's typical speed in a liquid line is 5% slower than fluid velocity, which is around 1 m/s (i.e. 4 km/h or a gentle walking pace).

Pigging a gas pipeline

There are two main uses for pigs in gas pipelines. The first is a moving block separating liquid (e.g. water) from gas (e.g. nitrogen) during the initial fill or decommissioning phases. The second is during operation: a condensate remover pushes condensed liquid out, in quantities as large as 100 tons of liquid at once. This liquid must be contained in a slug catcher, and properly sizing these slug catchers often requires sophisticated multi-phase simulation.

As the pressures and flow rates at any gas pipeline's inlets and outlets vary over time in an imbalanced manner, delivering gas nominations is done by exploiting how its linepack is allowed to change over time. A simulation is the only real way to reach a good estimate of any arrival time, including a pig's arrival time. If we send a pig through for the first time through a simulator, the estimated arrival time is typically much better than anything a team was using before without a pipeline simulator. Even then, it can usually be made even more accurate. A pig moves through the pipeline slower than the fluid velocity by some slippage, and that slippage must be learnt. Atmos SIM can learn a pig type's slippage factor as well as a pipe section's slippage factor, in a real-time deployment or as a replay run. These learnt slippage factors then generate more accurate predictions in offline scenarios when planning future pigging operations.

In a gas pipeline, we want to avoid a pig moving too fast. We could drop the boundary condition at the supply or demand nearest a fast-moving pig, and replace that boundary condition with a new controlling constraint at the pig: a velocity setpoint. We can then deduce how to control the original constraint at the relevant supply or demand, in order to keep the pig's velocity within some chosen limit. It is somewhat surprising that this works – one would expect enforcing a setpoint in the middle of a pipe to be unstable – yet this scenario is simulated successfully. To date it's not been widely used in practice.

8.7 – Logical control schemes

Realistic simulation beyond idealized control

So far we've discussed *ideal* control: the idea that model control items such as supplies, demands, pumps, compressors, regulators and valves always switch to reach their target setpoints, immediately and exactly, as if by magic. Assuming idealized control lets a user say things like *"have this regulator's position do whatever is needed to hit this specified outlet pressure"*. This approximation is well-suited for many applications. Although some users choose to configure logic signals to control the operating speed of a pump or compressor, an easier approach is to use idealized constraints.

However, it's not how a real pipeline works, especially at short time-scales. For example, a real pump or compressor has inertia. During spin-up and spin-down, it has a speed-vs-time profile. A simulator abstracts this to spin-up and spin-down durations: speed just changes at a constant angular acceleration.

Real world behavior is much messier, with slow response times, overshoots, oscillation about the target, and so forth. In a real pipeline, control schemes are implemented in programmable logic controllers (PLCs). They read a measurement and adjust the control variable (say, the speed of a pump or compressor, or the position of a valve) in a gradual and somewhat complex manner to neatly deal with the aforementioned messy, real-world behaviors.

For applications like setting up the emergency shut-down (ESD) procedures of a pump or compressor station, the relevant timescales are the ones where the PID controller is acting, so it becomes necessary for us to simulate that. To this end, Atmos SIM provides logical blocks to configure logical control schemes.

Simple logical blocks

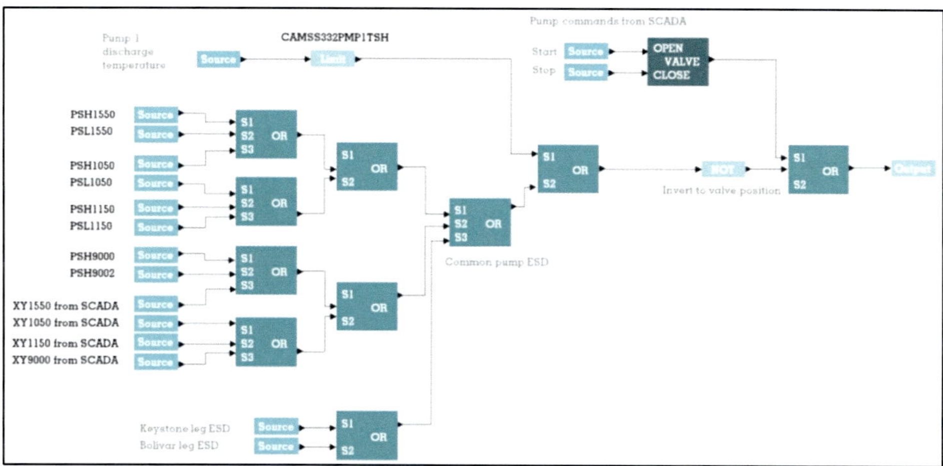

Figure 168 – Non-transient blocks include logical OR gates.

Some logical blocks perform calculations instantaneously. Perhaps the most rudimentary sort of logical control attempts to meet its target immediately. We can build a simple control system out of just these instantaneous or *non-transient* blocks: after each model step, take the difference between the measured pressure and the setpoint, multiply that by a constant and add that to the current pump speed. That just needs blocks to perform subtraction, multiplication, and addition. It is known as proportional control and is a bad idea for controlling something like a pipeline pump or compressor.

- Converting (offset and scale $A \cdot x + B$)
- Adding (summing $x + y + \cdots$) or subtracting (difference $x - y$)
- Taking a maximum or minimum (select largest or smallest of $[\mathrm{x}, \mathrm{y}, \ldots]$)
- Flagging a limit (true if exceeded, false otherwise)
- Comparing (If $x < y$ then A, else if $x = y$ then B, else if $x > y$ then C)
- Multiplying ($x \times y$) or dividing ($x \div y$)
- Selecting (If selector=0 then A, else B)
- AND (for example if $x \neq 0$ and $y \neq 0$ and $z \neq 0$ then 1 else 0)
- OR (for example, if $x = 0$ or $y = 0$ or $z = 0$ then 0 else 1)
- Instantaneous averaging (at a given instance $(x + y + z + \cdots)/n$)

Physical control of a pipeline is not the only application for logical blocks. One valid use case for logical blocks never assigns them as constraints of any kind in the model. These might be referred to as non-physical non-transient logical blocks. In this use case, they can be thought of as performing data processing, or a kind of *scripting*. This feature comes in useful for purposes such as simple arithmetic operations (e.g. add these together) to output some custom results. Another is for validation checks (e.g. is this greater than a limit).

Logical blocks with an internal state

Some logical operations retain some memory of preceding steps. This type of block can be used to perform differentiation, for example, calculating the rate of change of some quantity by looking at the amount of variation over the last model time-step. This kind of logic block can only ever be simulated with fixed time-step, and cannot fundamentally work with variable time-step such as that caused by cut-back.

Logical blocks that have an internal state include:

- Totalizing over time until reset or rollover; equivalent to integrating
- Counting; incrementing every time-step for a set count
- Rolling average over a given period as a *"moving window"*
- Proportional – integral – derivative control (PID)
- Rate of change
- Any block involving slope

These are distinct from simple logical blocks in that they depend on their history, and on step-length. This requires some special handling in the model, which Atmos SIM does automatically. For example, if a step is cut back to 40% of its original length because a check valve closes mid-step, a totalizer block should only add 40% of its input value to its running total.

Rate of change and one-shot

A rate of change (RoC) is a logical block with an internal state. It keeps a rolling window of N input values for the last time-step. On any step it outputs the most recent value minus the oldest value all divided by the elapsed time between those two values. So it is a finite difference first time derivative calculated over a window potentially longer than a single step.

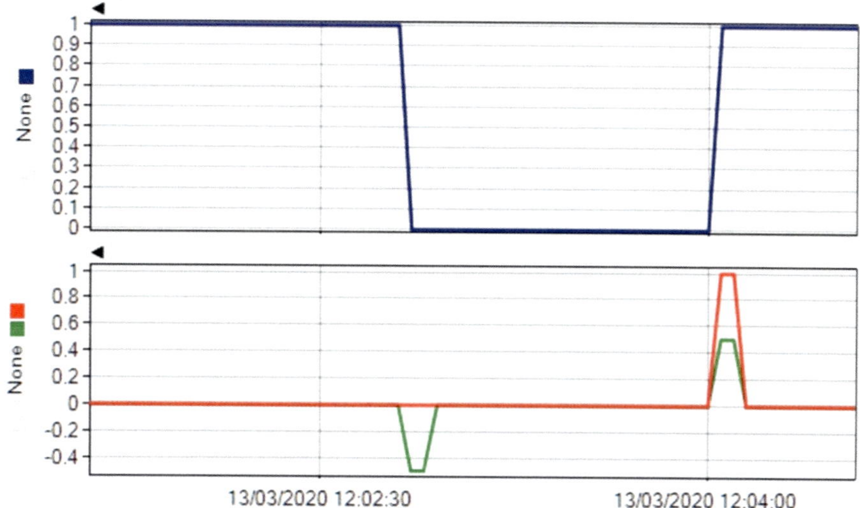

Figure 169 – A request signal (top) triggers a rate of change block (green) on both high-to-low and low-to-high transitions. This RoC block outputs the most recent value (0) minus the oldest value (1) divided by the elapsed time between them (in this case, 0.5). The red profile shows a one-shot, that triggers on rising but not falling input signal.

It is possible to use this block as part of wider logic to achieve one-shot (ONS) behavior, where an output signal is always triggered to a value of unity (1) on the transition of a request signal low-to-high but not on a transition high-to-low. In the above diagram, this one-shot response is shown in the red trend.

Rolling average

One type of state-preserving block is the averaging over a window of data. A gas pipeliner might wish to take a one-day window's rolling average of flow rate, resetting once per day.

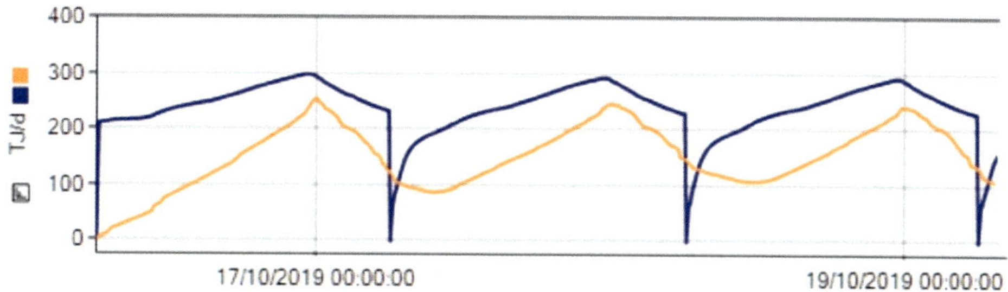

Figure 170 – A rolling-average logic block implemented within Atmos SIM to calculate the average energy flow rate over a gas pipeline's preceding 24 hours, resetting daily. The underlying value is shown in blue, and the rolling average is shown in orange.

Totalizer

In contrast to an adder ($F_{\text{ABC}} = F_{\text{A}} + F_B + F_C$), a totalizer keeps a rolling sum of any selected quantity over a set number of steps. It is an integral disguised as a sum divided by the time-step length. This is useful in pipeline simulation for converting flow rates into inventory. It can also be used for tasks such as computing the average flow rate over the past hour as a custom result output.

$$\text{IF } \Sigma F_{\text{ABC}} > F_{\text{Max}} \text{ OR IF stepcount} > \text{maxsteps THEN reset}$$

1. It has a reset input – if that goes from 0 to 1 it resets its running total to zero, and does no more totalizing until the next time-step.
2. If it exceeds the configured maximum steps since a reset, it resets, then continues totalizing during that step.
3. It has a maximum value input, which if exceeded by the totalized value, has the block reset and do no more totalizing until the next time-step.
4. It can be configured to reset at a certain time or a regular time interval.
5. It has an input that tells it whether it should be totalizing (i.e. integrating) right now; if it got past steps 1 through 4 then it adds this time-steps input, integrated over the step, to its output.

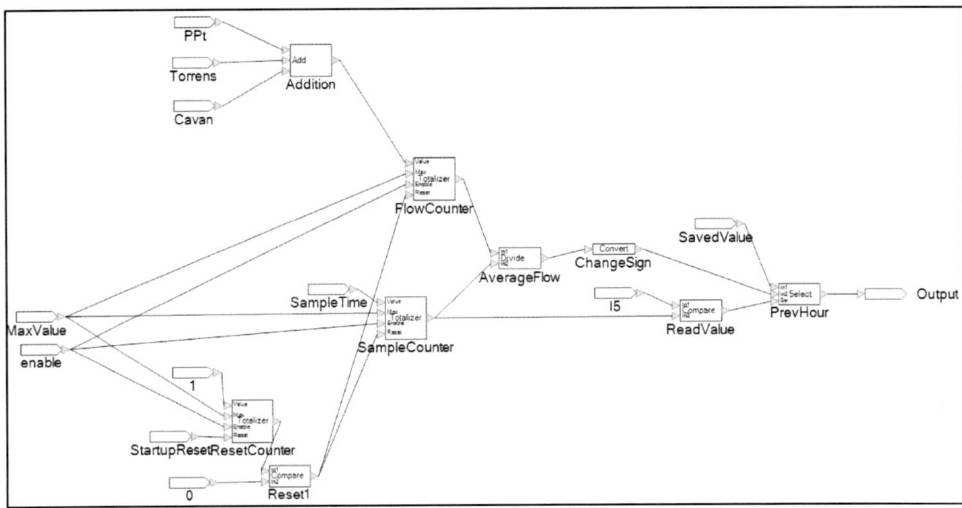

Figure 171 – A totalizer block in a logical control scheme within an Atmos SIM project

Proportional-integral-derivative

When we consider feedback, we note that there are still dynamics coming from the physical system. One cannot simply adjust a knob to reach a setpoint. We need one or sometimes two further considerations to decide what to tell our actuator to do. In addition to a multiplier, the proportional (P) term, there are integral (I) and derivative (D) terms, set so as to combine with the dynamics of the physical system in as inoffensive a way as possible. A proportional–integral–derivative (PID) logic block is the most commonly used feedback controller. It is a logical block with an internal state.

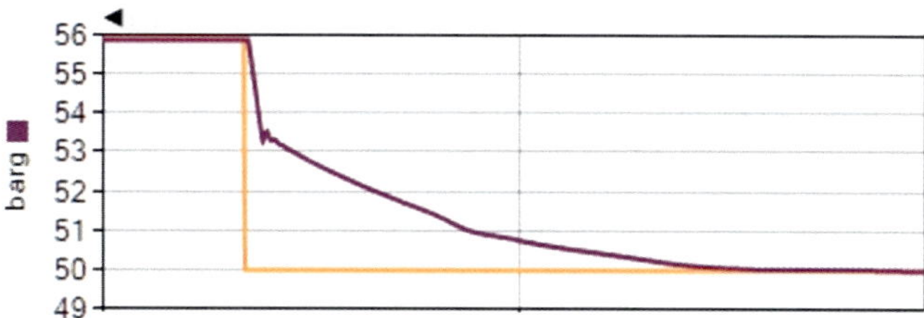

Figure 172 – A PID controller takes gradual action to adjust pressure (purple) when we ask for a sudden change in setpoint (orange), shown here as a trend from an Atmos SIM scenario

A PID block violates the basic concept of the steady state model that nothing is changing; its behavior is entirely driven by change. In a steady state, Atmos SIM sidesteps the PID block by using a user-defined override value. Once the transient model starts, Atmos SIM begins simulating the block as well.

The PID controller calculates a *controller error* (ε) as a difference between a measured *process variable* (PV) and a desired *controller setpoint* (SP). The controller attempts to minimize the error by adjusting the process control inputs. The user configures a PID controller with three distinct parameters: *proportional* (P), *integral* (I) and *derivative* (D):

- A *proportional gain* (K_p) acts on the error signal to provide the driving input to the process, adjusting our speed of response
- An *integral time constant* (T_i) implemented by an integrator, tuned to make the long-term response accurate so we eventually hit the setpoint
- A *derivative control* (T_d) for damping the system, rarely used in most industrial applications – people joke that it stands for *"do not use"*

Where used, the derivative control variable is applied to the *process variable* (PV) to counteract its rate of change (RoC). This is so that a sudden change in setpoint (SP) doesn't set off some violent oscillation due to the large time derivative of a process error. When the controller setpoint is unchanging, the derivative of the process error is the same as the derivative of the process variable.

A typical logic scheme including this kind of control might include overrides of the PID block in certain circumstances. Within a pipeline simulator, provisions can be made for initialization, alongside the setpoint coming from a SCADA system, and perhaps a manual override. This can become elaborate. The PID block in a typical logical scheme accepts a number of inputs, including a controller setpoint (SP), a process variable to be controlled (PV), and parameters for each of the proportional (P), integral (I) and derivative (D) actions. There is also a way to manually override the block with a user-specified value say for start-up.

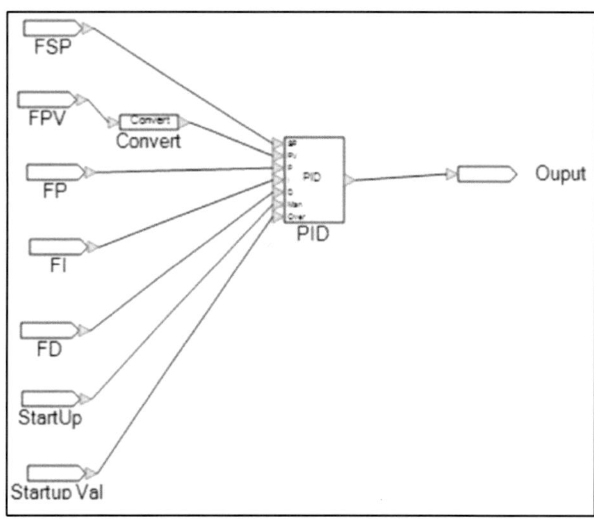

Figure 173 – The PID block in an Atmos SIM logical scheme can have seven user inputs

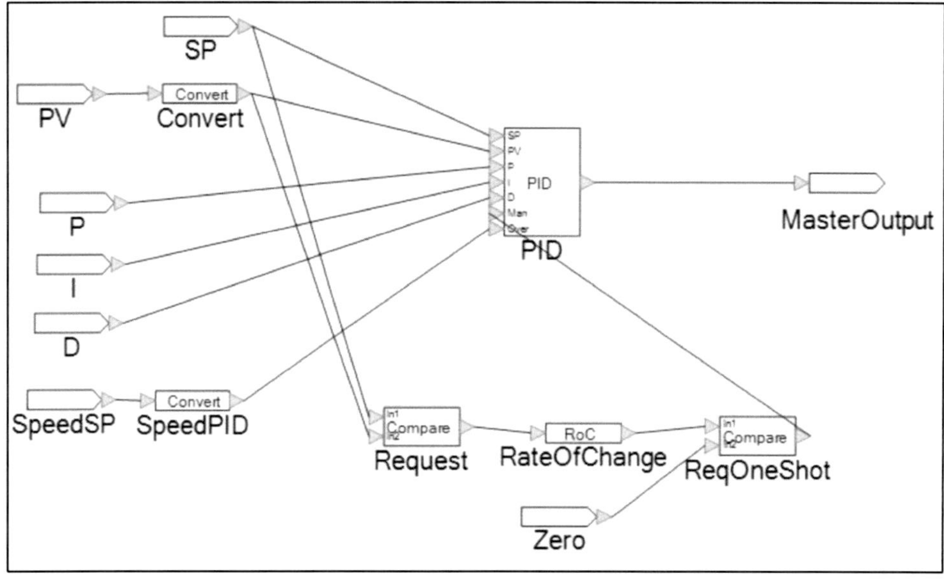

Figure 174 – A PID scheme in Atmos SIM to control the discharge pressure on a water pipeline

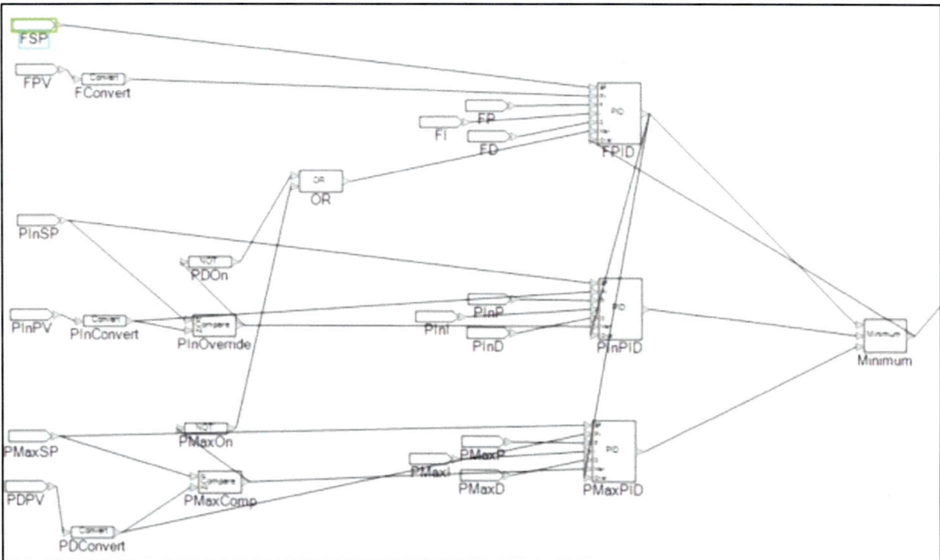

Figure 175 – A scheme within Atmos SIM involving several PID controllers, implementing flow control with both low- and high-pressure overrides to control a compressor's speed

Feedback in logical control schemes

These same simple logical blocks are also allowed to be used for physical tasks; the outputs of logical blocks can be fed into the setpoints of ideal control. For example they can directly control the speed of a variable-speed drive on a pump or compressor, or the position of a valve. Under the hood, this physical use case requires them to be treated differently if a non-constant time-step takes place (in case of a cut-back step).

Dependencies can become quite involved, when one control scheme feeds a signal into a preceding one. But these loops are allowed and are used in reality! Real control systems accomplish their most impressive feats with the use of feedback. This can be configured in Atmos SIM's logical control. Often the feedback is mediated through the pipeline itself. For example, a logical scheme might be driven by a measured pressure to set a valve position, then the hydraulic model – when solving a steady state or a transient step – might alter the measured pressure based on what the valve has done.*

It is also possible to put a feedback loop entirely inside the logical control scheme. Since these simple logical blocks we have discussed so far execute their function

* Indeed, idealized control embodied in offline scenario constraints – such as maximum discharge pressure – already implicitly uses feedback to perfectly hit their setpoints.

instantaneously, it becomes necessary to come up with some scheme to prevent infinite loops when trying to simulate them. Atmos SIM does this by only letting each simple control block execute once on each model step.

One should note that this can make the behavior of a logical control scheme with an internal feedback loop dependent on the simulation time-step. It is better to avoid this sort of feedback if possible.

Custom scripting languages

We have discussed control on paper, control in practice, as well as control in simulators using setpoints applied as ideal constraints and trip limits. The complexity of the control systems governing pipeline equipment can be captured to the appropriate level of detail if we apply the right degree of abstraction. Sometimes, custom control logic is needed – instructing equipment perform a certain action whenever a certain event takes place in the pipeline. This custom logic can be applied using logical control schemes, which Atmos SIM offers in an intuitive graphical format known as *control diagrams*. Or, the same custom logic can be applied using user-configured code, often called *scripting*.

Making a custom scripting language available to the user opens up the realm of possible instructions to be almost limitless. It can be convenient for asking a simulation to *perform tasks*. For example: *"keep checking for W+X, and if they exceed 20, then do Y, and then do Z"*. It's often possible to adapt logic blocks for this purpose, but the scripting approach can sometimes be more versatile or at least allow us to express the task intuitively.

Scripting languages might be able to leverage existing libraries (e.g. Python). They can perform the same sort of actions as the logical blocks we've been looking at. It can sometimes be quicker and easier to create a script for a given function than to assemble it from blocks. This is especially the case since the blocks have very limited ability to store data.

On the other hand, this does open an avenue for a different kind of challenge. Although all the vendor needs to do is test the fairly simple places where the scripting language calls their simulation package, with great versatility comes great latitude for something to go awry. Letting a user put together what is essentially a computer program raises questions about how to ensure it will work! A scripting language is code written by the user rather than a developer, and opening up this flexibility to users is powerful if – and only if – it is done right.

9 – Rapid transients and surge analysis

9.1 – An introduction to rapid transients

Aren't rapid transients just operations?

Rapid transients can be regarded as a specialty in their own right, in the sense that they involve changes happening on very short timescales, scales of typical pipeline distances divided by the speed of sound – so seconds or less, up to maybe minutes on a gas pipeline. Expert consultants dedicate their careers to these subjects, working on surge analysis and fast-acting control strategies.

In a versatile pipeline simulation engine, such as Atmos SIM, rapid transients are not treated in a fundamentally different way to slower pipeline operations. There are just a few special allowances made in the underlying code to handle the peculiarities of simulating a sudden transient such as a surge in pressure.

From the user's perspective, the timescale at which these rapid transients take place, and the way in which we look at them, are quite distinct from the operating scenarios we looked at in the previous chapter. This is why we'll now spend a dedicated chapter to closely examine the speed of sound and rapid transients.

Safe operating limits

It goes without saying that everyone agrees a pipeline should operate safely. *"If you think safety is expensive, try having an accident!"* In every jurisdiction, a set of regulations place the burden of proof on the pipeliner to demonstrate that all their operations remain safe. In the UK, *Pipeline Safety Regulations* (PSR 1996) stipulate that *"the pipeline operator shall ensure that no fluid is conveyed in a pipeline unless the Safe Operating Limits (SOL) of the pipeline have been established and that a pipeline is not operated beyond its SOL."*

Short excursions beyond the maximum allowable operating pressures (MAOP) might be permissible, but the stringent *safe operating limits* (SOL) must never be exceeded under any circumstances. How near we set MAOP to SOL is a multiplier (X) defined within the operating philosophy, typically 110% to 115%:

$$SOL = MAOP * X$$

Knots, time-steps and numerical dispersion

Although it may appear that our peak pressure never exceeds a maximum allowable operating limit, our choice of restriction on adaptive time-step might mask the reality. Simulating the same scenario again at shorter time-steps shows us the reality of any fast transient. Just as a succession of steady states misleads us by ignoring transients, simulating a rapid transient with too coarse a time-step might mislead us by ignoring the full extent of what's happening. Exploring how a pipeline responds to fast-acting control strategies is a specialist task. A consulting engineer is tasked with looking into a number of extreme transient scenarios, including surges, blowdowns and ruptures.

Provided the time-step is briefer than the timescale taken by an action such as a valve closure, there is no danger of a long time-step hiding fast transients. A different concern preoccupies those working with pipeline simulators, and that is numerical dispersion. Any time-step and distance step will be adequate to get the transient behavior right at the source of the transient, but as the time and distance step get coarser, it washes out more quickly due to numerical dispersion as it moves away from the source of the transient. So if we are counting on a reflected wave or the negative pressure surge from an upstream valve closure doing something at a particular valve, we must be sure that transient didn't get non-physically washed out by our model.

Liquid transients are fast

When pipeliners say incompressible, they really mean *almost* incompressible. A *truly* incompressible liquid's density stays the same no matter what pressure we apply. One consequence of this is that the force of pressure is immediately transmitted everywhere throughout its entire volume, so that in a liquid that is truly incompressible, the speed of sound is infinite ($c \to \infty$) such that there are no transients at all, just one steady state going straight to another steady state.

In a real almost incompressible liquid, the speed of sound isn't infinite, but it is very fast indeed. As liquids are nearly incompressible, disturbances transmit quickly. A liquid, being relatively incompressible compared to a gas, has no *"cushioning"* action. So when a container of liquid is subjected to some force (F), it compresses less than a gas would before reaching equilibrium with an internal pressure (P) that matches that force to its container's area (A): *

$$P = F/A$$

* youtu.be/O_HQklhIlwQ – Walter Lewin, 8.01x – Lecture 27 – Fluid Mechanics

This makes for exciting experiments. Suddenly stopping a flowing liquid means its kinetic energy turns into static head. One manifestation of this effect is that water pipelines below a street can – and do – blow manhole covers, spurting geysers of water high into the air. But nobody wants an *"exciting"* pipeline. To understand how these conditions affect a pipeline and how to avoid them in the field, we must predict the behavior of rapid transients in liquid pipelines. That means we must capture rapid pressure fluctuations, especially in severe abnormal scenarios. There are also normal but infrequent transient conditions that are less extreme. And, of course, routine scenarios. We want to be certain all of the possible scenarios always remain well within the realms of the mundane in their operating conditions throughout a pipeline.

Liquid transients are challenging to simulate, so much so that it has historically been a struggle to always keep up with real-time. This has been overcome with computational power and clever code. We are restricted to short time-steps, only ever getting away with something like five second time-steps before losing important details in the simulated results.

Liquid operations like batching require frequent changes, such as valves being opened and closed. Some of these in turn give rise to fast transients which rapidly transmit throughout the pipeline. Conservation of mass and momentum describes the resulting pressure surges and their knock-on effects, as well as how fast-acting logic control responds: Atmos SIM works well here. What is special about fast liquid transients is that they present a calculation engine with practical difficulties in solving its discretized equations. In some offline studies for surge analysis, important dynamics occur on a scale of milliseconds, and to capture these our simulation time-step must then be chosen to be that short.

9.2 – SPEED OF SOUND

The speed of sound is different in a pipeline

The speed of sound through a fluid contained within a pipeline is *slower* than it would be if that same fluid were *"free"* – at the exact same conditions but not contained within a pipeline. That free sound speed (c_{free}) depends only on the properties of the fluid, namely its isentropic bulk modulus (B_s) and density (ρ):

$$c_{\text{free}} = \sqrt{\frac{1}{\rho} \cdot B_s}$$

The density (ρ) is found via an equation of state. For liquids, reference tables usually contain the isothermal bulk modulus (B_T). The isentropic bulk modulus (B_s – also called the adiabatic bulk modulus) is the product of isothermal bulk modulus (B_T) times the ratio of heat capacities ($\gamma = c_P/c_V$):

$$B_s = B_T \cdot \gamma$$

For an ideal gas, this isothermal bulk modulus by definition equals pressure ($B_T = P$), so for a real gas that product ($B_s = P \cdot \gamma$) is a reasonably close first approximation of isentropic bulk modulus. A more exact approach is to calculate isothermal bulk modulus (B_T) via a suitable equation of state, as the derivative of infinitesimal pressure rise ($\mathrm{d}P$) with respect to rise in density ($\mathrm{d}\rho$) taken with temperature (T) held constant:

$$B_T = \left.\frac{\partial P}{\partial \rho}\right|_T$$

The speed of sound (c) is therefore faster in a fluid whose isentropic bulk modulus (B_s) is greater, or whose temperature (T) is warmer (because of its effect on density), or whose molar mass (m_r) is lighter. The speed of sound in *"free"* air is about $300\ \mathrm{m/s}$, in hydrogen it's $1\ 300\ \mathrm{m/s}$, and in water $1\ 500\ \mathrm{m/s}$.

The speed of sound in a pipeline is slower than the free speed of sound. Some make a distinction, calling it by a different term (such as wave speed), but this is misleading. It really is, honestly, speed of sound through fluid in the pipeline: if we hit the pipeline with a hammer (not advisable), a clang propagates through the fluid slower than it would through the same fluid at the same pressure and temperature had it been *"free"*. Pressure waves propagate through a pipeline at this slower *corrected* sound speed.

This is because it additionally depends on pipe properties – inner diameter (D), elastic modulus (B) and wall thickness (δ) – and a restraint factor (ϕ) accounting for how the pipeline is supported in place. This corrected speed of sound (c) can be calculated as follows for thin-walled pipes:

$$c = \left(\rho \cdot \left(\frac{1}{B} + \frac{D \cdot \phi}{Y \cdot \delta}\right)\right)^{-½}$$

Other physical effects of pipe expansion

Pressures and temperatures cause a pipe to expand and contract. Pressure in a pipe causes more *hoop stress* (σ_c) to arise within its wall, in the direction of its circumference as discussed in the *Primer* appendix. This is a force per unit area. If it changes, there is a corresponding elastic deformation called *strain* (ε), which lets us calculate a change in diameter due to the change in hoop stress. The *elastic modulus* (Y) * is a property of a solid material within its linear elastic region. It measures *stiffness*, the ratio of tensile stress (σ) to the strain (ε) that results in the same direction. It is typically a very large value compared to pipeline pressures, expressed in units of gigapascals (GPa):

$$Y = \frac{\sigma}{\varepsilon}$$

A longitudinal strain (ε_x) always generates a lateral strain (ε_c) in the direction of a pipe's circumference, as per Poisson's ratio (ν): †

$$\nu = \frac{\varepsilon_c}{\varepsilon_x}$$

Pipe material	Elastic modulus Y Gpa	Poisson's ratio ν
Steel	Between 190 and 210	Between 0.2 and 0.3
Polyethylene ‡	Typically less than 1	0.45

Table 19 – Elastic modulus and Poisson's ratio of steel and polyethylene pipe

Diameter varies with pressure as an elastic deformation. Depending on a pipe's wall thickness (δ) and its elastic modulus (Y), we can calculate its diameter (D) at some

* The elastic modulus is referred to as Young's modulus as he popularized it.

† This is Greek letter *nu* not Latin letter *vee*. Unfortunately they look similar (ν v).

‡ These values are taken from a vendor of high-density polyethylene (HDPE) pipes. Values for reinforced high-density polyethylene pipes can vary by a lot.

fluid pressure (P) relative to the diameter at standard conditions (D_{std}) and the pressure at standard conditions (P_{std}):

$$D = D_{\text{std}} \cdot \left(1 + \frac{P - P_{\text{std}}}{2 \cdot \delta \cdot Y}\right)$$

This is an approximate equation. There is also a further effect, resulting from the Poisson effect we just described. It says that a material tends to become narrower perpendicular to the axis it is being stretched in. Because of this, pipe expansion is also a function of longitudinal anchoring ($X = 1.25$ if the ends are free and $X = 1$ if the ends are fixed) * and the wall material's Poisson's ratio (ν). An exact equation has an $(X - \nu^2)$ factor as follows: [26]

$$\left.\frac{\partial A}{\partial P}\right|_T = \frac{D}{\delta \cdot Y} \cdot (X - \nu^2) \cdot A_{\text{std}}$$

With reference to the pipe's cross-sectional area at standard conditions (A_{std}). The Poisson's ratio of steel is such that the overall effect of this $(X - \nu^2)$ factor makes a negligible difference to the calculated diameter ($< 0.1\%$). For a small high-density polyethylene (HDPE) pipe it has an effect of perhaps 1%. The effect of temperature on pipe expansion by radial deformation is another term referring to standard temperature:

$$A = A_{\text{std}} + \left.\frac{\partial A}{\partial P}\right|_T \cdot (P - P_{\text{std}}) + \left.\frac{\partial A}{\partial T}\right|_P \cdot (T - T_{\text{std}})$$

Pipe material	Coefficient of thermal expansion (α) $°\text{C}^{-1}$
Steel	$10 \text{ to } 17 \times 10^{-6}$
Polyethylene	$100 \text{ to } 200 \times 10^{-6}$

Table 20 – Coefficients of thermal expansion for steel and polyethylene pipe

Cross-sectional area (A) varies from its value at standard conditions (A_{std}) with temperature (T) as a function of the pipe's coefficient of thermal expansion (α):

$$\left.\frac{\partial A}{\partial T}\right|_P = 2 \cdot \alpha \cdot A_{\text{std}}$$

For a $50°\text{C}$ change in temperature, this thermal expansion is small in steel pipes, something like a 0.2% expansion in cross-sectional area. It is, again, more pronounced

---

* Unburied exposed and undersea pipes don't use constrained supports. This leaves the ends of those pipe sections free to move. In contrast, buried pipes are constrained by the backfill, and so their ends are essentially fixed in place.

in pipes made of flexible, low-bulk-modulus materials such as high-density polyethylene (HDPE), something like a 2% expansion in area.

Ignoring pipe expansion in a pipeline simulator has noticeable consequences. Waves propagate in the pipeline fluid at a sound speed (c) which would be slightly wrong in that case. The actual expansion and contraction of the pipe's diameter in conjunction with the pressure wave alters the sound speed. In turn, if the sound speed is wrong, the wavefront of a hydraulic transient like a pressure surge arrives early or late, and has the wrong magnitude too – being smaller or bigger than reality. This affects the performance of a leak detector. The inventory of a pipe is also slightly affected by this expansion, so for any application such as leak detection that relies on modelling small changes in inventory it is important to get this right.

This equation for modified sound speed isn't directly applied by Atmos SIM because the same result automatically falls out of correctly accounting for pipe expansion in the fluid mass, momentum, and energy conservation equations. It is presented in an analytical form to serve as a handy check on any simulator.

9.3 – HYDRAULIC SURGE

What is water hammer, surge, or hydraulic shock?

There are many ways to break a pipeline. When a fluid is subjected to an inappropriate change in velocity, a phenomenon takes place known variably as *hydraulic shock*, *water hammer* or *pressure surge*. It has caused major accidents, [60] so surge analysis is conducted to ensure that operational controls avoid creating any hydraulic transients that exceed allowable limits anywhere. International conferences are held annually on the topic of surge analysis, which has become its own niche topic. Specialist consultants are employed to size surge relief equipment, and assess fast-acting logical control strategies to protect against worst-case scenarios. Their work is facilitated by pipeline simulation, with closely spaced knots and time-steps suitably short in duration.

The phenomenon in question, whatever we choose to call it (we'll call it surge), is considered an abnormal operating condition, and is typically attributable to poor control. A number of important scenarios give rise to surge conditions. And an incorrect speed of sound has serious implications in surge modelling.

Surge scenarios in a liquid pipeline

Surge scenarios arise on both liquid pipelines and gas pipelines. One is an *emergency shut-down* (ESD) that could kick in for many reasons. After a power failure, how quickly can we cease flow by closing valves? How would the entire pipeline system respond?

Causing flow in a pipeline to cease requires us to impose a net force from some source. In normal operation the force of friction balances the pressure gradient; once the driving pressure is released, that same friction gradually slows the fluid to a stop. To stop flow more abruptly by, say, closing a valve, more force is applied. This force ends up coming from a reverse pressure gradient – and that can be a concern. Reverse flow can be extremely costly, especially at pump discharge as it could cause a pump to spin backwards. A fast-closing and tight shut-off valve is necessary to prevent this problem.

A change in velocity happens in numerous liquid pipeline scenarios, including:

- valve movement – for example:
  - a block valve going from open to shut
  - or a regulator changing its position
- pump speed – for example:
  - a pump gradually spins up during a start-up sequence
  - or a pump trips to avoid exceeding some limit
- oscillations in tank levels
- collapse of vapor pockets formed at any point along a liquid pipeline
- sudden fluid blockage due to equipment malfunction
- sudden change in flow rate to meet an increased demand at a delivery
- a change in the incoming fluid's composition during injection of a new batch
- a batch of a different density hitting a pump operating at fixed speed

Some of these scenarios are, naturally, more dramatic than the others. A surge analysis considers all potential scenarios and unique hazards associated with a pipeline. Identifying and mitigating surge requires hydraulic analysis of the entire network in every routine and non-routine scenario. Surge mitigation measures are assessed for every section, condition, and scenario. Short lines within terminals are simulated with a very tight mesh of knots and time-steps. Offline transient scenarios can investigate events that have happened in the field. Perhaps a leak could cause a pump trip, and that might then cause an emergency shut-down. Simulating this series of events helps figure out what happened and how best to deal with similar occurrences in the future. This can then be incorporated in a simulator trainer-trainee application.

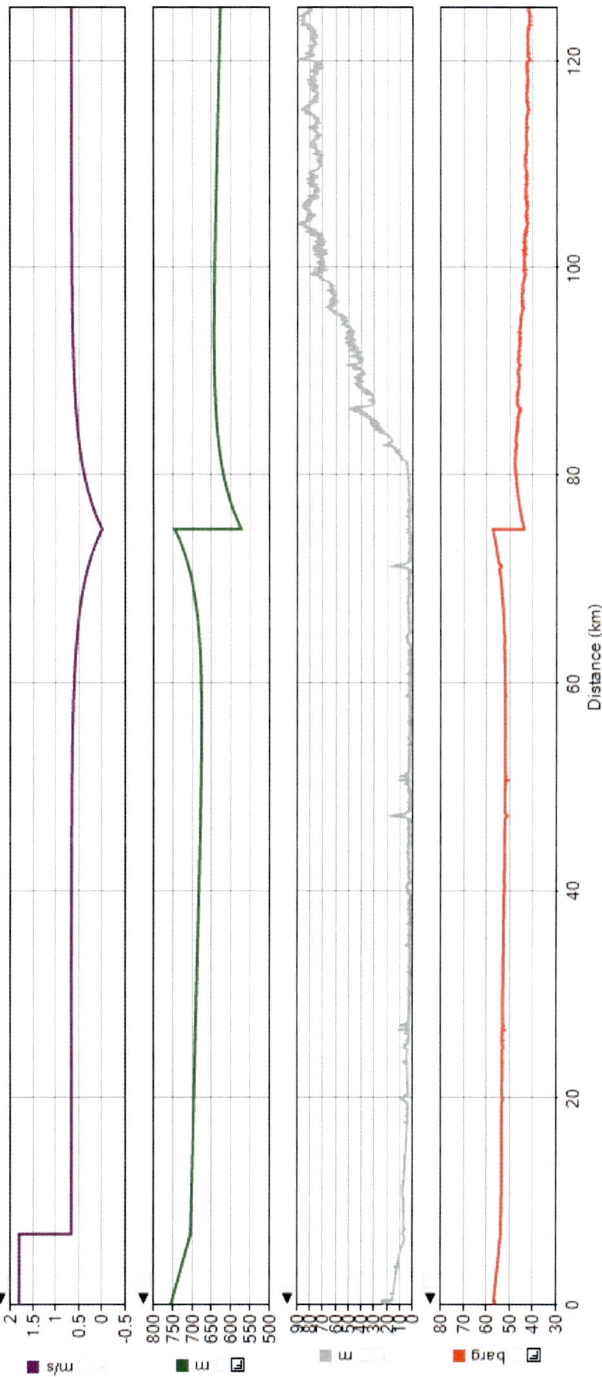

Figure 176 – The instant of a valve closure within a flowing liquid pipeline. A surge travels in the upstream direction and another travels downstream. The top profile is of flow velocity, the second is of head, the third is elevation and the bottom is pressure.

Transient simulators during the design phase help ascertain the safety of pipeline operation during uncontrolled pressure transients. The economics of surge control systems can be assessed versus thicker-walled pipe. Water hammer calculations help with the sizing and choice of location of air vacuum and positive pressure surge relief valves. Station control valves can be designed with pressure override logic to protect the pipeline from over-pressure.

High speed data collection allows us to compare actual data of surge waves with their simulated values. It is not typically offered in off-the-shelf commercial simulators, and although it's a new and interesting avenue to gather fast scanning data that might help with validation, more ordinary data collection intervals have proven adequate for usual simulation tasks.

Very fast transients in liquids have complicated effects going on, where our assumption that the flow rate is uniform across a pipe diameter does not hold. There are specialized versions of the one-dimensional plus time (1D+T) mass and momentum equation that account for this.

Rapid transients in a gas pipeline

Surge scenarios are assessed for gas pipelines in much the same way as they do in liquid pipelines. The compressibility of a gas, even when it is very dense at the pressurized conditions of a pipeline, means the speed of sound is lower. Taken with the fact that a pressurized gas has a lower density than a liquid, the Joukowsky pressure rise is lower in a gas pipeline than in a liquid pipeline. But this doesn't mean that a gas *"absorbs"* a surge: gas pipelines are still liable to excessive up-surge and down-surge if operations are performed improperly.

In a liquid pipeline a big pressure change is needed to affect flow, whereas in a gas pipeline even a slight pressure change causes a big change in flow. The reason for this, as we'll see, is the Joukowksy equation. It states that a change in pressure is proportional to density, sound speed and a change in velocity. In a pressurized gas pipeline, density is typically a factor of two to ten smaller than in a liquid pipeline, and sound speed is three to five times smaller than it is in a liquid pipeline. The same change in pressure affects gas velocity more.

What this means in practice is that a sufficiently large pressure change in a gas pipeline, such as that seen during a rupture or blowdown, creates a shock wave with choked flow. The flow rate in the pipe reaches a limit – sound speed – at the point of choking. At that point, there is a stationary shock in the pipe. The shock is located at a local narrow spot, such as a leak. Atmos SIM simulates choked flow if it occurs, either inside the pipe – typically either at the downstream end or at a leak – or in the leak itself. Choked flow in a pipe rather than the leak itself may happen in the event of a long crack in the pipe opening up, such that *"leak area"* is bigger than the pipe's cross-sectional area. Then we might get choked flow in the pipe coming from either direction.

Many things cause these abrupt changes in flow rate within gas pipelines. Sudden shut-down of compressors, sequential valve closures along several stations, or an abrupt increase in gas withdrawn from the mainline are all routine operations that significantly affect the pipeline. They cause pressures to rapidly fall at a rate comparable in magnitude to that associated with a large leak.

Unacceptable rapid transients are those that give rise to extreme pressures, adversely interfering with equipment. Dynamic components (compressors) might fail, leading to high loads on other devices. Static components might fail through fatigue. The pipes themselves might fail due to the extreme pressures.

By understanding the detailed characteristics of unacceptable rapid transients, we can differentiate between normal routine operations such as compressor shut-downs or valve closures, and abnormal conditions such as leaks and ruptures. This allows an operator to tune automatic controls to enact measures via an appropriate response. In the case of a rupture, a pipeline should automatically shut-down by closing valves around the ruptured pipeline segment to isolate it from the rest of the network. Whereas following a compressor shut-down, a pipeline might safely remain operational, running at a reduced flow rate. [61]

Emergency shut-down in a gas pipeline

Following a leak or rupture, it is essential for emergency shut-down controls to respond correctly, avoiding unnecessary leakage of harmful emissions to the environment, which if ignited could explode. An uncontrolled stoppage can cause costly disruption by damaging pipeline equipment such as compressors.

An emergency shut-down (ESD) is a procedure carried out using specially installed emergency shut-down valves (SDVs or ESVs) present in all pipelines. These are designed to automatically close in case of a major pipeline event to avoid a line break. Interestingly, these complex control systems can be built to operate independently of SCADA systems in an entirely mechanical manner that is driven by pipeline pressure, so that it doesn't require power to operate. They rely on automatic mechanisms that fall into two categories: those that trigger on low pressure by measuring a differential between the pipeline and a reference reservoir, and those that trigger on high rate of pressure change. Modern emergency shut-down valves incorporate both approaches to trigger a closure. Emergency shut-down valves are calibrated to detect a sudden transient event with particular features (e.g. a surge) and can respond to a number of possible scenarios. Their duty is not restricted to *"just shut down if there's a leak here"*.

Automated responses can be problematic during routine operations. Automatic shut-down valves close main line block valves if pressure sensing mechanisms detect an unexpected fall in pipeline pressure. This can lead to incorrect actuation of valves during normal operating procedures, such as a planned compressor shut-down, resulting in the unnecessary closure of pipeline mainline block valves. That then involves a costly restart procedure, combined with loss of earnings while the pipeline is out of service. Automatic shut-down controls therefore must be adequately planned, and the equipment calibrated to avoid these scenarios.

An emergency shut-down may be automatically triggered in the field based on the slope of a wavefront, i.e. based on the rate of change of pressure at that point. But a typical simulator's numerical dispersion might wash out this slope to the point that the simulated valve doesn't trigger. It is therefore especially important to use appropriately short knots and time-steps for this application.

Figure 177 – A pneumatically actuated shut-down valve

To minimize potential flow interruptions, highly accurate modelling of pipeline dynamics is used to tune the behavior of automatic shut-down valves. Unless they are optimally calibrated, these line break mechanisms could be improperly actuated, resulting in unwanted interruption of flow during routine operations.

Joukowsky's magnitude of a peak surge pressure

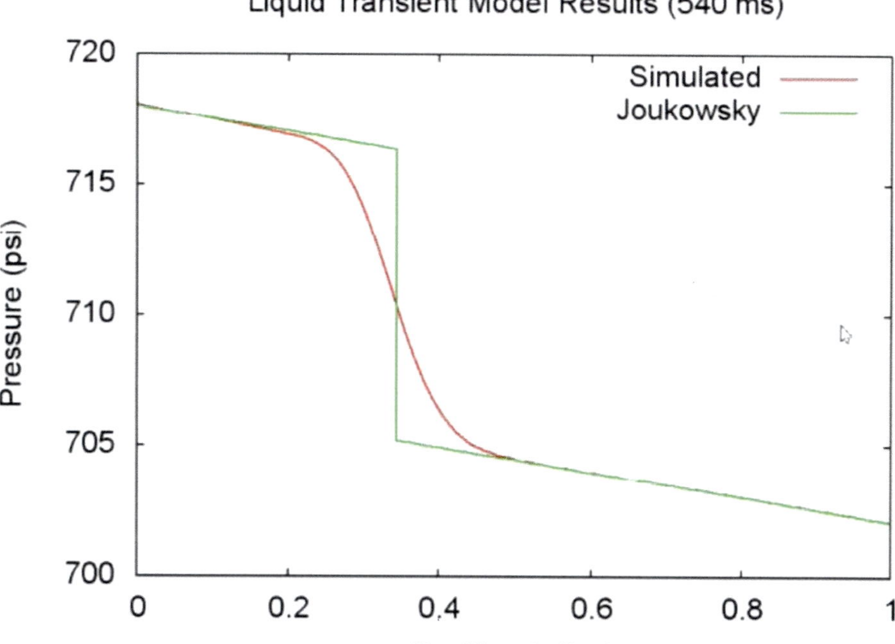

Figure 178 – A simulated pressure surge wavefront and the analytical Joukowsky result. This simulation has produced some numerical dispersion, which results in the smoother profile.

Sound speed, by definition, affects the arrival time of a wavefront. It also affects the *magnitude* of the pressure wave associated with a given change in flow rate. A change in velocity (Δv) causes a change in fluid head (ΔH) whose magnitude depends on sound speed (c) and the acceleration due to gravity (g):

$$\Delta H = -\frac{c \cdot \Delta v}{g} \cdot \left(1 + \frac{v}{c}\right)$$

This is derived from conservation of momentum ($F = m \cdot a$) coupled with the fact that a disturbance propagates down a pipeline at the local speed of sound.

In most rigid pipes, sound speed is far greater than fluid velocity ($v/c < 0.01$) so the head increase due to a change in velocity reduces to *Joukowsky's Law*:

$$\Delta H = -\frac{c \cdot \Delta v}{g}$$

Multiplying head by density gives us pressure. So the head increase manifests as a pressure surge (ΔP) whose magnitude depends on the fluid's density (ρ):

$$\Delta P = -\rho \cdot c \cdot \Delta v$$

Pipeline fluid	Density	Speed of sound	Flow velocity	Pressure surge
Water	1000 kg/m^3	1485 m/s	1.6 m/s	+24 bar
Crude oil	840 kg/m^3	1250 m/s	1.6 m/s	+17 bar
LPG at 50 bar	540 kg/m^3	880 m/s	1.6 m/s	+8 bar
Natural gas at 150 bar	210 kg/m^3	430 m/s	10 m/s	+9 bar

Table 21 – Peak pressure surges in fluids at pipeline conditions, as per Joukowsky's equation

The importance of speed of sound in these calculations should be appreciated.

Is the reported sound speed (c) consistent with how a simulated surge wave progresses through a pipeline?

If, in the case of a plastic pipe, expansion due to operating conditions reduces sound speed by 10%, that reduces the estimated peak pressure during a surge event by 10%. For the same reason, if we performed a transient simulation with an isothermal model, the simulated pressure surges would be way off: an isothermal model means that the isothermal bulk modulus is effectively used in wave propagation. Though for water this is a safe enough approximation, for many hydrocarbon liquids it might result in a 10% error in sound speed, and for a gas it might result in a 20% error in sound speed.

An isothermal sound speed might be 20% off, owing to how an isothermal and an isentropic bulk modulus differ. This causes a calculated pressure surge to be off by 20%.

Active surge mitigation strategies

Calculating a pressure surge allows pipeline designers to put in place devices (passive measures) and procedures (active measures) for mitigating hydraulic surges. Such systems might be safety critical, relied upon to perform as a layer of protection against what might otherwise lead to catastrophic consequences.

Demonstration: closing a valve ruptures a pipeline

A valve is slammed shut at the outlet of a short crude oil pipeline. The safe operating limit is a maximum design head of $H_{\text{max}} = 1\ 000\ \text{m}$ for the crude oil to be transported.

Here we shall perform a manual surge analysis to demonstrate the procedure. In a user-friendly pipeline simulator such as Atmos SIM, simulating a surge scenario becomes trivial. The equations are stated alongside typical values to give the reader an idea what sort of magnitude one might expect in their result.

We calculate the Joukowsky head via the sound speed and fluid deceleration. The sound speed (c) at which waves propagate depends on diameter (D), pipe restraint factor (ϕ), bulk modulus (B), elastic modulus (Y) and wall thickness (δ):

$$c = \left(\rho \cdot \left(\frac{1}{B} + \frac{D \cdot \phi}{Y \cdot \delta}\right)\right)^{-\frac{1}{2}} = 1\ 027\ \text{m/s}$$

For fluid deceleration – a change in velocity over time (Δv), we might assume fluid velocity decreases linearly over valve closure time. This assumption could give inaccurate calculated values, depending on the type of valve being used:

$$\Delta v = -\frac{q}{\pi \cdot r^2} = 8.06\ \text{m/s}$$

Joukowsky's law assumes the valve closes instantaneously, approximating head increase (ΔH):

$$\Delta H = -\frac{c}{g} \cdot \Delta v = 843\ \text{m}$$

Initial head at this location (H_{initial}) plus head increase (ΔH) calculates the maximum head (H_{max}):

$$H_{\text{max}} = H_{\text{initial}} + \Delta H = 1\ 440\ \text{m}$$

This exceeds the line's maximum design head. We need a surge mitigation procedure or a surge relief device.

Joukowsky's law can be modified for short pipes to include valve closure time, chosen to be sufficiently longer than wave reflection time. An approximation for head increase (ΔH) at valve closure times of 1, 2, 3, 4 and 5 seconds is:

$$\frac{\mathrm{d}H}{\mathrm{d}t} = -\frac{c}{g} \cdot \frac{\mathrm{d}v}{\mathrm{d}t}$$

Rearranging gives a valve closure time. In this case, choosing 2.8 seconds ensures the pressure stays within acceptable levels following valve closure, with a safety factor of 10%. If we slam the same valve shut over only 1 second, this scenario would give rise to an extreme surge exceeding the maximum allowable operating pressure, causing the pipeline to rupture.

The length of the pipeline section immediately upstream of the closing valve to the next upstream constraint or boundary condition also considerably affects maximum head; this is also not accounted for here. In practice, the safe valve closure duration is entirely dependent on the wave reflection time. There is no simple formula for calculating that: engineers need a pipeline simulator.

Figure 179 illustrates how simulated surge pressures initially match those predicted by the analytical Joukowsky equation, and how differences arise later in that scenario. There is a bit of numerical dispersion which is not physical, but which also has little meaningful effect. There is also a larger amount of change in the upstream frictional loss, which is physical, demonstrating the benefit over simpler methods of conducting surge analysis via pipeline simulation using a complete physical model.

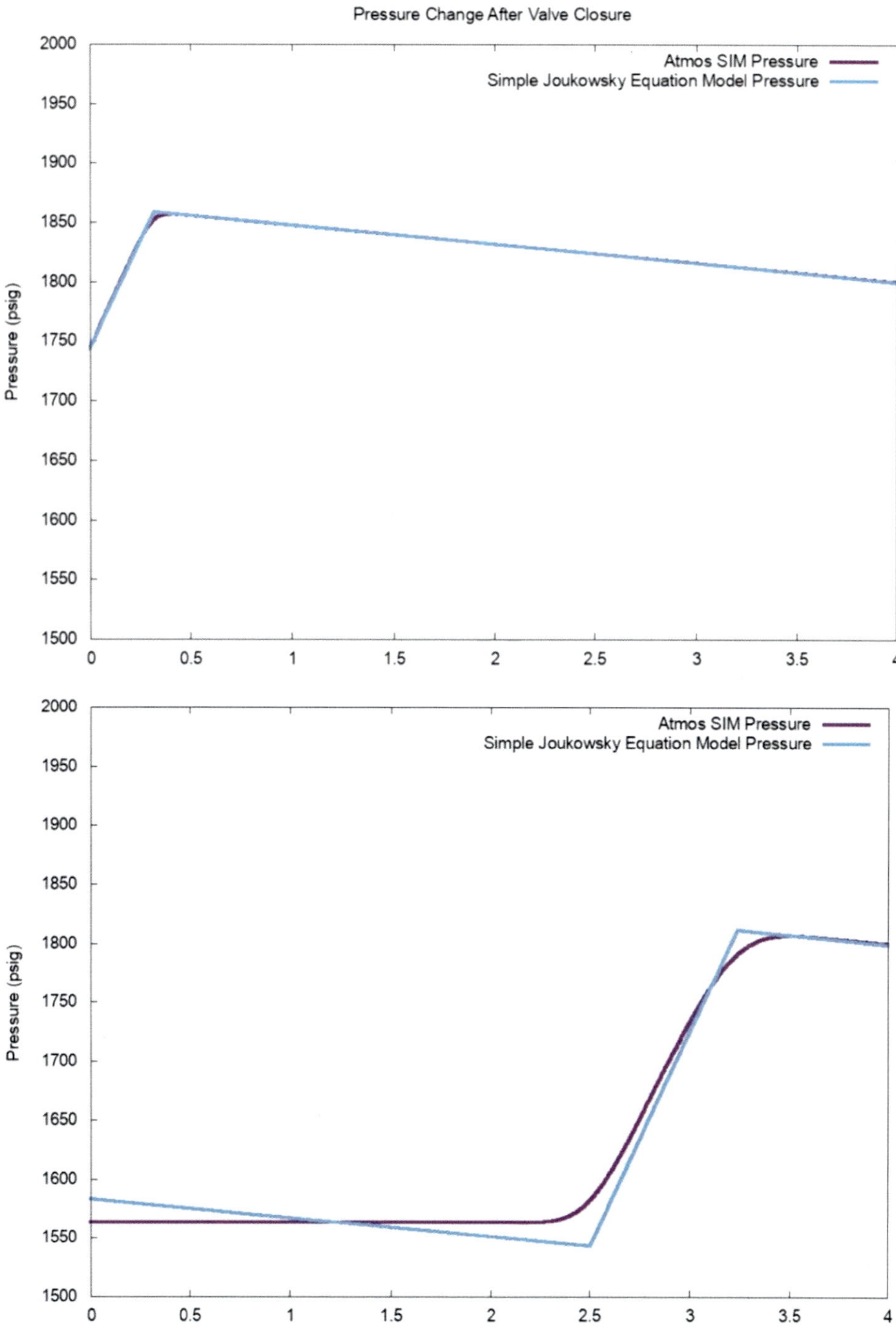

Figure 179 – The surge pressures simulated by an Atmos SIM transient scenario initially agree closely with those predicted by the Joukowsky equation (top). Differences arise later (bottom) partly due to non-physical numerical dispersion, but mostly due to physical frictional loss.

Valve timing: slower closure won't always mitigate a surge

Figure 180 – What happens if we shut this valve suddenly? What if we shut it gradually?

A valve must close slowly enough to avoid what engineers call *"valve chatter"*. Some valves have a special *"silent"* design so they can be faster-closing, but there is always a limit to how fast it is feasible to close a valve. In the case of a check valve or a non-return valve (NRV), the closure time must be quick enough to avoid reverse flow. Backflow might be occurring whilst a check valve is in the process of closing, but closing it too suddenly (a spring or weight slamming it the instant flow ceases) creates shock waves. Instead, check valves in normal operation are regulated by a damping mechanism that is tuned to bring flow to rest over a carefully calculated valve closure period. How quickly a particular valve in a particular pipeline should be closed is a question that is complicated to answer, falling within the remit of surge analysis studies.

Another matter to consider is that we cannot assume a fluid decelerates linearly throughout a valve closure. A gate valve, for instance, is only 2% to 5% open before significantly reducing flow velocity; each valve type behaves differently.

In one example, the operating procedure for an emergency shut-down was taking over 3 minutes, but it was possible to do something much safer within 30 seconds. A better recommendation for this valve closure scenario might be:

1. Move the valve towards a closed position until pressure reaches some value
2. then wait for the pump to trip
3. then wait for the trip pressure drop to arrive at the valve
4. then close the valve the rest of the way until it is completely shut

If we close this valve over just 20 seconds in this way, it would be adequate.

> A surge study recommended to *"close the valve over 208 seconds"*. It assumed a linear valve profile with almost no resistance for the first 200 seconds. It was really saying *"wait 200 seconds then close the valve over 8 seconds"*!

For now, we shall leave the matter of valve maps aside for the purposes of this discussion, and consider a valve closure as simply being a linear reduction in flow rate over the specified period of time, enforced using Atmos SIM's regulator model item. This lets us focus on the main consideration limiting on how quickly we shut a valve: the peak pressure spikes due to hydraulic surge.

When liquid stops or reverses because of a valve slamming shut in a pipeline, a hydraulic surge ensues. The tremendous racket that accompanies such a surge isn't a mere nuisance: the noise and vibration might cause substantial damage. In fact, every valve movement inevitably creates a pressure surge of a certain size, which then propagates through the pipeline. A valve *closure* increases the pressure upstream, and decreases the pressure downstream. The magnitude of this pressure surge depends on how quickly a valve closes, the properties of the fluid, and the pipeline's restraint and elastic properties.

In most applications of pipes, it is excessively fast valve closure that causes upstream pressure to become too large. There have been cases of pipes being ruptured by these associated pressure spikes. This means that in most applications, a surge mitigation measure is simply to have a longer valve closure period, slowing the fluid more gently, lowering the peak pressure. The only way to avoid *at least* the Joukowsky pressure rise happening at the downstream side of a closing valve is to close it so slowly that the reflected wave returns back before it's done. This is not usually feasible on a pipeline. Contrary to what is suggested by the idealized analysis that is popularly taught, with instantaneous valve closures along frictionless pipelines, it is possible that on a real pipeline, closing a valve slower may make a surge *worse* not better.

Moreover, the maximum pressure might not be occurring at the valve. It might be occurring at the reflection point, because the operating pressure there is higher in the first place. We can mitigate that a little by closing the valve slowly, because there the reflected part of the wave only has to go the length of the wavefront, not the length of the pipe. Modelling that accurately within an error of 1 bar might necessitate modelling the control logic associated with the closing valve.

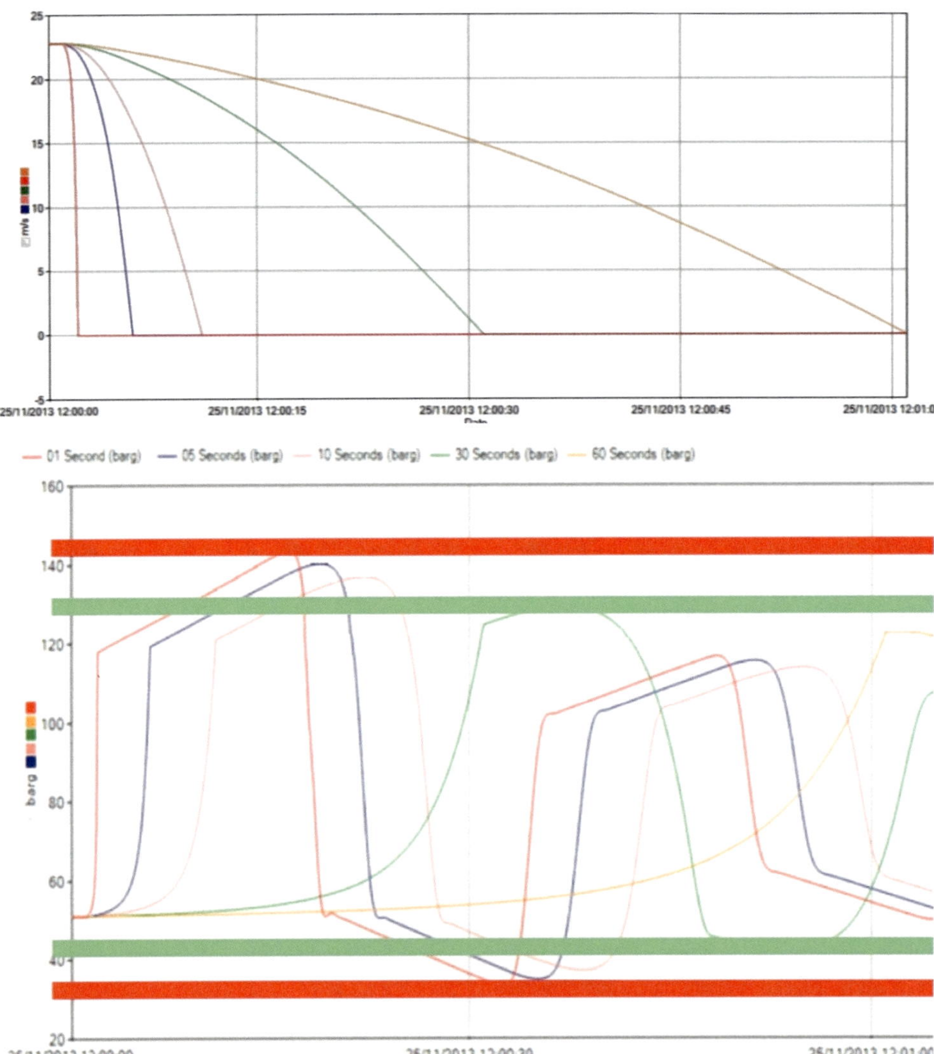

Figure 181 – Flow velocity (top) and upstream pressure (bottom) for a linear valve closure over 1 second (red), 5 seconds (blue), 10 seconds (pink), 30 seconds (green) and 1 minute (orange).

More importantly, closing a valve more slowly won't always reduce the surge pressure! Pipelines, as it turns out, are a little different from most applications of pipes, because they are so long and at such high pressures. The slowest practical valve closure time must be weighed up against a need for valves to close quickly enough during the worst-case scenarios, according to emergency shut-down procedures. In a liquid pipeline, for example, we must close block valves quickly enough to maintain enough pressure that is sufficient to avoid fluid cavitation (slack or column separation).

Pressure wave propagation and reflection time

Any pressure wave propagates along a pipeline at the speed of sound (c). It traverses the length of the pipeline (L). The wave subsequently reflects off the other end and then comes back. The period elapsed until the reflection of an outgoing wave is seen coming back towards the original location is therefore the time taken to traverse the section of pipeline twice if the origin is at either end. The reflection time is shorter if the pressure wave is initiated in the middle of the section.

$$\Delta t_{\text{reflection}} = \begin{cases} \dfrac{2 \cdot L}{c} & \text{if origin is at the end} \\ \dfrac{L}{c} & \text{if origin is in the middle} \end{cases}$$

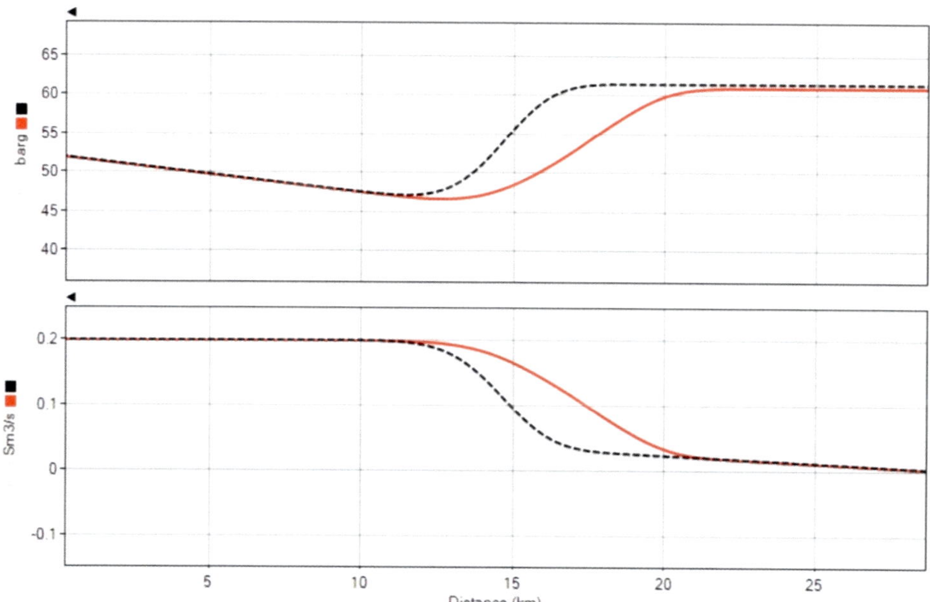

Figure 182 – A hydraulic surge propagates a pressure wave along the pipeline at sound speed, after a valve linearly reduced flow rate to zero over 5 seconds (solid) and 1 second (dashed). The peak pressures are the same because this water pipeline's reflection time is 40 seconds.

This happens both upstream and downstream of the location at which a surge wave originated, with the potential to cause over-pressure conditions upstream and under-pressure conditions downstream.

We need to consider whether the valve closes slowly enough that the time-period elapsed is greater than this reflection time ($\Delta t_{\text{reflection}}$):

$$\Delta t_{\text{closure}} > \Delta t_{\text{reflection}}$$

A typical stretch of pipeline transporting fluid at typical pressures has quite a prolonged reflection time. This is one peculiarity of performing surge analysis in a pipeline.

Fluid	100 m	1 km	10 km	30 km	100 km
Water	130 ms	1.3 s	13 s	40 s	2 m 15 s
Crude oil	160 ms	1.6 s	16 s	48 s	2 m 40 s
LPG at 50 bar	230 ms	2.3 s	23 s	1 m 8 s	3 m 47 s
Natural gas at 150 bar	470 ms	4.7 s	47 s	2 m 20 s	7 m 45 s

Table 22 – Wave reflection times in pipelines of length 100 meters to 100 kilometers for a selection of pipeline fluids will vary from milliseconds to the best part of ten minutes

No emergency shut-down (ESD) procedure keeps a pipeline flowing for the best part of ten minutes until a valve travels from open to closed! On a pipeline, closing a valve slowly doesn't necessarily help with overpressure. The peak pressure stays the same.

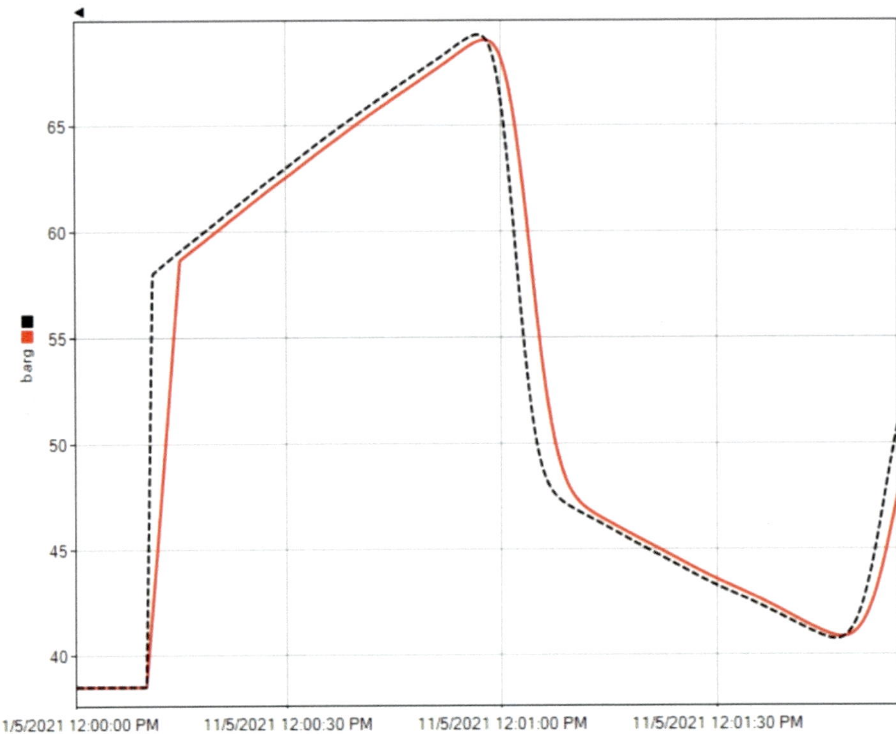

Figure 183 – The peak surge pressure is the same for a linear reduction in flow rate to zero over 1 second (dashed) and 5 seconds (solid), because this pipeline's reflection time is 40 seconds

If we choose a slower valve closure time that is still briefer than the pipeline's reflection time, that might make the surge *worse*. We might get the same pressure rise (as per the Joukowsky relation), as well as a region of reduced flow extending upstream, which means reduced losses from friction, and therefore a higher base pressure at the inlet to the closing valve (which is the outlet of that upstream section of pipeline).

For this reason, a pipeline's active surge mitigation strategies may involve waiting for a pressure wave to arrive from some upstream operation before taking an action that would create a pressure spike at a certain location. Figuring out the timing of this sort of operation relies on sound speed being very accurate. For example, if the inlet of the upstream stretch of the pipeline we just examined – instead of being constrained on pressure – had a valve been shut by the time the wavefront arrived, we could time the down-surge to cancel out part of the reflecting up-surge.

This raises the issue of how to shut a pipeline down. There is no single way of doing this across the board. On a gas pipeline, a routine shut-down might close downstream valves first, in order to keep the pipeline packed with inventory. An emergency shut-down might close upstream valves first, in order to drain the section in which a leak is detected and reduce the quantity of gas released. The effects of elevation, length of a stretch between block valve stations, and behavior of pumping or compressor stations comes into play. On a liquid line that crosses some hilly terrain, downstream valves might be closed first to hold enough pressure at the high points along the route to avoid column separation. It's the case-by-case nature of surge scenarios such as shut-downs that makes surge analysis and transient simulation a fascinating, extensive endeavor.

The peak pressure can be lowered using advanced control logic which trips upstream pumps or compressors at exactly the right moment as soon as anything triggers a valve closure along the route. The negative pressure surge moving downstream from the tripped pump or compressor is carefully timed so that it adds linearly to the positive surge moving upstream from the recently-closed valve.

Controlling how pumps trip

Even a routine procedure to shut down a series of pumping stations requires us to investigate what happens when, and what happens next.

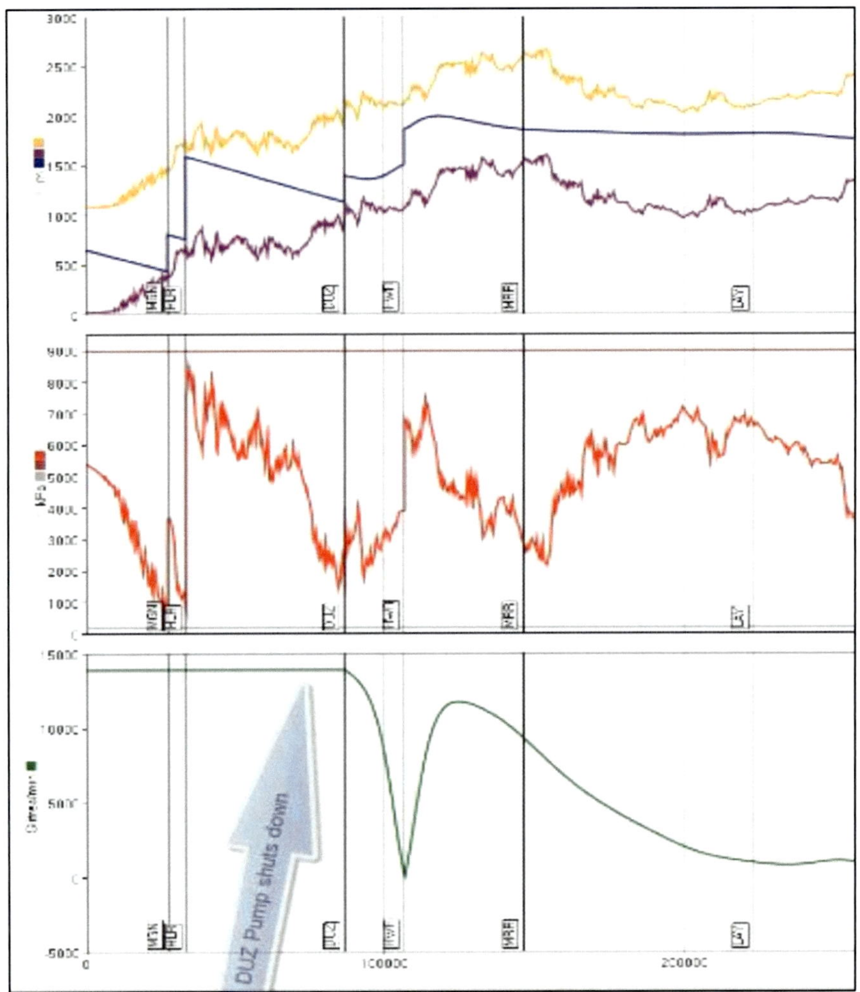

Figure 184 – A series of pumping stations shutting down in succession on an oil pipeline, remaining within allowable operating limits throughout the entire route at all times

If too little or too much is asked of a pump, it violates an *operating limit*, and will accordingly *trip* if configured to do so. This operating window might be rather tight. While the consequences of a trip might be bad, they ought to be tolerable. We must be confident that we can always trigger a pump trip to avoid something worse, like melting the pump. Tripping a pump can cause disastrous hydraulic surging along a pipeline, particularly in a pumping station just before a stretch of pipeline climbing up a rising elevation profile, the kind of pipeline system we say has a high *static lift* – the maximum elevation increase the fluid is going to obtain in the pipeline downstream of the pump.

Discharge pressure falls rapidly after a pump trips. A robust, defined shut-down procedure is vital to safely operating any pipeline, with thorough consideration given to how valves are controlled. We want to plan it so this hydraulic transient settles without slack. If the pressure drop is enough to induce slack, then this propagates, causing vapor pockets to open up. This triggers the check valve on the delivery side to shut, to prevent flow reversal which might damage the equipment. In turn, this sudden valve closure could itself cause rapid flow reversal (again, *hydraulic surge*), particularly if only one of several units trips. Unless such transients are properly controlled, the system is liable to over-pressure, particularly near the pumps.

The following figure details a pipeline with multiple pumping stations. Stopping a pump suddenly – unless properly planned and studied – has the potential to disrupt other pumps causing further surges to propagate through the pipeline. This has the potential to exceed safe operating limits and rupture the pipeline. A surprising number of surge scenarios need considering when we consider the various configurations in which pumps and valves might be operating, even on a simple pipeline like this one.

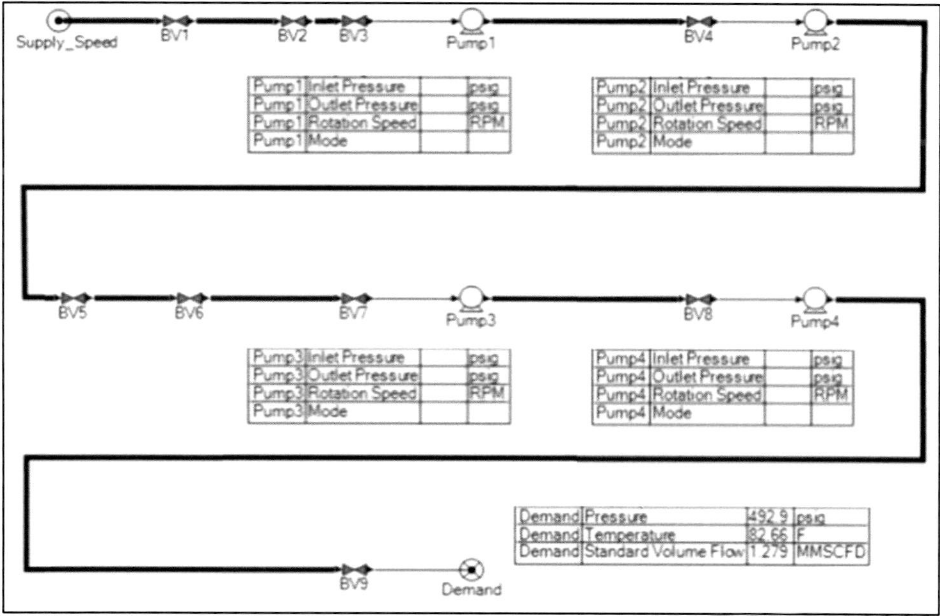

Figure 185 – Four pumps and nine valves along a pipeline model. As the equipment is operated in various configurations, one might not realize how many surge scenarios arise – even on a simple pipeline like this one, there is a surprisingly large number of surge scenarios to consider.

Controlling how a compressor trips

A compressor stop is of paramount importance and must happen quickly.

A variety of reasons can cause a sudden shut-down of a compressor unit or a compressor station. These include scenarios such as a power outage or a defect. While running, the discharge pressure of a compressor on a large transmission pipeline is significantly higher than it would be if the compressor is stopped. A sudden compressor shut-down causes an abrupt reduction in discharge pressure. This sudden fall in pressure creates a *rarefaction* wave, a negative pressure wave that propagates down the pipe in exactly the same manner as a pressure surge.

Compressor stations integrate a selection of safety systems and practices, including automatic shut-down systems. These systems can detect abnormal conditions such as unanticipated pressure falls, leaks and ruptures. They automatically stop the compressor units and isolate the section of the pipeline to limit environmental damage in case of a rupture. Regulations require operators to periodically test compressor stations and maintain the emergency shut-down system to ensure reliability. [61]

This presents us with a challenge. We must discern between falling pressure caused by routine operations, and events caused by leaks. A line break mechanism must be sufficiently sensitive to identify and act in the event of a leak, without promoting improper closure of valves during routine operations.

Small transients don't really disperse as they go down a pipeline, so any dispersion we see is necessarily an artefact of our simulator. We can get an accurate simulation of any fast transient by looking at its initial length in the pipe and just assuming it always has that length, while its height (in pressure, flow, or whatever it may be) evolves in such a way as a simulator says it does.

Hydraulic simulation of these rapid transients requires a mathematical model capable of modelling the detailed rarefaction wave that arises. On that basis, we can tune emergency shut-down valves by modifying the simulated results.

Case study: controlling an emergency shut-down valve

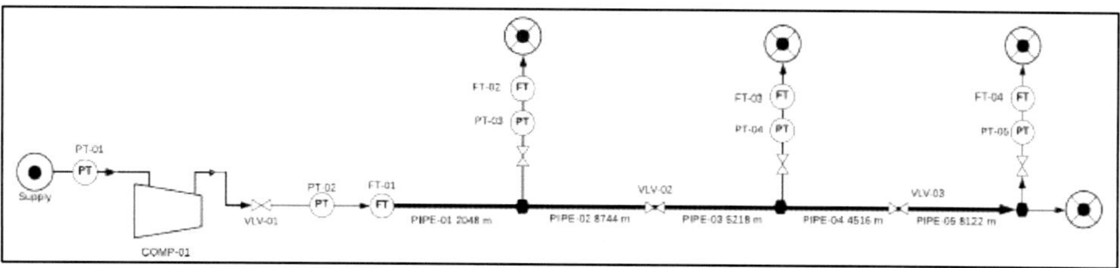

Figure 186 – The region of a gas pipeline selected for this case study

The physical case study pipeline is a 29 km, 24-inch segment of a natural gas pipeline. At its inlet is a compressor station with a shut-down valve immediately downstream. Metered off-takes are located at the 2 km, 16 km and 29 km points along the mainline. At the 11 km point between first and second off-take is an automatic shut-down valve.

On this pipeline, Atmos SIM was used for its planning capabilities. This assists with suitably calibrating the emergency shut-down controls. By taking an actual operational state of the pipeline, saved from an online real-time model, then restoring it to launch an offline simulation, detailed studies can be conducted accurately. Operational scenarios such as compressor shut-downs can be simulated, allowing us to analyze the resulting pressure waves. These operations are then evaluated against simulated leak data, to help differentiate operational scenarios from pipeline leakages, and identify desired calibration ranges for each automatic shut-down valve.

High-resolution data from the physical pipeline was measured at the location of the automatic shut-down valve. This data was sampled at a rate of 0.1 seconds. The typical pipeline measurement data received from the pipeline instrumentation for the network has a sampling rate of 10 seconds. The objective of the case study was to obtain a clear reference of the actual rate-of-change of the pressure at the location of the shut-down valve when a compressor upstream tripped, so that the valve isn't activated unintentionally. We wish to ensure that the emergency shut-down valve (ESV) doesn't trigger on events like a compressor trip, but automatically triggers on a rupture event.

Simulations using Atmos SIM investigated whether the shape of a moving wavefront is retained during its journey through a gas pipeline. The shape of a pressure or flow wavefront is characterized by a height and a width, as shown in Figure 187.

Two scenarios were considered. The first was a compressor trip scenario in which boundary pressure falls by 5 bar over 0.1 s, which ideally should not trigger the emergency shut-down valve (ESV). The second was a rupture scenario in which boundary pressure falls by 16 bar over 0.1 s, chosen to be just short of what would cause choked reverse flow at the upstream end, as in the event of a real rupture it is likely that flow would choke at the rupture point. Within 10 seconds of the event, the pressure wave travelled 3.2 km. Within 80 seconds, it had travelled 26 km.

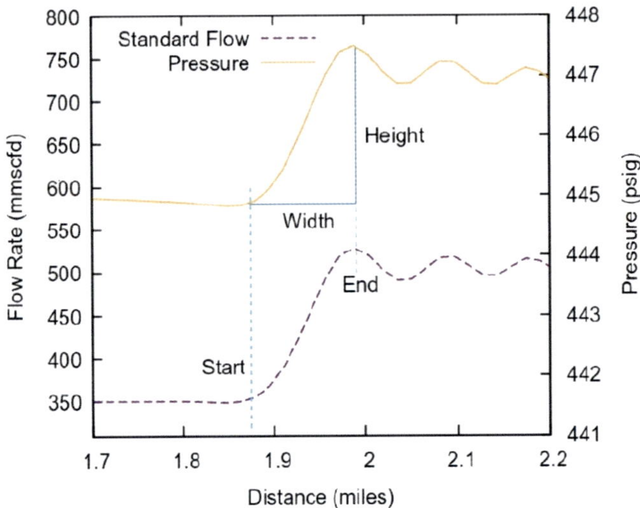

Figure 187 – A definition for the width and height of a pressure or flow wavefront [32]

The trip scenario is different enough from the rupture scenario that the arrival of a wavefront at the emergency shut-down valve some distance away can be attributed to one scenario or the other. Moreover, unlike in a compressor trip scenario, after a rupture the entire pipeline upstream of the wavefront is unpacking.

One question was: does the wavefront retain its shape in these two scenarios? It was determined that there was no real dispersion of a sharp but relatively small (in delta-pressure terms) wavefront as it moved down the gas line. Any wavefront spread seen in simulation results is due to numerical dispersion. This means the simulated wavefront from the trip scenario is quite spread out compared to the real wavefront, whereas the wavefront from a rupture scenario – which gets spread out naturally while it is formed and so is initially less sharp – is correct. We want to use a box scheme simulator with a fairly coarse grid of knots so it solves quickly, and to have that produce the correct shape of wavefront from the compressor trip scenario, we need a solution.

The true *width* of a wavefront is unchanged since the time of its creation, and only its *height* really changes as it travels along a pipeline. Since the pressure before and after the wavefront are the same – and therefore presumably correct – in all of the different simulation approaches, we should extend the pressure profile in the downstream and upstream regions into the dispersed region to produce a correct profile. The width of the wavefront is taken to be the speed of sound multiplied by the time it takes the compressor to trip. The position of the wavefront in the pipe is taken to be the original dispersed profile's center point. And the pre- and post-wavefront linear pressure profiles are extended up to the true start and end times of the pressure surge. The result of the corrected pressure surge is shown in Figure 189.

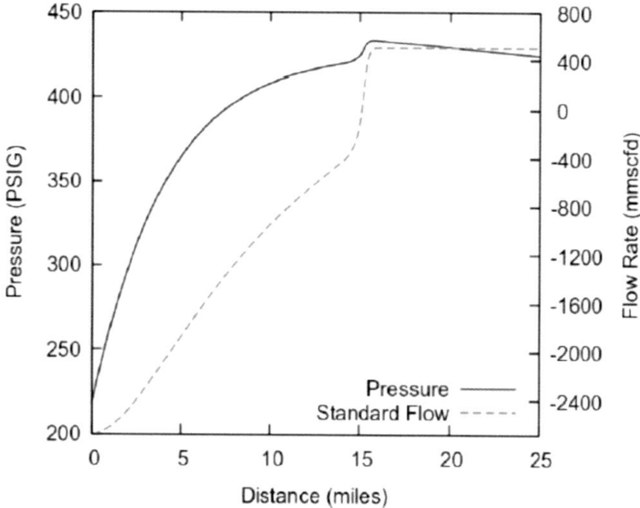

Figure 188 – A pressure wavefront and a corresponding flow wavefront are visible along the profile of this gas pipeline. This snapshot was taken 80 seconds after a rupture.

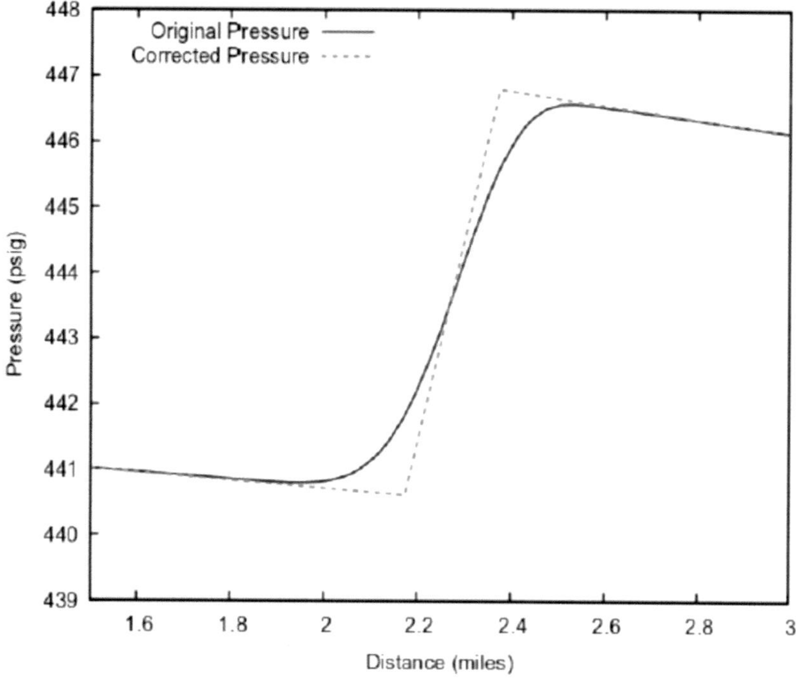

Figure 189 – Manually correcting a pressure wavefront that was dispersed during a simulation.

In this way, a transient pipeline simulator like Atmos SIM can be used to tune emergency shut-down valves so they are triggered only on a rupture, without triggering on a compressor trip. First, we run the simulations with the appropriate features such as a transient ground thermal model, a high-quality equation of state, etc., using a tight enough mesh that the wavefront is not completely washed out when it passes the valve location. Then, we calculate the wavefront width and height using the above procedure. By doing this for an expected compressor trip and again for the valve closure surge scenarios, we calculate tuning parameters for the emergency shut-down valves that prevent accidental closures. [32]

Passive surge mitigation strategies

Relief valves

Relief valves are safety critical, so pipeliners must be sure that they work. This is challenging because they don't often get used in operation. Good standards and practices ensure they operate at the right time and at the right pressure.

At Bellingham, [62] a relief valve that was improperly configured failed to operate as intended, causing a major disaster. It failed to open, and so failed to prevent a dangerous pressure surge following a sudden shut-down in the pipeline. The surge ruptured the pipeline, releasing gasoline. Gasoline vapors exploded in a fireball shortly afterwards, taking three lives. It led to the first conviction against a pipeline company in the United States under the 1979 Hazardous Liquid Pipeline Safety Act.

> Carefully selecting and placing a relief valve at a pipeline's inlet to deal with a sudden shut-down of a mid-line station reduces the peak surge pressure.

Pressure relief valves open fast once a predefined pressure is exceeded. They allow fluid to escape. This prevents a downstream velocity change from [fully] propagating upstream past the relief valve, because the relief valve opens, and the flow goes through it instead. A relief valve thereby significantly lowers the peak pressure upstream of a surge event such as a suddenly closing valve.

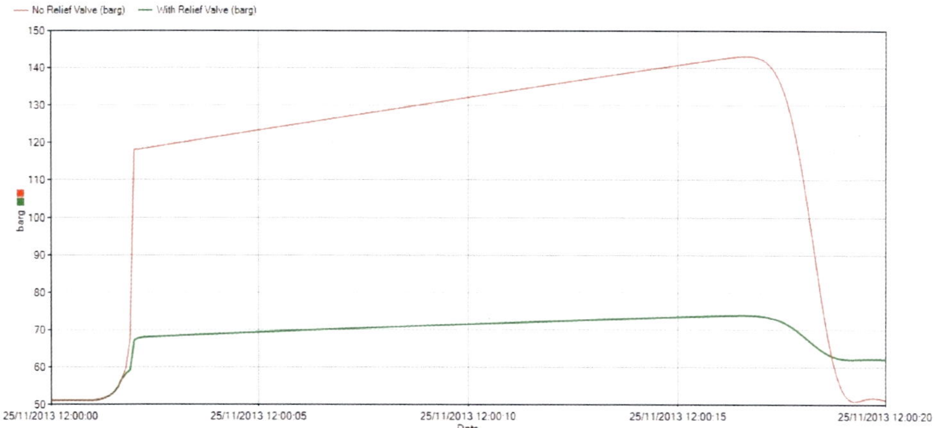

Figure 190 –Pressure upstream of a 1 second valve closure; with (green) vs. without (red) a pressure relief valve positioned immediately upstream of the closing valve.

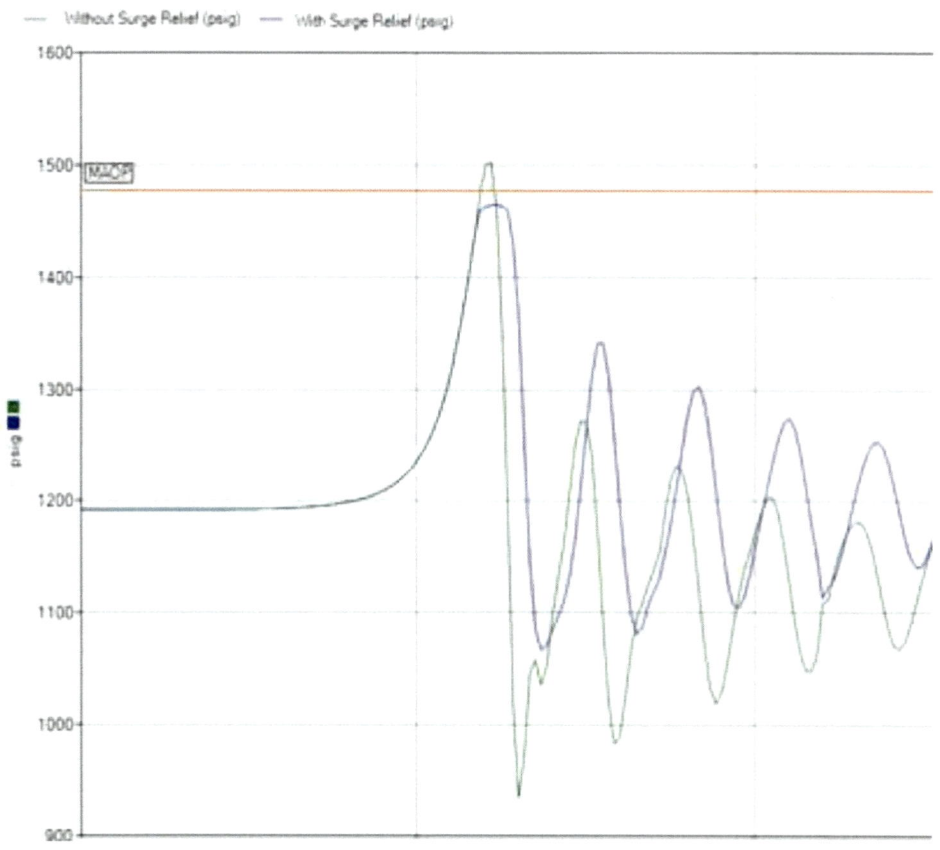

Figure 191 – A relief valve reduces pressure up-surge and down-surge to below max. allowable operating pressure MAOP (orange); without surge relief (green) versus with surge relief (blue).

Relief valves come in several types in terms of their mechanical construction. These range from inexpensive spring-loaded discs to complex systems that are capable of reacting within milliseconds.

Surge tanks

Surge tanks can mitigate both up-surges and down-surges in pressure. In the event of a line pressure exceeding a limit, liquid is diverted into the tank which reduces the height of the pressure surge; if pressure falls below the tank pressure, liquid flows out of the tank and increases the pressure at the trough. This serves to reduce transient pressure surges as well as dampening oscillations in velocity. As excess liquid is diverted from the mainline to a surge tank, it acts as a temporary storage device, thereby ensuring a velocity change is more gradual, and transient pressure waves smaller. If velocities fluctuate, it acts as a damping device, supplying liquid, thereby ensuring that deceleration is more gradual, in order to avoid excessively low pressures.

Pipe size

Increasing diameter slows the fluid at a given flow rate, reducing momentum change in bringing it to rest, reducing potential surge pressures. But too large a diameter impairs frictional damping of pressure fluctuations. A pipe's design pressure (P) is based on its yield strength (σ), wall thickness (δ), outer diameter (D), and multipliers for its design, location, longitudinal joints, and temperature derating (K):

$$P = \frac{2 \cdot \sigma \cdot \delta}{D} \cdot K$$

All other factors being constant, a larger diameter pipeline can only handle a reduced design pressure. As such, a larger pipeline must be thicker-walled and / or manufactured of stronger materials to operate at the same pressures.

In certain jurisdictions, there are stipulations that other factors go into this as well. For example, a recent hydrostatic test must be performed that goes up to the pressure at which the pipeline is intended to operate.

Pipe strength

Wall thicknesses selected during preliminary design might be adequate for steady operation, but might not safely withstand surge pressures resulting from rapid valve closure. Installing a surge control device might mitigate operational concerns, but if the surge overflow effectively becomes a spill or a release they may fail to meet environmental protection requirements. If other suitable surge mitigation techniques are not feasible, the pipeline might need to be stronger, e.g. via thicker walls. This incurs an initial capital cost but requires less maintenance and testing than other mitigation methods. It is only considered during the initial design phase, as this kind of modification is impractical to retrofit. To achieve the same end in an existing operational pipeline, we might have another pipeline connected in parallel to increase capacity – *"looping"*.

Monte Carlo simulation

Certain pipeline decision making tasks, such as the sizing of surge vessels, are about choosing one of many possible options. Wherever we find ourselves faced with an overabundance of possible choices, the power of computational statistics can be harnessed to our advantage.

A Monte Carlo simulation (MCS) is used to generate thousands of scenarios. For sizing surge vessels, this is a big set of simulations covering many cases of up-surges and down-surges. The results are then analyzed to develop a tailored model, which can predict the initial and expanded vessel's gas volume.

Developing these models can serve as a basis for selecting suitable sizes of surge protection in preliminary feasibility studies. They outperform traditional design charts as measured by several statistical indicators, producing more accurate results. Analyzing them provides qualitative insight into what affects a surge vessel's gas volumes, as the simulations vary parameters such as pipe diameter and static head. [63]

9.4 – AUTOMATED SURGE ANALYSIS

Key features of the Atmos SIM Surge Analysis Tool

Every change to a pipeline's operating characteristics, material specifications, or fluid properties warrants a performing a comprehensive surge analysis, involving copious calculations. This study covers the entire pipeline network, including branches, which becomes tedious. Atmos SIM includes an automated tool that automatically launches each scenario in succession and produces a report summarizing key results including the peak surge pressure and its location. In achieving this task, its key features include:

- Variable knot spacing, reducing interpolation for resolution
- Variable time-steps, for accurate pressure wave propagation
- Reverse velocity modelling for dynamic check valve behavior
- Pump modelling for accurate spin-up and spin-down behavior
- Control logic for simulating control during surge events
- Valve characteristic inputs for accurate valve closure

This expedites the verification of how to safely operate and control the pipeline. By automating the analysis of extensive pipeline networks in their numerous operating scenarios, it greatly facilitates how pipeliners work to prepare their submissions to regulators. Instead of spending tedious days on a task prone to human error, users set up a model in a few hours and simulate all the different surge scenarios in one click.

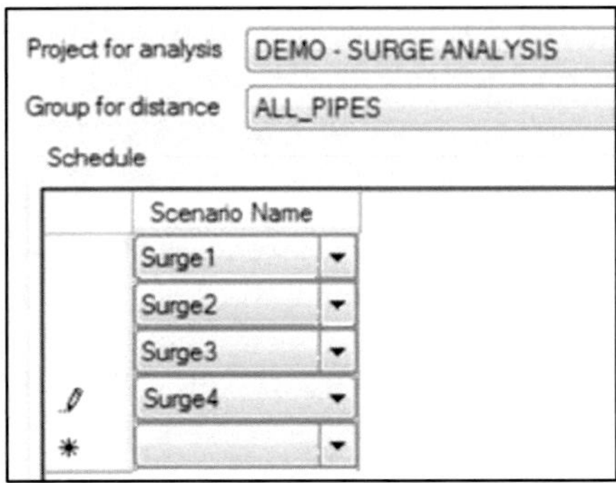

Figure 192 – Instructing Atmos SIM's Surge Analyser to launch four distinct surge scenarios.

Identifying the location of any violation, such as a violation of maximum allowable operating pressure (MAOP), the simulated pressure at these points is trended over time. The report outlines important results in a convenient format for scrutiny, including:

- Indicating any valve closure or pump trip, with its name and the time
- Indicating violation of MAOP, with its time, location and pressure

The report also generates trends for each pressure surge scenario including:

- A trend of pressure at the location of MAOP violation
- A trend of pressure upstream of a valve closure or a pump trip
- A profile of maximum pressure for each scenario
- A profile of minimum pressure for each scenario
- Trends of flow rate for each scenario

This data helps select a surge mitigation procedure or device to incorporate. Minimum and maximum pressure profiles may be used, during a new pipeline's design phase, for example, to help calculate the required wall thickness (δ) based on a suitable design pressure.

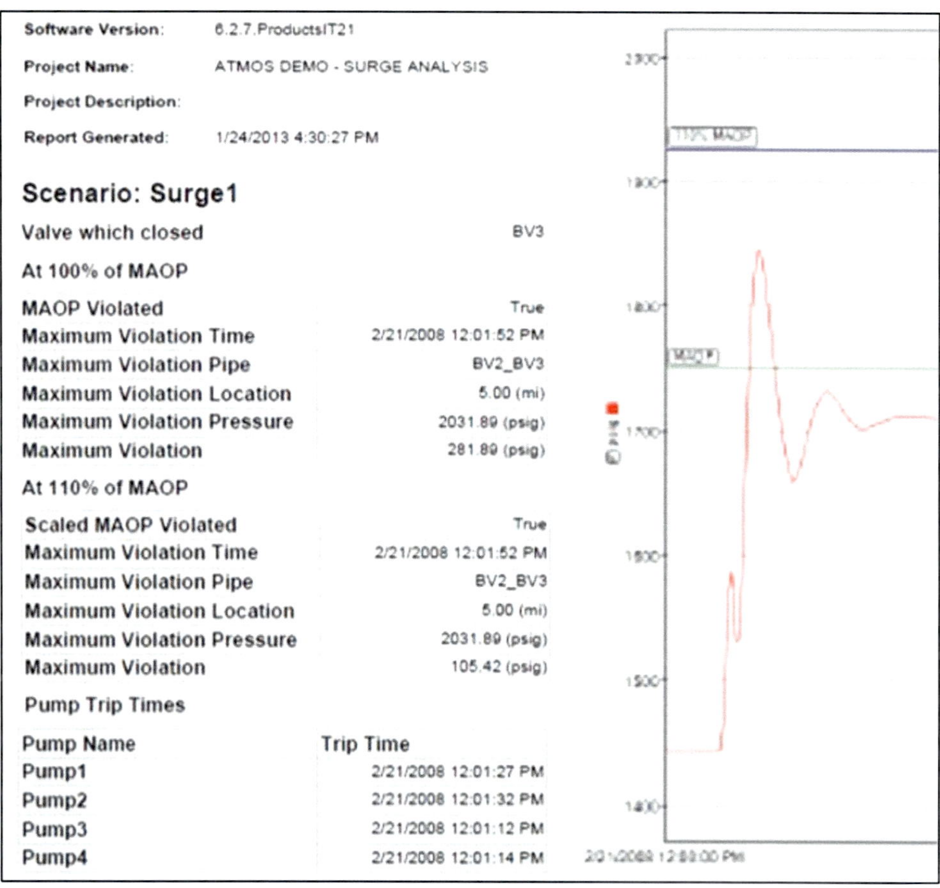

Software Version: 6.2.7.ProductsIT21

Project Name: ATMOS DEMO - SURGE ANALYSIS

Project Description:

Report Generated: 1/24/2013 4:30:27 PM

Scenario: Surge1

Valve which closed	BV3

At 100% of MAOP

MAOP Violated	True
Maximum Violation Time	2/21/2008 12:01:52 PM
Maximum Violation Pipe	BV2_BV3
Maximum Violation Location	5.00 (mi)
Maximum Violation Pressure	2031.89 (psig)
Maximum Violation	281.89 (psig)

At 110% of MAOP

Scaled MAOP Violated	True
Maximum Violation Time	2/21/2008 12:01:52 PM
Maximum Violation Pipe	BV2_BV3
Maximum Violation Location	5.00 (mi)
Maximum Violation Pressure	2031.89 (psig)
Maximum Violation	105.42 (psig)

Pump Trip Times

Pump Name	Trip Time
Pump1	2/21/2008 12:01:27 PM
Pump2	2/21/2008 12:01:32 PM
Pump3	2/21/2008 12:01:12 PM
Pump4	2/21/2008 12:01:14 PM

Figure 193 – An automatically generated report from the Atmos SIM Surge Analysis Tool

Peak pressure can be lowered to acceptable levels by implementing advanced control logic, so as to trip upstream pumps upon valve closure. Atmos SIM's Surge Analysis Tool captures the way a pipeline is controlled during operations by replicating this logic, with no surge mitigation devices or with any combination. It can be instructed to automatically re-launch the full gamut of scenarios afresh, incorporating the chosen surge mitigation procedures and devices. It generates an updated report for these modified scenarios, and a pipeline configuration report detailing properties of pipes along the length of the route – lengths, diameters, wall thickness – as well as pump and valve performance curve data and model boundary conditions for every significant model item.

Case study: removing a balance tank from the line

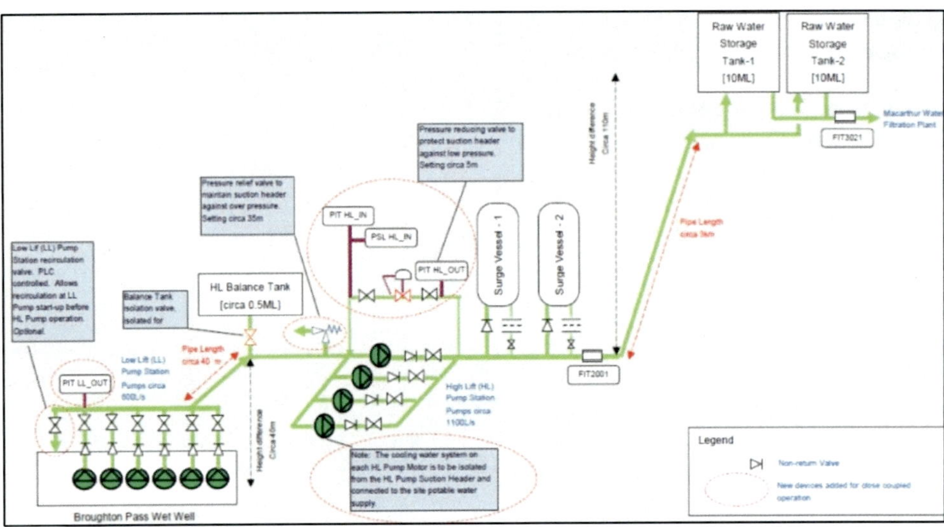

Figure 194 – Surge vessels, a relief valve and a pressure reducing valve mitigate surge effects according to this pipeline's control strategy. Can we remove the balance tank?

A balance tank at a high lift pump station for a water pumping system needed to be isolated for four weeks. The water supply is critical to a major Australian city. Taking the balance tank offline would mean the water system would be *closely coupled*: the flow upstream of the tank would always have to match the flow downstream of it. Six low-lift pumps would send water directly from the reservoir to the suction of the four high-lift pumps, which then send the water to tanks at a water filtration plant.

This change in configuration would significantly impact the high-lift station's suction header: surges would no longer be mitigated by going into the tank. The frequency of induced pressure surges would be exacerbated by the number of pumps, by planned starts and stops, and by unplanned stops. To protect system integrity, mechanical changes and changes to process control were therefore proposed.

Modelling the total system was instrumental in examining the proposed planned and unplanned scenarios, identifying if the mechanical measures and process control changes were adequate to protect the system from over-pressure and under-pressure in every circumstance. The two pumping stations and existing mainline water transfer system, with the balance tank in operation, was modelled on Atmos SIM, and tuned to maximize simulation accuracy using real-time data.

Time-based events for the pumps were assigned to represent each case of the study. The modified model simulated the system's response in each scenario. Simulated results provided confidence that operating the low lift pump rising main without the intermediate balance tank wouldn't result in the propagation of damaging surge pressures exceeding allowable operating pressures along the pipeline and pump stations, even during the most severe scenario of an instantaneous complete uncontrolled shut-down of all pumps at both stations. It also showed that without installing a back-pressure regulator (a pressure sustaining valve), damaging conditions could occur during a simultaneous trip of five low-lift pumps and two high-lift pumps.

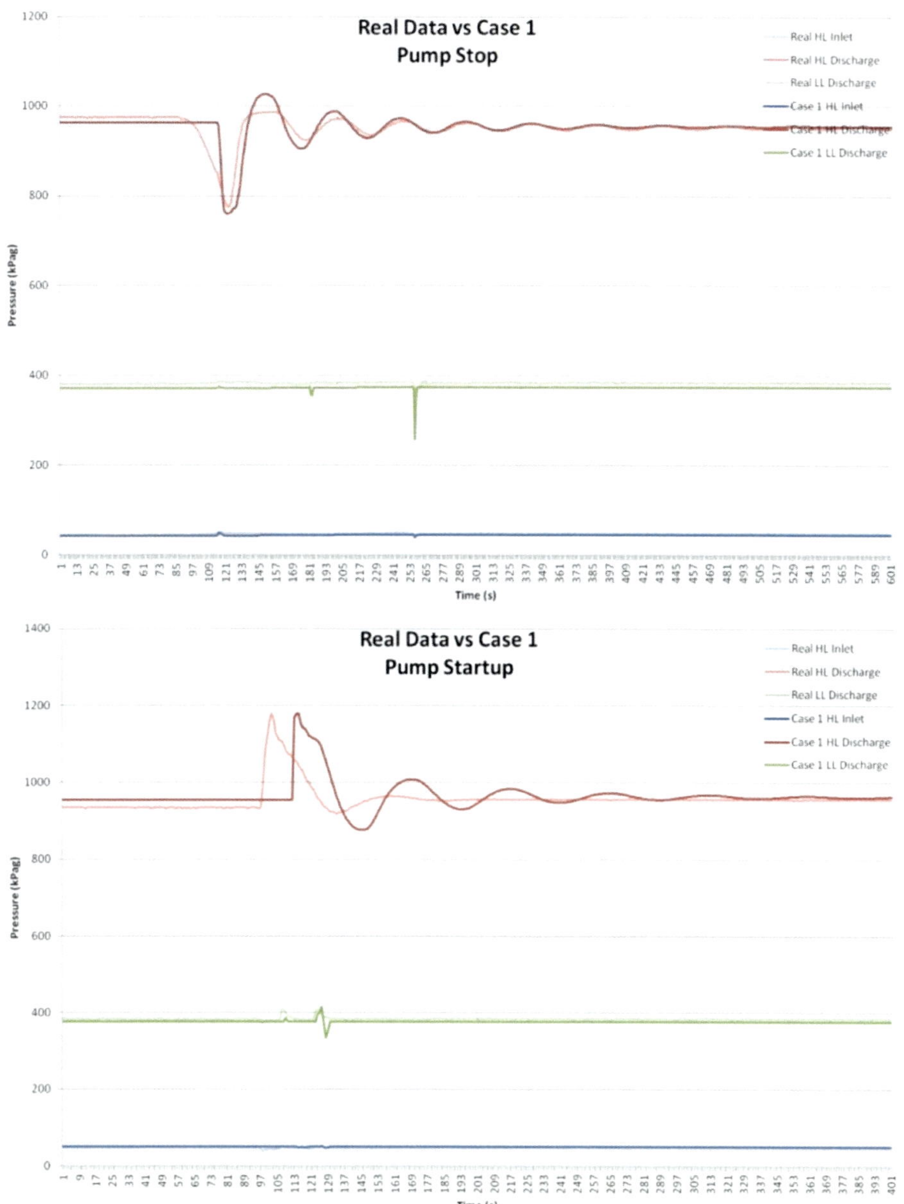

Figure 195 – Pump start-up and shut-down, simulated results closely matching real data.

The verified model was then modified to reflect the proposed changes in mechanical protection and process control. The balance tank was removed, and in its place a pressure sustaining valve and a 40 m length of DN150 pipe was added at the high-lift pump station. At the low-lift pump station, discharge pressure relief valves were added.

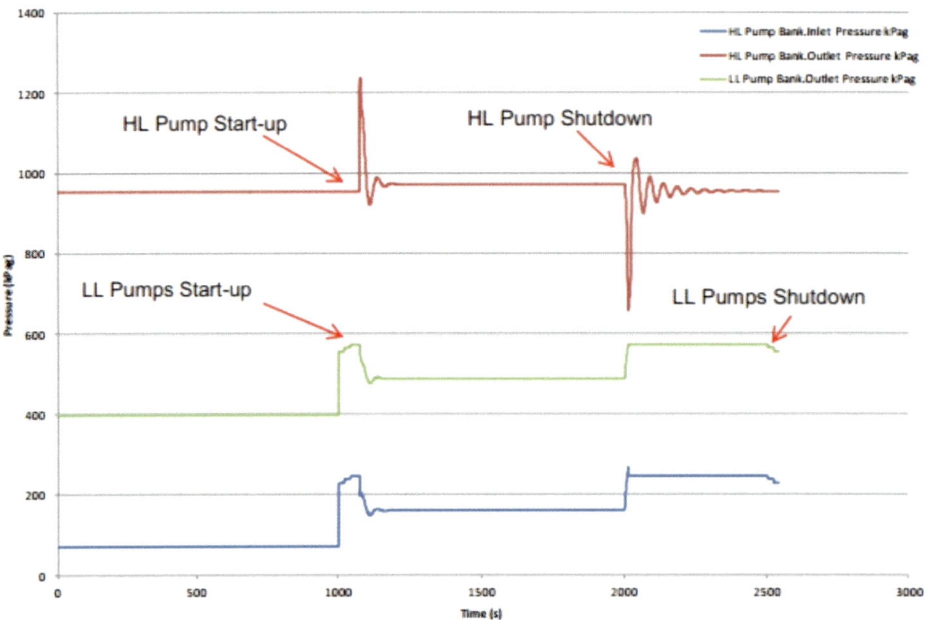

[kPag]	HL Inlet	HL Discharge	LL Discharge
Minimum	153	657	486
Maximum	258	1237	572
Minimum MAOP* at Measured Point	2800 DN1000 - Balance Tank	2800 DN1000 – Header	2500 DN600 – Header

Figure 196 – Simulated surge pressures associated with pump start-up and shut-down for one of the cases considered in the study.

In this way, Atmos SIM scenarios validated the study's recommendation to install the back-pressure regulator. It also recommended that a minimum of 10 minutes be allowed between shut-down and restart of the high-lift pumps, to dissipate the oscillating surge pressures during this type of operation.

9.5 – CHOKED FLOW, BLOWDOWN AND LEAKS

Choked flow in gas pipelines

In normal conditions, lowering downstream pressure causes an increased flow rate. A sudden discharge of fluid might, for some period, become *choked*. In that choked condition, lowering downstream pressure further doesn't affect flow rate. In choked flow, fluid velocity reaches the local sound speed (c) at the upstream side of the choke point. A disturbance can travel no faster than sound speed, so it cannot travel upstream past a choke point. Therefore, what happens downstream of a chokepoint, such as the pressure there, cannot affect what is happening upstream of the chokepoint. There is only one exception: if the downstream effects conspire to end the sonic flow condition.

Choked flow is considered to be a challenging edge-case that some basic pipeline simulators cannot handle gracefully; the moment our pipeline's flow velocity exceeds sound speed (c), a simple simulator may struggle to reach a good solution. Pipeline simulators that don't explicitly handle the point of choking flow often have trouble simulating it, because some of the quantities that appear in the model equations take on non-physical values if they are pushed to a higher gas velocity than the speed of sound, which can happen while solving the simulation due to numerical discretization. It has ramifications as choked flow arises in scenarios such as rupture and blowdown.

Blowdown or pressure relief: venting to atmosphere

If something goes wrong in a pipeline, presenting a risk of over-pressure or in case of a suspected rupture, an emergency shut-down is conducted, then a blowdown operation (BD) reduces the pressure driving accidental release by evacuating inventory until the section reaches near atmospheric conditions, and flow finally ceases. After venting, combustible fluid is directed via *"burn pit lines"* to be flared off. Pilot flames are monitored, as if they are extinguished, a dangerous cloud of flammable vapor might form. Dispersion is safety-critical.

How long does it take for a shut-in section to be *"empty"*? How much of its inventory must be released by that point? How does flow rate vary during the course of a blowdown?

Blowdown is conducted through a branch pipe which has valves. Blowing down any pipeline obliges us to think more closely about those valves. Their orifice, smaller in diameter than the surrounding pipework, is likely to be the location where choked flow first occurs. So we expect there to be choked flow in the throat of a discharging pressure relief valve (PRV), for example. Choking will occur at the throat of the narrowest valve at the low-pressure end of that pipe. Resistance downstream of the choke point doesn't have an effect unless it moves the choke point. Resistance upstream of the choke point has an effect, and can be inserted as a model item between the pipeline and a demand point.

A blowdown scenario may start from a shut-in pressurized condition. Suddenly a vent is opened to the atmosphere. Regulations stipulate these scenarios must be examined carefully at the outset, and revisited by the team every few years to assess if the surroundings have become more urbanized for instance.

A related case study is: *"What if a fire occurs just outside of the pipe?"*. Fluid temperature and pressure rise at constant volume until a pressure relief valve opens to protect the pipeline. Then the fluid expands at constant pressure. [64]

A boiling liquid blowdown scenario

In the extreme case of a liquid hydrocarbon in a scenario where there is a fire near the pipeline, a boiling liquid's composition changes during the venting process. Any simple manual calculation premised on an incompressible assumption here by applying Bernoulli's equation won't give the right answer. Indeed, modelling such a scenario hinges on a sophisticated simulator that moves well beyond single phase, which comes with its own special challenges. We touch on some of these challenges in the final chapter of this book.

Gas blowdown

A gas blowdown is more complicated to model than a liquid blowdown. As hydrocarbon inventory contained as linepack within a section of gas pipeline is discharged, the sudden change in pressure causes a change in *temperature*. Indeed, when blowing down a *wet* gas pipeline, this Joule-Thomson cooling may cause *liquid dropout*, and a poorly-characterized choked two-phase flow.

Case study: gas blowdown as a leak or rupture

In this scenario we see a blowdown can be represented as a leak point, with a somewhat different discharge coefficient (C_D) to represent a perfectly round valve throat rather than the longitudinal crack which might happen in a leak.

Figure 197 – A simple single-in single-out natural gas pipeline with a leak in a shut-in section.

This 18-inch natural gas pipeline is 25.7 km long. Its nominal flow rate is 70 MMSCFD. The time taken to reach 0.029 MMSCF of inventory in the pipe – equivalent to 0 barg, the point at which there is no more *"available linepack"* to unpack from the pipeline – depends on whether it is initially operating at higher pressure (in this case 26 barg) or at a lower pressure (4 barg).

Assuming the leak opens immediately rather than gradually, to ambient air at a temperature of 27°C, and that the block valves starts to close after 1 minute thereafter transitioning to fully shut over 1 minute, the duration of this process was calculated.

Orifice diameter	**Duration taken to release all the available linepack**	
	from 26 barg initially	from 4 barg initially
7 mm	3 days 11 hours	11 hours 30 minutes
22 mm	4 hours	1 hour 20 minutes
100 mm	1 hour	20 minutes

Table 23 – Duration of blowdown or leak until full inventory is released from a shut-in pipe.

Results like these have been validated against real experience in the field, as well as against an equation calculating gas blowdown time from the American Gas Association (AGA). There is also a manual method for estimating gas blowdown time listed in the references. [65] One advantage here of a pipeline simulation is it generates trends showing how inventory diminishes over time, and how that relates to the shut-in pipe's pressure, the leak flow rate, and whether the *"leak mode"* is critical or sub-critical.

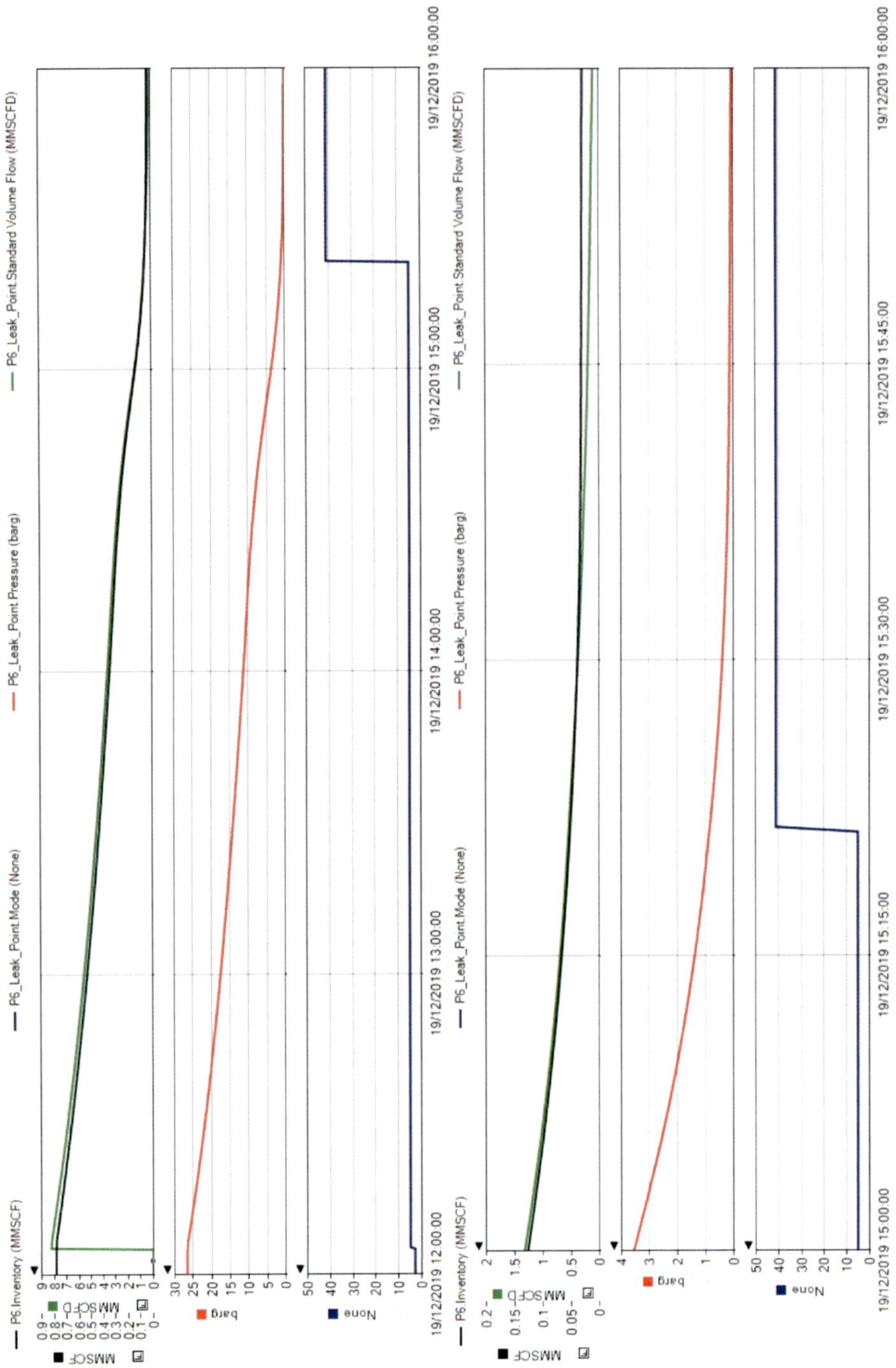

Figure 198 – Trends showing how inventory diminished during the gas blowdown operation.

10 – LEAKS AND RUPTURES

10.1 – WHY PIPELINES LEAK

Performance is difficult to measure for leak prevention. Leaks are events that we want to never happen, so speaking about improvements in the rate they happen or mitigating their effects can be thorny. Quantifying how much pipelines leak is not straightforward. We must define the question carefully, as not all leaks are equal. Water mains leak 200 000 times every year in the US. Hazardous liquid pipelines leak less often, about 300 times per year, but their impact is worse. An oil spill needs cleaning up. A volatile liquid poses an immediate threat if it forms a flammable gas cloud. Surprisingly, we cannot assume this risk is obvious to everyone. Disastrous spills have happened because *"people and organizations involved in design, operations and maintenance were unaware of similar [incidents]"*. [66] For this reason, pipeliners take leaks seriously.

Although natural gas is lighter than air, a natural gas leak is certainly a problem to avoid and mitigate. Gas leaks happen mostly in extensive low-pressure distribution networks, and are difficult to quantify: the industry estimates 0.8% to 4% *"lost and unaccounted for gas"*. This seems alarming, but it should be noted it is not known exactly how much of that is due to leaky pipes rather than poor metering. The twin topics of metering and operating conditions come up whenever we try to assign numbers to the issue of leaks.

Pipeliners won't intentionally violate safe limits and cause a leak, though they assess the consequence of these unlikely scenarios to be sure mitigation is in place. Leaks are caused by various reasons, some obvious and others more subtle. There are dedicated organizations in the United States,[*] the European Union,[†] and elsewhere who have categorized the causes of pipeline leaks. These may be typically listed as:

- Excavation and outside force
- Material and equipment
- External and internal corrosion
- Operational causes
- Natural causes
- Other causes

Most pipeline incidents are caused by external damage. [67]

---

[*] Pipeline & Hazardous Materials Safety Administration (PHMSA, pronounced *femsa*)

[†] Conservation of Clean Air and Water in Europe (CONCAWE)

10.2 – MIMICKING A LEAK USING A SIMULATOR

Is a leak or rupture similar to a blowdown?

At the start of a leak, a massive flow gushes forth. A rupture can be defined as being so significant that it causes reverse flow, such that the leak-flow is higher than the line-flow. The leak continues to flow until enough fluid has escaped from the section of pipeline that its pressure has dropped substantially, and the leak sputters out into a mere trickle. More properly, during a leak, flow takes place in several successive *"modes"*. These classify the nature of flow through an orifice, including periods of *choked* flow or *critical* flow, and *sub-critical* flow. In a high-pressure pipeline, choked flow occurs, reaching the sound speed (c) of the fluid in the pipeline. This is the highest possible flow rate during a leak. [61]

Making a distinction between a large leak and a rupture is not clear cut: an orifice of half the pipeline's diameter is a rupture, whereas an orifice of one tenth the pipe diameter is referred to as a leak. Inside the pipe, a large leak or full-bore rupture is almost identical to a blowdown – a similarly violent process. If a leak is small, a flowing pipeline doesn't really *"blow down"*. It keeps flowing but just loses some of the fluid along its way. A shut-in pipeline will eventually drain completely if there is any leak, even a leak that is very small.

Leak points, leak coefficients and leak scenarios

A pipeline simulator characterizes a leak point as an orifice with a few simple parameters: orifice coefficient, effective diameter and how fast it opens up. We can simulate scenarios such as a thief gradually opening an illegal tapping valve.

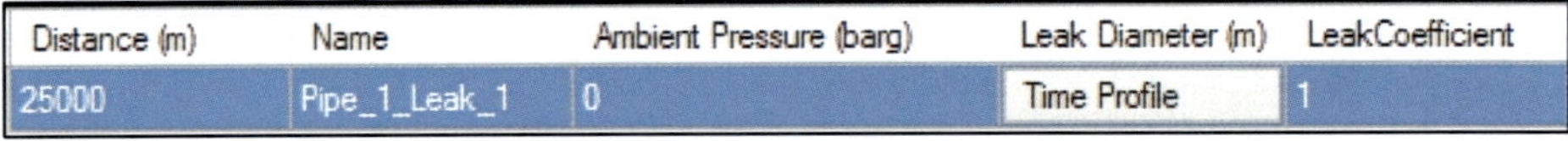

Distance (m)	Name	Ambient Pressure (barg)	Leak Diameter (m)	LeakCoefficient
25000	Pipe_1_Leak_1	0	Time Profile	1

Figure 199 – The diameter of a leak point is set to vary over time in an Atmos SIM scenario.

A leak point simulates a leak by imposing the effect of an artificial demand at that point along the pipeline. A leak coefficient captures the nature of the type of orifice through which this leak occurs. A leak happens through a longitudinal orifice as a pipe cracks along its length. A theft happens through an illegal tapping point, as an unmetered offtake. For all intents and purposes, the same *"leak point"* model item suitably represents a leak, rupture or theft if we simply assign its orifice coefficient a suitable value. Orifice coefficients are well-established, and have wider application in engineering beyond leaks. A typical value is 0.67 for a leak's orifice coefficient – somewhere between sharp-edged and rounded.

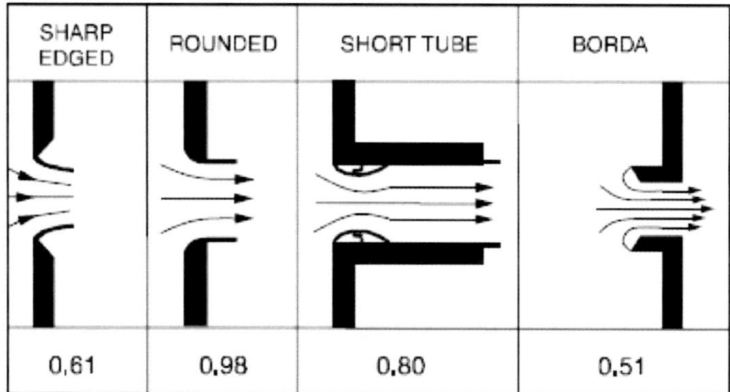

Figure 200 – Orifices and their nominal discharge coefficients.

An offline scenario in Atmos SIM can mimic the effects of a leak over time. There is much more to consider in doing so – leaks are far more involved than adding a discharge coefficient (C_D) through an orifice. For a leak at a high point or a low point, what is the peak flow rate, mass / volume released, and their effect on the line? Where is the *"worst"* location for a leak in terms of fluid released? What about the worst location in terms of the resulting surge? Where are the sensitive areas, populated areas or river crossings? If a line commences its shut-in procedure 5 minutes after a leak, what quantity would have been released?

A leak or rupture in a liquid pipeline

A leak in a liquid line has a characteristic leak signature in its flow rate profile.

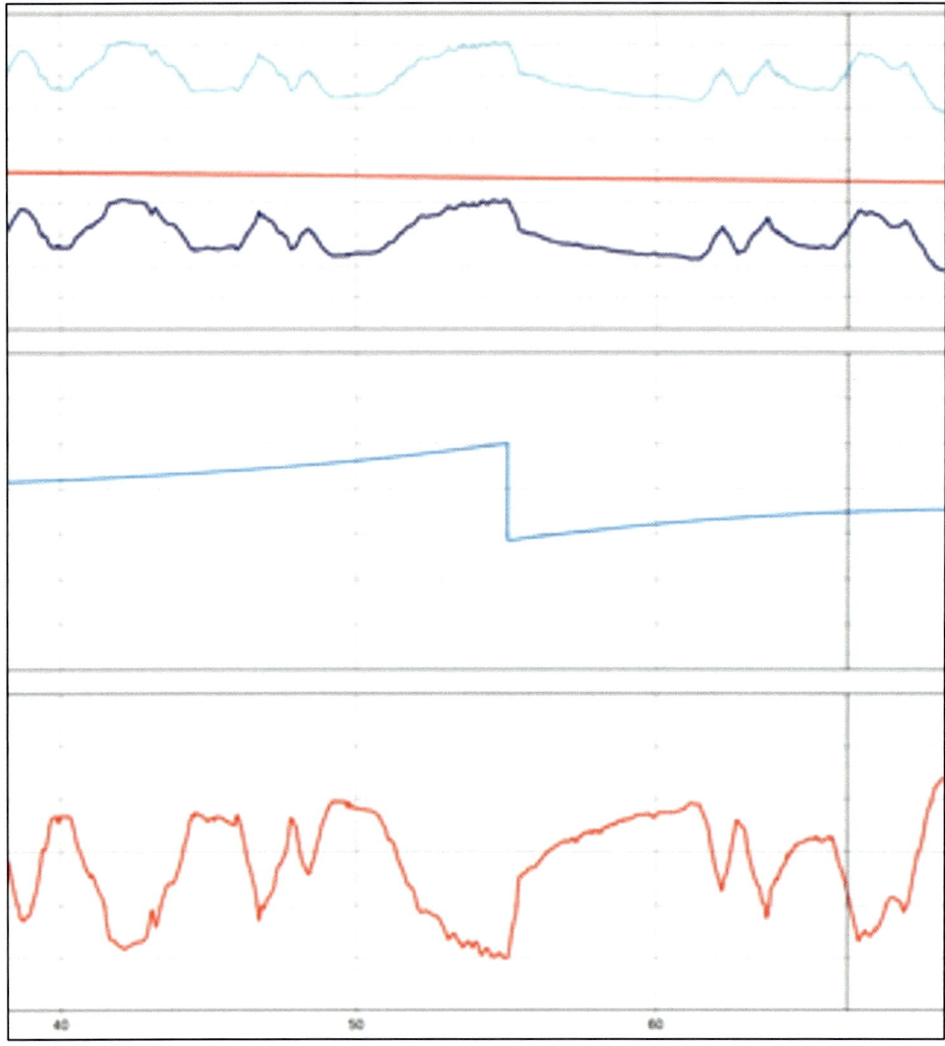

Figure 201 – Simulating a small leak in a liquid pipeline. This offline scenario's results are shown as profiles along the route, of head (top), standard flow (middle) and pressure (bottom).

To imitate a leak or rupture by simulating it in an offline scenario, constraints at nearby model control-items, like supplies, demands, pumps and regulators, should be *"loosened"*. Otherwise we cannot expect to imitate a leak signature if the offline scenario is meeting setpoint constraints exactly at the same point as the meter is located! Loosening a controlling item (such that the constraint is not immediately met exactly once the leak signature arrives) can actually be achieved by a workaround, using a length of dummy-pipe or a point resistance.

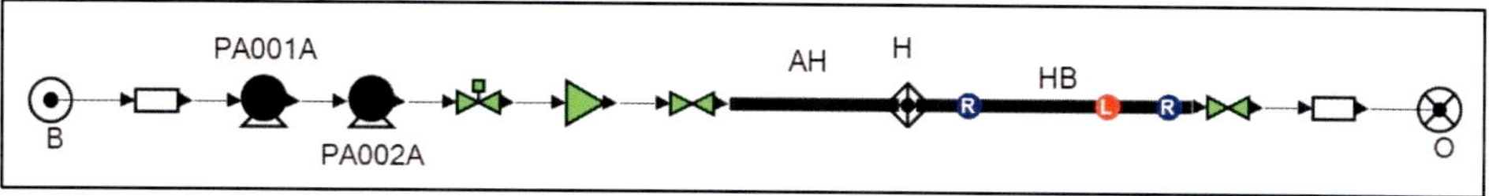

Figure 202 – A leak points and reporting points are directly placed anywhere along a pipeline.

A leak or rupture in a gas pipeline

A leak or rupture exposes the pipeline to atmospheric conditions at the location of the leak. The pressure within the pipeline falls at the leak location, trying to equalize with atmospheric pressure. This pressure fall propagates along the pipeline as a pressure wave. Eventually, if the leaking section has been isolated, its pressure falls to match atmospheric pressure. The last amount of a leaking gas takes the longest to escape through the leak orifice. But most of the gas departs very quickly indeed: from the onset of the leak at what is probably a substantial pipeline pressure on a transmission pipeline, down to as low as twice the ambient pressure, the flow through the leak orifice is very high indeed. It is as high as physically possible, a condition known as *choking*. Extreme flow rates cause extremely low temperatures in a high-pressure natural gas pipeline, due to the Joule-Thomson effect manifesting as cooling. This risks causing a brittle transition in the steel pipe, which means that someone will end up having to replace significant lengths of the pipeline.

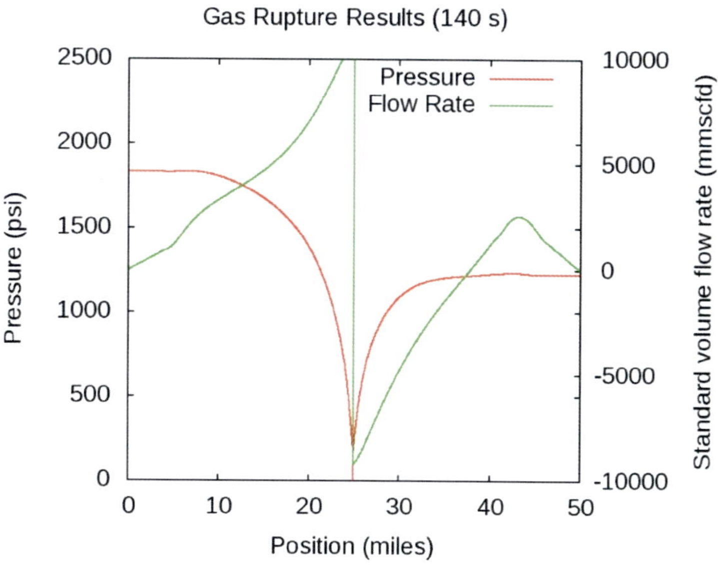

Figure 203 – Simulating a rupture scenario within a gas pipeline in Atmos SIM.

Demonstration: a rupture in a gas pipeline

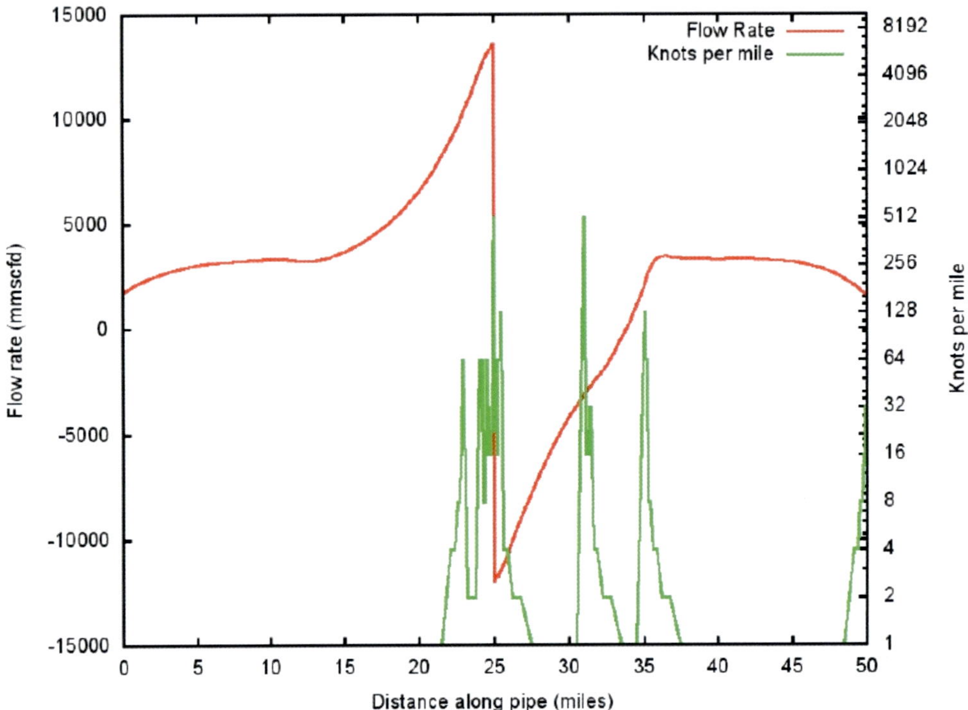

Figure – A snapshot of flow rate and knot density along a pipeline one minute after it ruptures.

This demonstration covers a 36" gas pipeline, 50 miles long. It initially operates from a supply at 2 000 psig to a single delivery at 1 000 psig. A rupture is introduced at the 25-mile point, to atmospheric pressure. It is assumed larger than pipe diameter, so flow is choked in both directions. This simplifies a real rupture, but what matters here is it causes rapidly changing fluid properties. Valves are immediately closed at both ends of the pipe. It is run for 8 hours, at 10-second fixed time-steps. Temperatures get so cold that ethane should start liquefying. We ignore this, choosing an equation of state to calculate vapor-like densities.

Fluid flows from both directions towards the rupture. Within the flow rate profile above, bends reflect transients, which move over time. Knot density is varied along the line to adapt automatically: knots are densest near the rupture, and at clusters around transient bends. Elsewhere we see 1 knot per mile – the coarsest we allowed.

10.3 – THE CONSEQUENCES OF A LEAK

What happens beyond a pipeline is outside the scope of our simulation, but our results (including the leak flow rate over the course of the leak's duration, and total quantity released until flow ceases) can be useful for the consequence analysis.

When a hazardous liquid or gas is released due to a leak, the consequences can be classified using an event tree like the one shown below. It might not ignite. It might immediately ignite and burn as a fire. Or it might form a combustible cloud of vapor in air. If a release of a highly volatile liquid such as propane does not burn slowly as a flash fire, the delayed ignition becomes a vapor cloud explosion (VCE). [12] A gas leak doesn't look like an oil spill, but is no less serious. The term *"boiling liquid expanding vapor explosion"* has arisen because it has been happening in industrial accidents for decades, and has sadly continued to happen recently. The notoriety of these events is attested by the widespread use of its acronym (BLEVE). This particularly violent type of explosion happens when a pressurized liquid such as propane is released to atmosphere, and suddenly finds itself to be a liquid at a temperature above its new boiling point. It therefore boils very rapidly; if it also ignites, ensuing heat release accelerates the evaporation of liquid and this continues until it combusts all of the liquid.

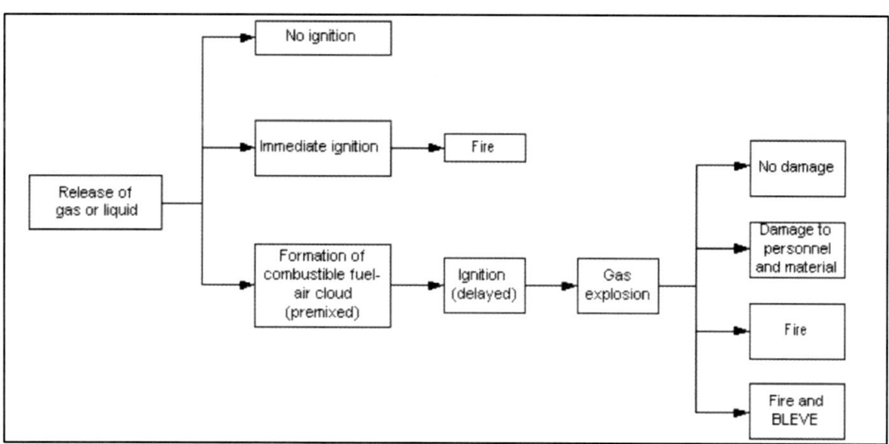

Figure 204 – Event tree showing typical consequences after release (Credit: UK HSE). [68]

Of course, fires are also very serious; if the reader is at all in doubt about that, we would point out that while this book was being written, a gas leak set part of the Gulf of Mexico on fire. Pipeline leaks have many other consequences, for example: damage to people, the environment, property and a company's reputation. Loss of revenue, cost of cleaning up and financial penalty are the financial consequences. It is essential that an effective and reliable leak detection system is implemented on any operational pipelines. We simulate leaks offline. We can detect, locate and quantify leaks online.

10.4 – Model-based leak detectors

Is a leak detector needed?

In 1977, Davenport examined the origin of leaks causing vapor cloud explosions, and found the leading culprit to be pipelines, valves and flanges. Though his study wasn't specific to the pipelining industry, including all kinds of process pipework, it serves as a worthwhile reminder of how significant the humble pipeline can be in the grand scheme of a major enterprise going awry. A leak might cause significant upsets in a pipeline that nonetheless don't make it obvious to a human operator that a leak is taking place. A major event could happen – a pump could trip – yet the operator may be none the wiser as to its cause! [69]

> *"Many publications about pipe failures attribute them to causes such as fatigue or inadequate flexibility. This isn't helpful. It is like saying a fall was caused by gravity"*

While leak *prevention* strives for zero incidents, leak *detection* is for consequence mitigation. All it can do is potentially reduce the size of a release and help plan actions to minimize its consequences. A leak detector asks: if a leak occurs on site, how quickly can we detect it, how closely can we locate it, and how well can we estimate flow rate?

Is a model-based leak detector needed?

The consequences of a leak depend on the quantity and type of fluid that is released, and the location of a leak. How do we estimate them? Accordingly, what procedures should be put in place to respond to a probable leak? How do we say it's probable? Do we need to go to the effort of a model-based leak detector to answer these questions?

Often the answer is no. The bread and butter of our success is originally Atmos Pipe, a *statistical* pipeline leak detection system, a great leak detector that isn't model-based.

Many methods and contraptions detect leaks. What's harder is detecting leaks *without false alarms*. We must reliably trigger an alert followed by an alarm, for all leak sizes, during all pipeline operations, in a timely fashion. But false alarms affect the psychology of the operating staff, and might instill deviations from procedures that could cause a major incident. The investigation into the Deepwater Horizon disaster revealed safety-critical alarms had been silenced because senior managers did not want the staff to be disturbed by false alarms.

For some applications, pipeliners find a model-based approach essential. A leak doesn't only occur when flow rates disagree, especially in a compressible fluid. And conversely, flow rates could be imbalanced if there is a leak-free transient taking place, as happens during all kinds of pipeline operations. Packing or unpacking, unit start-ups and shut-downs, or setpoint changes at injection or delivery points; all these events cause upsets which disturb the line. For this reason, primitive leak detection methods were a major cause of false leak alarms. To avoid them, engineers would introduce some form of intentional degradation during certain pipeline operations. This desensitizes the leak detector, or even deactivates it altogether: an undesirable approach. During those transients with drastic changes in pressure, a pipeline is precisely most susceptible to developing a leak! So how does anyone get away with it?

How are Atmos Pipe – a statistical volume balance system – and Atmos Wave – a negative pressure wave system so good in the real world? Yes, a well-tuned physical model would make an even better leak detector, but if a simpler approach works, it is better to opt for that. The answer is *time-scale.* On gases the time-scale at which hydraulic transients occur is similar to the timescale at which we are doing leak detection, so we can't filter them out. Whereas liquid transients don't persist, so we can filter them out for the purpose of leak detection. In a buried liquid pipeline, thermal transients are extremely slow due to the buffering effect of the soil, and hydraulic transients are extremely fast. Not lightning fast, but sound speed fast (c). A low pass filter takes care of the thermal effects, and a high pass filter deals with the hydraulic transients. There are a lot of tricks involved in compensating for a new batch going through, for instance, and hydraulic transients run into the faster leak detection periods. But in essence, we rely on leaks taking place on their own timescale.

An important facet of Atmos Pipe and Atmos Wave being so successful is their low cost of upkeep. A well-maintained pipeline model is significantly more expensive than non-model-based solutions – it is deployed only where needed. If well-tuned, a model-based leak detector could outperform other methods particularly in gas pipelines which undergo frequent transients. They extend the old principle of flow-imbalance with a far more sophisticated *inventory-change* capturing linepack via a physical model. Some call this approach an extended real-time transient model (E-RTTM). Regulations such as API-1130 in the US oblige operators of certain pipelines to deploy real-time leak detection systems based on computational analysis of flow, pressure and other measured data. Model-based leak detection is one of the methods included in API 1130. This is an umbrella term: model-based leak detectors and locators don't all use the same method. Each software vendor takes their own unique approach. But the general principal remains reliant on a broad principle of missing mass or flow.

Figure 205 – *"Contributing to the delay in recognizing the release were limitations of pipeline SCADA to detect small leaks."* – National Transportation Safety Board (Credit: NTSB)

Is a model-based leak detector possible?

To determine the *smallest detectable leak*, we consider the fluid's velocity and pressure throughout every section. A pipeline with a very low flow rate for its diameter (a low *velocity*), or a very low pressure, is challenging to tune in general, and particularly hurts the performance of any leak detection system. Some pipelines are oversized by design to accommodate potential future capacity. Even in a well-designed pipeline during its routine operations, a low flow rate could manifest as it copes with various commercial drivers. And for any pipeline, a crucial stage of its lifecycle often involves low flow rates: commissioning activities. This challenging phase of work is also our first opportunity on a given project to validate the leak detection system in anger.

A model-based leak detector only works if a complete physical description of the pipeline is baked into our model. In addition, it requires that both pressure and flow rate are always metered at every inlet and outlet. It also demands an obsessively correct thermal model, so we need both inline and ambient temperature metering, processed by a dynamic thermal model incorporating transients. This means layers describing the pipe, insulation and surroundings.

In addition to typical conditions required to run any online simulator, a leak detector can only reliably monitor a given region if every one of its inlets and outlets has valid, good quality measurement of *both flow rate and pressure*. A section bounded by both pressure and flow meters at every inlet and outlet (a PF-FP section) is the ideal way of forming a leak detection region, determining our granularity. If this is not satisfied, instead of making our leak detector inactive for the affected regions, we use a sort of *"virtual flow meter"*.

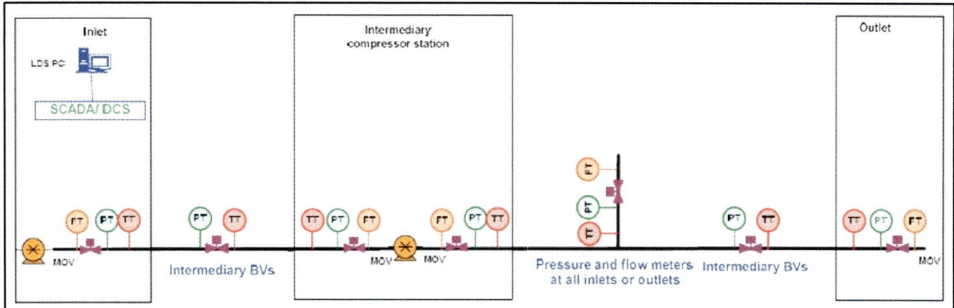

Figure 206 – The typical instrumentation required for a model-based leak detector.

Atmos SIM has detected leaks at a sensitivity of 1 or 2% of nominal flow rate. If a mid-line valve position was open and is then closed at a certain point in time, leak detection *sub-regions* might result, which potentially improves expected performance.

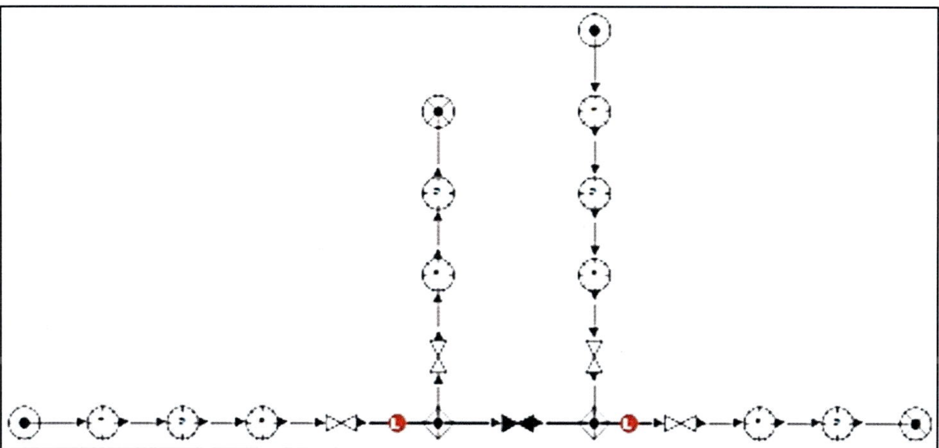

Figure 207 – This leak detection region splits into two sub-regions if the mid-line valve is shut.

But what if a mid-line flow meter becomes unavailable? Adjacent leak detection units could merge to form a single larger one, but this may degrade expected performance. In a real pipeline, we can't insist on having enough mid-line flow meters to bound leak detection regions of reasonable length: we might not have a flow meter for hundreds of miles! Luckily we can make do with just a pressure meter; we can deduce a suitable *"virtual flow meter"* by simulating the adjacent pressure-bounded region – the next *leak detection unit* (LDU) – in isolation to find its flow rate. This estimates flow rate independently of our own region's model, which is all that we need to perform leak detection.* Unlike other methods, model-based leak detectors unequivocally rely on good instruments to measure with not only repeatability, but also *accuracy*. Our

* This approach works so long as there is not also a leak in the adjacent region.

physical model must be the truest representation of reality at all times. Much like us, a model-based leak detector must be well-fed to perform adequately.

How to feed your leak detector

No pipeline packs or unpacks indefinitely; over a long enough period, net flow into or out of every subsection of a pipeline averages to zero. We find discrepancies implied by meters in our pipelines by taking advantage of this fact: over a very long period, the net flow into the system summed over all inlet and outlet points must be zero. We can thus very gradually automatically adjust flow meter readings. Note that this adjustment must be several times slower than our longest leak detection period, or it might also tune away a real leak! To avoid false alarms, persistent discrepancies are gradually learnt to bring their corrected readings into agreement, over a steady, leak-free period. This period is chosen to be far longer than the pipeline's transit time.

Imbalance in inflow and outflow changes the inventory. If this disagrees with what pressure meters imply, it's a leak.

Any imbalance in flow, that is inflow minus outflow, should cause a change in inventory. Our model converts pressures into modelled inventory, which is then compared to the last step's inventory (ΔM or ΔV_{std}) to calculate how it changed:

$$\Delta M = \text{function of (pressures and temperatures)}$$

When a leak occurs, that pipe section depressurizes. This change in inventory no longer accounts for measured flow imbalance, and we see a *discrepancy*. Pressure-deduced inventory falls below what we'd expect. But it's actually far more subtle and cleverer than that. By this method, we detect a leak even where inflow doesn't exceed outflow. There might be a leak whilst unpacking! What matters is that the flow imbalance ($F_{\text{in}} - F_{\text{out}}$) over a time-step ($\Delta t$) interval is fully accounted for by the pressure-deduced inventory change (ΔM):

$$\text{Discrepancy} = \sum F_{\text{in}} - \sum F_{\text{out}} - \Delta M$$

Any slight steady-state discrepancy is learnt during leak-free conditions. If instantaneous discrepancy deviates from its learnt mean, there may be a leak. Calculations can be done on either a standard volume (V_{std}) or mass (M) basis. So discrepancy, and the leak flow rate it implies, can be reported in either units. A possible leak detected can be verified by shutting-in the suspected section, and monitoring its pressure. Shut-in leak detection works by a principle that is similar in some ways and distinct in others. It has a strict requirement for metering temperature and pressure on both sides of each closed valve. If a region is shut-in, there is zero flow imbalance! The discrepancy is between pressure-based inventory at present temperature, and what it was on the previous step. During a leak, measured pressure falls faster than we expect. This shows up as a discrepancy in our formula, just like a leak in a flowing line.

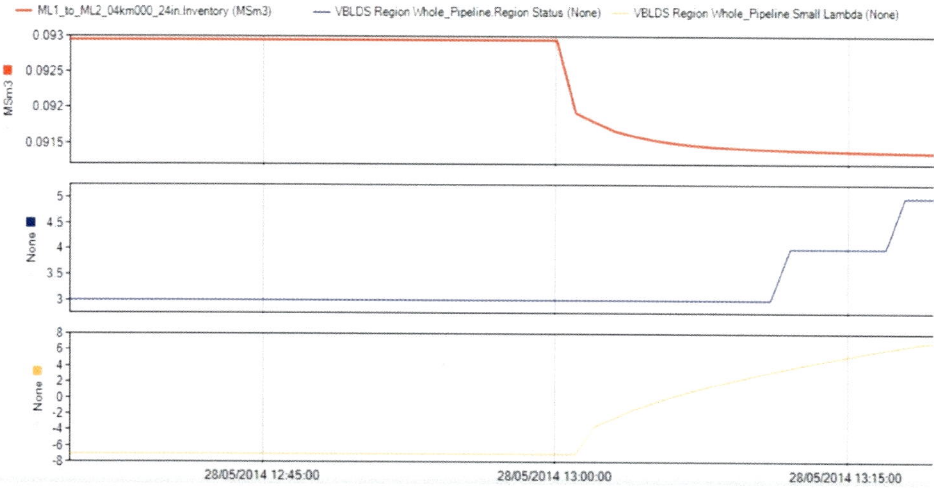

Figure 208 – This leak causes inventory to fall (top), but not all leaks do this. Almost immediately after the leak's onset, a region's leak alarm (middle) is triggered by a rising leak signal (bottom).

A pressure-deduced inventory model

A leak is detected whenever the relationship between flow rate and inventory appears to be suspicious. All the charm is in *how* we calculate this inventory.

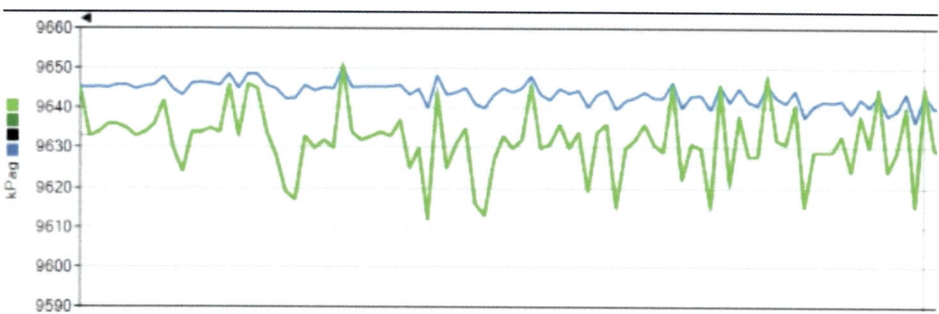

Figure 209 – A pressure-based model (LDU) is used by Atmos SIM's leak detector and locator, working according to a different principle than the maximum likelihood state estimator (MLSE).

Online models need some sort of *state estimator* to make best use of all of the available pressure and flow meters. Atmos SIM uses an algorithm called *maximum likelihood state estimation* (MLSE), which we shall explain in detail later. But if we were to calculate the inventory for leak detection using MLSE, we would to an extent be comparing metered flow rate to itself! So in Atmos SIM, neither the leak detector nor locator algorithms rely on the maximum likelihood state estimator (MLSE). A separate model, sometimes called the *leak detection unit* (LDU) model, is more appropriate for leak detection. It treats pressure meters as boundary conditions wherever possible and does not insist on conservation of mass at those points; it treats pressure meters as defining separate leak detection regions and doesn't require mass flow into a pressure meter to equal mass flow out in the way the MLSE model does. Between every pair of pressure meters (A and B, B and C, etc.) separated by a pipeline section (P-P section AB, BC, etc.) the LDU model deduces an inventory based as far as possible on the pressure meters. In Atmos SIM, this separate pressure-based LDU model runs independently and concurrently live on site alongside MLSE, dedicated to allowing the leak detector to compare measured flow rates to *pressure-deduced* inventories.

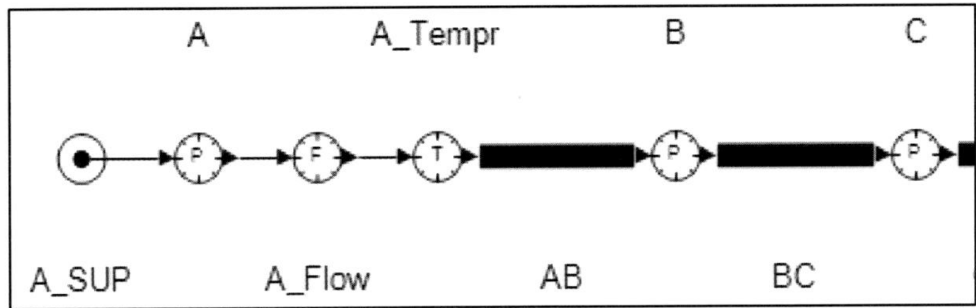

Figure 210 – Flow, pressure and temperature metering along an example pipeline.

Sigmas, lambdas and statistics

Our offline data's leak-free total imbalance is zero, so learning it is pointless as it does not use instrumented data. But on-site data, even during nominally *"steady"* periods, has biased meters, causing leak-free discrepancy to be non-zero at all times. Learning tunes a leak detector, correcting imperfections in both meters and the calculation itself. To minimize false alarms, we have incorporated the proven statistical algorithm from Atmos Pipe into Atmos SIM. The deviation of the discrepancy from its mean is fed to the *sequential probability ratio test* (SPRT):

$$\lambda_{i+1} = \lambda_i + \left(\text{demeaned discrepancy} - \frac{\text{size}}{2}\right) \cdot \frac{\text{size}}{\sigma^2}$$

Where sigma (σ) is the standard deviation that impacts on the detection time, and *"size"* is the leak size this particular test is looking to detect. Several lambdas (λ) can be defined, each performing an SPRT calculation for its own leak size and sigma parameters. Smaller leaks are configured with longer detection times. This is all specifically tailored for every pipeline, to optimize sensitivity, reliability and robustness.

The SPRT method analyzes any leak signal, establishing the probability of a leak. Here it reliably triggers true leak alarms whilst avoiding false alarms.

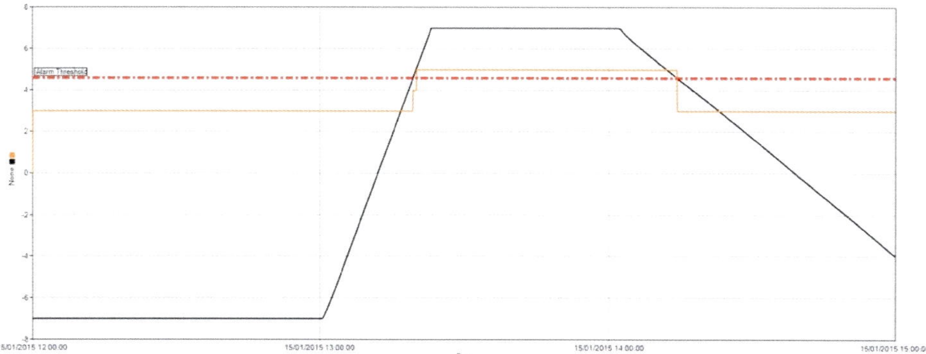

Figure 211 – When any one of the lambdas (black) reaches its threshold (red), a leak alert is triggered. When it stays above the threshold for a configured period, a leak alarm is generated. This is cleared when all lambdas drop down below the threshold.

The smallest detectable leak size depends on flow meters' performance and, for regions bounded by mid-line pressure meters, the model's general transient behavior, as well as the pressure meters' performance. A typical way of configuring a system for some throughput is for a small lambda to detect a 2% leak within 30 minutes, a medium lambda to detect a 10% leak within 10 minutes, and a large lambda to detect a 30% rupture within 1 minute. When the statistical calculations show the probability of a leak is above 99%, an appropriate leak warning is raised. A tag indicates which section is leaking and the leak size estimate is also given. In addition, the leak start time and total volume leaked since its onset are provided.

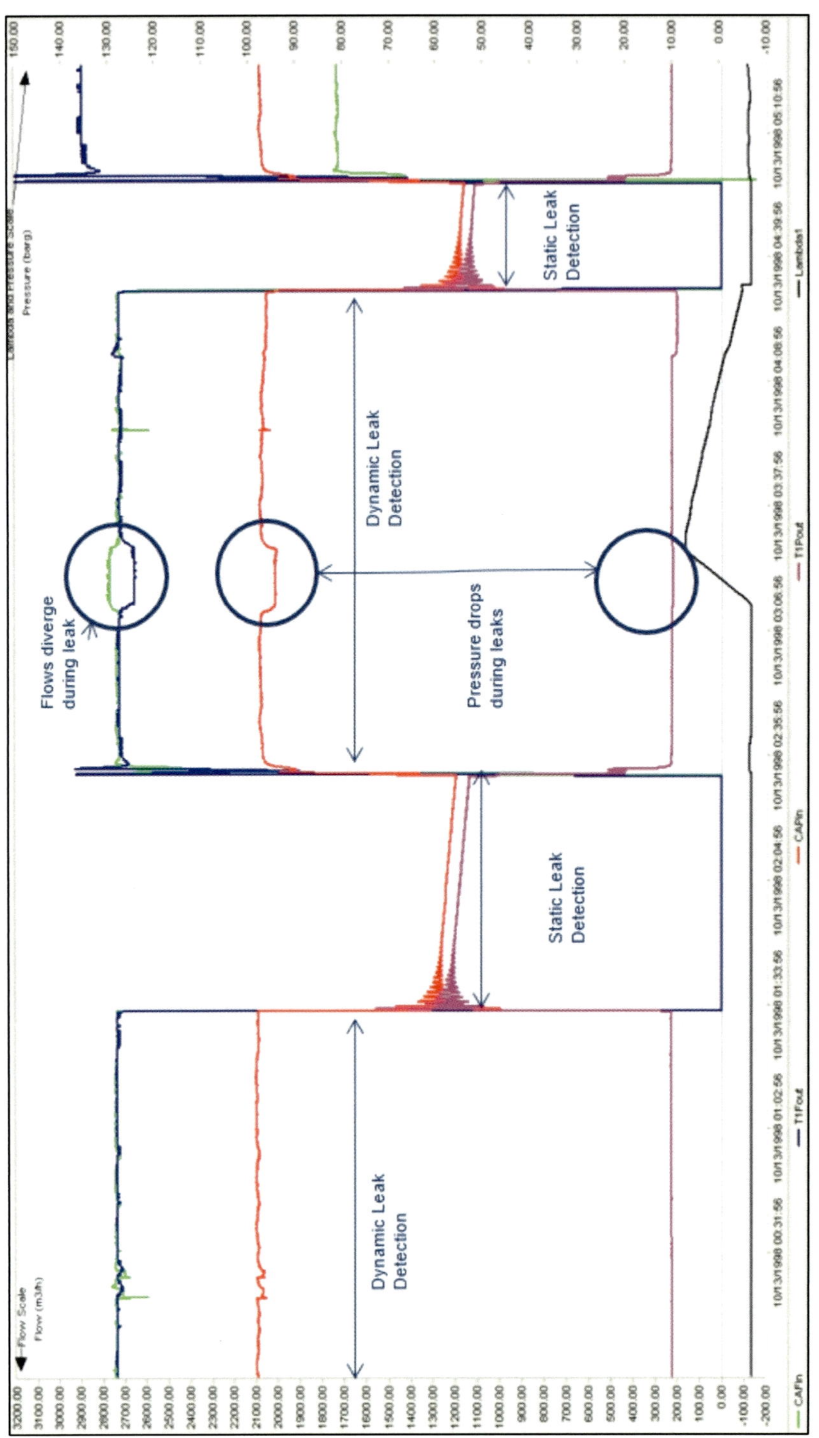

Figure 212 – Atmos SIM's dynamic leak detection and static leak detection in action. The leak detection signal (grey) is shown at the bottom.

10.5 – MODEL-BASED LEAK LOCATORS

Are leak locators needed?

Gas leaks whistle and on a high-pressure line, they can usually be heard for miles. This is not the case for most liquid leaks. To look for a leak, common practice was to go out and walk the line, perhaps with sniffer dogs. This doesn't scale up well for, say, long-distance subsea and cross-country pipelines – but the principle remained the same: a boat or helicopter would routinely patrol along the route to check for leaks. Methane from a large leak can affect the buoyancy of the water or air, so to be safe, vessels checking a gas pipeline for leaks deliberately avoid sailing or flying directly overhead.

Sniffer dogs, boats and helicopters are slow and expensive. It matters a lot that a model-based leak *locator* provides an accurate location estimate within a known time-frame. Both the response time and location accuracy are unique to every leak, depending on the length of a section, and the quality of instrumentation.

Can we locate leaks without a model?

There are plenty of leak location methods which don't involve going to the trouble of building, maintaining and simulating a full model. The time-of-flight method works well works well but requires we identify the arrival of the leak onset signal at two different pressure or flow meters, which is often impossible using just SCADA instrumentation.

It is also theoretically possible to locate a leak based on its effect on the frictional losses upstream and downstream, but this needs the line to be in steady state and a steady state may take a long time to appear after the leak starts. So we do not use this method.

In recent years we have seen Atmos Wave locating tapping points to an accuracy of a few meters in crude and multi-product pipelines where thefts are detected. It uses negative pressure wave technology and high speed data sampling of dynamic pressure sensors. Atmos Pipe uses both time-of-flight and friction method for locating leaks within one to a few kilometers depending on the data sampling frequency and pipelines.

The special pressure-only model for leak location

Atmos SIM's leak locator uses the same inputs as its leak detector but works in sections bounded by pressure meters (a $\mathrm{P}-\mathrm{P}$ section). A leak locator is more sensitive to fluctuations in metered pressure than a leak detector. Not only must it deduce inventory in these sections, it must also deduce a *flow rate.* This is possible without a flow meter: it calculates a flow rate based on the metered pressure drop. Deduced-flows from a pair of neighboring pressure-pressure section are acting as if we have placed a pair of virtual flow meters, telling us more about the leak if we look at how they disagree with one another. At steady conditions, a well-tuned pressure-based model estimates *the same flow rate* through every one of its many sections. That is to say, at every pressure meter, there is zero reported discrepancy in pressure-deduced flows as implied by the two adjacent pressure-pressure sections.

A leak locator algorithm long seemed elusive and finicky to get just right. Atmos SIM uses a method that has been successful at leak location during hydraulic transients on liquid pipelines; our experiments found it works with slight modification for gas as well. In brief: it takes a ratio of flow imbalances. The worst flow imbalance is presented by the pair of pressure meters bounding a leaking section. For mid-line pressure meters without collocated flow metering, we deduce flow rate from neighboring sections' pressure drops, *"virtual flow meters"* provided by the pressure-based LDU model. If the leak is immediately adjacent to a pressure meter, we expect the flow imbalance at the other end will be zero – to the extent that flow imbalances are zero in the absence of leaks. We can interpolate between the cases where the leak is at either end of a section by estimating the leak location based on the ratio of flow imbalances at the two ends. *

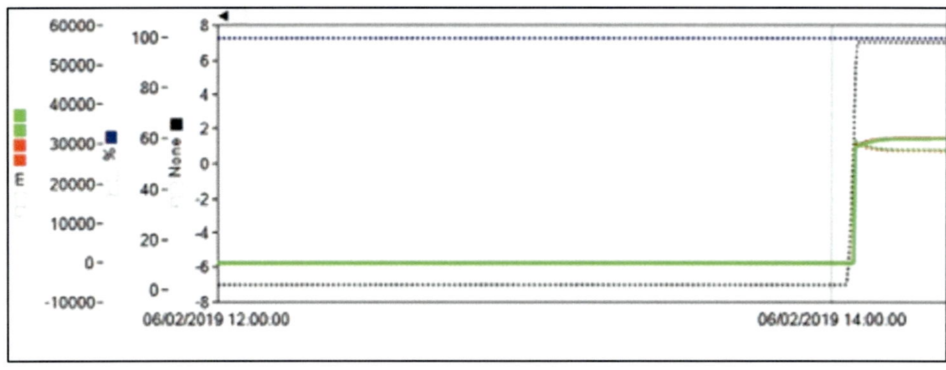

Figure 213 – A leak locator estimates where a detected leak lies within a pipeline.

The performance of a leak location estimator is highly variable and strongly caveated. A location can be identified within a surprisingly accurate distance, and within a surprisingly short timeframe. An accuracy of a few percent is quite achievable, especially in favorable conditions. If an unmetered stretch of pipeline is particularly long, we understandably must wait for a *"leak signature"* to appear at meters at both ends. Once the transient resulting from a leak hits the *"neighboring section"* beside the true location, it can infer a good initial estimate of leak location. Transient operations

---

* As far as we know, this was invented in the 1980s by Ed Nicholas at Modisette, Inc.

are particularly challenging on gas pipelines. If flow rate changes significantly during a leak, which it often does, the initial flow imbalance used for leak location could be *"corrected"* at each subsequent step.

We check a region's calculated flow imbalance by comparing the pressure-based flow to metered flow. The flow imbalance ought to be zero for offline-generated steady-state leak-free data.

Shut-in leak location is not particularly accurate unless there are pressure meters available at both ends of the shut-in section, but what a pipeliner usually needs to identify is *which* shut-in section is leaking.

10.6 – Estimating leak flow rate

Using the same discrepancy, we calculate how much fluid has gone missing, and how quickly it does so. The leaking flow disappeared in the difference between flow imbalances at either side of the leaking section, yet it hasn't been accounted for in the pressure-deduced inventory change within the leaking section of pipeline. Leak flow rate is instantaneous discrepancy divided by time-step (Δt).

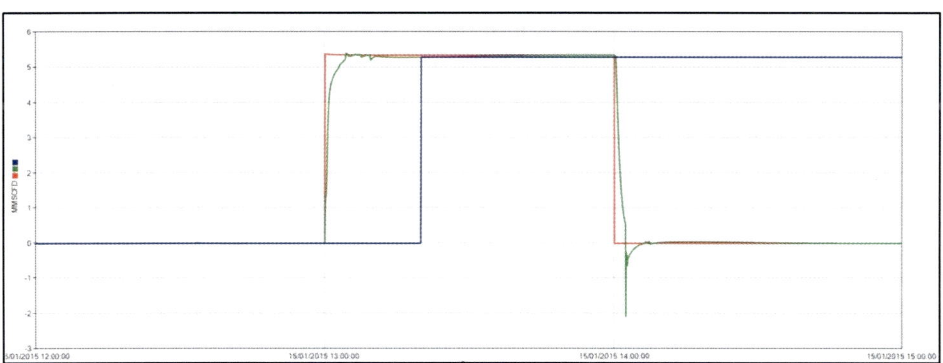

Figure 214 – A gas leak appears (true flow rate in red), giving rise to a flow rate discrepancy (green) which is used to accurately estimate the leak flow rate (blue).

If the section is underwater, it might be that the surroundings are at a higher pressure than the local line conditions, so a leak might result in ingress of seawater rather than an escape of pipeline fluid.

10.7 – Performing in anger

Is there a way to locate leaks in the middle of a gas network, even in a long unmetered subsea stretch of pipeline? The belief was that the difference seen at the meters was so tiny, that any distinction between leaks 100 km apart would be impossible, clues being swamped by discretization error, imperfect tuning or inevitably *something else*, perhaps uncertainties in the parameters of the pipeline itself, especially those affecting thermal behavior – for example, if a subsea pipe is buried. All of this meant that until the early 2000's, some gas pipeline operators still did not want to buy a leak detector.

If meters are 100 km apart or 50 km apart then we can do much better than if they're 1 000 km apart. Performance doesn't scale linearly; there doesn't exist a system locating all leaks to within, say, 2 km in any conceivable network. In a 50 km section, we locate a leak very closely. But in a 1 000 km section, locating a leak within 20 km is difficult or maybe impossible. And by *"meters"*, do we mean both flow and pressure meters, or just pressure meters? Atmos SIM's leak location is driven by pressure meters, so theoretically it is enough to use shorter locator sections defined as pressure-pressure (P – P) bounded. But this brings tuning into the discussion. The algorithm, in the absence of a real flow meter, relies on a virtual flow meter reading the flow rate of the adjacent pressure-pressure section. So if tuning is wrong in that adjacent section, flow is wrong, and so leak location is very wrong! Similarly, if a leaking section is poorly tuned, its own calculated flow rate is wrong, and so again is the leak location estimate.

Atmos SIM's model-driven leak locator, when it's driven by perfect noise-free simulated data, estimates leak locations miraculously well. The errors seen amount to nothing, dwarfed by errors associated with real instruments. Therein lies the puzzle. In a real system, noise and meter errors come into play. Leak locations are incredibly sensitive to everything. And the real world indeed throws everything at us. For a leak locator to not get inundated by imperfect site data, it must tackle the real dreadful data elegantly.

10.8 – Correcting inventory after a leak

If a false leak alarm has occurred, we don't need to do anything. This is because the model Atmos SIM uses for leak detection doesn't remove inventory from the pipeline, even when it's estimating a leak's flow rate it continues to use a leak-free model.

But what do we do when a valid leak alarm has occurred? Perhaps the operator instigated an unmetered offtake to prod the leak detection system in order to verify it's working as intended. Going back to a pre-leak state will introduce inaccuracy, because the pipeline has moved on since then. Carrying on as we are also incorporates an inaccuracy, because now some inventory is missing and conditions near the leak's location are very different. It would be helpful, in this unusual case, to allow the user to somehow manually confirm that a leak alarm is valid, and for the system to duly remove the correct amount of inventory from the pipeline at the correct point.

11 – LIVE ON SITE

"Keep calm and carry on"

11.1 – THE ONLINE SIMULATOR: A DIGITAL TWIN

The investment in a strong underlying physical model pays dividends throughout the pipeline's lifetime. Having built and calibrated an underlying model for offline studies, a qualified team can adapt it for deployment live on site, in an *online* environment. This adds value to a smarter operation to keep it flexible, safe and profitable. Auxiliary software validates incoming raw data, and displays calculated results and alarms for interaction with the operators. Automatic learning of parameters keeps the model well-tuned. Virtual meters, inventory calculations, composition trackers and other features such as leak detectors inform the operators about the present state of the pipeline system. Look-aheads anticipate operations into the future to help plan what to do next.

Figure 215 – Controllers in an oil pipeline control room must monitor operations in real-time.

Deploying a simulator to a live site environment is an application that demands fast-solving, seamless performance at all times. More than ever, abiding by our underlying principle stands us in good stead: great models should be kept as simple as possible.

For pipelines carrying batches of most liquids, provided that they aren't drastically compressible or multi-phase, an *incompressible model* is suitable for most applications. It keeps track of batches and their custody as they move through the network. It does not attempt to deduce flow rate, but instead follows a fiscal meter. As every batch is detected passing by a mid-line station or arriving at its destination, *"approximating corrections"* are made to fudge-factor parameters, improving the accuracy of the batch tracker.

For other liquid pipelines and all gas pipelines, a *real-time transient model* (RTTM) provides planning and operations professionals insight into what is truly happening throughout the system at all times, including deducing physical conditions at unmetered locations. It evaluates measurements against calculated flows and pressures. A state estimator is complemented by learnt tuning corrections that maintain the model's agreement with valid measurements. It also helps deduce possible causes by working *backwards* from consequences, diagnosing symptoms in real-time, a process known as *abnormal situation management* (ASM).

States are saved at regular intervals every few minutes. If the simulation fails, perhaps due to unflagged instrumentation failures that are presented to it as good-quality data, upon restart the last saved state is restored. The simulation catches up by linear interpolation, if instruments were lost during downtime. Alternatively, some simulators use the historic logged data for catching up.

Alongside real-time modelling, predictive capabilities are deployed as part of these live systems. Using a current state saved online as a starting point, *look-ahead* scenarios are launched to anticipate what will happen in the future.

Nothing can be achieved by a pipeline simulator unless it can prove itself to be a capable system. Its value hinges entirely on being a model the users can trust. This confidence is built on a foundation of remaining robust and resilient in the face of real-world conditions, validated using metrics such as transit time.

11.2 – Communicating with the user

Who is the user?

Pipeline simulators are now interacting with a much broader cross-section of a pipelining team than in the past. It is not only simulation specialists doing offline studies, or day-to-day operators who rely on a modern pipeline simulator. There are all kinds of technical and commercial tasks to which the model and its auxiliaries add value.

Properly integrating a pipeline simulator with a SCADA system enables information from both systems to be clearly displayed to everyone who needs it. Multiple users in the control room and the office can take advantage of this information, combining real-time data at virtual meters and throughout the pipeline with predictive look-aheads and the historian's logs.

This diversity of use cases also makes it imperative that due attention is paid to access and permissions. This is a more sophisticated matter than keeping track of separate logins and recording who made a change: perhaps somebody, for instance, needs read-access to some facet of the system, but is not allowed to enact a change at all.

At the same time, it can be concerning that the department responsible for supporting the digital infrastructure, usually referred to as information technology (IT), might get jittery about allowing something like remote access for a software vendor, although this means they will be running out-of-date unpatched software – a situation which comes with its own organizational risks.

The human-machine interface

Once our simulator deduces what's happening, a suite of auxiliary software interfaces with the end-user. There is a lot more to this than merely displaying results to a dispatcher. An operations team consists of several complementary roles, which might be interested in different live and historical information, and are responsible for several kinds of reports. The system often involves communication back to the simulator, affecting interventions and corrections, such as adjusting the head position of a liquid product batch, or retrospectively issuing a pig launch command. A single console may manage many pipelines.

The human-machine interface (HMI) is a set of screens and views that keep an operator informed and in control. These are tailored for the operations team's needs, complying with stringent standards covering everything from colors to logging and access permissions. There are obsessive studies into the psychology of a control room; layout, procedures, shift rotations and so on. As we wrote this book, in the midst of the COVID-19 pandemic keeping much of the world confined, operators of many pipelines had to operate their infrastructure entirely whilst working from home. More than ever, the impact of the look and feel of a human-machine interface cannot be understated. An HMI (referred to commonly by its abbreviation) is difficult to define. Strictly speaking it refers to any human-machine interface. Some in the pipeline industry only use the term for specific screens that accommodate information which cannot be readily handled via the broader SCADA system. It has to integrate seamlessly with the SCADA system, or link from it. The specifics of what a HMI shows, and how, are customized to every project's needs. It is likely that more than one type of user interface is used: SCADA for controllers and applications interfaces for engineers.

11.3 – PROCESSING SITE DATA

No simulator is an island

Even when a pipeline simulation system is performing well, the onus of having one's wits about them always lies with the operations team. A pipeline simulator is not subject to the same scrutiny as, say, a self-driving car: its purpose is advisory, and even where it becomes so useful to an operation as to be considered mission-critical, we must bear in mind that it is not audited to be *safety*-critical: simply because that is not its purpose!

That being said, software commissioned by a capable team supported by developers is almost always robust enough to require minimal intervention in service. Simulated results are capable of impressive accuracy, provided the system is set up adequately and is fed validated data that isn't so unexpected as to somehow sneak bad inputs into the underlying physics engine. Once cold-started, a healthy simulation system gradually comes into alignment with reality, then regularly saves a state to restore in case of a warm or hot-restart.

SCADA

> *"In 1999, an employee was performing maintenance on a SCADA data acquisition server [controlling] a gasoline pipeline in Bellingham. Database maintenance caused the model-based leak detector to not be working at the instance of the rupture, allegedly, and prevented them from shutting down the pipeline in a timely manner. This contributed to the rupture not being detected during a leak. Gasoline ignited and burned a two-mile section of the creek. This incident resulted in loss of life. The SCADA system had no security features that prevented a maintenance procedure from affecting the operation"* [70]

Operational technology (OT) has come on in leaps and bounds. Professionals dedicate careers to an array of technologies, such as ruggedized general purpose industrial computers called *programmable logic controllers* (PLC), *remote terminal units* (RTU) and open-standard *process automation systems* (PAS). Most pipelines use a *supervisory control and data acquisition system* (SCADA, pronounced as a word), whereas process industries have tended to prefer proprietary *distributed control systems* (DCS) dispersing autonomous controllers throughout an industrial plant, with on-board monitoring and control. In practice, these are a family of related technologies.

The boundaries between a connected PLC system and a DCS are blurring today, as features are converging. It is a disservice to refer to all this by any umbrella term, though pipeliners often just call it all SCADA. As it is not our focus, we won't digress.

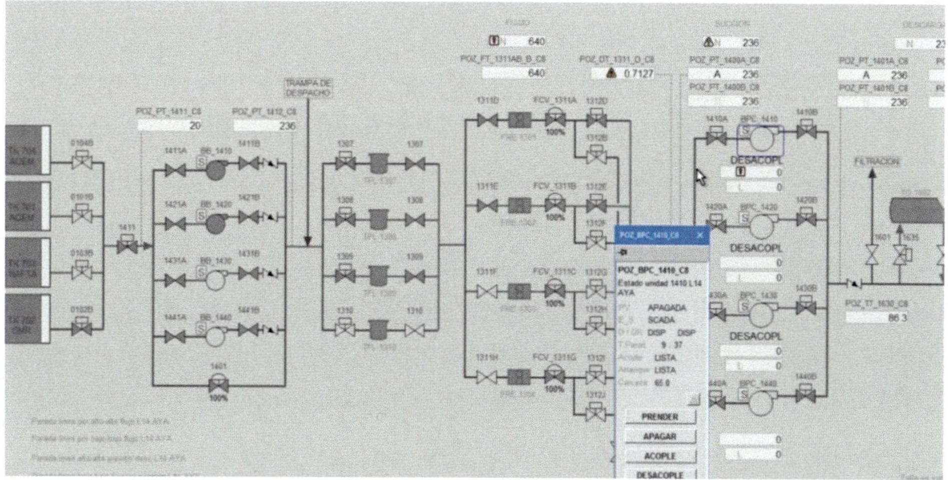

Figure 216 – A pumping station along a pipeline as shown on a typical SCADA screen.

"SCADA" is a centralized system that sits between our simulator and field instruments, data historians, and operator screens allowing users to respond. It involves many operational technologies, gathering data from remote devices and making it available at a single centralized location. It handles raw data, covering every variable informing us about the pipeline's ongoing operation – conditions at metering stations, valve positions, pig launch events and so on. Data comes in from instruments, and all manner of field equipment, reporting on measurements and the *quality* of each measurement.

It might be the case that incoming information is inconsistent. Helping a simulator make sense of conflicting information it is a significant task for a team executing a project or maintaining a commissioned system. Auxiliary software processes these incoming raw data, as even the most reliable SCADA system could be flawed for understandable reasons. The chosen polling strategy, for instance, might cause some tags to lag behind others. Oftentimes, a project engineer digging into a historian's logged dataset will find flaws in at least some of the incoming tags, and seek ways to deal with them.

A host of supporting digital infrastructure and auxiliary software is instrumental in commissioning and maintaining any accurate, robust simulation suite. The online model is merely one component of this system. It must be fed only those datapoints pertinent to what the pipeline simulator has been tasked with, and those must first be carefully validated. How we set up a system, tailor it for a unique site, and monitor it during operation will determine how well it performs.

Communicating with the world

The Open Platform Communications protocol (OPC) has become ubiquitous not just for software interacting with pipeline SCADA systems but across industrial automation. A protocol is a commonly agreed way to communicate between very different software, which means more than just transferring data. In some senses the protocols in use today are being pushed near their limits.

Positive changes have unfolded. In the 2000's, typical SCADA scan rates were 60 seconds for a gas pipeline, and 10 seconds for a liquid pipeline. In 2020, we are exchanging more data and processing it an order of magnitude more often. More than ever before, a lot is expected to happen *"now, without delay"*. This has pushed forward innovations in many sectors, and pipeline simulators are slowly catching up as pipeline SCADA systems supply us with better data.

Good or bad – quality, range and error

It might be unclear if a meter is isolated from the rest of a pipeline. If some decision depends on a valve being closed, do we trust the valve position tag?

How do we validate a meter's range? Is it *"everything the instrument can measure"* or an *"expected practical range based on the operating philosophy"?* Bad meters can dip in and out of being good-quality, so we use a validation hold count before trusting a *"bad-quality"* meter truly returned to being good!

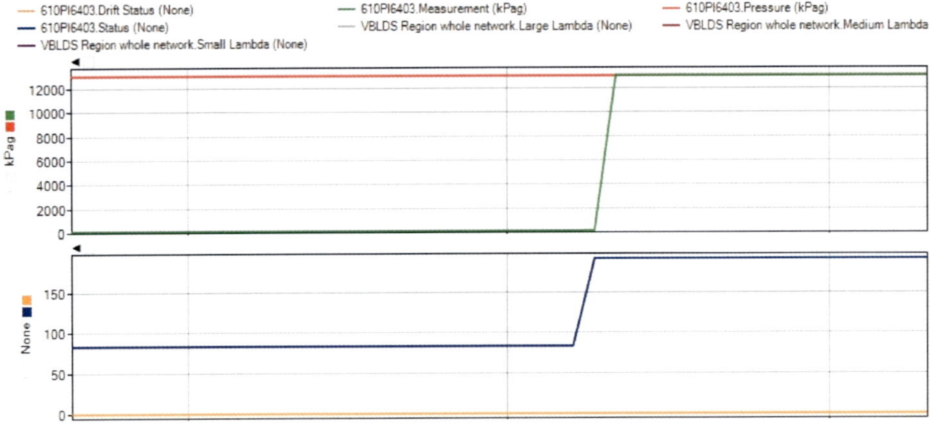

Figure 217 – In the top trend, a pressure reading (green) resumes, having been zero while bad. In the bottom trend, that meter's quality tag (blue) becomes good, indicated by a value of 192.

Numerous signs of bad metering on liquid pipelines are not obvious on gas pipelines. In general, one would check that incoming data is free of major blips or plainly wrong measurements. It would be tough to spot flow repeatability issues, as the system is not particularly sensitive to flow measurements alone.

How well a live model performs depends on the digitization being employed by a SCADA system and what sort of noise is present. The digitization or quantization scheme is apparent from inspecting logged data, as all meter changes will jump by an even number of digitization increments. Also report-by-exception presents a problem if the definition of an exception is too weak. If an exception is any change in a meter, even in its last significant digit, then it is identical to report-by-polling. But generally it is typical to see it set less sensitive than that.

The data manager

To deploy our simulator live on site, it needs valid data. As the famous adage goes: *"rubbish in, rubbish out"*. The SCADA system itself might possibly perform some kind of range-checking or even more sophisticated tests of data quality, but this is rarely adequate to drive a simulator, for which a single pressure meter reading 0 barg with a *"good"* data quality can look like a rupture. Rather than burdening the physical simulator with these non-physical tasks, we delegate the processing of data to happen separately beforehand. Each incoming datapoint is first processed via a *data manager* (DM).

Figure 218 – A fibre optic cable rack. Live measurements from a pipeline make their way as *"site data"* across vast distances to be processed via a SCADA system and reach a simulator.

The Atmos Data Manager is auxiliary software dedicated to performing data validation. It is capable of deciding an incoming tag which claims to be good-quality is in fact bad, flagging it as such and holding the last known good value instead of passing on the suspect value. This is customizable: for example, we might check if one tag goes out of range, but for another tag a tell-tale sign of bad quality might be the rate of change.

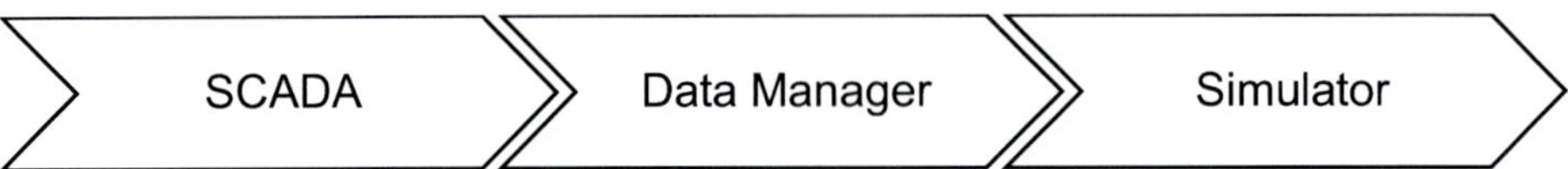

Figure 219 – SCADA fetches raw data from site instruments at somewhat arbitrary intervals. A data manager (DM) pre-processes it to look for obvious problems and converts it into a regular series of updates for each instrument. This is then passed on to drive the simulator.

There is a whole host of approaches to help us deal with all kinds of *"flaky"* data! Where the simulation software cannot yet handle a new situation that a project engineer encounters, custom scripting inside the data manager itself lets them do whatever it takes to prepare suitable inputs to feed the simulator.

Tag	Value	Unit	Quality
Valve position	Transitioning	None	Good
Outlet pressure	9 918 203 287	BARG	Bad (Invalid)
Inlet flow rate	0	GPM	Good
Pump status	Stopped	None	Bad
Pump speed	3 000	RPM	Bad

Table 24 – An example of tags in the data manager, with respective values, units and qualities

The data manager can also perform other useful functions, such as logging data in its own historian, in an amenable format to feed a simulator in the future. This is handy for replaying data which was found to have challenged the live simulator in some way, troubleshooting bugs and failures, or perhaps even for feeding a trainee scenario and seeing how operators would have responded to a real situation which occurred on site.

The Atmos Data Manager (DM) uses standard protocols to gather, validate and record real-time and simulated data at configurable intervals, typically working in a redundant environment so it can seamlessly switch from a primary server to a secondary server.

11.4 – METERING AND INSTRUMENTATION

Virtual meters

Even a steadily flowing pipeline is likely to have certain locations along its route where operating limits are likely to be violated because of the terrain. It's worth monitoring these key locations with what Atmos SIM calls *reporting points,* which are essentially virtual meters: the model tells you what a meter would despite no meter being present. This informs design decisions during offline studies, or acts as live virtual meters within a live simulation – vital for remote unmetered sections. But no model can run and deduce these virtual meters unless enough good live data is available at all times. A host of pipeline instrumentation is used to measure our operating conditions in the first place: we ought to be familiar with how it works.

Pressure meters

Measuring pressure

It is common to see co-located pressure meters on a pipeline. On a piping and instrumentation diagram (P&ID), they are typically labelled as either pressure transmitters (PT) or pressure indicators and transmitters (PIT). For example, there are pressure meters on both sides of a valve in case the valve closes, and on both sides of a regulator or a pump or compressor, to measure what they are doing. If the valve is open, or the pump or compressor is off, then the pair of pressure meters located at each side are both measuring essentially the same pressure as one another because the head loss across these devices is typically negligible.

Each kind of pressure meter works according to its own principle and has its own utility. Depending on how we perform our measurement, we might be observing any one of several different kinds of pressure. Meters typically only indicate pressure with respect to a reference pressure – typically atmospheric pressure, and so *"it goes without saying"* that what they are reporting is a *"gauge pressure"*. This is important to be aware of because oftentimes, people do neglect to say it (e.g. 25 barg wrongly written 25 bar).

Mechanical methods for pressure measurement have been known for centuries. U-tube manometers were among the first pressure indicators. Originally, these tubes were made of glass, and scales were added to them as needed. But manometers are large, cumbersome, and not well suited for integration into automatic control loops. Therefore, manometers today are usually found in the laboratory or used as local indicators. Depending on the reference pressure used, they could indicate absolute, gauge, and differential pressure. Differential pressure transducers often are used in flow measurement where they are designed for differential pressure measurement across a venturi, orifice, or other type of primary element. The detected pressure differential is related to the flow velocity and thereby indicates volumetric flow rate. Many features of modern pressure transmitters have come from the differential pressure transducer. In fact, one might consider the differential pressure transmitter to be the model for all pressure transducers.

A pressure transmitter is a standardized pressure measurement package consisting of three basic components: a pressure transducer, its power supply, and a signal conditioner / re-transmitter that converts the transducer signal into a standardized output. The sensing element reacts to the force or pressure of the process, creating an output signal that can be interpreted by a read-out device or a data-collection device. The sensing element, therefore, is the heart of the transducer or load cell.

The first pressure gauges used flexible elements as sensors. As pressure changed, the flexible element moved, and this motion was used to rotate a pointer in front of a dial. In these mechanical pressure sensors, a Bourdon tube, a diaphragm, or a bellows element detected the process pressure and caused a corresponding movement.

Bourdon gauge

The most common pressure meter in practical use is the Bourdon gauge. Ubiquitous, cheap and robust, it is so easy to use that it attracts curiously little interest – so much so that when pipeliners say *"metering"*, they invariably refer to flow metering rather than to pressure metering. Its principle of operation relies on the elastic deformation of a solid: a curved tube with an elliptical cross-section. The device is constructed such that one end of the tube is closed off and free to move, while at its other end it is fixed in position and open to the fluid whose pressure we are measuring. When the tube's pressure exceeds the internal pressure, its cross-section becomes more circular. This makes the curved tube uncurl, moving its free end thus indicating the pressure.

Figure 220 – The mechanical inner workings of a Bourdon gauge, and its external appearance.

After the 1920s, automatic control systems evolved, and by the 1950s pressure transmitters and centralized control rooms were commonplace. Instead of connecting the free end of a Bourdon tube (bellows or diaphragm) to a local pointer, pipeliners configured it to convert a process pressure into a transmitted (electrical or pneumatic) signal. At first, the mechanical linkage was connected to a pneumatic pressure transmitter, which usually generated a 3-15 psig output signal for transmission over distances of several hundred feet, or even farther with booster repeaters. Later, as solid-state electronics matured and there arose a need for signal transmission over longer distances, pressure transmitters became electronic. Early designs generated various DC *voltage* outputs (10-50 mV; 1-5 V; 0-100 mV) and eventually standardized as a DC *current* output signal (4-20 mA).

Types of pressure transducer

The search for improved sensors for pressure measurements over the years has given us several types of transducers based on a number of different physical principles, including the *strain gauge*, *capacitance*, *potentiometric* and *piezoelectric* principles.

A strain gauge-type pressure transducer uses a strain gauge as a component to measure the deflection of an elastic diaphragm or Bourdon tube. These strain gauge pressure transducers are widely used, and depending on their application, they are capable of measuring gauge pressure, differential pressure or absolute pressure.

A *capacitance*-type pressure sensor uses the change in capacitance resulting from the movement of a diaphragm element. The diaphragm is usually made of metal or metal-coated quartz and is exposed to the process pressure on one side and to the reference pressure on the other. A capacitive pressure gauge can be configured to measure gauge pressure, differential pressure or absolute pressure.

A *potentiometric*-type pressure sensor is a simple method to obtain an electronic output from a mechanical pressure gauge. The device consists of a precision potentiometer, whose wiper arm is mechanically linked to a Bourdon or bellows element. The movement of the wiper arm across the potentiometer converts the mechanically detected sensor deflection into a resistance measurement, using a Wheatstone bridge circuit. The resonant-wire pressure transducer was introduced in the late 1970s. As a digital counter circuit can detect a change in the resonant frequency precisely, it can be used for low differential pressure applications as well as to detect absolute and gauge pressures.

A *piezoelectric* pressure sensor exploits the fact that when a force is applied to a piezoelectric material, an electric charge is generated across the faces of the crystal.

This can be measured as a voltage proportional to the pressure. The output signal will gradually drop to zero, even in the presence of constant pressure, but the device is sensitive to *dynamic* changes in pressure across a wide range of frequencies and pressures. This means that while a piezoelectric pressure sensor is unsuitable for measuring static pressure, by using Quartz crystals it measures dynamic pressure in a stable and repeatable fashion. The robustness, high frequency and rapid response of piezoelectric pressure sensors make them suitable for a wide range of industrial and aerospace applications where they are exposed to high temperatures and pressures.

Flow meters

Measuring flow

On a piping and instrumentation diagram (P&ID) a flow meter is typically labelled as a *flow transmitter* (FT), or as a *flow indicator and transmitter* (FIT).

On a gas pipeline, custody transfer demands strict standards of accuracy that render many equations of state unsuitable for metering gas flow rates, despite performing adequately for other purposes. We can only easily measure the actual volumetric flow rate being delivered (unless we use a relatively expensive coriolis meter), but what is billed is proportional to the number of molecules delivered (and in some cases their quality too). So we need to use an equation of state along with a pressure and temperature measurement to calculate billed flow from measured flow. Errors of 1.0% in the gas flow metering may amount to millions in losses to a seller or a purchaser.

On a liquid pipeline, fiscal metering is so important that it forms the basis forms the basis of Atmos SIM's real-time incompressible model (which, as we discussed, is suited for many applications where SIM's full real-time transient model would be overkill), taking precedence over any flow rate implied by pressure meters. The business of trading crude and products is complex. The API publishes a Manual of Petroleum Measurement Standards (MPMS), which specifies matter such as *flow conditioning* – placing straight pipe before and / or after a flow meter depending on its type to ensure good accuracy. Broadly speaking, liquid flow meters can be categorized as being either *direct volume* meters that separate the fluid flow stream into distinct segments, or *inference meters* that deduce flow rate by way of measuring other properties. [58]

In both gas and liquid pipelines, the stringent requirements make flow meters eye-wateringly expensive. Flow metering is therefore uncommon on a number of gas and liquid pipelines, except where absolutely necessary. A certain maximum and minimum measurable flow rate is always specified by the manufacturer of any given flow meter.

Flow meter error

Usually, a raw measurement is corrected to correspond to standard pressure and temperature, so a transmitter reports standard flow rate rather than actual flow rate.

At very low flow rate, the error is constant. This means the fractional error increases as flow declines. This means that pipelines often have a minimum flow rate for deliveries to ensure that the amount delivered is metered with sufficient accuracy. In the regulated aviation industry, API 1540 (2004) stipulates that flow meters *"should perform in service with a maximum tolerance of 0.2 percent at flow rates between 20% and 100% of rated flow unless [other] regulations are more stringent".* [71] A flow meter's accuracy is inherently rather poor at low flow rates. The way we capture this is to say a flow meter is subject to an *absolute error* as well as a *relative error*. Overall error is the larger of the two, so at low flow rates the absolute error becomes relatively significant.

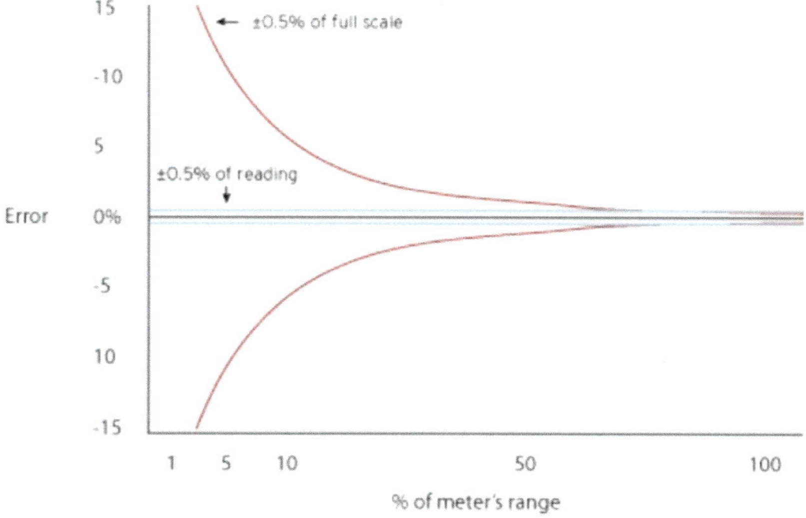

Figure 221 – A flow meter is subject to both absolute error (red) and relative error (gray)
Flow Expert / Wikimedia Commons / CC BY-SA 3.0

Co-located flow meters are rare on a pipeline. Devices (like valves) and equipment (like pumps) conserve flow. So there is no reason to place flow meters on both sides of them in the manner that we place pressure meters on both sides of a valve or a pump.

The inherent accuracy of a meter will deteriorate over time, an effect captured by what is known as a meter factor. Because this error is of a predictable and consistent nature, *"meter provers"* can be used to re-calibrate a flow meter by flowing a known volume through it. Indeed, any meter must be regularly proved on a routine basis to maintain its accuracy. Whilst being proved, a meter quality should indicate *"bad"* or *"unavailable"* as its reading varies over the scale. Hopefully the SCADA system is informed of this;

sometimes this human process involves a phone call, which might not always happen, leaving a *"bad"* meter marked as good quality within the SCADA system.

Positive displacement flow meters

Positive displacement (PD) meters work like a turnstile to directly measure actual flow rate (Q) by counting the chambers as they rotate. They measure liquids accurately but can't measure gas flow rates – as a gas passes through, the pressure drop causes the gas to decompress, reducing in its density, thereby causing the meter to spin too fast. For the same reason, positive displacement meters on oil or hydrocarbon product lines must be operated at enough pressure to avoid the fluid falling below its vapor pressure at any point within the meter. In some cases, this requires a back-pressure valve. With these caveats, *"PD meters"* are standard for custody transfer on oil pipelines. [58]

Turbine flow meters

A turbine meter is based on the flow velocity of a fluid turning a rotor (a *"turbine wheel"*) at a proportional rotational speed. These are inference meters which were introduced into the pipelining industry in the 1960's. They are intolerant of debris, so are widely used for refined products and gas but not crude oil. [58]

Coriolis flow meters

Coriolis meters are popular on oil pipelines and are also used for natural gas. The coriolis principle measures mass flow rate using *"an oscillating loop"*. This is a tube which is twisted by the energy of a fixed vibration. This deflection indicates the amount of fluid passing through because the vibration depends on the momentum and therefore the mass flow rate (Q^{mass}). This is converted to actual volumetric flow rate (Q) by dividing by density (ρ) in a flow computer, or to standard volumetric flow rate (Q^{std}) by dividing by standard density (ρ^{std}). Coriolis meters are the only type of meter described in widespread use that directly measures the mass flow rate rather than the actual volumetric flow rate, and thus don't require a flow computer implementing an equation of state to determine the actual billable amount of natural gas delivered.

Orifice plate flow meters

The orifice plate flow meter is commonly used in clean liquid, gas, and steam services. The fluid passing through the orifice plate creates a pressure drop which varies with the flow rate. Orifice plate calculations used today still differ from one another, although various organizations are working to adopt a universally accepted orifice flow equation. Orifice plate sizing programs usually allow the user to select the flow equation desired.

Ultrasonic flow meters

Ultrasonic flow meters use high-frequency waves to sense the flow velocity of a liquid or gas, which a flow computer then uses to calculate volumetric flow rate. It has no moving parts, which practically eliminates pressure loss – a drawback common to other types of flow meter. Both inline and clamp-on ultrasonic flow meters are available. For custody transfer and fiscal oil and gas measurement, in-line ultrasonic flow meters are typically used. They come in a range of different options, from single-path to 8-path, the number of paths representing the number of times a sound wave is transmitted through the fluid.

There are two types of clamp-on (non-intrusive) ultrasonic flow meters: *transit-time* also known as *time-of-flight*, and *Doppler*. The transit-time flow meter works by looking at difference in transit time for a sound wave transmitted across the interior of the pipe at an angle so that it's moving with the flow, and for one moving against the flow. Doppler flow meters look at the Doppler shift of sound waves reflected off particles or bubbles in the fluid, and so they won't work for gas or for liquids that don't contain small bubbles.

A transit-time flow meter uses two transducers, working at a frequency of 1 to 2 MHz, or higher for smaller pipe installations, hence the need for several different transducer sets. A Doppler flow meter uses a single transducer operating between 0.5 and 1 MHz.

Flow computers and equations of state

When a car is filled up at a filling station, the fuel pump measures actual volume without correcting for temperature. A hydrocarbon expands when warm, so the same mass of fuel costs less on a cold day! This discrepancy is not substantial at such a small scale. On a pipeline handling millions in valuable commodities, the sums involved motivate pipeliners to correct for these operating conditions.

For this reason, custody transfer is – rightly – based on standard volumetric flow rate (Q^{std}), or mass flow rate (Q^{mass}). If the chosen meter is measuring the actual volumetric flow rate (Q), then a *flow computer* is required to be able to deduce either of these. This flow computer incorporates an equation of state which it uses to find the density at the specified temperature, and this principle is expressed via calculating a supercompressibility (Z, often referred to in this context as *"Fpv"*).

Historically, quite unsuitable equations of state (e.g. AGA NX-19 for gases) were built into the meter itself. Today's flow computers implement much more accurate equations such as AGA-8, which was first presented in 1985 and revised in 1992. This is all done before flow signals reach a pipeline simulator.

Valve positions

It is easy to dismiss a valve, especially a block valve, as a straightforward piece of equipment. It is not. The design, selection, and operation of a valve employs a fair number of intelligent professionals. Every pipeline simulator needs to know what a valve is doing. This, in practice, is more than a data tag reporting its position. Is the valve in transit? Is the reported position uncertain?

This is a good place for an aside on report-by-operator valves. Because often, operators don't! Relying on an operator walking to a computer to tell us what a valve is doing a few minutes later is not a hallmark of accuracy. If such a valve is important enough in a pipeline to be monitored using an online model, we usually need some automatic way to infer its status from other measurements. Although it is far from ideal for a model to have to deduce a valve status at all, Atmos SIM has been used to deduce whether a block valve is open or shut.

Figure 222 – Manually operated valves might not transmit their own position

Status and speed of compressors

The challenge of poorly instrumented equipment is rather disappointingly not restricted to the realm of valves. We have encountered compressor status tags that were so often wrong that operators were routinely unsure which of their many compressors were actually running, and what operating speed, and thus how much fuel they were consuming. Atmos SIM provided them with that valuable information to handle temporarily unreliable compressor status tags. It deduced which compressors were on at a given instance based on their maps, by analyzing operating conditions including the suction and discharge pressure. Of course, a feature to perform such a circuitous deduction is strictly a fall-back workaround of last resort. The pipeliners we worked with in that instance fortunately did get their compressor status tags fixed and working.

That particular situation was doubly unusual because an online model is not usually expected by users to do much with compressor statuses. They don't matter for linepack management, leak detection, or estimated times of arrival. We mention it because some pipeliners are interested in the fuel consumption, which may not itself be subject to an (expensive) flow meter.

What is perhaps a more common situation is that a pipeliner cannot always assume that a compressor *"status on"* tag in their SCADA system actually corresponds to when a compressor is on. It takes perhaps 15 minutes for a compressor to start up, and during that period its status tag might not be telling us that it's on. The system might raise alerts to say that a compressor has started, but not provide a live trend to show a particular compressor is operating.

Liquid level measurement

Several options are possible to determine how much liquid is in a tank, such as radar level sensors or float gauges, and specialist software relates liquid volume (V) to tank level with reference to what people in the industry call *strapping tables*. Measurements must compensate for temperature to indicate inventory as a standard volume (V_{std}) and compensate for density to indicate it as a mass (M).

Other measurements

For operational purposes, it is useful to have instruments measuring many parameters, including pressure (P), actual volumetric flow rate (Q) and temperature (T), as well as:

- *Transmission amplitude* for pig detection in gas or liquid pipelines
- *Speed of sound* (c) for batch detection in multi-product pipelines

The above can all be measured and transmitted by Atmos Eclipse. Further measured quantities on a pipeline might include:

- *Density* (ρ) and perhaps also *standard density* (ρ_{std}) in liquid pipelines – and gas pipelines, too, may be equipped with densitometers for flow metering
- *Viscosity* (μ) in liquid pipelines, using a viscometer
- *Composition* in gas pipelines, using a gas chromatograph (GC)

Many liquid pipelines lack in-line viscometers. One application of a real-time model is to deduce a fluid's viscosity by calculating it from other measured values or applying blending equations to calculate the overall viscosity of an outgoing mixture based on the viscosities of the incoming streams that form it. A liquid pipeline carrying hydrocarbons as batches is equipped at strategic locations with a batch detector (BD). This works by any of several principles, one of which is to shine a light through the liquid so as to measure its opacity.

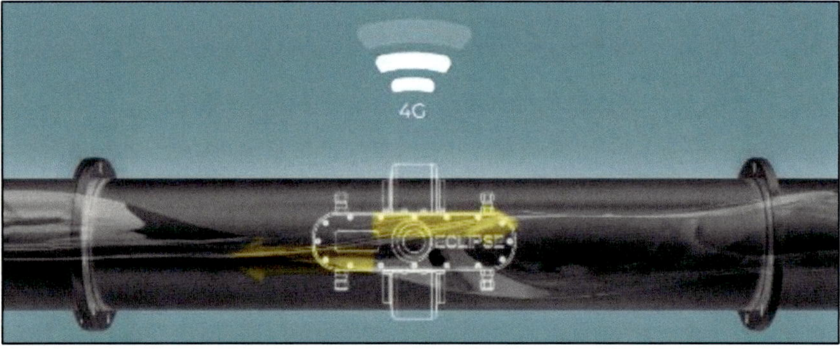

Figure 223 – Atmos Eclipse is a device to measure pipeline pressure, flow rate, temperature, transmission amplitude and speed of sound. It handles bi-directional flow and shut-in.

Flaws in instrument data

Errors can creep into site data from many sources. State estimators, as we will see shortly, can remedy normally distributed noise and quantization error. These operate by assuming that *instrument noise* to be normally distributed and uncorrelated from step to step, even though it probably isn't. Instrument round-off error is meant to represent the limited precision of each instrument available through a SCADA system. *Quantization error* arises due to a limited number of bits being available in any analogue-to-digital converter. *Instrument drift* is compensated by automatic tuning parameters learnt in real-time, to the extent that it can be handled in software at all. Tuning out this sort of error is only possible if a pipeline has a lot of extra instruments; most do. Even then, it is premised on a few assumptions. If, say, every flow meter is reading 1% higher than reality, there is fundamentally no way for any online model to correct this error, no matter how well it is configured and tuned.

Simulating flawed instrument data

Artificially noisy measurements in simulated runs are useful for verifying that an online simulator will perform well in the field. They also make for a more realistic experience when we come to set up trainers. Normal-distributed noise and other effects can be superimposed onto the model results from an offline constrained simulation.

11.5 – REAL-TIME INCOMPRESSIBLE MODELS

Why would anyone ignore any of the physics?

Simulating liquid pipelines online can often be done using a straightforward *incompressible mode.* This approach relies entirely on a chosen fiscal flow meter, following it exactly at all times. It is convenient for many batch tracking tasks, and has been deployed with success for a long time. It is also adapted to handle compressible liquid batches and volatile batches liable to slack-line. Instead of calculating density at operating conditions based on an equation of state, density is taken to be completely constant within each batch. This sets the transient density term in the mass equation to zero, so the density (ρ) of that moving point never changes over time (t):

$$\frac{\partial \rho}{\partial t} = 0$$

A scenario with no hydraulic transients is equivalent to a fluid with infinite isothermal bulk modulus (B_T) and thus infinite speed of sound (c). All hydraulic transients in incompressible mode immediately vanish to a steady state, so this approach is not always taken literally in a physical sense. It is only suitable for batch tracking, pig tracking and simple hydraulic profiles. For example, it simply models pumps by looking at the pressure meters at their suction and discharge sides. For other liquid pipeline applications, or for gas pipelines, we use a transient physical model. An incompressible approach doesn't *ignore* physics altogether, but focuses on the special case of fluid physics that applies to these almost-incompressible liquids. It helps liquid pipeliners answer many important questions, including:

- What is the inventory and line-up?
- Is the pipeline fluid nearing a slack condition anywhere?
- Is it near maximum allowable operating pressure (MAOP) anywhere?

The third of these questions – i.e. whether conditions approach an operating limit – can only be addressed using an incompressible model by providing an extra conservative safety margin, because by definition, such a model misses any surges that might take place during the pipeline's operation. Compared to a fully-physical transient approach to simulating a pipeline live on site, the choice of an incompressible model can to an untrained eye seem rather basic. However, these sweeping assumptions don't mean that the incompressible model isn't subtle and sophisticated. Incompressible models lend themselves well to many liquid pipeline applications, particularly those where commercial aspects of the operation rely firmly on well calibrated fiscal meters.

Shipments of batches have long been tracked on the basis of *"what goes in immediately comes out of the other end"*, and until quite recently teams performed this task using methods akin to *railroad charts*. The challenge that arose is that real pipelines – with terrain, weather, fluid properties and so forth – are not really like railroads. An incompressible model takes the core principle and then runs with it to meet that challenge by doing a whole lot more.

A simple pressure profile for incompressible liquids

An incompressible model computes a special pressure profile, capturing the effects of densities, elevations, and diameters. It makes two big assumptions: that the pipeline is always at a steady state everywhere, and that the friction factor is constant throughout each section bounded by pressure meters. On this basis, the *"incompressible"* approach essentially performs a sophisticated interpolation between pressure meters.

It applies a simplified version of the momentum equation which has been adapted for incompressible liquids by dropping a term that is small for such fluids in steady state:

$$\frac{\partial P}{\partial x} + \rho \cdot g \cdot \sin[\theta] + \frac{\rho \cdot f \cdot v \cdot |v|}{2 \cdot D} = 0$$

We elucidated each of the terms in this equation in the chapter on conservation laws. Using this equation, we assume that in addition to the density (ρ) remaining constant within each batch, the friction factor (f) is also constant throughout the section. We solve for friction factor, deducing a rudimentary indicative pressure profile everywhere. This simple approach is typically accurate to within around 50 psi (3.5 bar) for liquid products, and 100 psi (7 bar) for crudes and light fluids. Deviations from this accuracy are observed to be short-lived in *"normal operation"*, so the approach is suitable so long as there are no substantial transients ongoing.

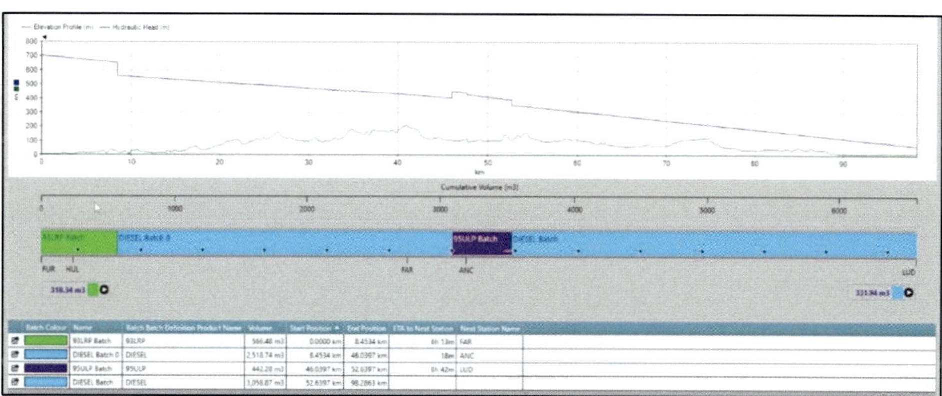

Figure 224 – The head profile is calculated when simulating via incompressible-mode, capturing the effect of batch densities as they move along the route of the pipeline.

In this example, there are no mid-line pumping stations: the jumps in head along the hydraulic profile reflect the difference in density of inline batches.

Batch tracking of liquids: a domain unto itself

Tracking the contents of a liquid line is a deceptively complicated task even on a purely logistical level. It involves a kind of *spatial awareness* around how a line is operating and how it's planned to operate next. A lot of information is considered by those responsible for decision making or weighing up options.

	93LRP Batch	93LRP	Injecting
	93LRP Batch 0	93LRP	Delivering
	95ULP Batch	95ULP	Inline Unscheduled
	AVTUR Batch	AVTUR	Delivering
	AVTUR Batch 0	AVTUR	Unscheduled
	DIESEL Batch	DIESEL	Delivering
	DIESEL Batch 0	DIESEL	Inline Unscheduled
	DIESEL Batch 1	DIESEL	Injecting
	LIGHT CRUDE Batch	LIGHT CRUDE	Unscheduled
	LIGHT CRUDE Batch 0	LIGHT CRUDE	Unscheduled
	LIGHT CRUDE Batch 1	LIGHT CRUDE	Unscheduled
	LIGHT CRUDE Batch 2	LIGHT CRUDE	Unscheduled

Figure 225 – Batches listed along with their status in the Atmos HMI connected to Atmos SIM.

When a batch enters a pipeline, its *"custody"* is transferred to the pipeliner, but often its ownership is retained by a different party. This means that a pipeliner transporting multiple products needs to always know the location of the head and tail of every batch for commercial reasons. A commercial team can use an accurate visual display of precise batch locations to optimize sales revenues. There are, of course, many technical considerations too. An operator who is well-informed on where batch interfaces are in real-time can prepare to swing the valve at the exact moment a batch arrives at a station, delivering *"product"* to the right storage tank or to an end-customer with minimal contamination.

Batch Colour	Name	Volume	Start Position ▼
	95ULP Batch	125.59 m3	42.3122 km
	DIESEL Batch 0	2,518.74 m3	4.7259 km
	93LRP Batch	316.69 m3	0.0000 km

Figure 226 – Batch colors, names, volumes and start positions in the Atmos HMI.

It is relatively easy to track multiple batches in a fixed-diameter pipeline across flat terrain. However, when there are drastic changes in elevation and many different internal diameters, the task becomes far more complex. Multiple batches may be present with very different properties, and it is challenging to reconcile flow rates with inventory to perform physical composition tracking, even with a well-tuned and well-fed thermal model.

It is often best to dispense with any physical model for throughput altogether. Devices called inline batch detectors (BD's) are installed not just at origination and delivery stations but also at certain intermediate stations. These send signals that can be used to automatically adjust the positions of batch heads and tails. The widespread installation of these devices in modern batched product pipelines means that pipeline simulators can perform batch tracking very well without implementing a full physical model with all the thermal properties and so forth. This meets the needs of the teams operating this kind of pipeline in that the simulator follows a fiscal meter exactly via an incompressible-model approach. The focus in such an application is for batches to be manually adjusted at will, which is occasionally needed, and for routine reporting tasks to be automated.

Batch Detector Type	Inline	Duration Threshold (s)	0
Next Name Tag	<none>	Arrival Distance Threshold (m)	0
Scheduled Volume Signal (m3)	0	Upstream Threshold (m3)	0
Scheduled Flow Signal (m3/s)	0	Delivery Threshold (m3)	0
Fixed Anomaly Signal	0	New Process On Delivery Threshold	☐
Growing Anomaly Signal	0	Batch Colouring Type	Product colour
Interface Signal	0	Colour Cycle	<none>
Product Signal	0	Future Schedule Process Batch Name	<none>
Flow Type	StandardVolume...	Future Schedule Process Volume (m3)	100
Totaliser (m3)	0	Future Schedule Process Flow (Sm3/h)	359999.9999999997
Totaliser Maximum Value (m3)	1000000	Future Schedule Process Product	<none>
Totaliser Drop Limit (m3)	100	Future Schedule Process Start Time	<none>
Totaliser Jump Limit (m3)	100	Trigger Schedule	<none>
Volume Alarm Threshold (m3)	0	Fungible Session Trigger	<none>

Figure 227 – An Atmos SIM batch detector (BD), typically configured by a project engineer. It has many properties because such devices are not perfectly reliable. We specify how the model behaves when a batch detector misses a batch transition, or fails to register one that we expect.

Typically, a liquid product batch originates from a tank farm, and its destination is another tank farm. While tanks are their own specialized domain, dealt with separately, a pipeliner needs to consider *blending* of, say, different grades of gasoline in order to satisfy a quality specification. Certain liquid batches might be fungible with some but not all others, providing a degree of interchangeability. It is crucial at all times to have a method of origin tracking: identifying the origin of a given batch. A further complication is the formation of interfaces as adjacent batches intermingle during a journey, even those separated by spherical pigs.

Operators in the field might make cuts at various mid-line delivery points based on opacity or density. As the *"batch train"* is sent, one after another, there is mixing and contamination. When a pure batch arrives at its delivery location, we provide the operator with a good arrival time estimate so they prepare to get everything ready well in advance. They switch that valve to send the incoming product to its corresponding tank – jet fuel contaminating crude is acceptable, whereas crude contaminating jet fuel is not! Many factors impact arrival time by a few minutes, and it adds up over the length of the line.

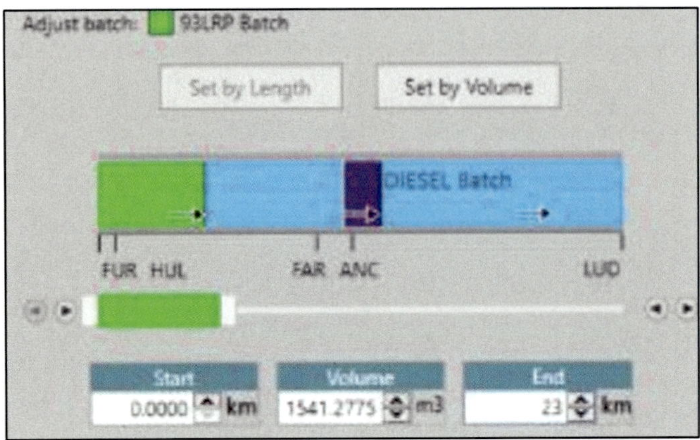

Figure 228 – A batch adjustment dialogue in the Atmos HMI, connected to Atmos SIM.

Most liquid product pipelines don't separate most of their batches with spheres, simply allowing the head of one batch to partially mix with the tail of the other. This means that a decision must be made about how to cut a batch. The industry refers to the mixed volume as an *"interface"* while it is in the pipeline. If a cut is made before and after this mixture and it is transferred to its own storage tank, it is referred to as *"transmix"*. Other options are to cut only at the start of the mixture, at the end of it, or part-way into it. Over a thousand-mile pipeline, a batch interface might grow to roughly a mile between the 1% and 99% mix points. If the flow is ever laminar or only partially turbulent, then much more mixing is likely to occur, and spheres are needed. Such a situation might occur during shut-down, especially if a heavier batch is uphill of a lighter batch on a slope, a situation in which the batches might form layers in the pipe. Then when the pipeline starts up again these layers form a long interface.

There is an entire domain of reporting and logic involved in batch tracking, via a human-machine interface (HMI). As the pipeline is operated, it is imperative that an audit trail is created and retained. Custody reports are produced, describing where batches of fluid are, and when they are expected to arrive. Automating how these reports are generated saves significant manual rework, but there seems to be no single shared standard in use by all liquid pipeliners.

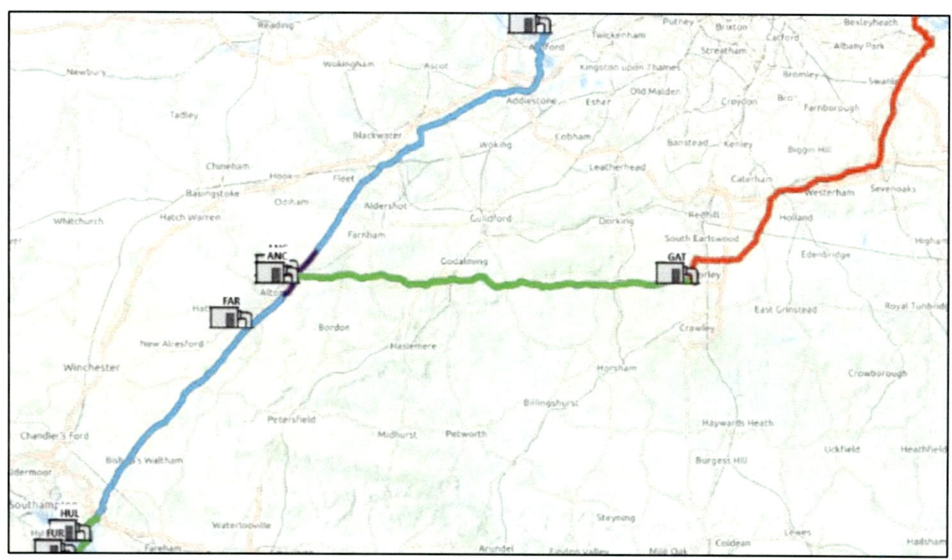

Figure 229 – Batches displayed on an Atmos HMI map view connected to Atmos SIM.

Case study: blended batches

A pipeliner is transporting different grades of crude oil. In addition to standard requirements of providing volumes, start and end interface positions and estimated arrival times (ETA) at each station, they need to track how much they blend together along the line, and to recalculate viscosity on that basis.

The blended batches are themselves tracked, their make-up described by the ratios of their *"component batches"*. Batch tracking has been particularly tricky historically when dealing with long pipelines, especially those across rugged terrain. Manually tracking all these variables is time-consuming and complex. The Atmos SIM batch tracker, integrated within a pipeline simulator, proved to be a high-performing system, producing excellent results.

The ability of a pipeline simulator to accurately calculate these parameters automatically during changing operating conditions lets a batch tracker predict the arrival of a batch within a narrow timeframe, so operators are ready to receive a batch and route it to the right storage tank. This avoids unnecessary tank mixing and wastage.

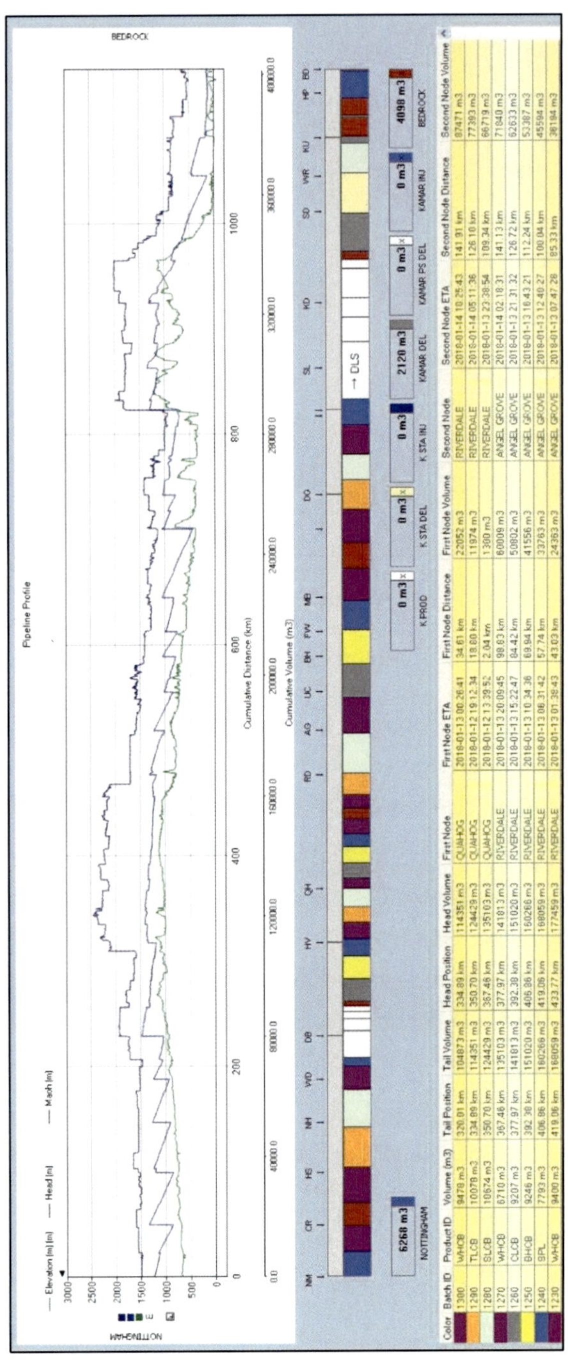

Figure 230 – A pipeline's customized human-machine interface displaying simulated results from Atmos SIM in a real-time application. At the top is a live profile of maximum allowable operating head (MAOH), dynamic head and elevation. In the middle is a batch train showing the head and tail positions for each batch. At the bottom is the corresponding batch list with detailed information on the status and progress of each batch.

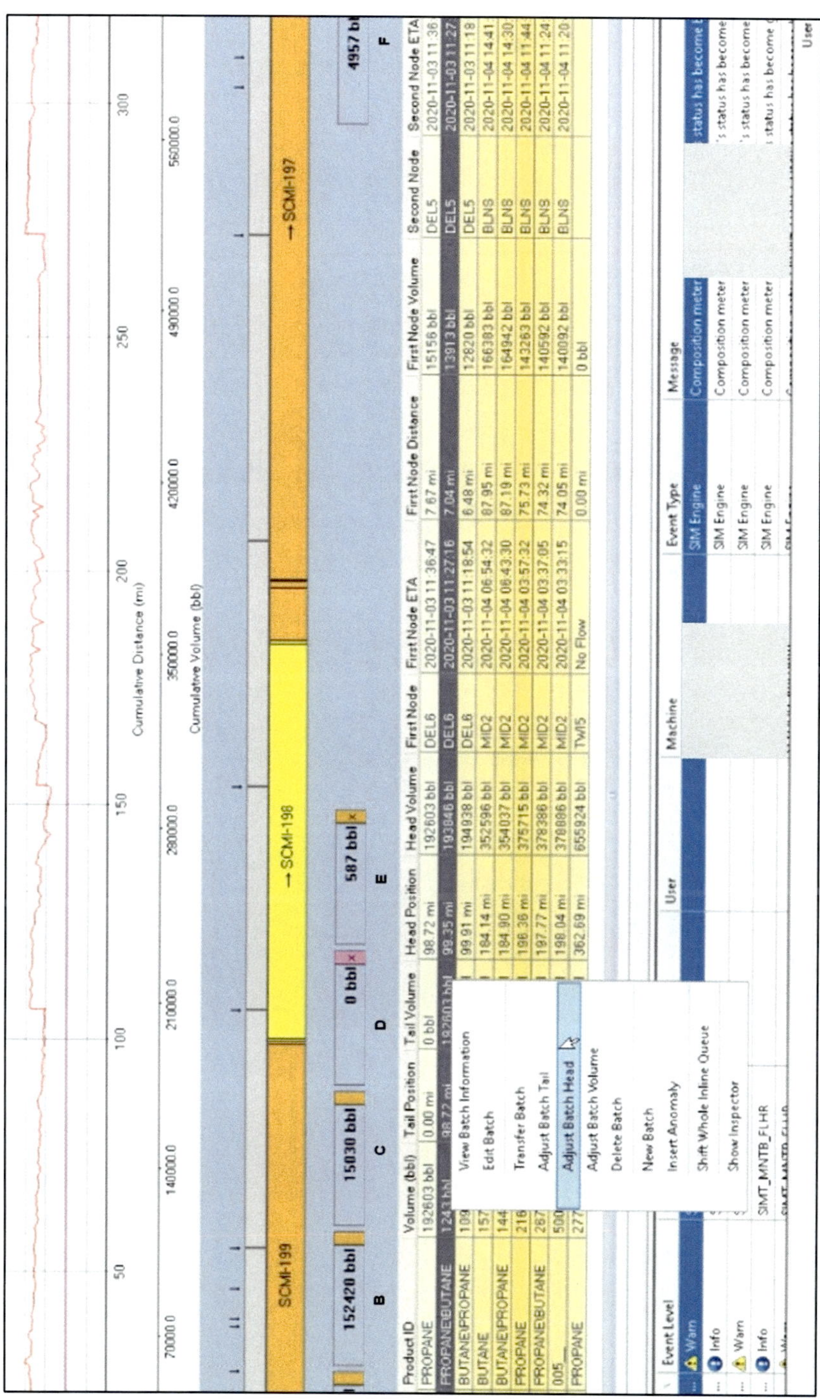

Figure 231 – Another view (known as the mimic in the HMI) showing an evolving pressure profile. There is a context menu on right-clicking a batch which allows batch transfer, or batch tail adjustment, batch head adjustment, batch volume adjustment, batch deletion, adding a new batch, inserting an anomaly, or shifting the whole in-line queue (ILQ).

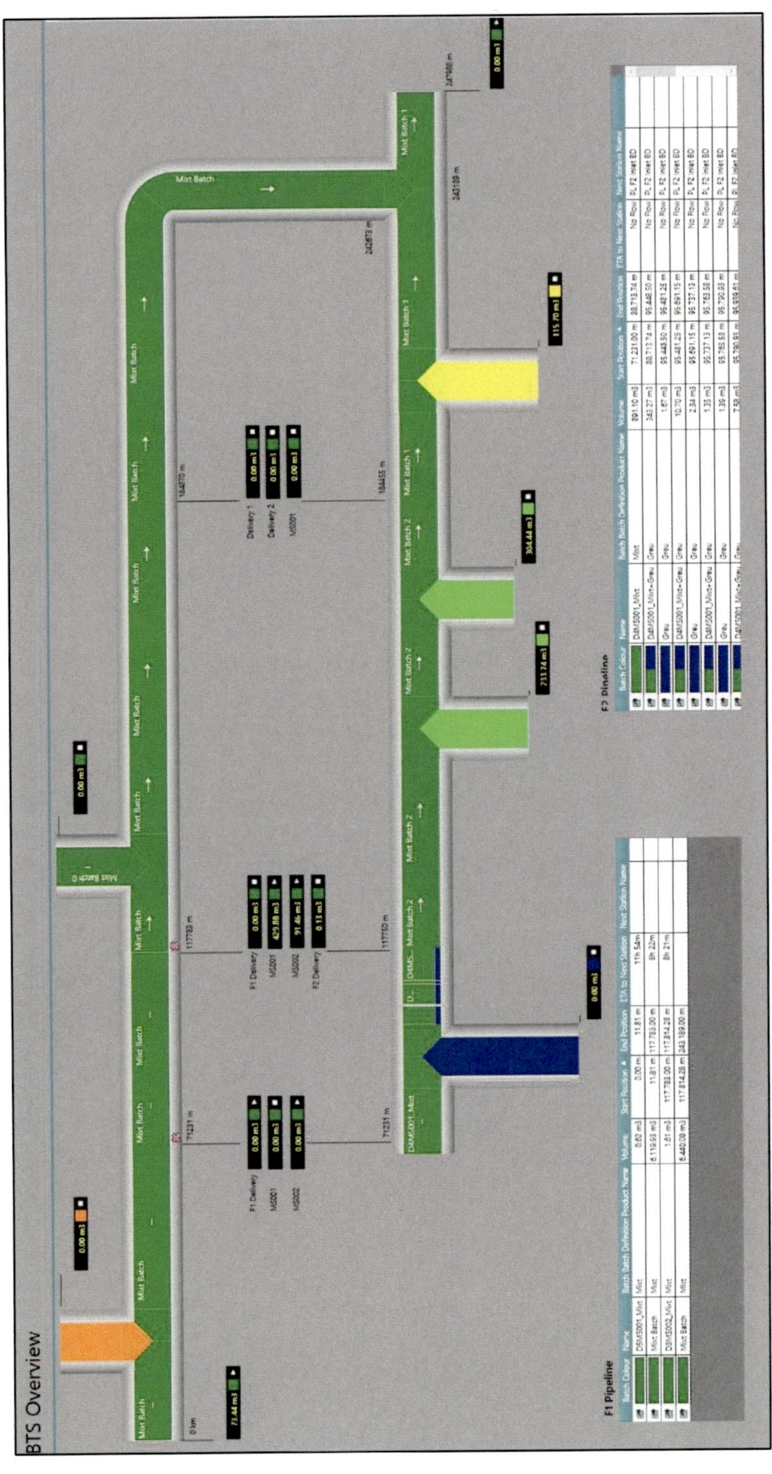

Figure 232 – A batch tracking overview screen, showing batches and blends being tracked, and their properties updated in real-time including estimated time of arrival to the next station.

Case study: an empirical model for slack-line flow [94]

The Trans Mountain pipeline in Canada carries crude oil and refined liquid products from Edmonton, Alberta westwards to Vancouver on the coast of British Columbia, traversing the rugged Rocky Mountains. Within some of its sections, the pipelined fluids exhibit significant compressibility. More importantly, a section of the pipeline is routinely operated under slack-line conditions. Atmos SIM's batch tracker was adapted with a specially designed *"drain-and-fill"* module to make accommodations for these physical phenomena while retaining all of the advantages of a fiscal incompressible model. [94]

Figure 233 – Liquid storage tanks at the western terminus of the Trans Mountain pipeline, significant infrastructure that has been critical to exporting Canadian oil to the Pacific. Codex / Wikimedia Commons / CC BY-SA 4.0

Packing and unpacking – also called filling and draining – is possible for a section containing a compressible fluid (that can be squeezed) or a vapor region (that can be collapsed). Delivering more than what is injected in this way, or vice-versa, affects the location of a batch head and a batch tail. Column separation (slack) occurs when the operating pressures and temperatures conspire to drop below a liquid's phase transition curve at any point within a pipeline. The temperature is especially liable to drop during a shut-in condition. In this situation, some of the liquid turns to vapor. These vapor pockets change the volume of a batch in a pipeline, moving the physical position of its head and tail interfaces, causing the estimated times of arrival to be less accurate.

In this live environment, Atmos SIM was set up to track the volume entering and leaving each segment. This empirical approach bypasses the physics entirely, using the volume contained within each segment to estimate arrival times. This is found to be accurate within 15 minutes, after a batch travelled over 1 150 km in a pipeline with drastic elevation changes along its route. This method of batch tracking has been proven to be highly accurate and reliable without resorting to fully physical modelling of the underlying liquid hydraulics and the complexity that would entail for the commercial aspects of a liquid pipeline. Tailoring the incompressible model in this way has allowed the controllers to operate with significant slack-line as well as draining and filling operations. [94]

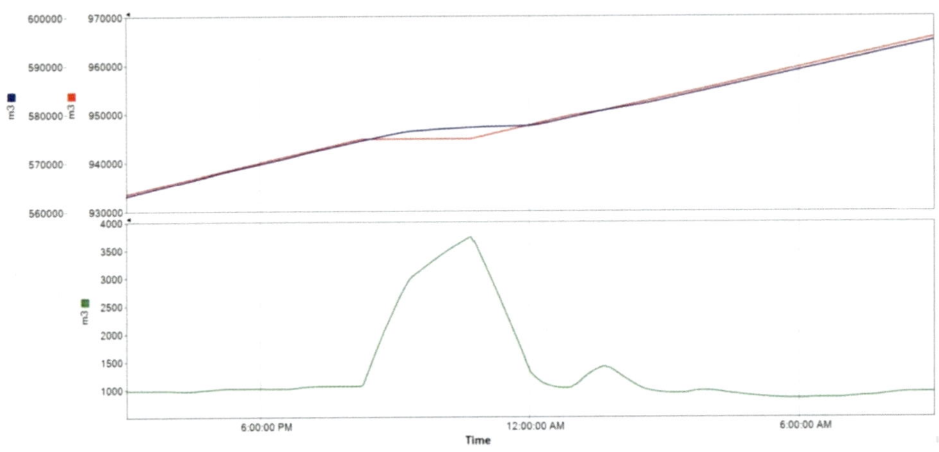

Figure 234 – The empirical approach known as a drain and fill module has proven successful as a feature of the Atmos SIM incompressible model that many batch pipeliners prefer to use. The accumulated inflow and outflow is totalized as liquid volume into and out of a region (top) and that is used to deduce the *"empty volume"* (bottom) which is in reality occupied by vapor.

In this way an incompressible model, as favored by batch pipeline operators, can be modified to handle the packing and unpacking which occurs when a region of their liquid pipeline has become partially or completely occupied by vapor due to its low pressure. This lets it calculate estimated times of arrival. [94]

Base Volume	Current Volume	Empty Volume	Unit	Fluid Percentage
306820.00	309064.42	17.32	m3	100.0
71615.21	71744.01	1342.99	m3	98.1
18229.12	18217.00	-18.23	m3	100.1

Figure 235 – *"Empty volume"* is reported for each region and can be adjusted by operators

Pig tracking in a liquid pipeline

It is desirable that a pig along a liquid pipeline should be tracked in tandem with the batches either side of it. In a typical multi-product pipeline, this is achieved within the same real-time incompressible model live on site. Pigging is a routine operation, and simulators are key for preparing to receive a pig.

Duration Threshold (s)	0
Volume Threshold (m3)	0
Distance Threshold (m)	0
Signal	0

Figure 236 – A pig detector in Atmos SIM has properties that tell the simulator how to handle unexpected pig signals or expected pig signals that did not occur.

If the next pig's estimated arrival time anticipates that it is expected to arrive sooner than a *duration threshold*, a *volume threshold* or a *distance threshold,* a warning is raised. Similar thresholds are also available for a pig parker – a facility to receive and hold a pig until it is re-launched – and for a pig receiver.

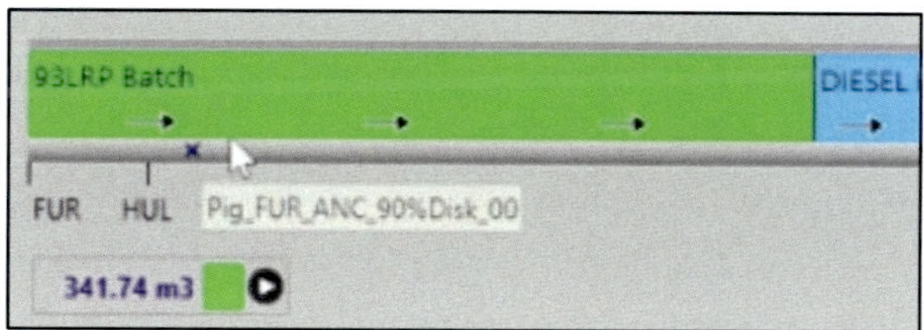

Figure 237 – A pig is tracked by Atmos SIM concurrently with liquid product batches, and its position and parameters are displayed for the user on the same screen within the Atmos HMI.

A pig being tracked on a line transporting various batches ought to agree with estimated batch positions, particularly if the pig is a sphere whose purpose is to separate two batches thus reducing the interface which forms between them. However, whether this will happen depends on the slippage of each type of pig and each section of pipeline.

11.6 – REAL-TIME TRANSIENT MODELS

Deducing a pipeline's physical state as it operates

The real-time transient model (RTTM) approach experienced teething issues in its initial decades, but has today become robust, rapid and ubiquitous through continuous improvement based on its performance in the field. It handles uncertainty in instrument data to deduce the true physical state, easily keeping up with real-time.

Complex software of this nature can only be effective when an organizational culture exists to best exploit it; model-based leak detection requires systems and processes in place for users to take advantage of its capability. Operators invest substantially in ongoing support for the technology, people and resources to maintain a system. The cost to configure, tune and run such simulations is justified for long (>150 km) gas pipelines without extraneous mid-line instrumentation, or for any pipeline that presents particular complexity.

A real-time transient model describes the current state and predicts its future trajectory. This offers several opportunities to add value. It can be used to test leak detection systems before installation, then to perform model-based leak detection in operation. It can be used for training operators, analyzing proposed or possible scenarios, assessing handling of previous situations. It can form the basis for look-aheads, aiding with *what-if* calculations and production planning. These look-aheads can also simulate what a leak or sudden failure of stations or equipment would look like if such an event were to happen, to anticipate risks as they develop in real-time, assess their consequences, and take non-disruptive evasive action.

Real-time transient models are particularly crucial for gas transmission pipelines, where accurately computing the inventory in a long section without pressure or temperature meters informs how the pipeline should be operated. This helps operators cover demand fluctuations by using linepack, so that supply production facilities can operate more constantly for example.

In an online environment, if some *abnormal operating condition* (AOC) causes a pressure surge, readings from instruments at each end of a line might not reflect the absolute maximum pressure along the pipeline, in unmetered locations. This is particularly challenging in a situation where product density or elevation is changing along the line. To use only the end pressures and interpret what's going on within the domain is a challenge. And while a flow meter on land is expensive, any metering at the bottom of the ocean is prohibitively expensive and so generally won't be present except where absolutely required for operations. Models are crucial for these pipelines.

Real-time transient models allow operators to monitor areas where there is no instrumentation, such as long unmetered underwater sections, providing real-time hydraulic profiles, temperature distributions, and composition. Simulated data is made available as virtual instruments displayed on read grids, trends or datapoint outputs. Results can also be viewed throughout the length of a pipeline, as profiles extending along its route.

As it describes the fluid mechanics and other physical phenomena related to pipeline operations, a real-time transient model relies on correct physical models, sound calibration and continual tuning: automatic adjustment of learnt parameters to remain accurate. To maintain model accuracy, this live learning accounts for variations in gas quality, operating conditions and meter errors, adjusting them as it simulates.

Pipeliners rely on their live systems being available at all times, with elaborate measures in place to ensure redundancy, fail-over, and recovery of a real-time transient model for pipeline simulation. The traditional server architecture brings with it a number of challenges for the SCADA system, around such matters as data latency. In recent years, cloud computing has been used to host certain services on the internet, and this has complemented or replaced traditional servers in a range of applications. It makes a whole host of exciting opportunities a reality for modern pipeline simulation.

Figure 238 – Pipeline enterprises traditionally maintain their own server rooms, typically at two separate locations so that if one has an issue such as a power outage there is resilience in the redundant backup being unlikely to face the same issue.

11.7 – State Estimators

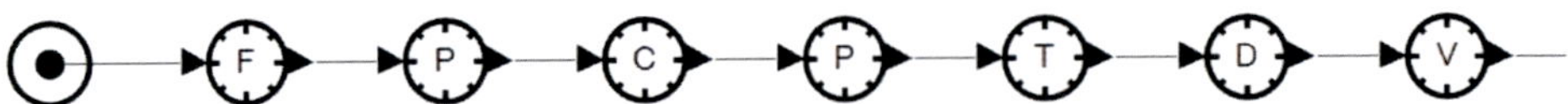

Why do we need a state estimator?

To use data coming in live from meters via a SCADA system, a real-time transient model must somehow handle an overabundance of boundary conditions, some of which are wrong. If multiple flow and pressure meters are located at the same point, what constraint applies there? If an entire network says a pressure or flow somewhere disagrees with a non-co-located meter that claims to be good quality, do we still trust that meter? Rather than ignoring useful data, a state estimator reaches a *consensus* in some way.

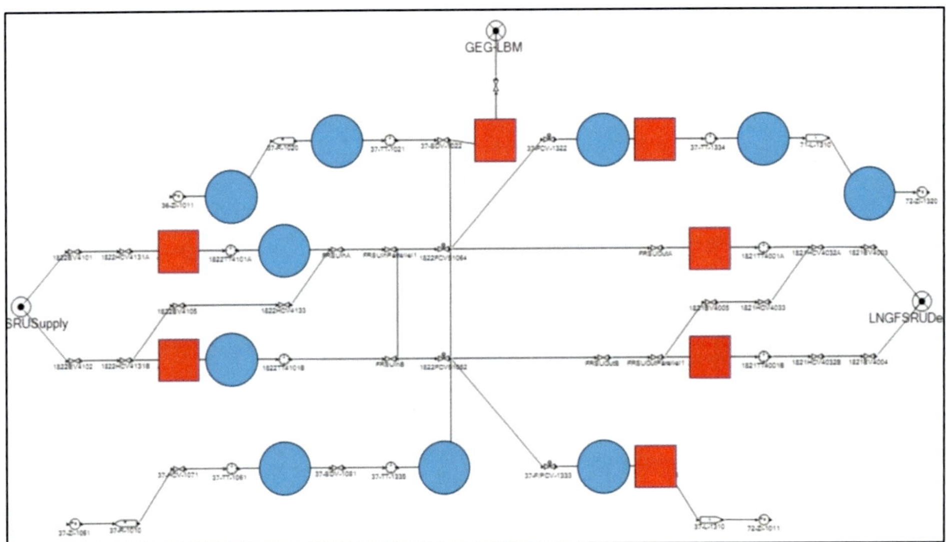

Figure 239 – A typical gas pipeline station. The redundant pressure (blue) and flow (red) meters must be handled by a live simulator, including when complex valve arrangements are switched.

A hydraulic model of a pipeline requires that one boundary condition, either pressure or flow, be specified at every point of entry or exit. Pipelines often have more instrumentation than this. There may be both pressure and flow measurements available at supply and delivery locations, and there may be mid-line metering stations. The extra data from these additional instruments can improve modelled estimates of the line's hydraulic state, for more accurate pressures and flows throughout the system.

There are clear, demonstrable benefits to using a state estimator, rather than just running a model with some of the meters as boundary conditions and ignoring the rest.

By and large, hydraulic models themselves now belong to one of two schools: *method-of-characteristics* (MoC) models, or *implicit finite difference* (FD) models. We discussed both of these earlier in this book. But there is no industry standard technique for state estimation. There are almost as many types of state estimator around as there are pipeline simulation experts! What we can all agree on is that there are several desirable properties of a state estimator. It's not enough to simply say that *"it should produce the best possible estimate of the state"*, as a method that works well in steady state might do something spectacularly wrong during a transient. Or a method that works well for certain types of instrument errors might have trouble with other sorts. Also, it's nice if a state estimator produces some form of output capable of driving *hydraulic tuning* mechanisms, learning corrections to parameters in real-time live on site.

In the following subsections, we take a look at a very basic form of each type of state estimation algorithm in use today (that the authors are aware of). All can be tweaked in practice and made more sophisticated. Here we just touch on the options that are out there, to understand what makes the maximum likelihood state estimator (MLSE) used by Atmos SIM great at compensating for our ever-evolving trust in each instrument, signal and meter.

The pressure-pressure state estimator (PPSE)

If we already have a functioning offline model, an easy way to implement a state estimator is to simply feed it some subset of available measurements at the boundary conditions as its constraints. After each model step completes, we somehow integrate the remaining measurements, and any discrepancies between the model results for different regions, into an estimate of the true state. This has the benefit that we re-use offline model without modifying it; additional measurements are included after the fact.

In a *pressure-pressure state estimator* (PPSE), the model is run with pressure boundary conditions taken from pressure meters at the ends of the pipe. * This is used to calculate results including flow rates at the ends of the pipe. These model flow rates are compared with flow meter readings, or lacking a flow meter, with the calculated flow rate from the other side of the pressure meter – which is usually not the same. If flow meters read higher than the model, this suggests true flow rate is higher than modelled, and vice versa. Given estimates of errors in flow meter readings and errors in modelled flows, we then come up with an overall best guess of the flow in the system. At its most straightforward (as outlined here) this can be done by solving for a *"consensus flow"* (F_i) at every location i that minimizes the weighted sum based on pressure-based flow (F_i^P) and measured flow (F_i'):

$$\sum_i \left(\frac{F_i - F_i^{\mathrm{P}}}{\sigma_i^{\mathrm{P}}}\right)^2 + \sum_i \left(\frac{F_i - F_i'}{\sigma_i^{\mathrm{F}}}\right)^2$$

Minimizing this sum with respect to the vector of consensus flow rate (F_i) is a system of linear equations. These can be solved exactly on each time-step. This consensus flow rate can then be our estimate of the true flow rate. In a network, we apply additional knowledge at this juncture: for instance, at a splitting point, consensus flow rates must add up to zero. This sort of thing can be included in the above optimization problem using the method of Lagrange multipliers; the problem remains exactly solvable.

One could implement this method using flow rate boundary conditions, but as a steady state can't be solved purely on flow rates, that requires special treatment for steady state. Moreover, if flow meters don't add up to a net inflow of zero, this eventually either drains or over-pressures the pipe, unless some pre-processing is applied to force them to sum to zero. Thus we have only seen this approach utilize pressure as its boundary conditions, except occasionally at small midline offtakes that lack a pressure meter.

* The reader might recall that this is exactly what the online LDU real time transient model, used in Atmos SIM for leak detection and location, does. However in Atmos SIM the state estimation is done in a separate model rather than using the LDU results.

Also, a closed valve must be treated as a zero-flow boundary condition if it doesn't have a pressure measurement on either side.

Although the pressure-pressure method is initially easy to implement, it presents potential problems. Overcoming these introduces a lot of complexity.

One problem is small mid-line deliveries. If there's a mid-line delivery that is small relative to mainline flow rate, the hydraulic model is more stable if a flow rate boundary condition is used at that location. Errors in flow rate would cause small errors in pressure, whereas if this boundary were pressure-metered, errors in pressure would cause substantial errors in flow rate. However, if we use flow rates as our boundary conditions at points of this kind, this scheme provides no direct way to use additional pressure data that is available there. It might be possible to incorporate that value into the automatic tuning, but it would be complicated.

A further problem is that if a section of pipeline operates at a relatively low flow rate, errors in pressure meters are significant compared to overall pressure drop across the section. So flow rates calculated by this scheme may turn out to be complete nonsense!

Yet another problem is that this method requires the user to find a weighting of flow meter accuracy relative to the accuracy of pressure-modelled flow rate. This latter term must capture errors arising from pressure meters themselves as well as those errors arising from poorly characterized pipeline parameters. This weighting is not straightforward. In our experience it involves guesswork.

Finally, it is aesthetically pleasing to nobody (developer, modeler, or user) for a model to compute flow rates which don't correspond to its own computed pressure drops. Relying on a system always being perfectly tuned up is naive.

Flow meter drift is handled by a pressure-pressure state estimator. Say there is a bad flow meter. The model sees a section of pipeline packing until the bad flow meter's difference with consensus flow balances the model not wanting to pack the line further, producing a *difference*, as it won't pack forever. Simple hydraulic tuning is possible with this approach, by examining the difference between consensus flow rate and modelled flow rate through each pipe. If consensus flow rate is lower than modelled, pipe efficiency or internal diameter is reduced, and vice versa. This corrects for a steady state error in pipe properties, provided that error stays the same at all the flow rates.

The performance of a pressure-pressure state estimator can possibly be improved by smoothing the pressure measurements with a low-pass filter, if there happens to be a large noise component.

The equal error fraction state estimator (EEFSE)

The *equal error fractions* state estimator (EEFSE) – for a simple pipeline – is easily implemented in an existing model by adding a new type of boundary condition. Unlike the pressure-pressure method (PPSE), EEFSE uses co-located pressure and flow meters on the same footing. It may seem complicated, but EEFSE just replaces PPSE's pressure or flow with a special EEFSE boundary condition.

If a pressure and flow meter are located in the same spot, then there is only one possible set of values for the modelled pressure and flow rate such that modelled pressure disagrees with measured pressure *exactly as much* as modelled flow rate disagrees with measured flow rate. Meters at the two ends of a pipe really aren't interchangeable. If a very accurate pressure and flow meter are located at the upstream end and a very inaccurate pressure and flow meter are located at the downstream end, the model still requires a downstream boundary condition, so the less accurate meters must still be used. So it's compelling from a physical standpoint to only treat collocated meters on the same footing, rather than all meters throughout the system.

For a pipe with both pressure and flow meters at both ends, this approach has the following boundary condition at both the upstream and downstream ends:

$$\frac{P_i - P_i^*}{\sigma_i} = \frac{F_j - F_j^*}{\sigma_j}$$

Where modelled pressure (P), measured pressure (P^*), modelled flow (F) and measured flow (F^*) are related by making the pressure error fraction equal to the flow error fraction – i.e. the model-vs-meter discrepancy divided by the respective error (σ_ι or σ_j). This tells us how to handle the end of a branch or the pipeline terminus if it has exactly one pressure meter and one flow meter. Unfortunately, EESE requires us to custom-derive a tailored boundary condition for every pipeline topography. This is burdensome. A mid-line delivery, mid-line pump or compressor station with suction and discharge pressure and flow metering, every meter arrangement, etc. – every single case must be treated differently if this method is selected as a state estimator.

Hydraulic tuning can be performed with this approach by defining regions of pipes bounded by pressure and flow meters. We then evenly distribute a field of pressures and flows in some manner that agrees as well as possible with these measurements. Differences in each pipe between measured pressures and flows and modelled pressures and flows are used to drive the tuning.

The diagnostic flow state estimator (DFSE)

Simulators that implement this method are complicated. A simple version of a diagnostic flow state estimator (DFSE) is to treat every meter as a boundary condition. Every interior meter is allowed to have a *diagnostic flow* – that is, we allow flow to be added or removed at these nodes. Allowing conservation of mass to be violated at nodes sounds outlandish, but the key here is how the entire network is solved. This approach is based on a sort of optimizer. If the network has trouble matching the meters, it will add or subtract flow at that node.

This essentially performs a sort of *flow meter offset tuning* wrapped up within the state estimator itself. Treating tuning at the same level as the state estimator is troublesome: when a meter goes bad, the effect can be felt as a diagnostic flow or spread out to many different spots, making it challenging to figure out what the source of a problem is and where in the system it originates.

Say that wrong viscosity data for a batch caused a false alarm. In that circumstance, disabling tuning would be useful to isolate our problem. It's good to have a state estimator that quickly points us towards the root of a problem, without involving other confounding factors that might mask it. The first few scans should tell us where an error appeared initially: errors propagate quickly, but a pipeline is so long that a typical scan-duration should capture those clues. [72]

A challenge with DFSE is it prefers the hypothesis that *"lots of different things are a little bit wrong"* rather than saying *"one thing is very wrong"*. This makes sense if we believe errors are normally distributed, but if one thing actually *is* very wrong, it is very hard to identify. A solution is that when it is clear one thing is very wrong because all sorts of different diagnostic flows and tuning started changing, we disable every sort of tuning except the diagnostic flows and see where they are changing. DFSE is used in commercial simulation, so it must be possible to configure it to overcome the challenge.

The maximum likelihood state estimator (MLSE)

Maximum likelihood state estimation (MLSE) is a statistical method to deduce the operating conditions throughout a pipeline without having to exactly follow any of the instruments. It lets a simulator apply its physical equations without insisting on any firm constraints. Instead, equivalent constraints are deduced from weightings, applied to account for how much we trust a meter's quality.

In this way, the maximum likelihood state estimator solves governing equations to best match all available flow and pressure data. It minimizes the weighted sum-of-squares of model vs meter difference at every pressure or flow meter, subject to all the physical constraints. These weightings then bias the state estimator to balance the effect of meter errors, for a more stable online model. The user configures an error to represent each meter's performance, relying on its specification, or better still on its proven track record as logged by the historian. The relative magnitudes of these errors, and the number of meters in indirect agreement with one another, are what makes this method powerful.

If every meter in our entire pipeline suggests that there is a certain pressure at some point, and one pressure meter alone says *"no, that point is at a different pressure!"*, the principles of statistics dictate we ought not to trust that meter! In fact, if there is a significant persistent difference between modelled and measured values at a meter, that actually indicates that something about the meter is probably suspicious. This acts as a sort of final layer of data validation.

The maximum likelihood state estimator treats meters *throughout the pipeline* on the same footing (not just collocated meters as is done in other methods), whilst still enforcing the model physics. This approach has a nice property. One set of equations works for every pipeline configuration, so we aren't forced to customize special configurations of meters (as was required by the EEFSE). This trade-off between all pressure and flow meters, based on each meter's associated error, utilizes all available data without disposing of any good data. This can maintain stability if the quality of several meters becomes poor.

If two pressure meters are at one location, and only one is at another, we are more confident in the former, but not *twice* as confident. Two identical meters are equivalent to one whose uncertainty is divided by $\sqrt{2}$ (which is 1.41 – 41% more sure, not twice as sure). Three are equivalent to dividing an individual error by $\sqrt{3}$ (which is 1.73). This holds for both the absolute and relative errors.

Say we're calculating a pressure (P) from two contradictory measured values P_A^* and P_B^* with associated errors σ_A and σ_B. MLSE minimizes the difference between modelled pressure and measured pressure via objective function (ψ):

$$\psi = \frac{(P - P_A^*)^2}{(\sigma_A)^2} + \frac{(P - P_B^*)^2}{(\sigma_B)^2}$$

This has one independent variable, namely model pressure (P). To minimize this objective function, we seek the minimum point by differentiating the objective function with respect to that independent variable (zero gradient):

$$\frac{\partial \psi}{\partial P} = 0$$

$$\frac{2 \cdot (P - P_A^*)}{(\sigma_A)^2} + \frac{2 \cdot (P - P_B^*)}{(\sigma_B)^2} = 0$$

$$P \cdot \left(\frac{1}{(\sigma_A)^2} + \frac{1}{(\sigma_B)^2} \right) = \left(\frac{P_A^*}{(\sigma_A)^2} + \frac{P_B^*}{(\sigma_B)^2} \right)$$

If we define an alpha (α) as follows, then we can rearrange the equation to tidy it up:

$$\alpha = \frac{(\sigma_B)^2}{(\sigma_A)^2 + (\sigma_B)^2}$$

$$P = \alpha \cdot P_A^* + (1 - \alpha) \cdot P_B^*$$

Thus if errors on two meters are equal, the MLSE pressure takes the average of the two measurements. But if the errors on the two meters significantly differ, the MLSE solution will be close to the value of the meter with the smaller error. For example, if $\sigma_B = 10 \cdot \sigma_A$ then MLSE pressure boils down in this case to favoring the good meter (A) with this intuitively pleasing weighted average:

$$P = \frac{100}{101} \cdot P_A^* + \frac{1}{101} \cdot P_B^*$$

MLSE weights the results more strongly towards the better meter than does a method like equal error fractions (EEFSE). Assuming meter errors are normally distributed, the MLSE result is the most probable value for the actual pressure.

MLSE assumes firstly that errors in measurements are all normally distributed. Secondly, it assumes the underlying pipe model captures the physics correctly. And thirdly that errors in measurements are not correlated between time-steps.

Now consider a model section with n pressures and flows and m model equations. In this approach, the normal model equations other than boundary conditions are treated as constraints. MLSE treats the m model equations as constraints on an optimization problem that minimizes the weighted sum-of-squares of difference between modelled and measured values at each meter. Indexing pressure meters by i and flow meters by j, and writing measured values at pressure meter i and flow meter j as P_i^* and F_j^* respectively, and writing modelled values as P_i and F_j, and writing a pressure or flow meter's error as σ_i or σ_j respectively, the objective function (ψ) being minimized is:

$$\psi = \sum_i \left(\frac{P_i - P_i^*}{\sigma_i}\right)^2 + \sum_j \left(\frac{F_j - F_j^*}{\sigma_j}\right)^2$$

The weights used are meter errors. But we are not quite done. Minimization occurs subject to all the physical equations of the (online) model as constraints:

- pipe conservation of mass (M), momentum ($\rho \cdot v$), and energy (e),
- conservation of standard flow (Q_{std}) through all zero-length objects,
- conservation of standard flow (Q_{std}) through all junction nodes.

There are n model equations which can be written as $f([P, F]) = 0$. This adds a final term for constraint equation (indexed k) with Lagrange multipliers (λ):

$$\psi = \overset{\substack{\text{Pressure}\\\text{Meters}}}{\sum_i} \left(\frac{P_i^* - P_i}{\sigma_i}\right)^2 + \overset{\substack{\text{Flow}\\\text{Meters}}}{\sum_j} \left(\frac{F_j^* - F_j}{\sigma_j}\right)^2 + \overset{\substack{\text{Model}\\\text{Equations}}}{\sum_k} \lambda_k \cdot f_k([P, F])$$

If the original model equations are linear in pressures and flow rates – or, as is the case in our implicit model, *linearized* in these – then this function is *quadratic* in all of its variables: the pressures, flows, and Lagrange multipliers. So it has a unique extremum. This objective function (ψ) is minimized by requiring it to be stationary with respect to independent variables (P_i, F_j), and all other pressures, flows, and temperatures in the model equations, as well as the Lagrange multipliers (λ_k). When we say *stationary*, we mean that its derivative with respect to each independent variable equals zero:

$$\frac{\mathrm{d}\psi}{\mathrm{d}P_i} = 0 \quad , \quad \frac{\mathrm{d}\psi}{\mathrm{d}F_j} = 0 \quad , \quad \frac{\mathrm{d}\psi}{\mathrm{d}\lambda_k} = 0$$

The derivative with respect to λ_k produces the kth constraint equation as an equality. The online model now solves this set of simultaneous equations as it does within an offline model. In fact, many equations are identical to offline. The full set of equations is non-linear, so the simulator linearizes them (as offline). Coefficients are numerically evaluated using the linearized form of this equation. A matrix of linear coefficients involves the Jacobian (J), its transpose (J^T) and a diagonal matrix (A) with non-zero entries corresponding to model pressures and flows associated with each meter:

$$\begin{bmatrix} A & J^T \\ J & 0 \end{bmatrix}$$

The order of variables here is first the pressures and flows, then the Lagrange multipliers. This means that the generated matrix is sparse and symmetric.

Like in the pressure-pressure state estimator (PPSE), flow meter drift can be handled by the MLSE. Although initially the model may pack (or unpack) a line somewhat, trying to honor the miscorrelated meters, it soon reaches a point where any further packing (or unpacking) would cause modelled pressures to drift too far from measured pressures. At that stage, it ends up accepting an error in flow meters in order to keep the linepack at a reasonable value.

On the other hand, MLSE doesn't inherently gain benefit from the known fact that over a long period the net flow in or out of the line has to be, on average, zero. Instead this is handled by a separate learning mechanism adjusting incoming flow data: flow meter offset tuning. MLSE results are used to drive tuning in the exact same manner as we had discussed for equal error fractions (EEFSE).

It is not really possible to assign an inherent accuracy to a state estimator algorithm. The performance of any state estimator depends on how well it's tuned up and on all sorts of auxiliary code, as well as the pipeline itself. With that caveat in mind, at Atmos we did investigate the matter further. We implemented all of these state estimation methods and subjected them to various sorts of noisy data to see which one performed the best. On our test data set, the results of which are shown in the next several pages, we found that MLSE matches or outperforms every other method of state estimation in almost every circumstance. It appears to be the most accurate estimator of a pipeline's state during transients. Although it requires solving a system of twice as many linear equations as simpler state estimators, this is fine as a pipeline simulator employs a sparse solver anyway, and it doesn't spend much of its time solving the linear system. The only obstacle to everyone adopting this state estimator is the complexity of its implementation, as it cannot be easily retrofitted to a pre-existing engine.

A well-configured maximum likelihood state estimator is robust and accurate, minimizing the effect of poor or missing meters, relying only on good meters. Simulated results are accurate within 1%, far better than the 3-5% accuracy of older approaches that ignored meter errors. Reducing operating costs or increasing pipeline capacity by 2% equates to significant additional revenue, providing operators with a rapid return on their investment. In addition, it stays accurate through transient pipeline operations, giving confidence in results at all times.

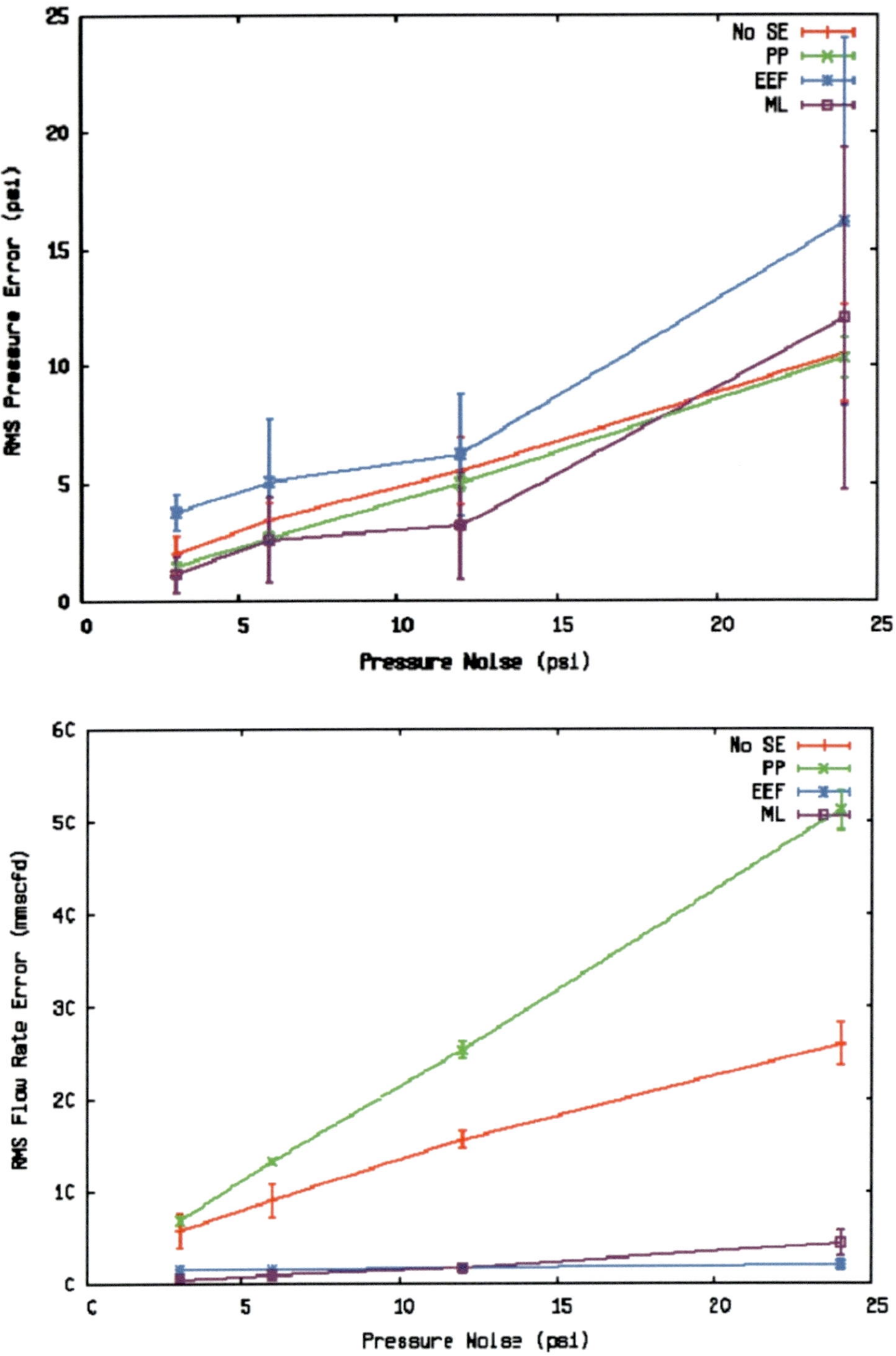

Figure 240 – A maximum likelihood state estimator (ML) outperforms all others in this test, in terms of RMS pressure error (top) and RMS flow rate error (bottom) due to pressure meter noise

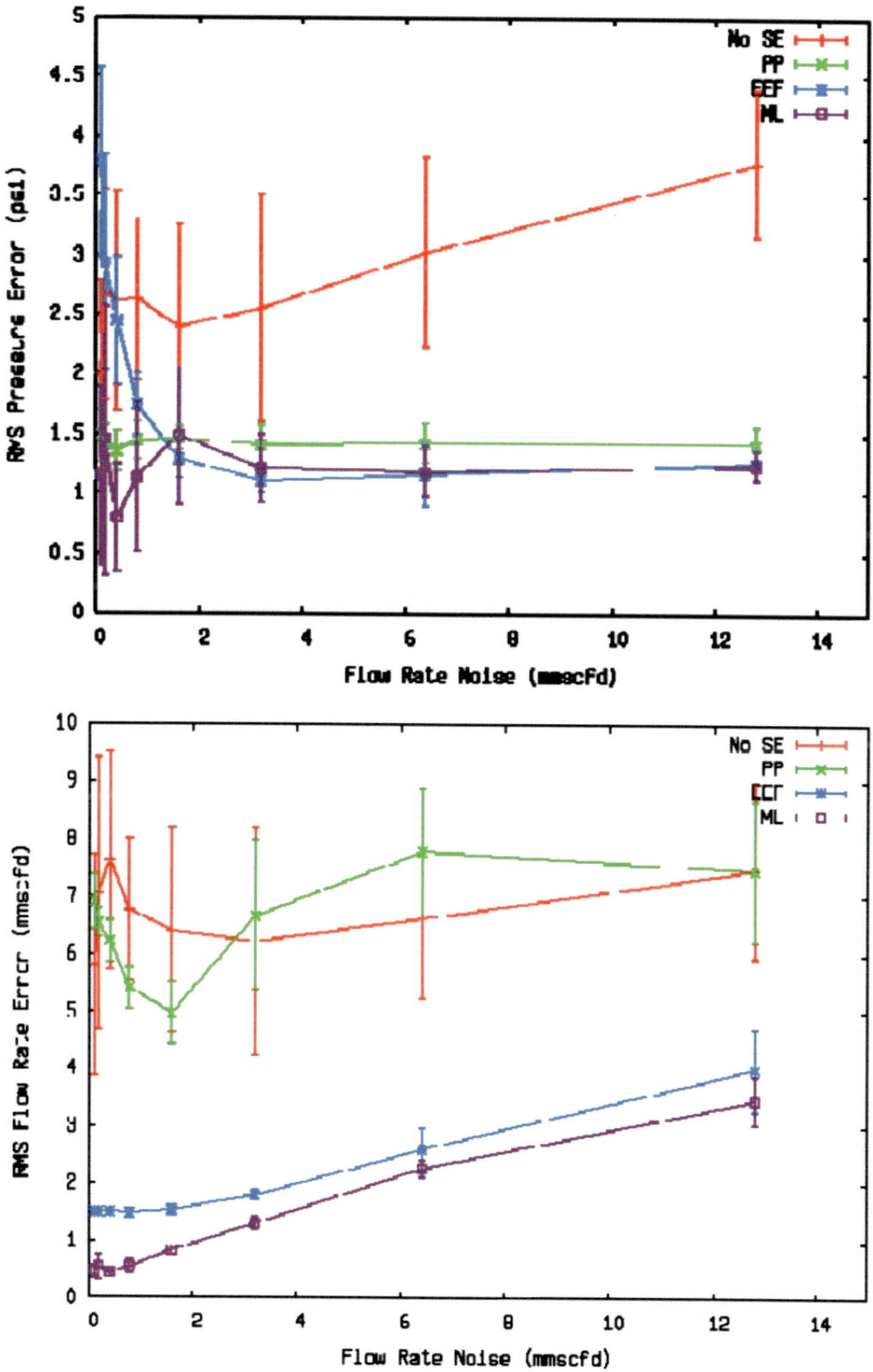

Figure 241 – A maximum likelihood state estimator (ML) outperforms all others in this test, in terms of RMS pressure error (top) and RMS flow rate error (bottom) caused by flow meter noise.

The model-versus-meter difference in MLSE

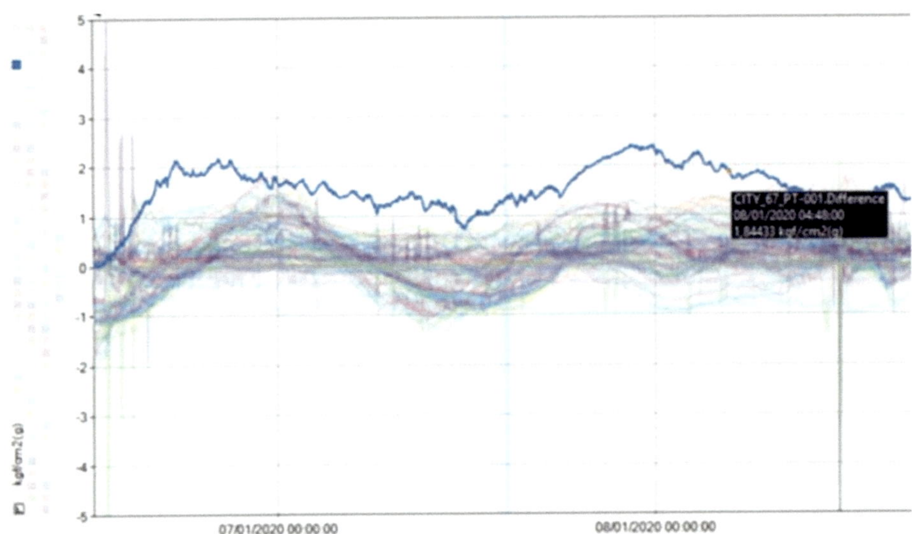

Figure 242 – This is a plot with very many traces representing the difference between model-calculated and measured values for all of the meters in a system. An outlier (in blue) is evident, indicating that that meter is less accurate than the others.

For each pressure or flow meter, a maximum likelihood state estimator (MLSE) quantifies a *difference* between calculated and measured values. If this difference becomes larger than usual for an instrument, *"usual"* being its historic performance and the performance of its peers, that may serve as an indicator that we ought to look at performing instrument maintenance. When a meter malfunctions, the MLSE model-vs-meter difference diverges from the norm, standing out in contrast to other surrounding meters and its own history. In this way, a real-time transient model can be leveraged to inform us of the health of our instruments, assisting with *condition monitoring*. On a batched liquid line errors in batch properties can give a false impression about apparent meter issues so we need to be sure a meter error persists over many batches.

Traditionally, maintenance teams carried out periodic maintenance by sequentially working through field instrumentation according to a schedule. A pro-active preventative maintenance approach anticipates instrument faults before they happen, but there is inevitable truth to the adage: *"expect the unexpected"*. By immediately identifying malfunctioning instruments, rapid response is possible, focusing the efforts to be where it is most needed to minimize any disruption.

MLSE keeps the wider model following healthy meters while we rectify a problem. A pipeliner may choose to display simulated results as a *virtual meter* on their SCADA screens, alongside the raw value reported by each instrument. This takes some stress off control room operators if a meter is bad-quality, by providing them with model results. This makes troubleshooting transparent, and helps them continue to operate their pipeline during a meter malfunction.

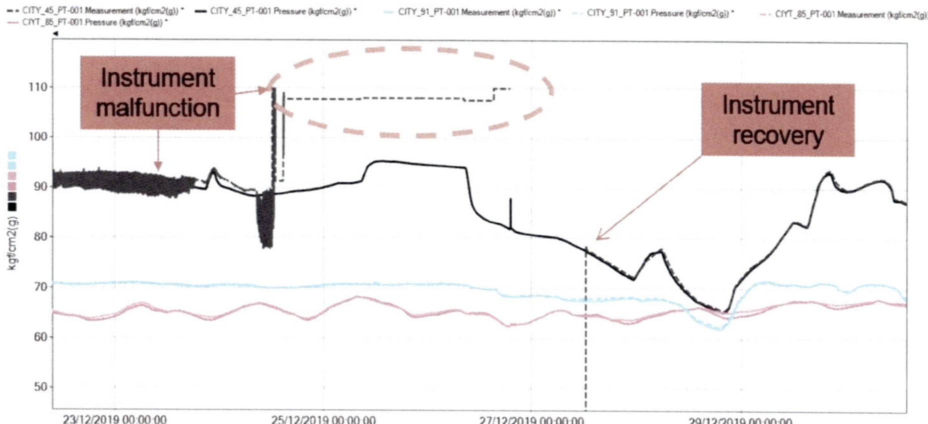

Figure 243 – Good instruments show the measured data (dotted lines) match model calculated values (solid lines). The black line indicates a meter failure for a period of time before it is fixed in the field.

Once a meter fault is rectified, MLSE sees that the problem meter has recovered, and smoothly eases the wider system back into trusting that meter, without giving rise to any false transients.

Questions arise here about normal operation too. When instruments perform well, what sort of model-vs-meter difference is acceptable? In real data, we do not expect a difference of zero. It is possible to achieve a difference of 1% using our model even if each meter has 1% error, especially if those errors aren't random and so can be tuned out. Even if the errors are random, provided there are a lot of extra meters, it may be possible. Two meters at the same place with a 1% random error are as good as one meter with a 0.7% random error.

Errors that are correlated over time

All the basic forms we presented of these state estimators exhibit a weakness. They treat each time-step in isolation from preceding steps. This discards a useful clue: how errors seen by the model *correlate* step-to-step. Such clues can be useful for tuning, specifically for distinguishing one type of error from another. If we observe a gas pipe imparting more resistance to flow than it did historically, is it a pressure meter drifting, or is it liquid hold-up accumulating?

Any state estimator is predicated on error measures being independent of one another, as its function is not primarily to handle sources of error that persist step-to-step. The correct way to deal with that sort of correlated error is to use a Kalman Filter. This is a mathematical technique which is fairly easy to implement on a linear system, albeit computationally intensive, but which is much trickier to implement on a non-linear system such as a pipeline model.

We did try it, using a non-linear extension: the Ensemble Kalman Filter (EKF). That worked as well as the MLSE method coupled with the various forms of automatic tuning featured in Atmos SIM, maybe proving to be slightly better. However, it required running fifty or more models in parallel (the *"ensemble"*). Computationally that was quite unfeasible, certainly not worthwhile to achieve potential benefits that appear minimal. It might be an interesting avenue to revisit in the future. The largest benefit of the EKF approach was that it wasn't necessary to come up with a clever way to simultaneously tune all of the unknowns, we just had to say *"X is unknown but probably varies like so; and Y is unknown but probably varies like so;... etc."*, and the EKF algorithm did a good job. Having already come up with the aforementioned clever techniques, or in many cases borrowed from the pioneers in the field, that feature of EKF didn't prove helpful in this application. [73]

Instead, Atmos SIM implements heuristics to deal with this, with automatically learnt parameters that people refer to as corrections or online tuning methods. Each of these is intended to operate completely independently of one another.

11.8 – AUTO-TUNING: LEARNT PARAMETERS

Learning on site

As time passes while a section of pipeline is in operation, its parameters change. A line carrying heavy crude oil is liable to deposits that eventually clog it up, warranting periodic pigging, but it is almost impossible to completely clean pipes that have contained heavy oils, so they sometimes have a smaller effective inner diameter than their nominal bore.

Automatic tuning is *"learning on the job"*. Learning mechanisms adapt the model's pipe parameters during operation by lumping all effects into a *"tuned efficiency"*. If that learnt efficiency becomes and stays significant, this indicates a departure from the original situation, and the line should be recalibrated for both inner diameter and / or roughness. Learning live-on-site might even provide early clues about potential issues, in the same fashion as condition monitoring for rotating equipment.

It is convenient for a simulator to incorporate enough automatic tuning that an engineer could take a month of operational data and just instruct the model to *"auto-tune"*. An auto-tuner that works is a matter of respect for the user's time.

There are only so many ways for a model to be in error. Enough tuning parameters are available that we should be able to fix all of them automatically:

- pressure drop is wrong at a given steady flow rate
- downstream temperature measurement disagrees with model
- upstream / downstream flow response to a pressure surge is wrong
- transit time of batches / composition peaks is wrong

That is about the lot, other than occasional situations like *"two pressure meters at a location disagreeing with one another"*. We could do a sort of simultaneous offline tuning of all of the above. A universal auto-tuner raises its own separate questions. If there is a big configuration error, or a bad tolerance, to give two examples, the auto-tuner can't be expected to fundamentally fix those. In fact, it might hinder our troubleshooting by masking them in some circumstances.

Gas pipelines are more difficult than liquids to get to a certain accuracy, but fortunately, gas pipelines happen to be more straightforward to tune. There are probably enough degrees of freedom to make a gas-line auto-tuner correct all four of the main accuracy problems we mentioned, even if it got the actual physical properties of the gas and pipeline quite wrong. That is to say, the model can find a combination of errors that produces the right answers in all circumstances, because there are more unknowns than there are independent tests we have for model accuracy. It is especially hard to test the heat loss in the middle of long pipes.

Tuning batched liquid pipelines can become complex, as the simulator might be fed the wrong properties for a single batch amongst many.

Pipe efficiency correction

Efficiency correction adjusts a learnt efficiency for each pipe section, so in the long term, pressure drop across that section matches the flow rate through it. Sections can be instructed to *"learn together"*, meaning all pipes within the section maintain the same tuned efficiency. SIM does this automatically when there is insufficient instrumentation present to justify tuning sections separately. The user can also choose to force certain sections to tune together, for instance if they don't believe that a particular mid-line pressure meter warrants tuning the sections on either side separately. The scale time for this learning is adjustable, typically set to something on the order of 12 to 48 hours. Simulators either express this idea as an efficiency or a pipeline resistance, the two ultimately being inversely intertwined concepts, or lump it in with a point resistance.

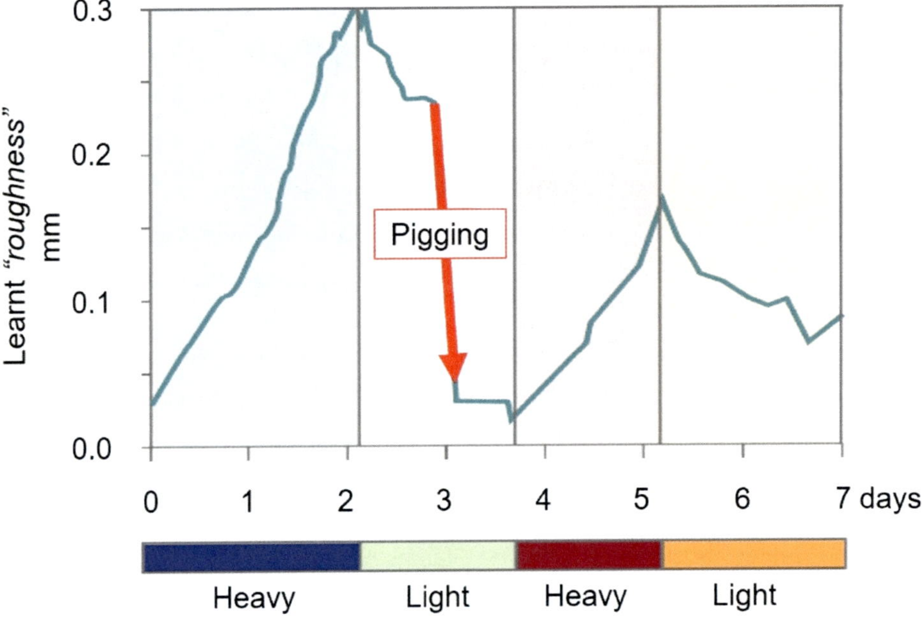

Figure 244 – The learnt efficiency of a pipeline varies over time as it carries crude batches. In this example, a pipe wall *"roughness"* is learnt. It increases when a heavy, viscous batch deposits wax, and decreases when a light, less viscous batch or a pigging run removes it. (adapted from original – credit: Kasch). [38]

When a section of pipeline carries a batch of heavy crude oil, heavy components are deposited, reducing the effective inner cross-sectional area thereby offering progressively more hydraulic resistance. When a light crude oil is batched, these deposits are dissolved, slowly reducing the hydraulic resistance. When the section is cleaned by pigging (green), the line is cleared and its hydraulic resistance quickly falls.

Efficiency tuning is only active if there is sufficient frictional pressure drop relative to meter error (typically a few atmospheres), and sufficient flow rate. The learnt efficiency range is restricted by default in Atmos SIM to between 75% and 125%, which we find covers the sort of legitimate roughness changes we see on most systems. Seeing a pipe clamped at these extreme values indicates the model requires recalibration, or often that something has gone wrong with an instrument somewhere.

Once a target efficiency is identified, the pipe is gradually eased towards it over time:

$$\eta_{\substack{\text{new}\\\text{tuned}}} = \left(\eta_{\substack{\text{old}\\\text{tuned}}} - \eta_{\text{target}}\right) \cdot \exp\left[\frac{-\Delta t_{\text{step}}}{\Delta t_{\text{scale}}}\right] + \eta_{\text{target}}$$

The efficiency tuning scale time (Δt_{scale}) should generally be set to something like something like a day on liquid pipelines or several days on gas pipelines. It must be longer than the time taken for a hydraulic transient to settle down, or else we would learn the wrong target efficiency in a futile attempt to follow each transient. We set the scale time to function as an averaging period that is long enough to filter out the effects of all transients. If we observe a larger pressure drop than expected, either the metering has a problem or that pressure drop is real and needs incorporating into the model. In a gas pipeline, sudden pressure drop would imply an associated Joule-Thomson cooling. Where this takes place is consequential. If the pressure drop happens suddenly at the downstream end, it has less impact on inventory than if it happened at the upstream end. So if we wish to model additional pressure drop without affecting our inventory, we introduce additional lumped resistance at the downstream end.

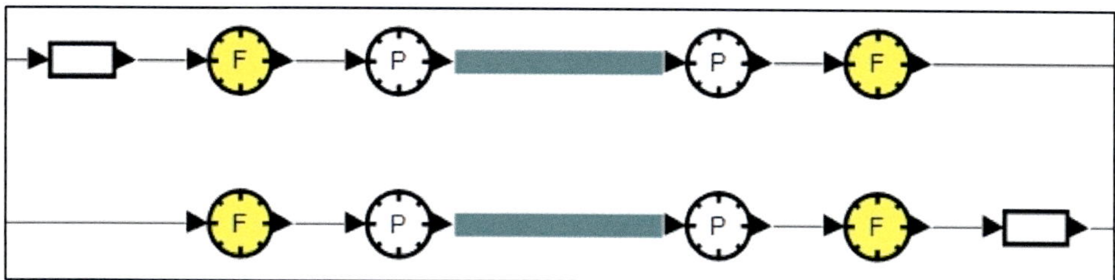

Figure 245 – To model additional pressure drop without affecting inventory, we should introduce additional lumped resistance at the downstream end rather than upstream.

Another consideration is how multiple sections in series will tune overall. If each pipe were tuned individually, transient noise would cause pipes near meters to adjust more than mid-line pipes.

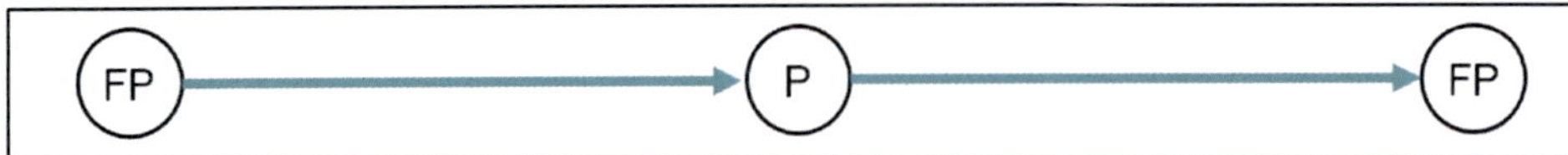

Figure 246 – Here there are two pressure-pressure regions forming a single flow-flow region.

Instead, a single tuning efficiency is targeted for each region bounded by a pressure or flow meter. A pipe is only tuned if it is within a pressure-bounded region and a flow-bounded region: i.e. both meters on all boundaries. Three types of region are considered by efficiency tuning. Two are automatically generated, and might change during a simulation (e.g. if a meter goes bad):

- Pressure regions are bounded by good pressure meters,
- Flow regions are bounded by good flow meters.

For example, flow meters at both source and sink, and a mid-line pressure meter makes two pressure-bounded regions but only one flow-bounded region.

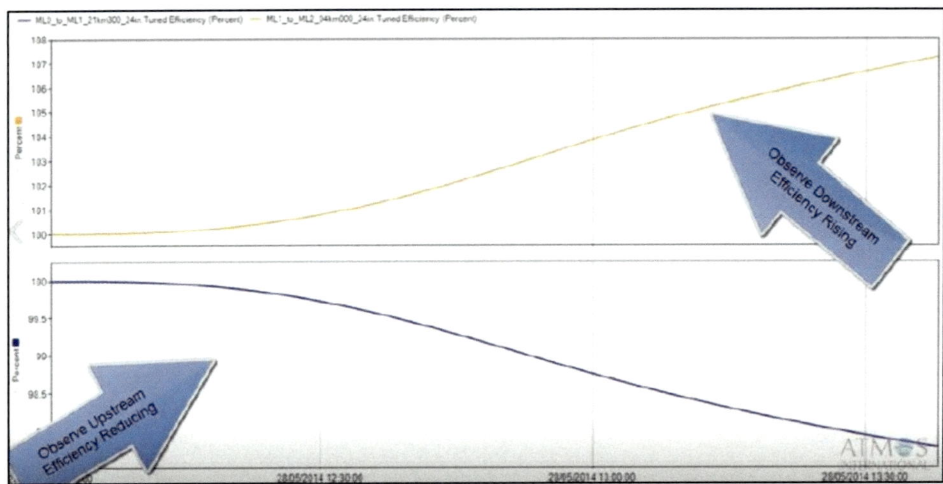

Figure 247 – A trend of pipe efficiency being learnt in adjacent regions: rising in a downstream region while falling in an upstream region. This may suggest a mid-line pressure meter is bad.

If we don't have enough instrumentation to justify tuning a number of sections independently, say because there is no meter between them, or just an offtake with a flow meter on it but no pressure, then we should tune them together. Atmos SIM generally figures this out automatically: it will only tune as many sections independently as it can justify. The user can define even larger groups as a third type of region, typically to prevent over-tuning of any very short pipes.

Flow meter offset correction

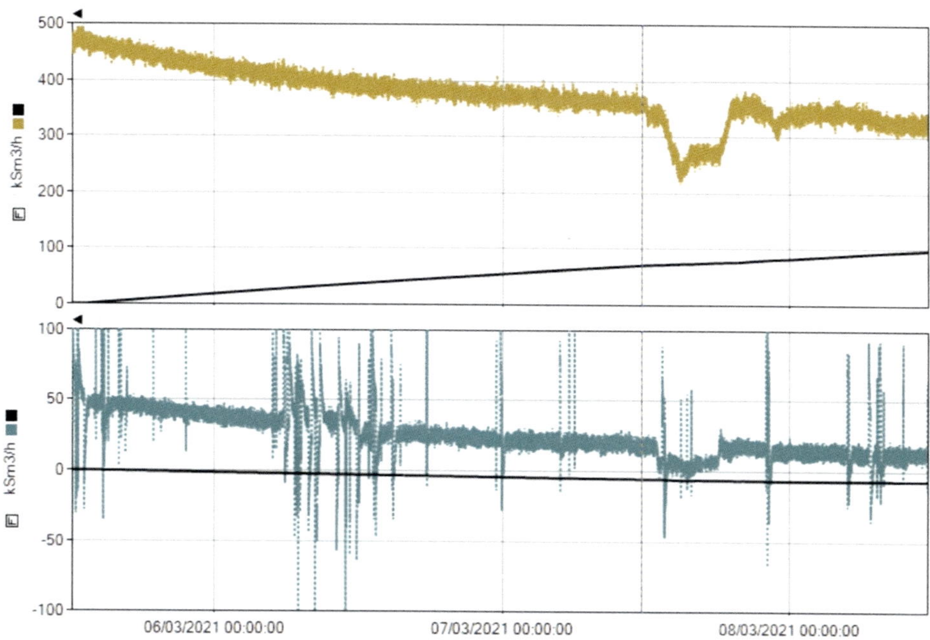

Figure 248 – Flow meter offsets (black) are gradually learnt after a cold start, reducing the *differences* (other colors) between measured and modelled flow rates at each meter.

Over the long term, a sustained net inflow into a pipeline section causes its pressure to rise, and a net outflow causes it to fall. A net inflow or outflow with no corresponding change in pressures suggests that flow meters have drifted.

Even calibrated flow meters drift over time. If an inlet and an outlet flow meter are each calibrated to an accuracy of 1%, the upstream of the pair might read 1% too high while the downstream one reads 1% too low. This results in an imbalance of 2% across a section of pipeline, which over days and weeks causes its calculated inventory to be inaccurate. Exactly how this manifests depends on the state estimation method in use. In maximum likelihood state estimation (MLSE), used by Atmos SIM, the modeled pressures would rise above the pressure meters to a certain point and then stay there; at that point modeled flows are balanced, disagreeing with the flow meters. Flow meter offset tuning corrects this imbalance, so that inflow equals outflow in the very long term.

We compare the change in modelled linepack in each model region with the net measured inflow from the flow meters bounding the region. If needed we apply a small offset to our flow meters. Meter errors are used to determine the relative offsets for each meter, distinguishing between an inlet flow meter that is reading too high and an outlet flow meter that is reading too low. For each flow meter, we then define an objective function coefficient (ψ_m) based on the modelled flow rate (F_m), metered flow rates (F_m^*), and the respective error (σ_m) on each flow meter:

$$\psi_m = \sum_{m}^{\substack{\text{Flow}\\\text{Meters}}} \left(\frac{F_m - F_m^*}{\sigma_m}\right)^2$$

The resulting equations are solved for a *target offset* for each flow meter.

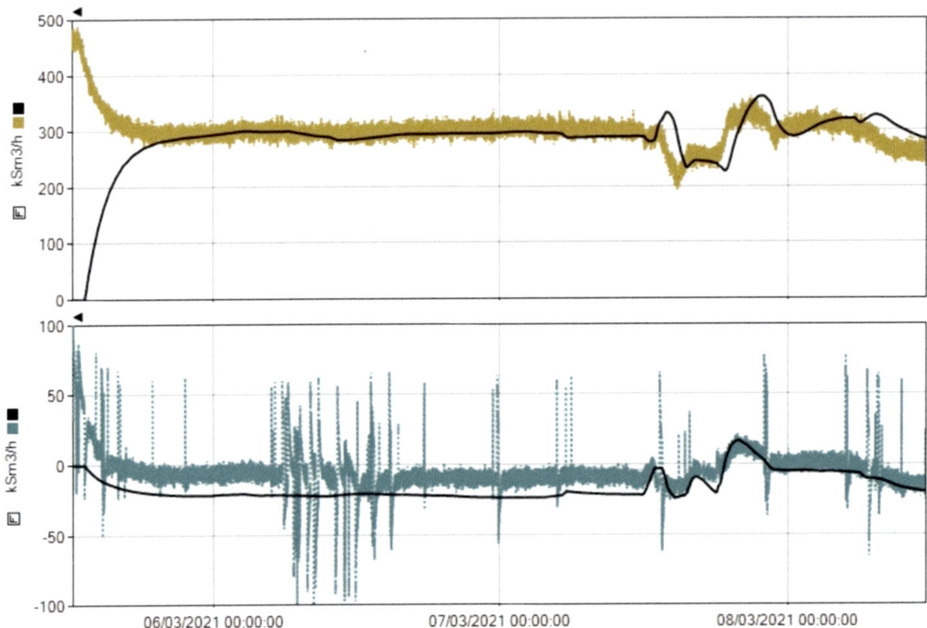

Figure 249 – An example of overly keen learning of flow meter offsets; the characteristic scale time should be considered so this learning is configured to happen slowly enough. It can be tempting to *quick-tune* in this way, and for the first couple of hours it may be useful, but overly keen learning might then over-react to normal operational changes.

The target offset is then averaged over a user-specified scaling period. The tuned offset is then applied gradually using the following equation:

$$F_{\substack{\text{new}\\\text{tuned}}} = \left(F_{\substack{\text{old}\\\text{tuned}}} - F_{\text{target}}\right) \cdot \exp\left(\frac{-\Delta t_{\text{step}}}{\Delta t_{\text{scale}}}\right) + F_{\text{target}}$$

Over time, this learnt offset should correct any flow meter drift by adjusting its bias for each flow meter. The scale period should be set much longer than that set for learning pipe efficiency, something on the order of a week.

What causes an arrival time to be wrong?

A well-known challenge in pipeline simulation is when a pressure or flow meter goes bad without getting flagged as bad quality. Once that is picked up it is accounted for in the subsequent steps, but it still takes a long time for the linepack to self-correct. A sensible way of rectifying this situation is to re-simulate with good inputs, one way or another. This can either be done by rolling back and running again, or automatically spinning off a second parallel model when the pressure starts changing drastically with the assumption that perhaps that meter is wrong, and if it does turn out that the meter was bad since that moment in time, switch over to that parallel model.

There is another, less common situation in which the model pressure is inaccurate due to heat transfer parameters being inaccurate. This problem too is evident to the user via arrival times. That is to say: if estimated arrival time is consistently 5% too soon, say, it may be due to one of two possible causes, depending on the scenario.

In the first scenario, it might be that every flow meter is 5% off in the same direction – which would presumably not happen if there were many well-calibrated flow meters, but in a section with only two flow meters is quite possible. * Indeed, if there are only two flow meters we don't have enough instrumentation to tell. Perhaps a downstream temperature meter will support one or the other hypothesis. If we had data sets at two different flow rates that might also help.

In the other scenario, it might be that the model linepack is 5% lower than reality, due to incorrect heat transfer parameters. In this case, it is appealing to correct the linepack by tuning the heat transfer parameters, as they are a plausible cause of the inaccuracy.

Heat transfer correction and linepack correction

It is rather difficult to learn heat transfer parameters perfectly. Nobody usually knows the exact values and they can vary over *"moderate"* timescales (weeks). Is the subsea pipeline buried? Is the shallowly buried onshore pipeline in wet soil or dry soil? In most typical situations pipelines encounter, we can do a lot – more than enough for a pipeline simulator to perform its duties well.

Heat transfer parameters are important. They affect the temperature profile and especially on gas lines this directly impacts the linepack, which is one of the key model calculated outputs. In many circumstances, a pipeline simulator can take a number of

* One may ask: if flow meters are wrong wouldn't pressure metering pick it up?
The answer is yes – but a suitable efficiency would then be learnt, masking that clue.

possible measures. There are only two or three obvious ways to get at the properties of the soil surrounding a buried pipeline.

The first is to apply thermal conductivity of the inner soil layer everywhere as a fudge factor. What are we tuning towards? Linepack correction tuning could try to match arrival times of parcels or batches, by figuring out how much linepack was off, convert that into a temperature offset, then adjust soil thermal conductivity to get steady state heat loss to go in that direction. That would happen very slowly. It would look at transits over the last month and hope to adjust these parameters so the arrival times over the next month were better. It would not cause fluid to disappear.

Another might be matching downstream temperature in steady conditions, but a lot of the time that doesn't tell us much. If heat transfer is significant, then the downstream temperature is either actually equal to the soil temperature near the downstream end, or very close to it. So it is essentially just sampling soil temperature immediately at the downstream end. On an extensive network that information is basically useless. If the pipe is short enough, flow rate high enough, and heat transfer insignificant enough that the downstream end hasn't reached equilibrium temperature with the surroundings, then obtaining a thermal steady state at two different flow rates by running at each flow rate for days without changing it, we could estimate the ground temperature and the overall heat transfer coefficient (OHTC). With a clever system like a Kalman filter we might even be able to tune that in real-time. But even that won't help us if the fluid's temperature profile is at thermal equilibrium with the environment.

On a gas pipeline, a third might be to tune soil properties around the inlet if there was a big hydraulic surge. We could do that based on the surge pressure and surge flow rate, applying Joukowsky's equation to work out speed of sound in the fluid, then use the temperature dependence of that sound speed to work out the temperature in the region just downstream of the start of the line.

If there isn't anything we can do because one of these situations doesn't occur in the dataset, then we might need to fall back on a different less physically meaningful tuning mechanism: linepack correction tuning (i.e. *"adjust directly"*).

Linepack correction tuning has been a feature in several pipeline simulators. Its purpose is essentially to handle the model pressures in a pipeline being wrong. Pipe efficiency tuning and flow meter offset tuning can't correct linepack errors arising in a situation where the pressure drop across a section is correct, but the pressures are all uniformly too high or too low along the profile.

If the only reason the pressure is off is heat transfer issues then tuning the heat transfer might help, but in practice the main reason we have needed this is that some pressure or flow meter went crazy without getting flagged as bad quality and then once it was corrected it took forever to fix the linepack. Tuning the linepack gradually makes estimated times of arrival match what is observed.

The complication of the reasoning above makes inventory correction tuning problematic unless it is implemented very carefully indeed, so Atmos SIM doesn't default to using it in every circumstance. If the model works correctly otherwise (temperatures about right, gas composition about right) then it isn't necessary. The spirit of the reasoning is expressed by the adage: less is more.

Taking learnt parameters too far

If we aren't careful we could be our own worst enemies. Every system looks a bit surprising sometimes, and online simulators are no exception. Those monitoring and maintaining it want to know what's happening. If the system is getting into trouble, it's their job to troubleshoot, and it's their job to fix it.

Even if nothing is going wrong, going over the top with our tuning mechanisms makes surprising behavior difficult to explain. Troubleshooting becomes exceedingly difficult if overly keen tuning mechanisms have stuck fudge factors into countless places all over the system. Weirdness of all kinds could get stuck in half a dozen different *"bins"*, and the team tasked with solving the mystery is left scratching their heads, unable to figure out what the source of the problem actually is because we keep tuning it out.

An important issue with automatic tuning is that a lot of it ends up being *non-local.* If some tuned parameter starts suddenly changing rapidly, we want to automatically disable its learning and figure out where the sudden strangeness is coming from.

To keep the learning fairly localized, we stick to as few tuning *"bins"* as possible, with clear boundaries for what each is tasked with. The fewer learnt parameters we can get away with, the better. This is true for all online simulators that update their learnt parameter via some simple one-step-at-a-time method, and that is the method employed by most if not all leading pipeline simulators.

Another way to ensure tuning doesn't interfere when we are diagnosing problems is to keep the tuning time-scales relatively long. If a problem appears in a much shorter time than any tuning mechanism's scale time, we know that it can't have originated from the tuning, and that tuning cannot do much to hide it or spread it around over a large area.

11.9 – HOW ONLINE SIMULATORS FAIL

What can go wrong?

Several substantial problems arise on-site in any live environment. These include how to weigh up flow meters against pressure meters, how to transition from a state quite far from real conditions to a state matching reality closely, and how to persevere through flaws in data or lapses in communication. A lot of work goes into ensuring a real-time simulator handles all of these gracefully.

Weighting for flow or for pressure?

In an offline model, we can't constrain both flow and pressure simultaneously at the same point. In an online model we can locate flow and pressure meters essentially at the same point. In Atmos SIM the maximum likelihood state estimator (MLSE) generates a consensus value of model pressure and flow informed by both meters. Flow and pressure at the respective adjacent meters are both involved in the constraint equations, or at least linked by some subset of the constraint equations. Thus, at a given location, if we set the error values on both pressure meters and flow meters to be comparable, the simulated value will be some trade-off between what all the different meters are implying.

Atmos SIM is not usually configured to have an *"even trade-off"* between pressure and flow meters, because it is hard to say exactly what weightings ensure MLSE produces an even trade-off between the pressure and flow meter. It's also difficult to say what should be considered an even trade-off in the first place. If two pressure meters measure the same pressure, and one meter has a much smaller error than the other, the larger-error meter is essentially ignored in favor of following the smaller-error meter. Thus, it's only sensible to pick relative values for pressure meters as compared to flow meters, so a model honors pressure meters above flow meters, or vice versa.

In practice, when a project team deploys Atmos SIM in an online application, each region is usually configured to either *"follow the pressures"* or to *"follow the flows"*. This is done by setting the error values on the meters-to-follow of roughly their repeatability, and setting the error values on the other sort of meter to roughly their range of variation, or possibly something like 10% of their range of variation if we wish them to serve as checks in case of a serious problem with the meters-to-follow. With this setup, our model results match the meters-to-follow well if it's possible to do so without violating the physics of the system, and the differences between the model results and the other meters will drive the automatic tuning.

Taking the first transient step

Conservation equations at steady state differ from those over a transient step. In an offline model, as long as the setpoints and control modes of the model don't change, the solution of the steady state model is also the solution of the transient model as it advances in time from the steady state. In an online model using a state estimator like MLSE where the model results are not in perfect agreement with meters, this is not the case; there can be a start-up transient. This is illustrated in the case of a straight pipe with flow and pressure meters at both ends, where the flow meters have smaller errors than pressure meters. If the pipe is in steady flow at $100\ \mathrm{Sm}^3/\mathrm{s}$, but an upstream flow meter reads $105\ \mathrm{Sm}^3/s$ and a downstream flow meter reads $95\ \mathrm{Sm}^3/\mathrm{s}$, then Atmos SIM comes up with a steady state solution with a flow of roughly $100\ \mathrm{Sm}^3/s$ all along the pipe. In steady state, conservation of mass (M) implies standard flow (Q_{std}) all along a single pipe * is the same, else mass accumulates so it is not steady.

However, on the first transient step, MLSE finds it can do a better job of matching flow meters by assuming a flow imbalance: model upstream flow being roughly $105\ \mathrm{mmscfd}$ and downstream flow roughly $95\ \mathrm{mmscfd}$. This almost exactly matches the two meters, but causes the line to begin packing. This packing causes the pressure meters to deviate, but if these pressure meters' errors are large, the model would proceed for many steps, matching the flows, before the pressure meters stop the packing.

Long-term, Atmos SIM addresses this via flow meter offset tuning and linepack correction tuning, but those processes are relatively slow compared to the quick hydraulic transients moving through the system. So another technique, start-up offset, reduces the transient associated with sudden flow jump – for example, from 100 in / 100 out in steady state to 105 in / 95 out in transient. Start-up offset is needed because boundary and mid-line meters don't initially perfectly agree with model results. To settle to a transient value gently, offsets are temporarily applied to all pressure and flow meters. During this period, all automatic learning mechanisms should be deactivated. These offsets equal the difference between modelled and measured values at each meter at the end of steady state; i.e. these offsets make metered and modelled values match perfectly. We then relax these offsets linearly to zero over a set start-up period (one hour by default). This ameliorates the start-up transient caused by suddenly having constraints change, though the model still makes its way to the new solution over the course of that hour.

* Standard flow is conserved in most fluids of varying composition, except those exhibiting shrinkage or expansion due to blending – as discussed earlier.

Time-varying biases

Biases might emerge and develop over time in certain instruments. Pressure and flow meters drift slowly over time, which a pipeline simulator accounts for using efficiency tuning and flow meter offset tuning, as we discussed earlier. This drift over time also means that a pipeline operator must calibrate their meters periodically. Calibrating a pressure meter entails running it over its range; this can create havoc in a model if the operator doesn't indicate the meter is bad. If this happens, which it frequently does, the simulator probably has to restore from a state and catch up to undo the effects of such an undetected period of bad data.

Atmos SIM has special heuristics to detect un-flagged pressure meter calibrations and accordingly treat the meter as bad. These heuristics must be carefully implemented, so as not to accidentally hide the initial effects of a line break. They rely on watching other meters at the same location for confirmation of a sudden fall in pressure.

Incorrect inventory or composition

Modifications of the physical pipeline model, such as adding or removing an item, or changing the pipeline's dimensions, can cause the current and saved states of a simulation to no longer match the simulation model. Thus, any such changes to the physical model of the pipeline requires special handling. Changes involving a single item of zero volume, such as a valve, can be made *"on the fly"*. Since the current model results will not exactly match the new pipeline configuration, a small false transient is likely to occur. When adding or removing items which contain a volume of gas, such as pipes, or more complex assortments of items, the model cannot adjust current values to give a reasonable representation of the new pipeline configuration. In this case, previous saved states become invalid, and a cold start or cold restart is required.

Another point to bear in mind is that although states are saved frequently, so that a recent one is always available to restore, the composition along the pipeline might have changed since then. Liquid linefill or gas linepack then has the wrong composition for it to move to the end of the pipeline, which takes as long as a week in longer pipelines. One way of addressing this is a pre-emptive replay of a preceding period of time to prime the system with the correct composition from the outset of a real-time simulation. Another is to correct it retroactively. Atmos SIM has an advanced capability to restore a state even if the composition of fluid along the pipeline doesn't match very well, but any new items with volume are assumed to contain default fluid if no information is initially known. On liquid systems the user can manually edit this during the course of a simulation in Atmos SIM.

Occasional flaws in data

To deal with bad-quality meters in an online simulation live on-site, there is quite sophisticated handling involved. If a pressure or flow meter is bad from an initial steady state, the model uses the SCADA value with greatly increased error applied to it, typically a hundred times the usual error. The purpose of this value is to ensure that the meter isn't used if there are other non-bad meters at that location. If a meter is bad, it is used with a much-increased error.

Plenty of instruments are flawed or downright wrong occasionally. Technicians might perform calibrations on a meter without telling us about it, for example. A live system automatically realizes that a meter has gone bad using data validation, as discussed in the section about the Atmos data manager.

What do we do about a bad meter? If an adjacent meter is still good, the bad meter can be ignored in favor of its good neighbor. If it is likely that only one meter remains while the others are bad, we might script logic to select one value among the meters of the same type on a single location, or to use an average of the good meters. The original error is applied to the resulting scripted meter.

A point worth noting here is that if a flow meter's quality suddenly drops, it is not advisable to immediately switch to relying on a co-located pressure meter. If the model isn't tuned up perfectly, suddenly shifting the model from a balance between matching the flow and pressure together to just matching the pressure meter alone may well produce a significant false transient. To avoid this situation, which could potentially interfere with applications like the model-based leak detector, Atmos SIM automatically applies an offset to the persisting meter of however much it differed from the model result before the co-located meter became bad.

What if we suddenly lose a station's sole pressure meter, and the location in question only has a single collocated flow meter? We suddenly trust that flow meter a lot more! When a meter goes bad, the most likely state estimator (MLSE) freezes the last known good value. Any values from the SCADA system when the meter's status is bad are ignored. Its error is then adjusted in a careful manner. We must ease the system out of trusting the bad meter, gradually phasing it out of priority until it is ignored entirely (a user-set *"final error when bad"*). To do so immediately would cause a sudden *"jump"* from one state to another, creating false transients in the simulated results which are not truly happening.

In the event of data outage, linepack uses *last-known* values to calculate inventory, mass, and calorific value, until incoming good data resumes. It does not attempt to recreate missing data, so there is no period in need of updating.

If a pressure or flow meter goes bad during a transient online run, Atmos SIM won't immediately switch to the *"bad"* error, as this would create a big transient surge. Instead, the meter linearly ramps its error from the original value to error_when_bad over error_transition_time_on_status_change (set to around 10 minutes typically), while continuing to read the last good value. This allows the model to switch to any other meters at that location in a quasi-steady manner, slow enough to make the transition without creating a false transient surge.

For pressure meters, a user-specified error is always used if the meter is good, with the good-to-bad transition being eased gradually as described above.

What do we do if it is a flow meter that has gone bad? For flow meters, the error calculation is more complicated. Atmos SIM multiplies a flow meter's measured flow by a user-set override_maximum_fractional_error , then chooses the larger of this alternative error and the base error. If the flow meter reads close to zero ($< 0.001\ \mathrm{Sm}^3/s$), we assume it's against a closed valve and set the error accordingly if the meter recently switched to a bad quality.

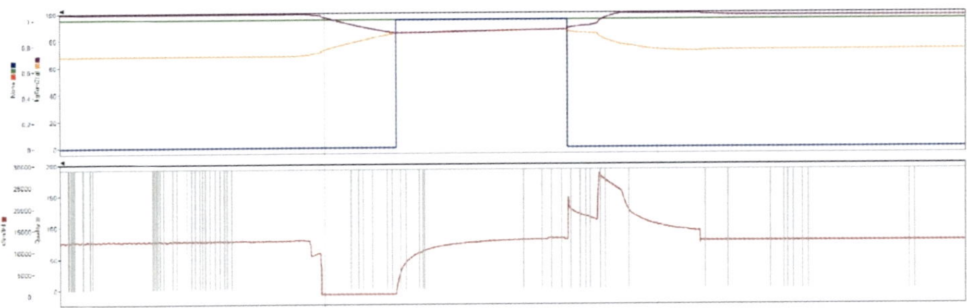

Figure 250 – Quality drops repeatedly (grey). Meanwhile, a compressor restarts (top). Going back to trusting a bad meter too soon creates a false pressure transient (red).

When do we start trusting a bad meter again? We can't immediately believe it's gone back to being good! Perhaps we apply a count as a form of validation, holding quality until *"new quality"* is seen to be stable for some number of steps. When a bad meter becomes good again, its effective error used in the state estimator starts ramping from error_when_bad to error over a suitable interval.

Major lapses in communication

It is not unheard of to briefly lose communications with an entire station at once. That is not to say that high-availability SCADA systems are flaky; it is just the nature of working with such systems. Even if we lose all telemetry to a single station, or indeed to the entire pipeline for a short period of time, we want our model to survive without any of its results becoming too badly degraded. If a whole station drops out, Atmos SIM keeps using all the meters with values frozen at the value they came in when their quality was last known to be good. Eventually, the state of a model doing this will start to deviate from the true pipeline, but if the telemetry system recovers in a short time there should not be any long-term degradation of model accuracy.

Over-constrained situations

The state estimator avoids setting any errors to zero, to avoid equations becoming singular. If we set pressure meter errors to be minuscule, and flow meter errors to be enormous, we nudge a simulator to almost exactly follow pressure meters in some circumstances, but not always. Mass is conserved at mid-line pressure meters, so it is impossible to follow mid-line pressure meters for extended periods of time. Moreover, if a simulator follows flow meters which say the line is packing while pressure meters don't indicate the packing, then over time the deviation at those pressure meters grows until eventually its effect equals the flow meter effect in the MLSE, causing the pipeline segment to cease packing in the model.

Many pipeline systems are *over-constrained*. For example, if we have a mid-line compressor, it typically has both a suction pressure meter and a discharge pressure meter. The MLSE algorithm effectively comes up with a solution such that each of these two pressure meters ends up with the same difference. This compressor prevents our simulator from enforcing pressure-pressure ($\mathrm{P} \rightarrow \mathrm{P}$) control on the adjacent pipes. The model compressor contains zero volume, so inflow equals outflow, enforced as a hard constraint in the engine physics.

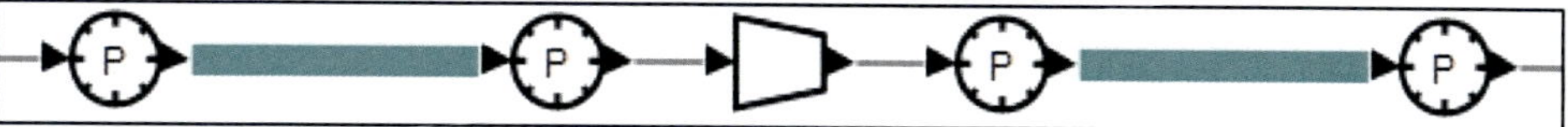

Figure 251 – The mid-line compressor causes this pipeline to be over-constrained. It is not possible for the adjacent pipes to be truly pressure-pressure controlled.

If we had a supply or demand with a weakly weighted flow meter at the same location, then the model would be on pressure-pressure ($\mathrm{P} \rightarrow \mathrm{P}$) control on both sides of the compressor. Therefore, any discrepancy between suction flow and discharge flow would be forced to flow into – or out of – the supply or demand.

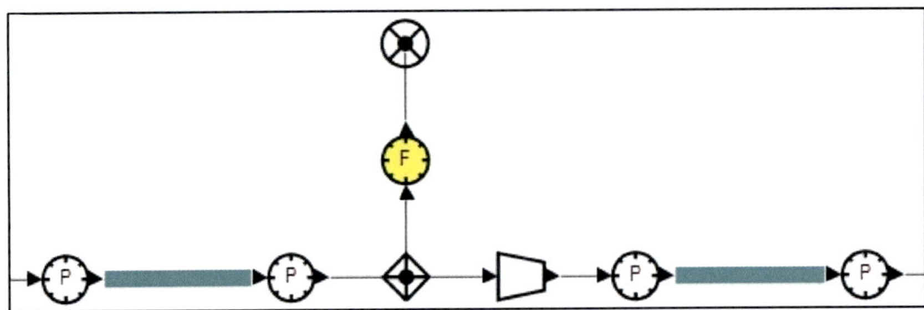

Figure 252 – A weakly weighted flow meter at the compressors inlet can accommodate any discrepancy that arises between adjacent pressure-pressure sections.

Numerical challenges and survival mode

Sometimes, there is no obvious issue with the incoming data, but the model struggles to solve a step for some other reason. In such a situation, instead of crashing, the failure to reconcile what's going on is handled in a more elegant fashion. If the model cannot quickly calculate a state within the allowed error boundaries, or if it cannot solve at all – an occurrence which primarily transpires if the incoming data happens to be atrocious on a particular step – Atmos SIM has a fallback mode which temporarily kicks in to keep the model limping along until everything eventually goes back to normal. This is called model *survival mode* – and it is entirely unrelated to a pipeline's survival time.

This situation usually arises when un-flagged bad data gets into the simulation, fouling up the model state. If the bad data causes numerical problems immediately, then as soon as the meter is corrected, Atmos SIM comes out of its survival mode. However, if the model is able to solve with the bad data for a few steps, and that *"bakes in"* bad temperatures and bad pressures to the internal state of the model, then even after the meters are corrected the model may still be unable to solve a normal transient step.

If Atmos SIM appears to be remaining in survival mode over a long period, it will try to resolve matters by taking a steady-state step of the hydraulic and thermal model – while retaining transients in features such as batch tracking and composition tracking, so that linefills are not lost. This has proven to be a very robust method of recovering from bad data with minimal disruption.

11.10 – OPERATING CONDITIONS

Point alarms and profile alarms

Simulated results are available at selected reporting points (virtual meters), as well as at each intermediate point along a pipeline route. We can detect and raise a notification, alert or alarm if any allowable range (of pressures, temperatures, etc.) is exceeded anywhere. Alarms are defined at items, reporting points, or accounting for variations in the maximum or lowest allowable operating pressure (MAOP, LAOP) along a profile.

Item Type	Item	Property	LoLo Limit	Lo Limit	Hi Limit	HiHi Limit	RoC Limit/s	Unit
Bank	D2PG001	Inlet Actual Flow	-	-	-	-	-	m3/h
Pipe	Barb - Hurezani	Inventory	-	-	-	-	-	Sm3
Centrifugal Pump	D02PG001A	Inlet Temperature	-	-	-	-	-	C

Item Type	Item	Profile	LoLo Limit	Lo Limit	Hi Limit	HiHi Limit	Unit
Pipe	Hurezani - Orlesti	Pressure	1	2	27.93	29.4	barg
Pipe	Barb - Hurezani	Pressure	1	2	27.93	29.4	barg
Pipe	F1 Oarja - Saru	Pressure	1	2	16.986	17.88	barg

Item Type	Item	LAOH	MAOH	LAOP	MAOP
Route	Barbatesti to Orlesti	☑	☑	☑	☑
Group	F1_PL_to_Petrobrazi	☑	☑	☑	☑

Figure 253 – Point alarms (top), profile alarms (middle) and dynamic profile alarms (bottom).

Being aware whenever we even *approach* a limit along the profile of a route can help operators take precautionary measures. Would an extra few atmospheres of surge from starting a pump cause issues? Might column separation arise along a liquid line? Could excessive liquids start dropping out and accumulating along our gas pipeline?

Survival time

Estimates of survival time on gas pipelines predict how long it would take to reach the lowest allowable operating pressure (LAOP) at any point whilst unpacking, or to reach the maximum allowable operating pressure (MAOP) at any point whilst packing. This can only be *strictly* calculated by running our full physical model. To this end, scenarios of this nature are spun off from the latest saved state and simulated alongside the live system as *look-aheads*. Alternatively, another, coarser estimate of survival time is computed at every step of a normal simulation, relative to a project-wide minimum and maximum pressure limit. At its simplest, we divide linepack standard volume (V_{std}) by net outflow to get how long we can operate for (at that net outflow) before we run out.

Northern Network	Survival Time	445.1	h
Northern Network	High Calorific Energy	271.7	TJ

Figure 254 – Atmos SIM's live calculated survival time and inventory of a gas network.

In fact, even the coarse estimate is a little more sophisticated than first appearances. For the chosen minimum pressure, we calculate the linepack it would have, and subtract it from the current linepack to deduce available inventory for unpacking. That is then divided by flow rate. The same is also done for the maximum pressure limit: we calculate available capacity to accommodate packing. This is actually quite a generic calculation that's not specific to pipelining; it is the same if we defined a least-allowable number of lollipops in a sweet shop, with a steady stream of customers walking out with them. It is only a useful assumption if we keep operating a stretch of pipeline at the same flow rates, but that is indeed what people sometimes do.

To allow for extra inventory making operating on linepack possible, a pipeline must be operated at a higher pressure than required for the ongoing demand flow. Packing more into a pipeline is commercially convenient. It consumes energy, but we get that energy back when we unpack the line, and frictional losses are lower while gas flows through a packed line. Overpacking is avoided as it leaves us less room to operate without violating MAOP, and more importantly, makes us less capable of responding to a loss of demand by packing the line.

11.11 – Pig tracking in a gas pipeline

As we saw in our chapter on steady hydraulics, the velocity along a steady gas pipeline is not constant. And as we saw in the subsequent chapter on transient operations, a gas pipeline is rarely steady in the first place. As the pipeline is packed and unpacked to accommodate variations in supplies and demands, the question of estimating the position and arrival time of a pig – which moves at some fraction of the gas velocity – presents a challenge to anyone brave enough to attempt pig tracking on a simple spreadsheet.

Luckily, we can apply the same principles of slippage that we discussed for pigging a liquid pipeline to any sound physical model of a gas pipeline. Each section of pipeline has its own learnt slippage, as does each type of pig. There is no difference in the underlying code, nor in how to configure it.

	Color	PigName	PigType	Slippage	EffectiveSlippage	Distance	Velocity	ETA	DistanceToReceiver	Comment
TANAP-GCS00 | Next pig name/type = /Intelligent										
Launch										
Edit		TANAP_GCS00_Intelligent_TANAP_GCS00_00	Intelligent	1	0.9025	0.87 km	3.74 m/s	2020-06-05 15:19:42	6.26 km	
Edit		TANAP_GCS00_Cleaning_TANAP_GCS00_01	Cleaning	1	0.9025	7.13 km	0.00 m/s	Passage not confirmed	0.00 km	
Edit		TANAP_GCS00_Intelligent_TANAP_GCS00_02	Intelligent	1	0.9025	7.13 km	0.00 m/s	Passage not confirmed	0.00 km	
GCS00-GCS01 | Next pig name/type = /Cleaning										
Launch										
Edit		GCS00-GCS01_Intelligent_GCS00-GCS01_00	Intelligent	1	0.9025	0.58 km	2.68 m/s	2020-06-06 22:41:13	269.52 km	
Edit		GCS00-GCS01_Cleaning_GCS00-GCS01_01	Cleaning	1	0.9025	29.51 km	2.66 m/s	2020-06-06 19:40:17	240.59 km	
Edit		GCS00-GCS01_Cleaning_GCS00-GCS01_02	Cleaning	1	0.9025	57.87 km	2.65 m/s	2020-06-06 16:41:36	212.23 km	
GCS01-ACS02 | Next pig name/type = /Intelligent										
Launch										
Edit		GCS01-ACS02_Intelligent_GCS01-ACS02_00	Intelligent	1	0.9025	0.47 km	2.23 m/s	2020-06-07 00:15:39	285.13 km	
Edit		GCS01-ACS02_Cleaning_GCS01-ACS02_01	Cleaning	1	0.9025	16.43 km	2.23 m/s	2020-06-06 22:16:33	269.17 km	
Edit		GCS01-ACS02_Intelligent_GCS01-ACS02_02	Intelligent	1	0.9025	49.22 km	2.33 m/s	2020-06-06 18:17:23	236.38 km	
ACS02-ACS03 | Next pig name/type = /Cleaning										
Launch										
Edit		ACS02-ACS03_Default_ACS02-ACS03_00	Default	1	1	0.59 km	2.84 m/s	2020-06-06 10:18:45	205.81 km	
Edit		ACS02-ACS03_Cleaning_ACS02-ACS03_01	Cleaning	1	0.9025	28.18 km	2.58 m/s	2020-06-06 09:26:11	178.22 km	
Edit		ACS02-ACS03_Cleaning_ACS02-ACS03_02	Cleaning	1	0.9025	66.49 km	2.75 m/s	2020-06-06 05:26:51	139.91 km	
ACS03-IPR01 | Next pig name/type = /Cleaning										
Launch										
Edit		ACS03-IPR01_Intelligent_ACS03-IPR01_00	Intelligent	1	0.9025	0.79 km	4.03 m/s	2020-06-06 00:37:33	118.63 km	
Edit		ACS03-IPR01_Cleaning_ACS03-IPR01_02	Cleaning	1	0.9025	119.43 km	0.00 m/s	Passage not confirmed	0.00 km	
Edit		ACS03-IPR01_Cleaning_ACS03-IPR01_01	Cleaning	1	0.9025	38.61 km	3.35 m/s	2020-06-05 21:40:52	80.81 km	

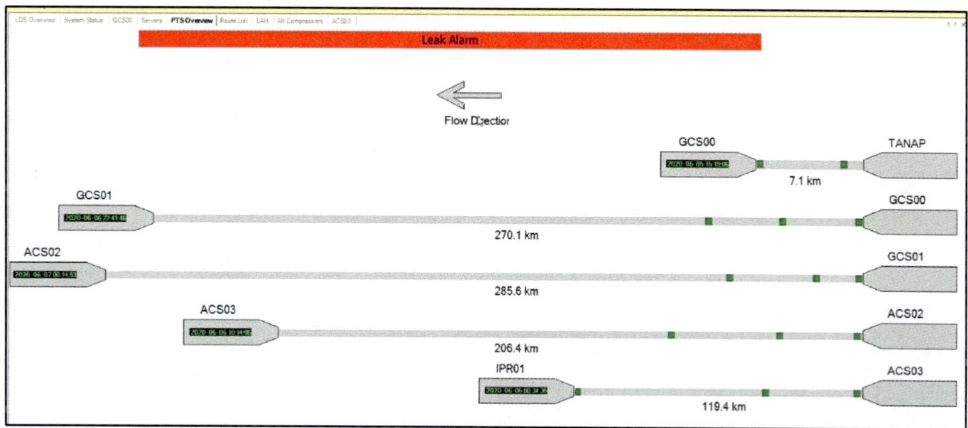

Figure 255 – Atmos HMI displaying pig tracking information as calculated by Atmos SIM, as tabulated values (top) and as a graphical representation referred to as a *"mimic"* (bottom).

However, there are big differences to bear in mind when one considers the very different set of challenges presented by the operating philosophy gas pipeliners are contending with. It is not uncommon for a demand point to draw gas out of the pipeline (at a variable flow rate) during some period of time, then cease to draw any gas out at all. That entails that saying we are *"within a few minutes"* in our estimate of arrival time can be misleading on a gas pipeline, as an approaching tracked pig on the final stretch may be within spitting distance of its destination, mere *meters* away, then the demand ceases to flow for days!

For pig tracking to be accurate on a gas pipeline, a dynamic, well-tuned thermal model is essential.

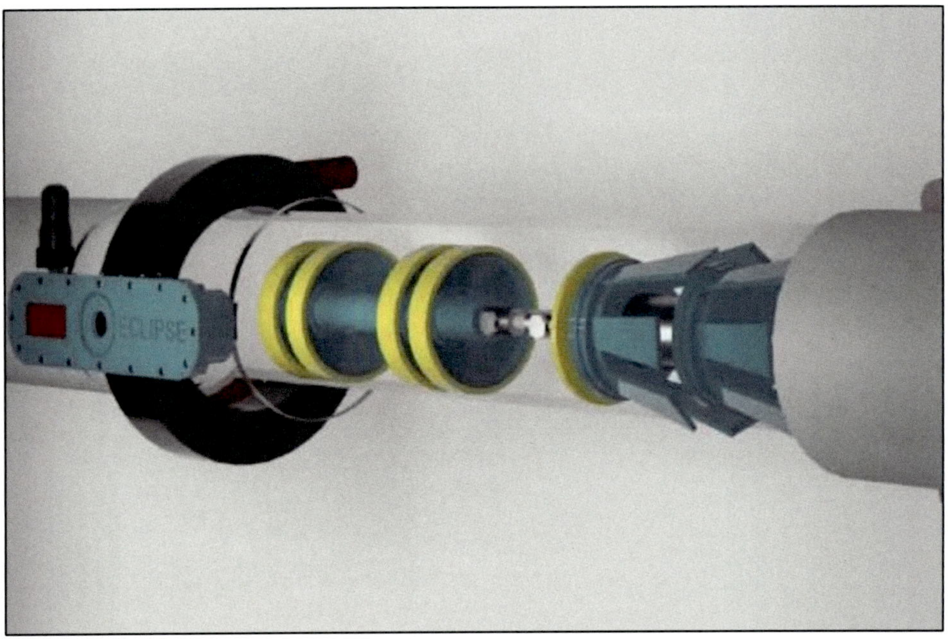

Figure 256 – A pig passes a non-intrusive meter which measures pressure, temperature and flow rate, as well as pig detection (Atmos Eclipse).

11.12 – Composition tracking in a gas line

It can be useful to define parcels of gas based on composition. These are sometimes called *"gas batches"*. Ensuring gas composition complies with specification throughout the pipeline, and especially at delivery to customers, might involve blending some higher-quality gas at strategic stations and at the right moment. A composition tracker is driven by the hydraulic and thermal model, and relies on them performing well. It then does some extensive bookkeeping throughout the pipeline. As a composition front moves through a gas pipeline, its arrival time is affected by the temperature along its route, just as a travelling pig would be. For this reason, a transient ground thermal model holds special importance for composition tracking scenarios.

Getting composition right is important in *all* pipelines. Even if a pipeliner is not interested in tracking specific parcels, all pipeliners are interested in accurate estimates of arrival time. This requires knowing the linepack very accurately, and linepack in turn depends on both composition and temperature throughout the entire pipeline. Composition tracking is not to be regarded as an add-on or an afterthought, but as an inherent pillar of any reliable online simulation.

Case study: real-time transient model at Gassco [39]

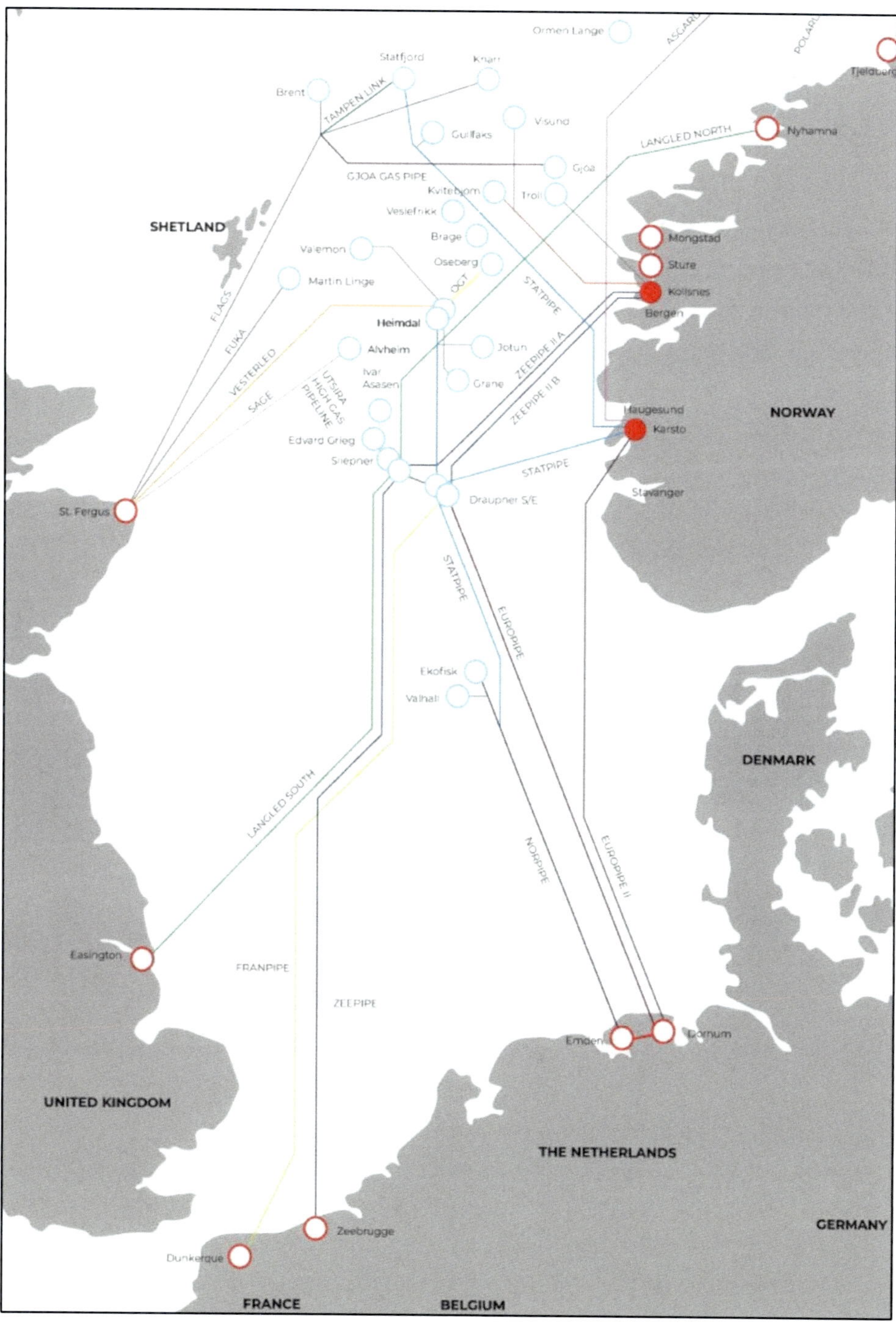

Figure 257 – The extensive Gassco pipeline network traverses the North Sea linking many gas fields, production facilities and customers across northern Europe.

Gassco is a Norwegian operator of a gas pipeline network extending 7 800 km. It links 2 processing plants, 2 riser platforms for blending, and 7 receiving terminals across the world's longest subsea pipelines, crossing harsh deep waters. It steadily supplies 100 billion Sm^3/year, 20% of EU and UK gas consumption. Since 1986 they have taken advantage of pipeline simulation to model their subsea gas network. Their experience has been improving in recent years. Given the long distances traversed across the remote depths of the sea floor, this provided the operator with *virtual metering* for the first time ever. Virtual instruments could be placed at inaccessible locations where a SCADA system could never otherwise measure operating conditions. [39]

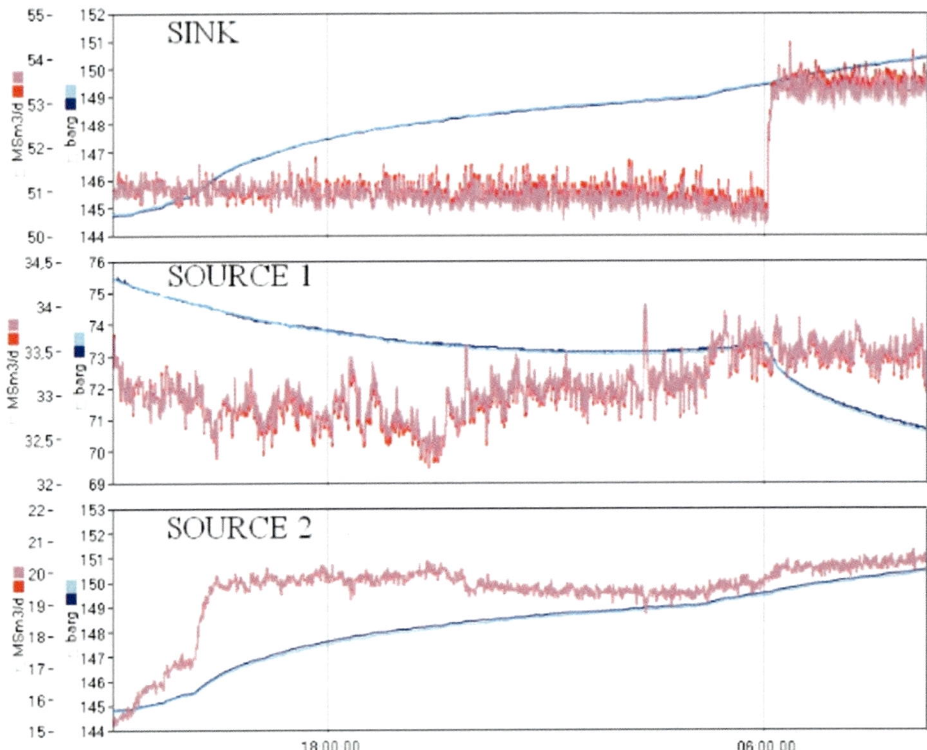

Figure 258 – Trends of measured (red and blue) versus modelled (pink and cyan) pressures and flow rates at several points show a very close match. [39]

The pipeline simulator deployed at Gassco is unusual in that they deploy their maximum likelihood state estimator (MLSE) with a strong weighting to favor their pressure meters. This lets us test the model physics in Atmos SIM – including the thermal model – by looking at historic data, particularly during transient conditions. Linepack fluctuations equate to substantial amounts of gas in a long section of pipeline, corresponding to major swings in operating pressure. To complicate matters, the degree of pipe burial under the sea floor make calculating the temperature a challenge because the thermal properties are always evolving. [39]

To validate simulated results, as the model tracks pressure at one end of a pipeline, we compare model-predicted flow rate to the flow rate measured by flow meters. This tells us if we've got it right. And we have! Even during transients, simulated pressures are observed to be accurate within 0.2 bar or better everywhere in this network. The discrepancy between modelled and metered pressures and flow-rates are only one amongst many checks for validation. Another important assurance, especially with regards to the thermal model and thus calculated density and inventory, has been the remarkable accuracy with which estimated times of arrival (ETA) predict when to expect a change in composition to reach a certain delivery point in the UK/EU. [39]

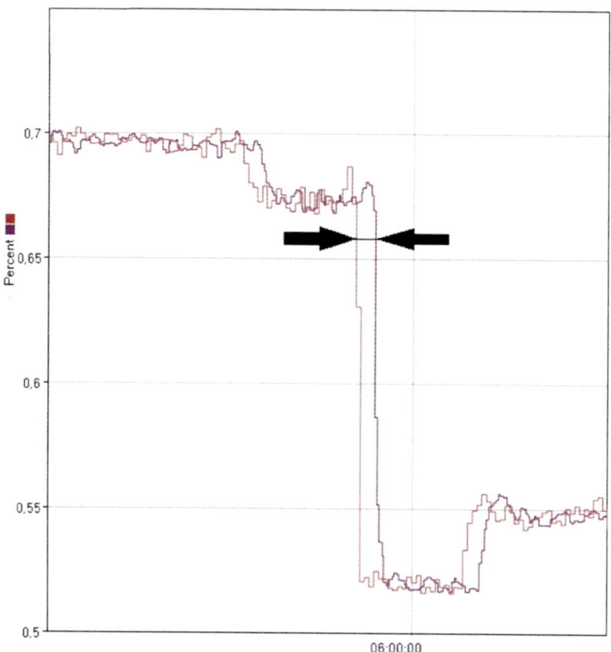

Figure 259 – A trend of a change in composition arriving at a demand after a 60 hour journey through a long subsea pipeline. Atmos SIM predictions (purple) are compared to measurements by a gas chromatograph (GC, orange). Arrival time is accurate to a few minutes on this line. [*] [39]

The trust attained through the rigorous validation of Atmos SIM over many years at Gassco provided the operations team with the confidence to rely on it to respond to opportunities, such as increasing flow during high price periods.

---

[*] It is interesting to note on this trend how *little* turbulent diffusion there is along a gas pipeline – much less than along a liquid pipeline. Tiny features are still visible in the measured composition even after traversing immense distances!

11.13 – SIMULATORS FOR OPERATOR TRAINING

"Practice makes perfect"

There are companies that suffered avoidable fatal accidents despite claiming a gleaming safety record, because in practice they paid little attention to people actually operating the facility. A healthy engineering team acknowledges that control-room operators are their clients, not just the managers at head office!

A good pipelining enterprise implements *standard operating procedures* (SOP) and guidelines that are well thought out. A pipeline simulator helps confirm these are fit-for-purpose, in the sense that an operator can execute them smoothly, knowing what to expect. This can be demonstrated and practiced by setting up a trainer environment, allowing for two-way communication and early feedback without waiting for challenging operations to happen in the field.

Training can be expensive. A tool that trains operators at their own pace, within an offline environment, helps reduce this cost. A trainer-trainee arrangement, where a skilled operator monitors the performance of trainees working with a simulated pipeline, is typically set up on a special console. The system mimics a pipeline's behavior and assesses an operator's response against operating procedures, allowing them to rehearse routine tasks, abnormal conditions and operations that are normal but infrequent. This assists the team and satisfies the regulator by building confidence that appropriate action plans are in place.

We can harness the power of dynamic simulators to mimic a pipeline's behavior in real-time. An operator training application of a pipeline simulator must appear realistic, in terms of timing, operational possibilities, and detailed responses to operator actions. Operators are conditioned by years of experience and training. To keep their skills up to scratch, theoretical learning must be supplemented by regularly trying their hand at performing every operating procedure. This hands on approach is insisted on by the regulators for a reason. It isn't just to assess an operator's competency: it's also to get the chance to practice certain rare operations, and demonstrate to themselves, their team and the regulator that they can handle anything that may come their way, reacting in a timely fashion in accordance with established procedures.

There are perfectly normal ways of operating a pipeline which are nonetheless unusual or extraordinary; rare in practice. For example, special operations that are done once a year when a portion of the network undergoes maintenance. A good trainer setup uses a simulator to mimic real behavior, allowing operators to stay familiar with processes, diagnosing and responding to infrequent events, or emergencies. Taking a simple action at precisely the right moment brings a situation safely under control, avoiding significant incidents.

The principle is simple: a trainer triggers an event, such as an instrument failure or a leak, which is simulated to happen in real-time, as and when he or she decides. A trainee is challenged to interpret the incoming data to deduce what's going on. An operator's competency is assessed based on how they execute each appropriate measure within its proper timeframe. In simulator terms what is special about a trainer model is that it is high-fidelity: it simulates all sorts of behavior beyond pipeline physics: emergency shut-down (ESD) sequences, pump start-up logic, PID control – anything implemented in a PLC. This level of detail is rarely needed for other modelling applications. When connected up to the SCADA screens everything feels convincingly similar to a real operation.

> *"Many pipelines use the Console Operator Basic Requirements Assessment (COBRA) to test controller candidates' ability to manage simulated operating tasks"* [58]

A trainer can even be used to rehearse real operations that have actually taken place. A particularly interesting use case of trainers is recreating historic events to learn from near-misses. This same principle is already very widely used by pilots in flight simulators. By recreating interesting events, we improve how our team would respond, how well our actions follow procedures, hone our *crew resource management* (CRM) and watch various outcomes play out.

ID	Type	Name	Data Point	Min Value	Max Value	Value Score	Start Time	End Time	Relative To	Time Score	Order Score	Missed Score
1	Valves	Valve	Position			0	0	30	Engine Start Time	12	8	10
2	Supplies	Supply	Temperature			0	0	40	1	12	8	10
3	Demands	Demand	Flow			0	10	50	2	12	8	10
4	Control diagr...	Average Diagram...	Value	0	5	3	0	30	Demand Pressure	5	0	15

Figure 260 – The Atmos SIM operator scoring module assesses the user-input value, the timing of the action and the order it was performed against the procedure.

12 – LOOK-AHEADS AND OPTIMIZERS

12.1 – GAS PIPELINE LOOK-AHEADS

Predicting the future

Look-aheads are launched and monitored alongside a live system to predict the future as it plays out in a variety of possible hypothetical scenarios. They forecast potential upcoming events, help spot anticipated risks and violations, and prompt timely action to be taken to avoid disruption. They are widely used to predict operating conditions based on nominations, enhancing how gas pipeliners exploit linepack to proactively avoid any inventory violations, delivery shortfalls or unexpected shut-downs, saving on costs and improving revenues. And they verify if operating at a given throughput is acceptable from a current state, helping deliver more during periods of high demand.

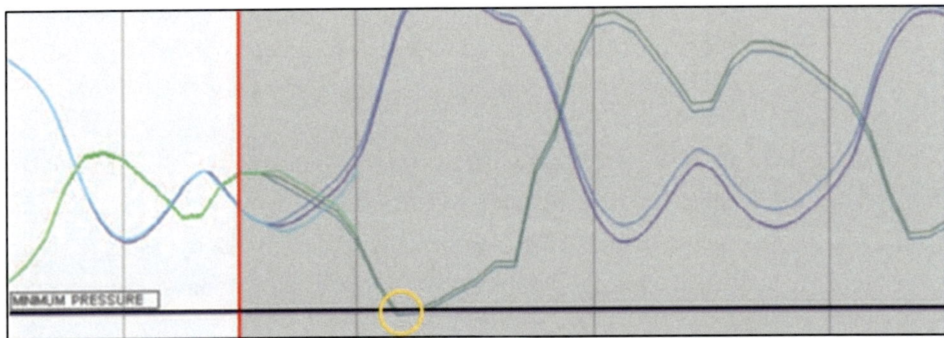

Figure 261 – A trend of a live pipeline's recent history (light background), alongside its predicted behavior into the near future (dark background). A minimum pressure violation is highlighted.

Current states are regularly saved on site, each capturing a snapshot of meter-deduced conditions and learnt parameters like hydraulic efficiency throughout a pipeline. A look-ahead identifies the latest of these *"online"* saved states, restores it, and uses it to initialize a special constraint-based *"offline"* scenario, bridging the gap between a live production system and offline studies. A look-ahead ramps from that most recent saved state's time-stamp to the instance its scenario begins. So if we ask a look-ahead simulation to predict what will start happening commencing at 4:13, having initialized from a latest state saved at 4:10, a three-minute ramp is required to precede that first step. Aside from the initialization which is adapted from a meter-based *online* state, the power of a look-ahead is in the user's ability to subject it to constraints on following steps much as in an *offline* scenario. We must smoothly ease the entire pipeline from one mode of operation to a very different mode without giving rise to false transients.

The parameters of a look-ahead scenario can be set like any other constrained *offline* scenario, except that we tailor certain parameters for speedy simulation, because the results from a look-ahead are only useful when available quickly. The adaptive mesh varies knot spacing automatically, up to a coarsest-knot limit, and it varies the duration of a time-step between the briefest and the longest user-defined interval, look-aheads conventionally only taking very long steps. Interestingly, allowing an adaptive time-step interval to be short if needed will sometimes achieve quicker simulation than forcing it to only use long steps. This is because excessively long steps miss some of the physics, so the latter approach makes a simulator spend time needlessly trying to solve difficult numerical equations which have arisen from a convolution of its own making. Unsuitably coarse steps have been known to slow a simulator down, especially if over-riding constraints differ significantly from values in the restored state. *

At every step, we simulate how the scenario plays out, affecting the inventory of the mainline and every branch in the pipeline system. For a gas pipeline, this can be reported as a standard volume, mass, gross energy and net energy. And, as with any Atmos SIM scenario, it only goes to the trouble of reporting the results along the entire length of every pipe section's profile if needed; the watchword is *speed*. As look-ahead scenarios are inevitably less certain than meter-based simulation, override alarms are set to be more conservative. If any alarm is raised part-way through a look-ahead, it immediately warns those monitoring the system, and in some cases it is sensible to cease the simulation at that point in time.

The purpose of a look-ahead scenario is embodied in the constraints specified to define it. We can instruct it to enable constraints, including multiple constraints simultaneously at the same point as flexibly as for an offline scenario. A dedicated look-ahead scenario can be defined for each of many possible operations the system might undergo in the immediate future. Commonly a case is defined for estimating survival time, forecasting arrival time, and planning nominations. To forecast how the pipeline will cope with nominations, we assign to demand constraints values which evolve over time.

A look-ahead can output any results we need. Gas pipeliners are interested in the inventory contained within our pipeline as linepack, the estimated survival time of the pipeline before a limit is violated, and the forecast arrival time of a pig or a composition peak, so something unexpected doesn't come hurtling in catching the team off guard.

---

* It might seem worthwhile to offer a *"whatever is quickest"* option. The problem is we don't know that in advance and so would have to try both short steps and long steps! Happily, computing power makes shorter steps fast-solving today.

Each pipeline enterprise uses its own sometimes convoluted terminology: *"projective"*, *"predictive"*, *"production planning"*, *"inventory management"*, etc. Thankfully, this is merely jargon: look-ahead scenarios are simple at heart. We shall speak about three main types: *Frozen look-aheads*, *What-if look-aheads* and *Nominations look-aheads*.

Frozen look-ahead scenarios

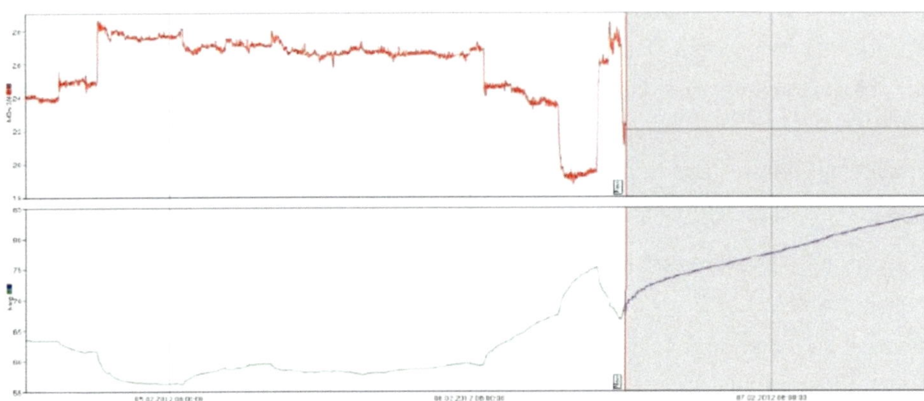

Figure 262 – A frozen look-ahead keeps flow rates as they were in the latest state and predicts how pressure, inventory and other parameters would vary if all flow rates continue as they are.

An online scenario doesn't have *"control modes"* like an offline scenario – it is therefore assigned whatever constraint mode the user specifies. Once a mode is specified, Atmos SIM deduces a setpoint value from the online results. *

A powerful feature of sophisticated look-ahead functionality as built into Atmos SIM is that *unlike* for offline studies scenarios, the user is not obliged to provide a separate instruction for each controlling item. Indeed, a *frozen* look-ahead scenario keeps everything as is; the user over-rides no constraints whatsoever. It defaults each controlling item to constrain in *flow rate* mode, deducing setpoint value from saved state conditions (the meter-driven model). Constraints can be set to control on other modes such as pressure if desired, maintaining the current value in the same manner. To control everything on flow, the pressures and inventory are allowed to vary. This frozen look-ahead predicts survivability if we keep everything constrained exactly as at the instance of the initial state. It can be launched automatically according to a pre-set schedule, running seamlessly in the background. If all flow rates are kept the same, how would physics evolve in time? How long could we continue to operate safely

* It would be possible to figure out constraint mode too, by analysing what has been constant over time. However, a need for this has not yet presented itself.

without intervention? Along with estimated survival time, a frozen look-ahead informs us which limit is violated, and where.

What-if look-ahead scenarios

Atmos SIM devotes a fair amount of effort to getting from current pipeline conditions to a new control situation without any false transient happening. This facilitates the use of long time-steps, thereby allowing a look-ahead to solve quickly while really doing exactly what the user has asked for. Achieving this entails a surprising amount of effort and code, as going from the solution of an online scenario to the solution of an offline scenario such as a look-ahead – even at the same physical state – is mathematically completely different. This feature has allowed pipeliners to do much more than *"freeze"* initial conditions.

A *what-if* look-ahead scenario asks what will transpire if an event takes place. What if there is an emergency shut-down? What if a compressor trips? What if a supply station becomes unavailable? Keeping other controlled items frozen at current conditions, the user specifies any such event, action or disturbance at control items. To instigate a station shut-down, a what-if case might override a valve or compressor status, or gradually ramp flow down at a given inlet or outlet. Everything else is, by default, frozen as captured in the saved state.

A project team might configure a few modelled supplies and demands, and then the operations team might create further what-if look-ahead scenarios capturing more such events. What-if look-aheads won't need modifying again: they are pre-configured once at the start of the project, and having been deployed to site, can be launched from the latest initial state. As inventory evolves during any pipelining operation, what-if scenarios are useful to launch on demand as look-aheads, rather than just a single *"one-off"* offline scenario. They typically project 24 hours into the future, with a 900 second step interval.

What-if scenarios may contribute to enhancing the safety of an operation. The results might suggest we perform a shut-down more cautiously, or take remedial action. They might be critical to maintaining the security of a supply, constantly checking our future capabilities, or helping us plan for contingency. The implications of an event can be assessed in advance. How long could we continue operating safely without making any intervention, if an event were to happen?

Nominations look-ahead scenarios

Varying flow rates over time

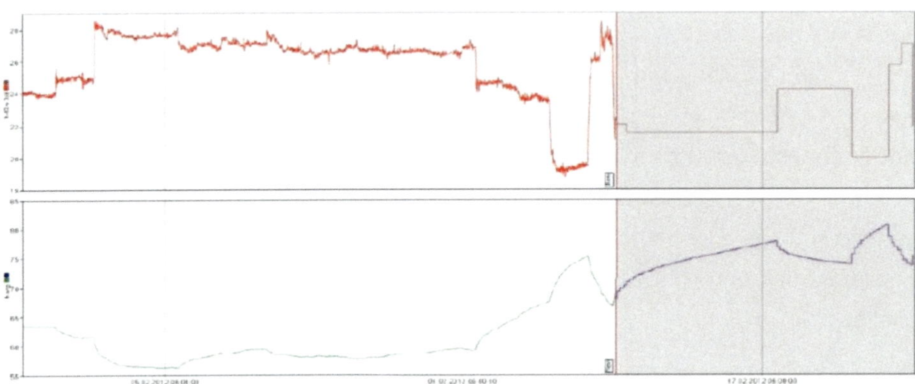

Figure 263 – A nominations look-ahead has a better certainty than a frozen look-ahead, allowing a pipeline to look further into the future.

A look-ahead can now routinely forecast pipeline conditions based on shippers' nominations. It uses exact nominations, varying upcoming flow rates over time at supplies and demands. The operations team updates overriding constraints on a routine basis, typically daily or weekly. The plan could be different but fixed: *"Could we cope with delivering 300 flow-units to that point over the next 48-hours?"*. Or it may be a time-series varying at every interval (e.g. 15 min).

A master production plan can be set up with default flow rates evolving over time for those pipelines without usable nominations. A *consumer-type* can also be defined to describe the pattern of variation over the course of a typical day as multipliers, so the user only needs to input the total delivered that day.

In this way, a look-ahead gives useful results for a long time into the future, and can be launched more often than was previously possible. Forecasting pressures throughout a pipeline based on loads and nominations facilitates production planning, helping to avoid overruns. The commercial team considers daily nominations which need checking if they're nearing a limit, to ensure we neither over-estimate nor under-estimate bookable capacity, taking full advantage of our infrastructure.

The importance of inventory in a look-ahead scenario

Inventory is used for hydrocarbon accounting in terms of total mass, volume, standard volume, gross calorific energy and net calorific energy. It is also used for applications such as estimated time of arrival (ETA), transit time, and settle-out pressure to name but a few. Such calculations are readily available to users via an online simulator without requiring manual calculations by the operators.

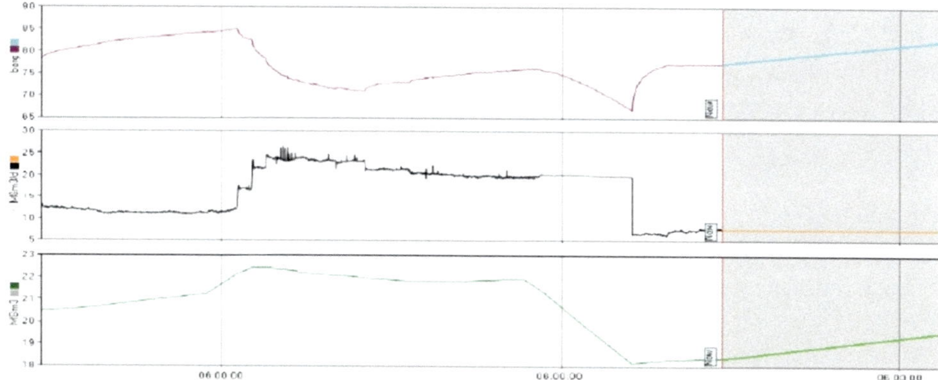

Figure 264 – A trend displaying pipeline inventory from a data historian alongside a forecast. The grey area to the right of the vertical line (the *"now"* line) indicates future projected results.

An accurate inventory is crucial to know the available capacity for packing and unpacking. It also contributes to the accuracy of estimated arrival time, which is important for tracking a pig, or of a peak of off-specification gas. This knowledge allows the control room operator to have a more informed view of the pipeline conditions, and is often used as a basis for manual calculations.

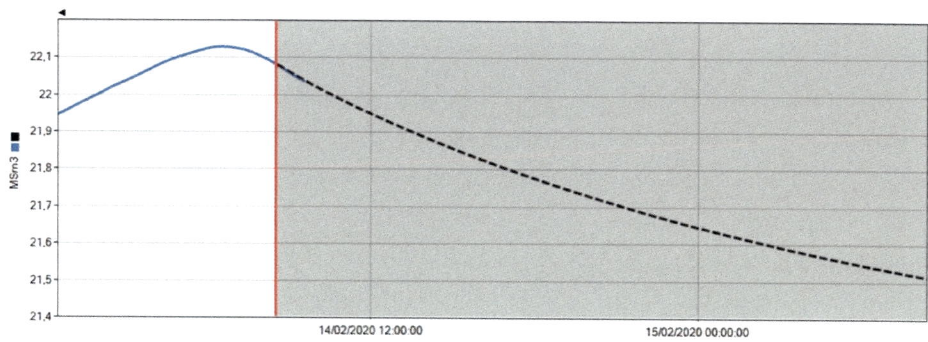

Figure 265 – A look-ahead predicting a pipeline's inventory during an unpacking operation.

Optimizing flow rate and inventory

Gas pipeliners wish to reduce fuel consumption in compressors, but maintain some excess linepack to accommodate a sudden increase in nominations. To represent this trade-off, some of them make use of *"nose curve"* type diagrams.

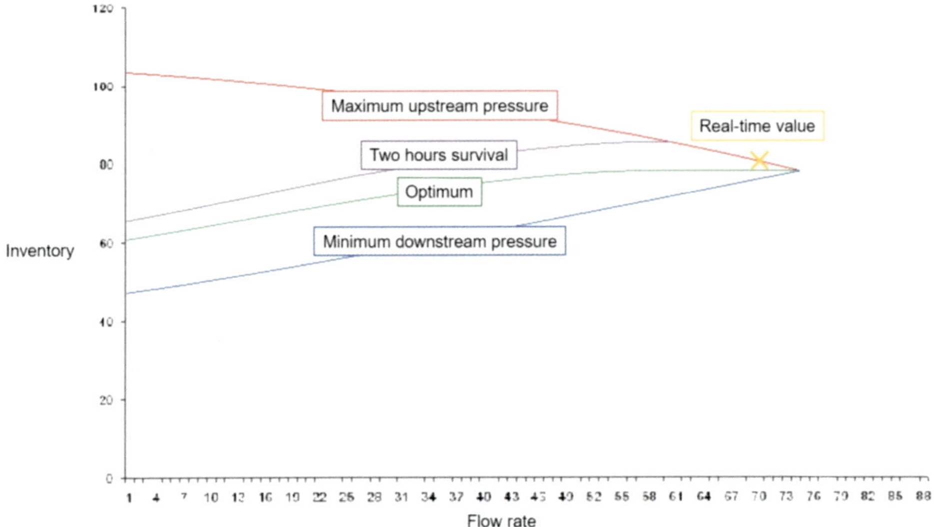

Figure 266 – A schematic of a gas pipeline *"nose-curve"* diagram. Real-time inventory (X) is bounded by upper and lower thresholds based on limits of allowable operating pressure.

This diagram instantly informs a gas pipeline operator where real-time linepack is within limits, as well as how to optimally adjust it over the following few hours.

Real-time inventory, a simulated result, is indicated by a floating *"X"*. It must lie between two inventory thresholds. An upper threshold is calculated based on a fixed maximum pressure at the supply point, and a lower threshold is based on a fixed minimum pressure at the demand. A further curve represents an optimal condition based on fuel consumption and operational flexibility during normal operation. A magenta curve provides two hours additional survival time over the usual optimal curve. This is useful during maintenance periods when fewer supplies are available, making the consequence of a supply shut-down severe. Here we see a big gap between the maximum pressure curve and the optimum curve. Typically frictional losses are minimal at maximum pressure i.e. at maximum linepack, so the gap would be small, unless perhaps we throttle at stations to get pressure down because we have fixed-speed compressors. It might also be that operational flexibility factors into defining *"optimum"*.

Tracking pigs and composition in a look-ahead scenario

Every Atmos SIM look-ahead is capable of handling pigs, composition fronts, component peaks and off-spec gas anomalies. If a pig or a peak is already travelling through the live system, it is picked up in the online saved state. A nominations look-ahead tells us how it will traverse the pipeline based on anticipated future conditions, not frozen present conditions. This is better than the traditional tracking approach, which assumes that a pig continues to travel at the exact same velocity that it happens to have at the initial instance in time.

A nominations look-ahead subjects a pig to the truly expected flow velocities, and applies pig-specific and section-specific slippages, providing an extremely accurate forecast of its arrival time. We can also launch a new pig during a look-ahead, to predict arrival time if we did decide to go ahead and launch it.

Anything else: ad-hoc look-ahead scenarios

The user has the freedom to build any look-ahead with the same versatility of control modes as would be offered for any offline scenario. For example, a special long-duration look-ahead might perform a prognosis into the average gross calorific value and Wobbe index based on upcoming compositions. To take advantage of this versatility, the interface must be easy to use. That feature should not be taken for granted as trivial; historically it did represent a barrier to pipeliners leveraging ad-hoc look-aheads to their full advantage.

Case study: look-aheads on gas pipelines [74]

Look-aheads at Gassco

Look-aheads are considered one of the prime application benefits of real-time modelling at Gassco. They cite at least five tangible benefits to using a better real-time model with better look-aheads: [74]

1. Blending or hold-back [*] passing mixing points
2. Blending after pressure harmonization [†]
3. Increasing capacity utilization
4. Increasing short term sales
5. Improving notification of off-specification gas

Gassco have many look-aheads set up, including one running five days' worth of nominations as a broad overview. Atmos SIM is integrated with Gassco's nomination system to automatically generate scenarios simulating planned future operation. Needless to say, a great deal takes place on any pipeline from now to five days' time!

Their survival time cases provide operators advance notification of issues that could impact security of supply. Projecting 24 hours into the future, they help to identify situations which may imply an interruption to routine operations. Anticipating a concern 24 hours in advance leaves the operators ample time to make suitable adjustments.

Contractual obligations specify the natural gas quality delivered by a pipeline. For example, it may stipulate that gas must include sufficient methane, and no more carbon dioxide (CO_2) than a certain limit. Say one field delivers good quality gas, while another delivers awful quality. Monitoring composition meters at that supply, we might suddenly see a low-methane *"off-specification"* batch entering. How long will it take for that parcel to reach the good-quality point? While it goes past, we could increase the flow rate from a good-quality supply, so that the resulting blend meets gas quality specifications.

Going past multiple points gives the operator several blending points at their disposal to get the gas to be *"on spec"*. The receiving terminal might run compressors off the gas being fed into it. A bad parcel can cause those to trip or bad implications for the equipment. The operator needs to anticipate the arrival of the off-spec gas, so they can

---

[*] Off-specification gas is "held back", i.e. not delivered, by diverting it away.

[†] The pressure of a high-pressure pipeline must be adjusted before a cross-over valve opens to send fluid to a separate pipeline or to receive fluid from it.

divert it and blend it to the required quality level. A penalty is incurred if off-spec gas is delivered to a customer, or to a compressor. If the calculated arrival times are only accurate to within two hours, that means two hours of good quality gas being wasted.

The annual benefit of a better look-ahead at Gassco is increased short-term sales and linepack utilization, amounting to 82.5 MSm^3/year in total. [74]

Model	Look-ahead case	Status	Next launch time
West-Axis	Frozen	Simulating	21 Feb 13:00
West-Axis	Planned nominations	Waiting	-
West-Axis	Projective	Waiting	-
Europipe	Production plan	Waiting	-
Europipe	Projective	Simulating	Immediately
Europipe II	Production plan	Simulating	21 Feb 13:15

Table 25 – A pipeliner's view of the status of their look-aheads, and the next launch time. Some of these are launched quarter-hourly, and others are only launched on demand.

A production plan look-ahead case with future nominations allows operators to fully utilize each pipeline's capacity, without fear of inventory violations, pressure protection shut-down or delivery shortfalls. For many of the pipelines in the network the control room operators have been able to better control the linepack using the results from the look-ahead. This allowed the utilization of an additional 2 MMSCMD of gas for a period of 12 hrs over 60 days a year, which translates to 60 MMSCMY of dry gas linked to swing field production. [74]

The overall impact of a more accurate online real-time transient model and predictive look-ahead simulations was a notable enhancement in short term sales. This pipeline enterprise increased their dry gas sales through two of their pipelines by 0.25 MMSCMD for 90 days per year, which is an increase of 22.5 MMSCM in annual sales of dry gas linked to swing field production. [74]

Gas parcels and peaks

Quality tracking is a feature facilitating blending fluids of different qualities. This helps maintain pipeline integrity by reducing concentrations of corrosive components such as hydrogen sulfide (H_2S) or carbon dioxide (CO_2). Accurate forecasting reduces the necessary period of flow rate hold-back for blending to happen. As a batch of off-specification gas moves through the network, there may be opportunities to blend it mid-line, to improve its quality. Estimating arrival time accurately can save valuable hours of under-utilized compressors. The operations team is ready to expect a gas quality change and divert off-specification gas for upgrading at just the right moment. This increases on-spec sales and reduces un-notified deliveries of off-spec gas. [74]

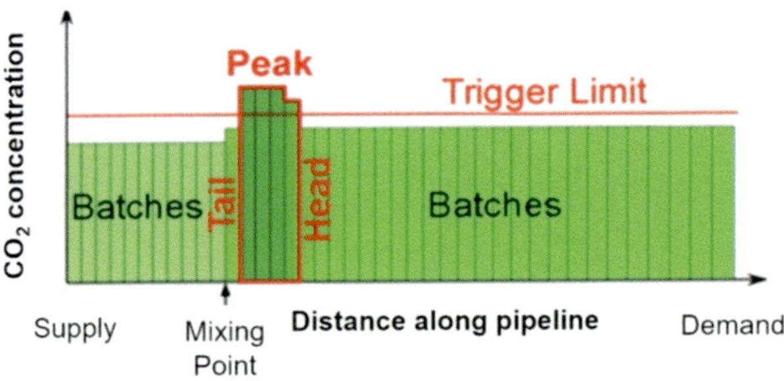

Figure 267 – A peak is a series of gas batches, also called parcels, exceeding a threshold.

A quality batch or *parcel* of gas is the smallest mass with a homogeneous concentration, within the resolution of each fluid component. A *peak* consists of a number of consecutive parcels which are all above a given threshold. [74]

Notification of off-specification gas to shippers

A key benefit of look-aheads on gas pipelines receiving gas supplies of varying composition is to notify shippers at the demand receipts about off-spec gas. If a volume of off-spec gas is delivered without due notification in advance, it calls for a large sales discount of a substantial value that is difficult to quantify. [74]

Location	PeakID	Description	RTM Status	Head KP	Tail KP	End KP	Vol (MSm3)	Sent (CET)	ETA (CET)	Duration	Min / Max	Avg
SKT	CO2-ASG #2021-00523	CO2 > 2,900 mol%. Automatic peak.	Exiting	1,30	0,00	1,30	0,251	12.11.2021 - 19:39	12.11.2021 - 20:12	02h 30m	3,205 %	3,159 %
NHT	CO2-SKT #2021-00528	CO2 > 2,900 mol%. Automatic peak.	Exiting	1,20	0,00	1,20	0,222	12.11.2021 - 21:55	12.11.2021 - 22:05	00h 27m	2,929 %	2,926 %
EAS	HCDP Trace.PMS Output-SLR #2021-00556	HCDP > -8 grader	Exited	544,20	544,20	544,20	8,000	11.11.2021 - 17:10	12.11.2021 - 22:08	03h 00m	-7,715 C	-7,886 C
NHT	CO2-SKT #2021-00531	CO2 > 2,900 mol%. Automatic peak.	Running	0,99	0,00	1,20	0,192	12.11.2021 - 22:34	12.11.2021 - 22:42	00h 07m	3,032 %	3,025 %
KRT	H2S Trace.PMS Output-KRI1 #2021-00528	H2S from KRI > 15 ppm. Automatic peak.	Exiting	0,36	0,00	13,65	0,011	12.11.2021 - 21:42	13.11.2021 - 00:02	01h 04m	15,657 ppm	15,615 ppm
EAS	HCDP Trace.PMS Output-SLR #2021-00557	HCDP > -8 grader	Exited	544,20	544,20	544,20	2,350	11.11.2021 - 21:07	13.11.2021 - 01:34	00h 53m	-7,849 C	-7,879 C
SKT	CO2-ASG #2021-00534	CO2 > 2,900 mol%. Automatic peak.	Exiting	1,30	0,00	1,30	0,242	13.11.2021 - 02:10	13.11.2021 - 02:32	00h 52m	3,377 %	3,308 %
SKT	H2S Trace.PMS Output-ASG #2021-00529	H2S from ASG > 3 ppm. Automatic peak.	Exiting	1,30	0,00	1,30	0,241	13.11.2021 - 02:16	13.11.2021 - 02:40	01h 05m	4,948 ppm	4,382 ppm
EAS	HCDP Trace.PMS Output-SLR #2021-00558	HCDP > -8 grader	Exited	544,20	544,20	544,20	1,105	11.11.2021 - 22:39	13.11.2021 - 02:55	00h 25m	-7,934 C	-7,935 C

Figure 268 – Peak tracking and batch tracking simulated in Atmos SIM and reported together.

Hydrogen sulfide quality tracking for mid-line blending

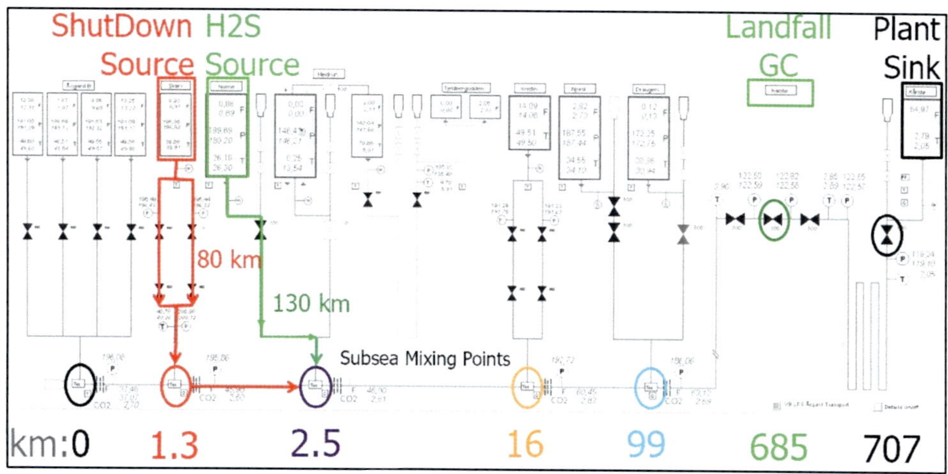

Figure 269 – Several color-coded unmetered mixing points along a subsea mainline. [74]

Our case study is a network with unmetered, unregulated subsea mixing points. There is a source of sour gas (green) containing excessive hydrogen sulfide (H_2S). It is upstream of a mixing point (purple) in the mainline. Another source even further upstream (red) shuts down, associated with a red mainline mixing point. The red source shut-down causes a surge at the purple mixing point, pulling sour gas containing excessive H_2S from the branch line into a peak in the mainline. This H_2S peak can be trended at the downstream mixing points (kilometer point 16 in orange and point 99 in turquoise), and compared to a landfall gas chromatograph (GC, green) at 685 km along the mainline. Finally it can be tracked all the way to the rich gas plant (black). [74]

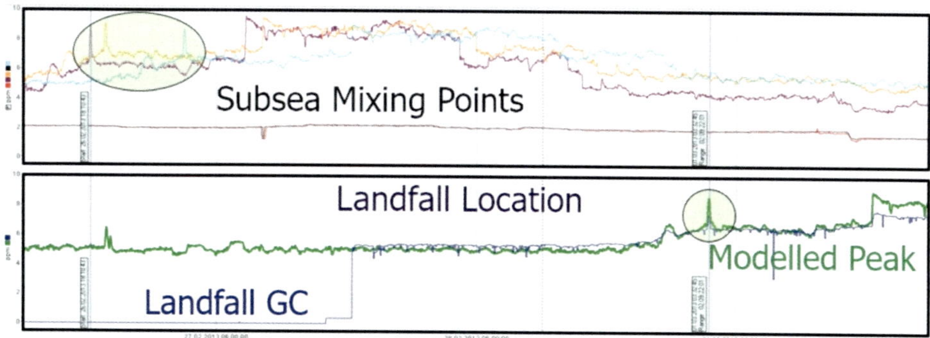

Figure 270 – A trend over time of a composition peak arriving at color-coded mixing points. This peak's arrival time estimate was found to be accurate within 60 seconds! [74]

In the upper graph, the hydrogen sulfide (H_2S) peak is first identified at the purple subsea mixing point. Then it is observed at the orange and turquoise mixing points. In the lower graph from the landfall location, we see the gas chromatograph being powered on (blue), It confirms the accuracy of the modelled peak (green) as it arrives at landfall after travelling 683 km over 2 days 9 hours and 22 minutes with an exact match. Here, the arrival time of the peak at the gas composition meter is computed to an accuracy of better than 60 seconds. [74]

An accurate arrival time estimate allows the operator to intervene by diluting an excessive peak. The way this is done is by injecting sufficient *"sweeter"* gas – that is lower in hydrogen sulfide (H_2S) – into a mid-line mixing point, blending the mainline gas into a resulting gas mixture that is more acceptable in composition. When calibrating this network using the Atmos SIM Tuning Assistant, we accept deviations in estimated arrival time within ± 45 minutes or 1.5% of travel time. Preferably the calibration is biased to arrive early. [74]

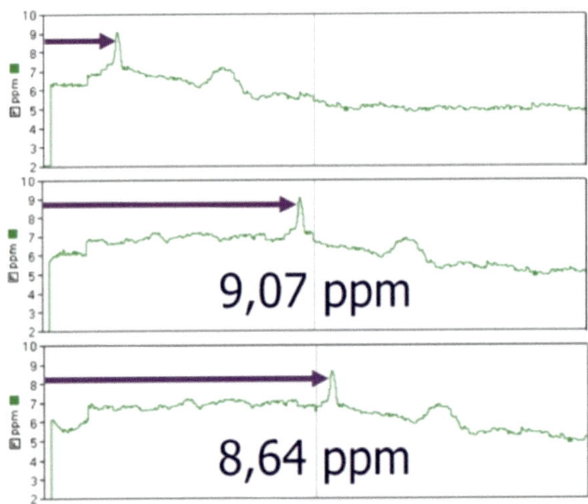

Figure 271 - Blending at the turquoise mixing point 99 km along the mainline, reducing the amplitude of the H_2S peak from 9.07 (8 hours, middle) to 8.64 ppm (9 hours, bottom). In this case, blending gas to reduce H_2S concentration by just 5% is enough to meet specification. [74]

Getting such an accurate prediction verifies that the model is calculating an accurate pipeline inventory and implies its parameters are set correctly. This gives confidence in other predictions for offline studies. [74]

Tracking carbon dioxide to reduce hold-back

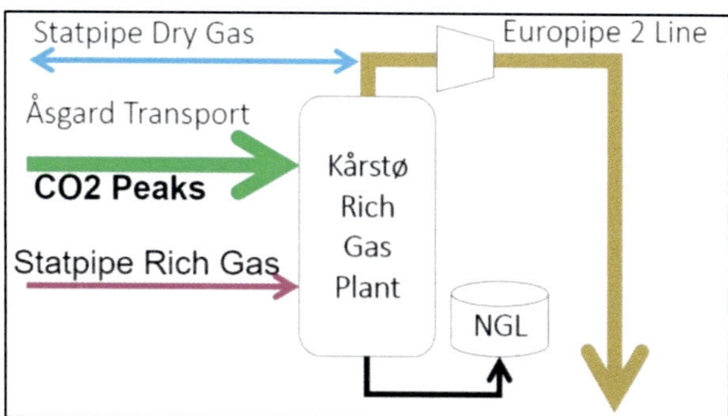

Figure 272 – Quality tracking for plant hold-back. Pipelines are color-coded. [74]

Three pipelines feed the Kårstø Rich Gas Plant. Åsgard Transport primarily delivers all its processed dry gas volumes into the Europipe 2 export pipeline. The two Statpipe pipelines are also used to bring gas in for export, as well as for blending across the plant. To successfully blend gas at subsea mixing points, flow rates must be controlled on the upstream boundary sources with consideration of the time lag to the subsea mixing point. A blending operation at a subsea mixing point due to off-specification gas may require one or more sources to reduce flow rate. This flow reduction results in lost profit during the blending operation which could typically last a few hours. Being able to accurately locate the position of the off-specification gas allows the control room operator to minimize the amount of gas to be held back for the blending operation. [74]

In handling carbon dioxide (CO_2) or hydrogen sulfide (H_2S) blending of rich gas at a subsea mixing point for one of the 17 sub-models comprising the total network, it is estimated a reduction in held back gas of 15 MMSCMD for 2 hours a day for 6 days a year due to improved system functionality and better quality tracking accuracy. This is estimated to result in 7.5 MSm^3/year reduced loss of rich gas production. [74]

Rich gas is converted to dry gas and natural gas liquids (NGL) for sale, and also impacts production of natural gas liquids at the delivering fields. Blending of the off-specification gas with the on-specification gas is calculated within the model for each individual component of the gas composition and for specified trace elements and properties. This allows composition to be estimated downstream of the mixing nodes where it is currently not viable to install compositional measurements, because these mixing nodes are along pipelines on the floor of the North Sea. [74]

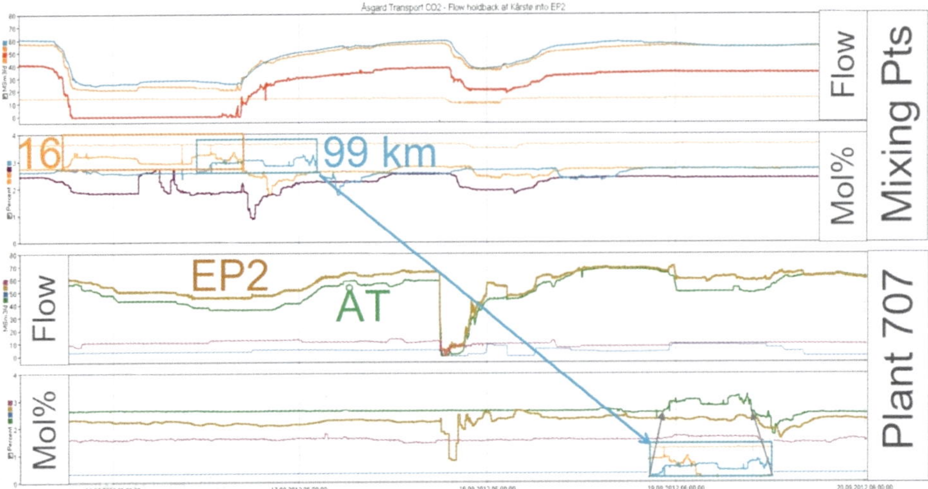

Figure 273 – Trends over time of flow rates and carbon dioxide (CO_2) concentrations (molar percent basis – mol%) at mixing points (top) and at a rich gas plant 707 km along the mainline.

The two upper trend graphs are from the subsea mixing points. A carbon dioxide (CO_2) peak exceeding 2.8 mol% is created at the orange mixing point, 16 km along the Åsgard mainline. The peak is slightly blended and compressed at the 99 km turquoise mixing point. It is copied into the lower graph for comparison. [74]

The two lower graphs are from the Kårstø Plant at the 707 km location. The peak is recognized a bit more compressed (green) due to flow reduction during travel. Åsgard Transport was delivering enough flow into Europipe 2, both before and after the peak arrived at Kårstø, but Åsgard volumes had to be held back to blend the peak down to acceptable Europipe 2 levels (2.4 mol%). [74]

The accumulated flow difference is found using the Atmos Trend statistics grid. The volume held back in the Åsgard Transport line at Kårstø is calculated by subtracting two numbers from the accumulated column. The annual benefit facilitated by this system's blending operations results in 7.5 MSm^3/year. [74]

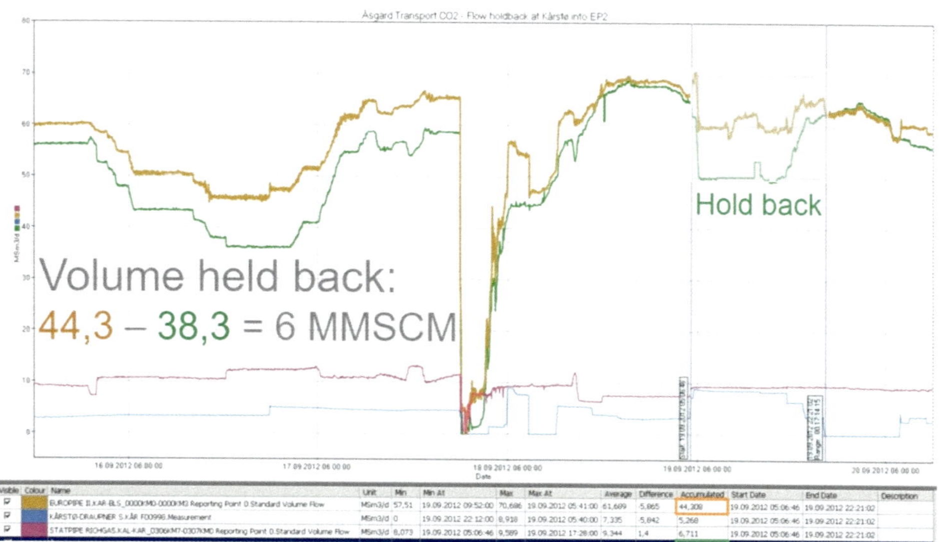

Figure 274 – Accumulated flow difference found using the Atmos Trend statistics grid. [74]

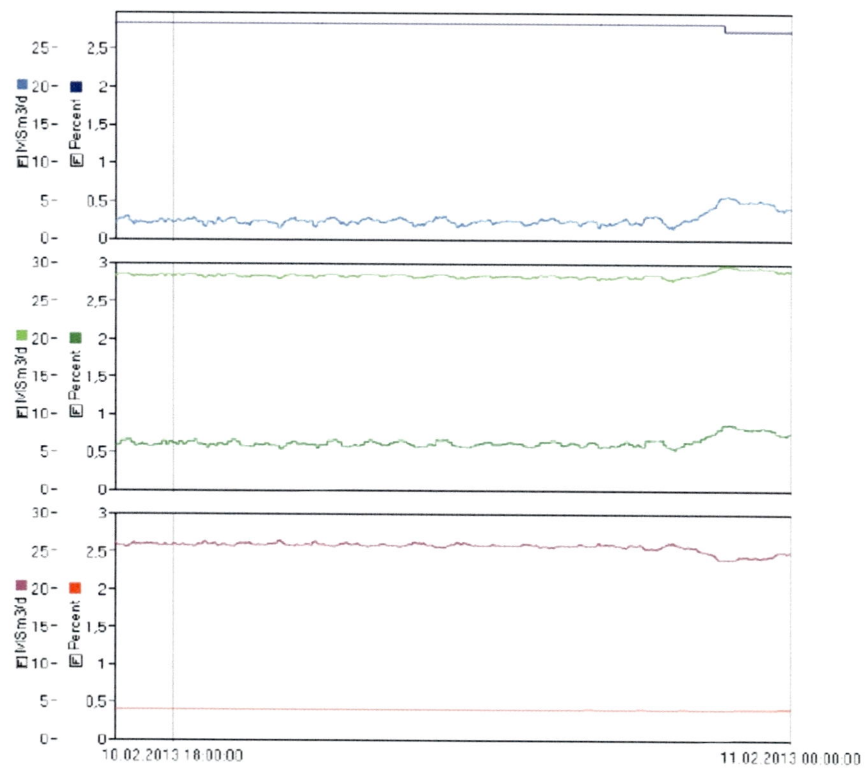

Figure 275 – Concentration of carbon dioxide (CO_2) before and after a mixing node. An upstream supply point (composition in red, flow in pink) mixes with a small incoming branch supplying off-specification gas (composition in dark blue, flow in light blue). This results in a mixed stream of blended natural gas (composition in dark green, flow in light green) that is on-specification. [74]

Pressure harmonization with blending

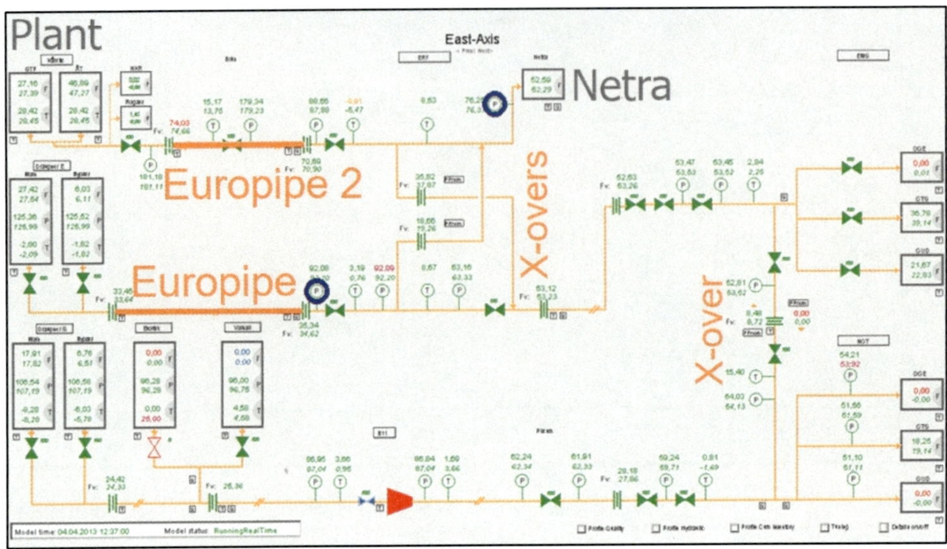

Figure 276 – *"East-Axis"*, a 4 in 1 network model, which makes modelling cross-overs possible for blending, as well as providing longer peak tracking routes. [74]

A few times a year Europipe may need to be packed up to deliver to the Netra exit point. The two pressures (blue) must be balanced correctly. Rich gas volumes at the plant may be held back. The benefit comes from reducing hold-back, adjusting linepack more accurately over a shorter period, and better quality tracking. The annual benefit of pressure harmonization with blending amounts to 1.25 MSm^3/year. [74]

Balancing pipeline pressures

On average twice a year the pressure level between two large pipelines must be balanced. This is so off-specification gas can be blended through a cross-over line. Given the large inventory, pressure drop and slow response from the source to the sink, operational procedures for balancing pressure conditions require extensive simulation. It is estimated that accurate arrival time estimates and look-aheads predictions optimized this operation such that the hold-back of rich gas on the upstream process plant was reduced by 5 MMSCMD for 3 hours, twice a year. [74]

If we only consider the four preceding types of repeating benefits to operations, in summary these represent 7.5 million USD in increased sales annually. [74]

Launching and monitoring look-aheads

A look-ahead case might be launched only when prompted, or automatically at a regular interval. If several are to launch at the same instance – say, those launching every 15-minutes coinciding with those launching daily – attention is paid to queuing. Do we launch the most-recent first, or the oldest-requested? Users can also manually request look-aheads on an ad-hoc basis. A watchdog monitors the progress of cases, confirming if they successfully reached completion without issue and without alarm. When a new look-ahead run is initialized, previous alarm statuses might be reset to *"off"*. Remedial action should be taken to mitigate what triggered the alarm in the first place, such as opening or closing relevant valves. That way, when the new case runs, the alarm, having been reset, is not raised again. It may be useful to log these events via SCADA so as to distinguish those cases where alarms were reset upon initializing a new look-ahead case, and those alarms which were intentionally cancelled because remedial action was taken to avoid the said violation. The way look-aheads are launched, simulated and reported is changing, and has brought about new use cases.

12.2 – ADVANCED LOOK-AHEADS

A new generation of ideas

In an age of unprecedented computing power, commercial paradigms, and disruptive technologies, it is worth stepping back and appreciating the bigger picture. In this chapter we take a fresh look at look-aheads, not as they have traditionally been used, but as an immensely powerful way of leveraging simulators to our advantage in promising new ways. We can go beyond validating nominations or verifying survival time to help with broader contingency planning, security of supply and decision-making.

Typically we wish to run a *survival time* case if a compressor trips, or a valve closes, or a supply changes by more than a certain rate of change. Could we trigger look-aheads based on a configured *"event"*? We've seen how look-aheads can serve as predictors of what will happen. But how do we decide whether we go ahead and do that in the first place, or what to do about a foreseen issue. Could this software be even smarter, acting as an advisor to suggest alternative ways of operating? As customers and regulators focus ever more on the environmental impact hand-in-hand with commercial considerations, might look-aheads help us to assess the carbon intensity of our pipeline? Or to perform the elusive tracking of green hydrogen or biomethane, distinguishing them from dirtier fossil fuels?

Look-aheads for liquid pipelines

Transient-model liquid look-aheads

Operators of a liquid pipeline are concerned with operating conditions along it. They swing valves open or shut, set pressure regulators, and execute practical procedures to inject batches, take off flow rate, etc. Whereas a gas pipeliner might launch a set of look-aheads every day or every couple of hours, a liquid pipeliner needs to launch them to happen more frequently. In the past there were significant challenges to overcome for look-aheads to adequately perform for liquid pipelines. They could not make timely predictions as they struggled to keep up with real-time! Some did consider devising a fundamentally different approach. For example, to have a liquid look-ahead solve quickly, they tried only asking it to simulate a time-step whenever a batch hits a pump station, since that is when the most substantial changes in pipeline behavior arise.

Incompressible mode look-aheads

Advanced prediction of estimated times of arrival (ETAs) of pigs and batches based on planned upcoming injection and delivery schedules are invaluable for scheduling personnel on a batch tracking system, and are quick to solve by using the simplified physics of a pipeline simulator's incompressible model. Even during steady state operations, hydraulic conditions such as flow rate may evolve over time as consecutive batches move along a pipeline. A look-ahead can alert the operator in advance to predict when a process condition is going to reach a certain threshold. Anticipating the exact instance that an event will take place allows the operator to prepare and take appropriate action: adjusting a pump speed or a regulator position for example.

The uncertainties involved in transporting many batches of distinct products from many injection points to many delivery points involve juggling a number of constraints such as terminal restrictions that affect flow rate. [75] Simulating how different operations play out helps operators make an informed decision.

Concurrent cloud-based look-aheads

Our discussion so far has touched on a multitude of ways in which different look-aheads assist or otherwise add value to a team operating their pipeline. Web based look-aheads offer options that were previously deemed impossible. Using multiple machines concurrently now allows launching, monitoring and analysis of many scenarios at once. When an alarm is violated, it is displayed clearly on a trend viewed from the same web interface. The results can be stepped through as an evolving profile.

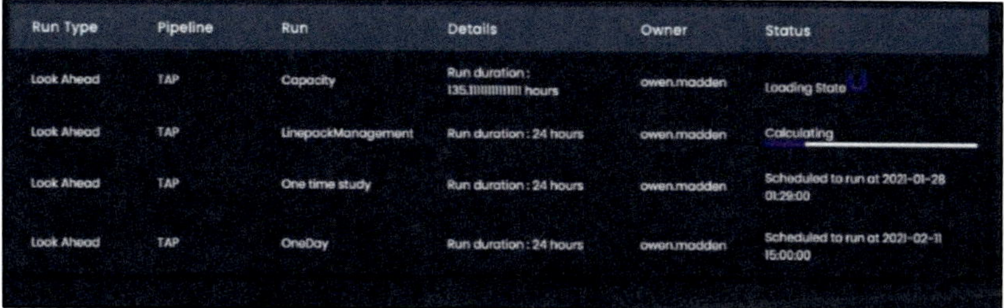

Run Type	Pipeline	Run	Details	Owner	Status
Look Ahead	TAP	Capacity	Run duration : 135.111111111 hours	owen.madden	Loading State
Look Ahead	TAP	LinepackManagement	Run duration : 24 hours	owen.madden	Calculating
Look Ahead	TAP	One time study	Run duration : 24 hours	owen.madden	Scheduled to run at 2021-01-28 01:29:00
Look Ahead	TAP	OneDay	Run duration : 24 hours	owen.madden	Scheduled to run at 2021-02-11 15:00:00

Figure 277 – A web interface for interacting with Atmos SIM informs us of the progress of cloud-based look-aheads. An ongoing look-ahead is loading a state, another is being calculated, and two look-aheads are anticipated. Details are displayed along with their scheduled times.

Pipeline	Run	Started	Finished	Owner	Result
Liquid line	Stop midline pumps	2021-03-23 11:00:00	2021-03-23 11:00:55	Employee 1	Completed with predicted alarms
GasNetwork	Close outlet valve	2021-03-23 11:00:11	2021-03-23 11:00:35	Employee 2	Completed with predicted alarms
GasNetwork	Close outlet valve	2021-03-23 10:45:11	2021-03-23 10:45:35	Employee 2	Completed
GasNetwork	New Case	2021-03-23 10:00:11	2021-03-23 10:00:12	Employee 2	Failed to find a latest state

Figure 278 – The web interface informs us about the status of look-aheads from several pipelines. This may, for example, be *"completed with predicted alarms"* or *"could not find a state"*, meaning there was no online saved state available because the online model just started.

Figure 279 – Results of a completed look-ahead run that raised an alarm. A trend of events (top), a slider to step through the simulation, and a hydraulic profile evolving over time. Here, the head has fallen below the lowest allowable operating head (LAOH), set as the elevation.

Case study: Teal Gas Pipeline (TGP)

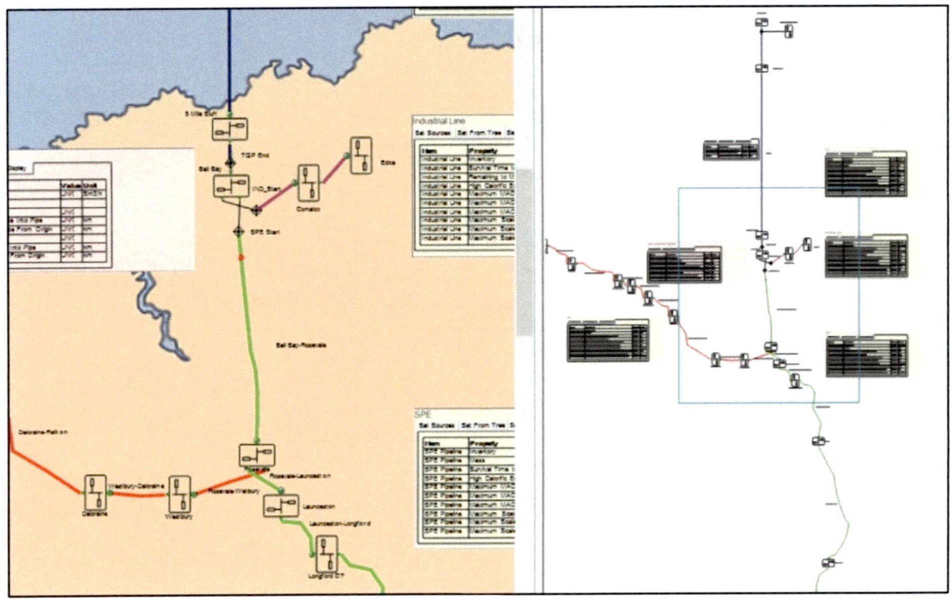

Figure 280 – A focused canvas view of the Teal Gas Pipeline's core sections (left) alongside an overview of the wider network (right)

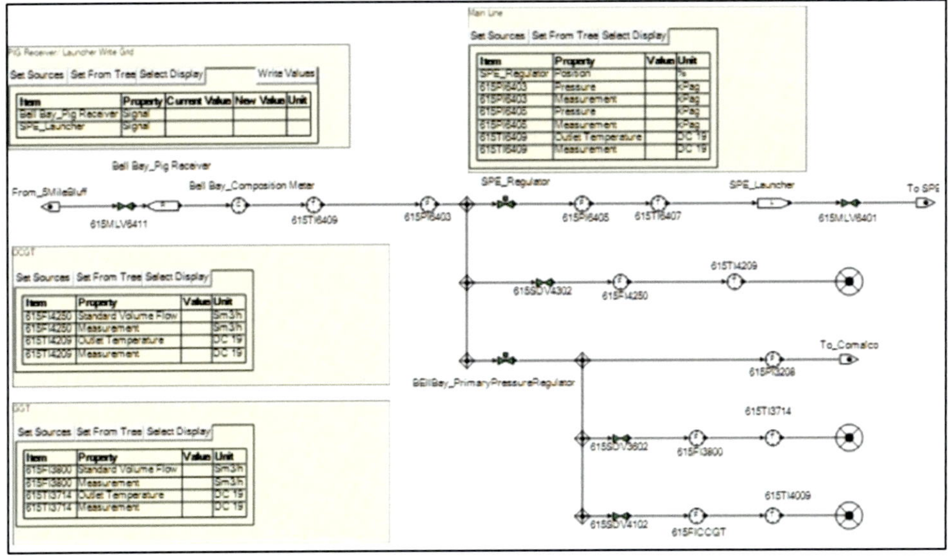

Figure 281 –The regulator station where the undersea pipeline meets a mainline, a branch line and several offtakes feeding gas-fired power stations.

The Teal Gas Pipeline (TGP) supplies an island with up to 130 TJ of energy per day in the form of natural gas from the mainland via an undersea interconnecting pipeline. Its bi-directional interconnection with the eastern gas pipeline and a hub injection point offers the opportunity for gas to be stored in the pipeline and injected into the main transmission system at times of peak demand. This makes it a part of a larger network of high-pressure gas pipelines, stretching over 20 000 km and serving several states.

As we explored in the chapter on energy, gas-fired power stations are relied upon as a versatile way to balance the electrical power grid. Here we see offtakes for open-cycle gas turbines (OCGT) and combined-cycle gas turbines (CCGT). These plants are especially important in regions which have seen a boon in solar- and wind-powered renewable generation, and where grid-scale energy storage has proven to be quite a major challenge.

For this reason, the team operating the pipeline routinely launches a number of tailored look-aheads on an automatic basis or manually by request. Look-aheads examine the impact of station outages in various combinations, and if a violation would eventually transpire. Others examine how the pipeline would handle ramping up the interconnection to transmit a higher energy flow rate for a few hours. This allows the pipeline to be operated flexibly and accommodate a change in nominations at short notice, in response to a sudden urgent need for gas-fired power generation to support the resilience of the electrical grid.

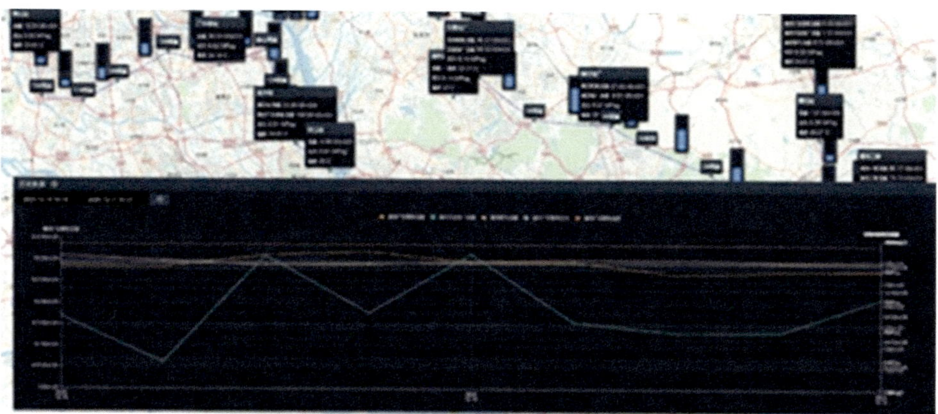

Figure 282 – Trends of pressure, temperature and flow rate at a selected point in the pipeline, as forecast by the results of a look-ahead. There are also screens for tabulated linepack and survival time results for each section of a pipeline, with consideration for the lowest allowable operating pressures (LAOP) along the length of the route.

Self-launching look-aheads

Event-triggered look-aheads

We might wish for certain events within the live system to trigger the launch of a look-ahead, such as a compressor failure, or sudden unpacking. We typically wish to run a survival time case if a compressor trips, or a valve position closes, or a supply changes by more than a certain rate of change. Could we trigger look-aheads based on a configured event taking place? *"Something has just happened, this is what it will mean for the pipeline"*

GasNetwork	Close outlet valve	Duration: 24 hours	Employee 2	Scheduled to run at 2021-03-24 11:00:10
GasNetwork	Compressor Tripped	Duration: 12 hours	Employee 2	Run when any compressor trips
Liquid line	Stop midline pumps	Duration: 4 hours	Employee 1	Calculating

Figure 283 – An Atmos SIM look-ahead scheduled to run at a set time, another look-ahead triggered whenever any compressor trips, and a third being calculated.

Peak-triggered look-aheads

There are sometimes opportunities for in-line blending of an off-specification gas peak before it reaches a delivery point. Look-aheads could be launched automatically as soon as a peak enters the pipeline to forecast when it is expected to reach a mid-line mixing point.

Leak-triggered look-aheads

If the pipeline simulator suspects there might be a leak, it could spin off one or several simulations to run concurrently alongside the real-time simulation. One of these could serve as a special survival time case. Another would predict the pipeline's response to an emergency shut-down and include the effect of a leak of the detected diameter at the detected location.

It has been tempting, for some, to consider the possibility that this may allow leak tests to be conducted via unmetered offtakes without disrupting the real-time transient model. They propose that if the model can back-track to the leak onset time (as detected) and account for the inventory of fluid released from the pipeline (also as detected, perhaps with an optional user-set override), then the model would remain physically correct following the leak test. Without this feature, they point out, we are left waiting for a long period of time so that our model's inventory is corrected – a characteristic scale time that may span days.

This wait is inevitable even if a state saved before the leak is restored, as it has the original inventory which is no longer in the pipeline. And it is also the case if we cold-start to a steady state, because again, a steady state has a different inventory to the real transient state. What may seem like a plausible alternative – directly modifying the linepack of a section of pipeline – would allow the user to set up a non-physical situation, a subtle difference that means it would be preferable to let the user inform the model about a leak's location, onset and diameter.

Historically, the omission of this feature may have been an intentional decision. The limited power and versatility of computing meant the focus was on how to keep real-time simulation persistent and reliable. Proposing to immediately and automatically launch a parallel simulation just in case a leak signal is legitimate seemed unfathomable. But today, it is not only perfectly fathomable, it is easily done by automatically configuring and launching a special *"look-ahead"* which follows live meters and adds the detected leak. If the user eventually confirms the leak is legitimate the real-time model could switch to continue on that basis.

Strategizing using look-aheads

Look-aheads can help implement a strategy to anticipate disruptive events, assess their risks, and respond in the most appropriate manner. In extreme cases, this may help avert a crisis. They help decide the best course of action, *in response* to a problem that appeared, or to *achieve* a target that is set.

An imminent emergency shut-down (ESD) ought to be flagged up urgently. Good *alarm management* alerts the operator to impending disruption of such magnitude more urgently than less concerning alerts, warnings and messages.

Predicting an avoidable violation in a couple of days' time typically gives an operator some breathing room to decide amongst a selection of possible alternatives. They have time to identify the best way forward, and take the most appropriate corrective action accordingly. Although there is often no one right answer, there may be a *best answer.* Deciding which is the best candidate amongst many possible operating strategies is the definition of optimization.

Automating certain tasks frees up operators time to focus on other duties. There are frequently occurring tasks that are amenable to automation using an online optimizer, like verifying if an upcoming operation will be smooth and efficient and advising on a better alternative if one happens to be possible.

12.3 – OPTIMIZERS

"Perfect is the enemy of Good" – Italian proverb

What is optimization?

To optimize is to find the best option amongst many possible alternatives. There is no universal optimizer. The best approach is tailored for each case. Someone might want an optimizer for *planning* their operation in advance. Another might want one for design studies to see how much a pipeline would cost to operate in typical conditions and what its capacity is. Another use case might be an optimizer to set targets for operating costs. But can we use an optimizer to help a team run a pipeline day-to-day? The brutally honest answer is it depends on *accuracy*, as discussed at length in the chapter on validation.

Often a problem statement is almost unique to a certain team working on a certain pipeline. The value added is as much about the user interfaces facilitating these unique inputs, and assisting with the analysis of results, as it is about the power of the calculations being performed. Making informed recommendations about the most influential factors, and the best way to modify an operation by leveraging them, is always going to involve a human in the loop. An optimizer is only as good as its interaction with that human.

Optimization is goal-seeking. It improves a model's parameters to achieve a specified task in the best possible manner. An optimizer advises how to best choose a parameter such as which pumps to run, or what pressure setpoint to use at a control valve, in order to optimize the wider overall operation. Optimization relies on defining an objective function, which is a measure of the quantity we want to improve. This might be a single number such as total cost of operation, or it might be a sum of terms to give, e.g. a least-squares measure of agreement between the model and reality. The target is to maximize or minimize this function. This might seem similar to calibration, but optimizers are much more challenging. It isn't a simple matter of choosing a new function to optimize. The algorithms involved, the parameters being adjusted, the nature of those parameters (discrete versus continuous variables, for example), and the sheer number of parameters an optimizer can adjust independently, are all completely different from those involved in a calibration problem.

Optimizers are in great demand yet have repeatedly proven to be challenging so far. Some early attempts fell short, partly because computing was not as advanced as it is today, and that made pipeliners cautious about optimizers. But it is unfair to give up on such a broad, useful concept. Teams expend untold effort finding the best way to design, modify or operate their pipelines. We can do better with our time than perform tasks at which computers excel. We won't digress to talk about error functions and linear programming – what should be appreciated is that optimizers are already very good if we fit the tool to the task.

Real-time optimization

It has proven surprisingly difficult to get pipeline optimizers into the control room. This has been true even when the optimizer was mathematically capable of producing valid operating plans. There are several reasons for this difficulty. One is that many pipelines are operating close to optimal efficiency in ordinary conditions, so an optimizer will just tell them to keep doing what they're doing, providing additional complexity but no added value. This is one place where accuracy matters: if an optimizer claims it can produce an improvement amounting to 3% of costs, but its underlying model is only accurate to within 2% on the running pipeline, it's hard to believe it really offers any benefit at all.

A second reason is that it's always going to take some back-and-forth with the operators to understand what all of the goals and constraints are that should be fed into the optimizer; and until this process is complete, any optimizer is not yet fit for purpose and so is going to produce visibly wrong and probably dangerous suggestions. This process is even more significant with optimizers than with online models, because an optimizer needs to get the model right and then also get the full set of operating constraints and goals right. A further problem stemming from this is that once any optimizer tells operators to do something unwise, they are unlikely to ever trust it again.

A third reason is that pipeline operations, especially gas pipeline operations, are highly coupled over space and time. So an optimizer will produce an entire operating plan for all stations and the entire schedule, but the operator may need to adjust it to account for an equipment problem or some unexpected event or nomination. This means that the optimality and perhaps even the feasibility of the solution go out of the window. Of course, the operator could run the optimizer again with their new conditions, if it's quick.

It is important to keep a highly skilled human in the loop, because if a pipeline operation goes wrong it can go very wrong. This means that a pipeline must always have skilled operators validating and signing off on an optimizer's plan. And so we might as well let those same operators actually operate the pipeline!

At this point it seems like an optimizer might be most useful in the control room if it could offer suggestions for small improvements that can be implemented piecemeal. It can help operators with suggestions such as: *"Instead of running compressors A and B, if we instead run compressors B, C, and D, the station overall efficiency would go from 40% to 60% at the current conditions."* To give another example of how optimizers can provide suggestions: *"Our incremental cost of head is twice as expensive at station A than at station B due to electricity penalty costs. If we could move some compression to happen at station B instead of station A, we can expect significant savings."*

Our optimization options

An *"optimizer"* might refer to a number of distinctly different concepts, expectations ranging from something as simple as a glorified scenario-launcher with algorithms to make progressive guesses, through to a general solver capable of exactly minimizing operating cost while considering everything a skilled operator might consider.

A simple *"pseudo-optimizer"* can be built without worrying about a *"proper"* exhaustive search algorithm. By launching a multitude of scenarios, either concurrently or one after the next, it reaches the best solution by trial-and-error. Many pipeliners and consultants already perform tasks manually in this way. Any iterative rinse-and-repeat procedure is an algorithm we can automate. Automated scenario-launchers, especially when complemented by automated guess-improvers, are an incredibly versatile tool. *

Trial-and-error (steady states)	Initial guess > steady state A > improve guess > steady state B > improve guess > steady state C > etc.... > *"optimal"* steady state
Trial-and-error (transient scenarios)	Initial guess > transient scenario A > improve guess > transient scenario B > improve guess > scenario C > etc.... > *"optimal"* scenario

Table 26 – Slow *"trial-and-error"* procedures to reach an optimal steady state or scenario

The trouble is: trial-and-error inevitably takes too long to answer all but the simplest questions. The effort expended developing an optimizer for a pipeline simulator is in figuring out how to take advantage of a problem's inherent structure to make a tailored optimizer solving it in a reasonable timeframe. Optimizing quickly is all about defining a problem, then choosing the right mathematical techniques and numerical algorithms to fit the task. Let's run through a few of the sophisticated tools used for this purpose.

---

* It is somewhat like the Atmos SIM Tuning Assistant, but instructed to do optimization.

Dynamic programming (DP)	DP breaks an optimization problem down into simpler sub-problems and utilizes the fact that the optimal solution to the overall problem depends upon the optimal solution to its sub-problems. It is the workhorse of liquid pipeline optimization. In many situations it can solve a steady state exactly and quickly.
Successive linear programming (SLP)	SLP is linear programming with successive linearization to solve optimization problems that are non-linear.
Non-linear programming (NLP)	NLP is the process of solving an optimization problem where some of the constraints or the objective function are nonlinear. Used in Atmos SIM's water pipeline optimizer and gas pipeline optimizer
Integer programming (IP)	IP deals with variables with discrete states, like whether a pump is on or off. *
Simulated annealing (SA) or similar methods	SA is used in the Atmos SIM liquid optimizer to deal with variables coupled between steps e.g. pump start / stop minimization †
Mixed integer non-linear programming (MINLP)	MINLP addresses nonlinear problems with continuous and integer variables. For example, the branch and bound method can deduce connectivity of valves and hoses in a ship loading terminal.

Table 27 – Tools and algorithms that have been used in the passes of pipeline optimizers

* The general integer programming problem is *"NP-hard"* which can be taken to mean that it cannot be solved in finite time for a real pipeline without doing something clever.

† An immense amount of expertise is needed to first get the problem into an amenable form to which simulated annealing can be applied.

Each optimization technique is best suited to certain aspects of a problem at hand, and so it is quite typical to use a different approach in successive passes.

Another promising approach is harnessing artificial intelligence: throwing large amounts of site data and / or simulation results at a black box. It learns to make predictions without our involvement – and without us knowing exactly how – using copious amounts of *"big data"*. It is one way other industries have answered questions about what an optimal arrangement would be. In the future, a machine learning algorithm could similarly help pipeliners tackle unwieldy problems, provided we properly hybridize it with our domain.

However, what most mathematicians mean when they talk about an optimizer is not a trial-and-error procedure (manual or automated), nor is it holding up our hands and asking an algorithm to treat our system as a black box. Rather it is a well-defined, overarching equation capturing the system's behavior, an equation we minimize or maximize – referred to as an *objective function* (OF).

Strictly speaking an optimizer must provably reach the absolute best possible solution, which can be mathematically demonstrated by reaching the minimum of an objective function that fully represents the decision space. In reality, there are many situations in which the problem at hand is not abstractable to a set of equations, but rather is more like the travelling salesman problem in that it becomes more complex as more decisions are made. This is rarely achievable; liquid steady state systems is one of the only places where we can guarantee an optimizer finds the true best answer. In all other applications we just want to get a really good answer, and for complicated problems, that's all anyone is ever likely to get in the best of circumstances.

A dynamic programming optimizer is fast – we can readily plug interesting problems into it. It's equivalent to a global search for a solution, if it does exist it'll find it. But often the domain requires some work in properly defining and adapting into a digestible form amenable to translation into a programming language. We need to be very careful about the question we are asking. Optimizers all work by the same components under the hood, but the problem statements are a broad family of equations that need to be deliberately defined.

In the next subsections we will look at how we perform optimization for liquid and gas pipelines. A simple approach is appropriate to answer simple questions, and advanced methods are employed for harder questions.

Liquid pipeline optimizers

Optimizing pipeline capacity

Even without thinking about time-varying, transient clutter, we can already answer some very interesting questions just by running many steady states consecutively (or in parallel!), each slightly different, then comparing results.

The problem statement of a water pipeline's operations seems, at first glance, to be deceptively simple. There are two reservoirs and a connecting pipeline. Yes, the reservoirs are large tank farms, and yes the pipeline crosses hills. But in the famous last words of the overconfident: *"how hard can it be?"* As it turns out, it is actually a difficult enough problem to be rather interesting.

The flow rate through any section of pipeline is constrained by operating limits.

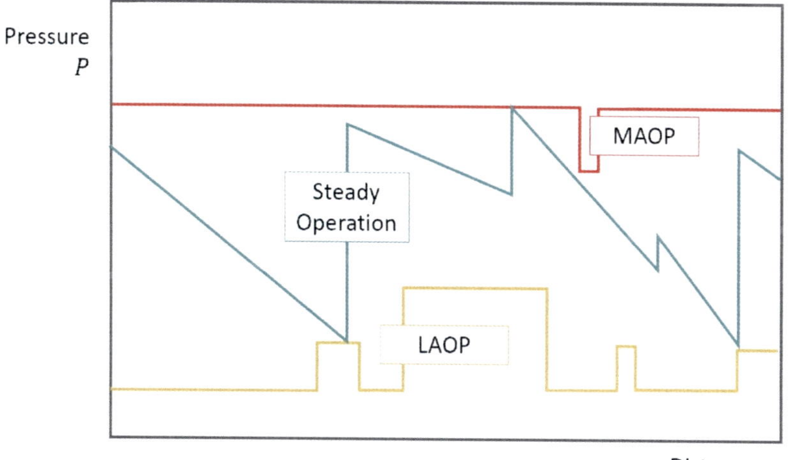

Figure 284 – A profile of pressure (teal), lowest allowable operating pressure (LAOP, yellow) and maximum allowable operating pressure (MAOP, red) along a liquid line flowing at capacity. A bottleneck is from where pressure touches MAOP to the downstream point it touches LAOP.

We can calculate a pipe section's maximum flow rate, known as its throughput capacity, with a simple type of bisection search. There are continuous control points in a pipeline: variable-speed drive pumps, throttle valves, or both. In each section extending from one continuous control point to the next, run as hard as possible without violating a maximum-type constraint such as maximum allowable operating pressure (MAOP) or rated pump driver power. We do this calculation down the entire pipeline, then ask where our closest approach is to a minimum-type constraint such as lowest allowable operating pressure (LAOP). If the pressure is not bumping against any minimum-type constraint, crank up the flow rate. And vice versa: if it is violating one then crank down the flow rate. This is a basic procedure: a doubly-nested Newton-Raphson iteration.

In throughput mode, an optimizer might tell us how to run at, say, 99% of its nominal capacity. It should also tell us why it cannot achieve a higher flow than that, specifying that the bottleneck is at a certain point in the line, or at a pump or a tank. For example: *"Because of [constraint X], the best possible outcome is to fulfil 43% of our nomination via [scenario Y]. Else [tank Z] would overfill".*

Optimizing pipeline capacity, pumps and tanks together

The Water Conversion Corporation (WCC) has used the Atmos SIM optimizer to recommend its pumping schedules. The network of pipelines is extensive, involving multiple branches linking numerous pumping stations and tank farms to take desalinated water from the sea coast to the depths of the arid desert, across a dramatic mountain range. Some pipeline sections are pumped while others are flowing gravity-fed. There are parallel pipes across the entire network, and pipes can be marked as unavailable for maintenance.

A pump's allowable operating range is defined as a window around the flow rate of the best efficiency point (BEP). Operation must remain on a pump map, but some regions are better than others. An optimum is sought based on defined constraints, minimizing pump-switching cycles and operational costs.

The model includes fixed-speed drive booster pumps, and variable-speed drive main pumps. Banks of parallel pumps are arranged in series within stations. An allowable operating range is defined as a window around the best efficiency point (BEP) flow rate, and operation must remain within the bounds of every pump map. We consider available suction head due to variations in tank elevation above sea level, plus the current height of liquid in each tank, corresponding to a difference of 20 meters of head available at the inlet to the pumps. This is reflected in the combined system resistance curve on the main pump performance curve.

A simple way to deliver a certain amount of liquid is to completely drain the tanks near the destination, which would leave the problem of refilling the tanks to another day. These tanks are used for buffer storage – balancing flow rates on an operational timescale – and also for strategic longer-term storage, so the user must specify maximum and minimum allowable levels in each tank.

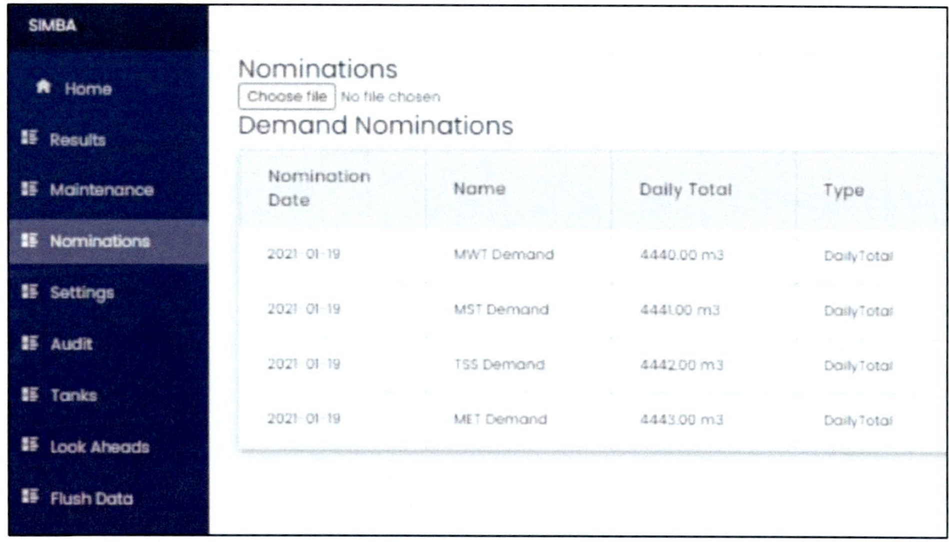

Figure 285 – Demand nominations can be updated to instruct the pump optimizer.

The inputs here are inventory of water contained within each tank, along with pump availability and consumer-side demand forecasts, expressed on a daily or hourly basis. While the live system is running, new overriding forecast or production values can be entered. The availability of tanks or pumps can be changed to account for failures or maintenance works. A user-configurable target stipulates tank levels to maintain at specified times, to ensure tanks are not completely drained at the end of each run.

Tank levels and the flow rates between them are physically coupled and can only be solved together, in what a mathematician would call a single global problem. This optimization weighs up many factors. For example, power contracts stipulate that electricity is cheaper during off-peak hours, so we should run the pipeline faster when power is cheaper. But simultaneously, more power is wasted overcoming friction in a scenario that flows faster for a while then slower during other periods, than in a scenario running the pipeline at a constant flow rate. So there is a trade-off here: we should also run the pipeline in as smooth a manner as possible to avoid excessive frictional losses.

A liquid pipeline's commercial flow rate is therefore constrained not just by what is physically possible but also by additional factors such as associated frictional losses. An optimizer determines the best scenario for operation. A 24-hour pumping plan minimizes operational costs, including cost factors associated with pump switching. Its outputs provide recommendations on which pumps to use and their setpoints in terms of flow rate, suction pressure and discharge pressure. It forecasts tank levels within operational limits, their fill or drain rates, the energy usage and its price.

This water pipeline optimizer has five passes each using a different approach:

1. A nested Newton-Raphson search calculates pipe segment capacity
2. Non-linear programming calculates flow rate schedules everywhere
3. Dynamic programming finds optimal pump combinations on each step for each segment
4. Simulated annealing to mix-and-match sets of pump combination options from step 3, to reduce maintenance costs due to pump starts and stops
5. A transient hydraulic model produces detailed results

If it cannot find a solution satisfying the requested demand schedule, first it relaxes tank limits early in the run, then it relaxes them throughout the run, all while remaining within the physical capacities of the tanks. If that doesn't do the trick, it starts reducing the delivered water, trying to minimize curtailment before minimizing operating cost. By looking at the location and time of the curtailment, the user can identify the bottleneck which is causing the problem.

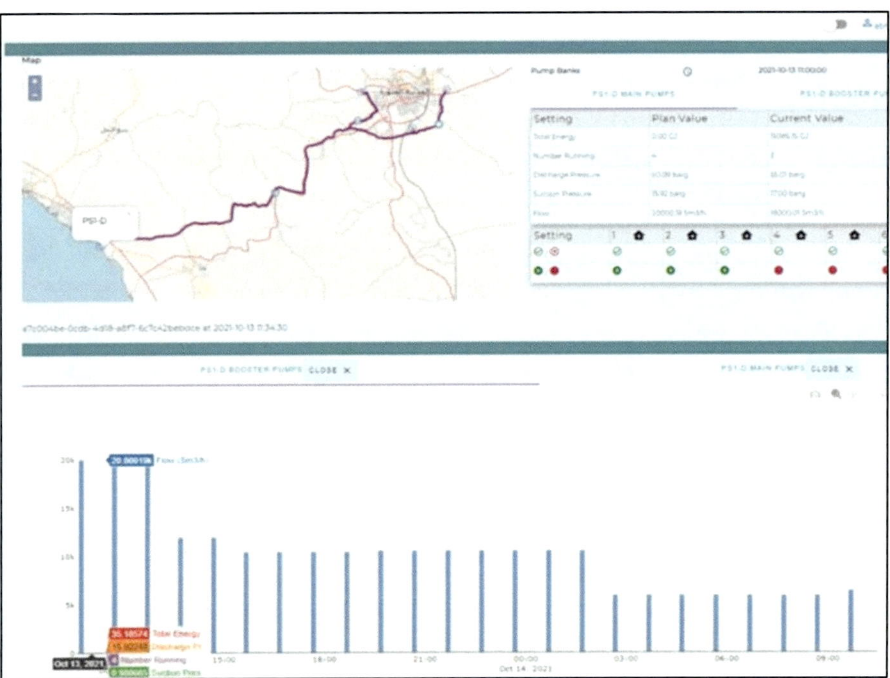

Figure 286 – The dashboards of the Atmos SIM Operational Optimizer. A geographic information system (GIS) view intuitively superimposes a pipeline's route and key locations onto a map, aiding operators spatial awareness when viewing results.

Optimizing all of the above with heaters and drag reduction

In the optimization of crude oil pipelines, the questions seen in the water industry around the operation of pumps, flow capacity along each section of pipeline, and levels in tanks are compounded by several further considerations that add several more operating parameters. The flow of a viscous liquid like crude oil can be supplemented by injecting a drag reduction agent (DRA). This could serve one or several purposes at once – the pipeliner might be seeking to:

- improve the throughput flow rate using the same pumps
- reduce the energy consumption for a given throughput
- extend the operating life of a de-rated stretch of pipeline

This optimization balances savings achieved by injecting DRA with purchase costs. Far more parameters are at play than just *"how much to add"*. There is also *"which batches to put DRA in"*, *"which stations to inject it at"*, and even *"which pumps to turn off and when, to let DRA through without destroying it"*.

The effectiveness of DRA depends on the type being used, which location it is injected, its concentration, the viscosity of the batch containing it, and the pipe's roughness. If flow needs increasing, we might want to ask: is injecting DRA a viable option, or do we need a further intermediate pump station? Such questions are today answered by launching and analyzing numerous offline study scenarios as successive simulations, but the problem statement is quite amenable to *"real"* optimization.

Another measure that can improve the flow behavior of viscous crudes is using in-line heaters, adding yet another parameter to the optimization milieu.

Optimizing batch schedules

The capacity of a batched liquid pipeline changes over time. It is a function of how we decide to transport our batches. So in fact if we transport our batches in an unwise manner, we might reduce the average capacity of our pipeline.

A shipper makes nominations; formal requests to transport a quantity from a receipt point to a delivery point. If nominations received by an operator exceed available capacity, they are reduced according to scheduling priorities stated in the pipeline's published tariff, which governs movements and deliveries. An enterprise typically includes a dedicated team to do this production planning.

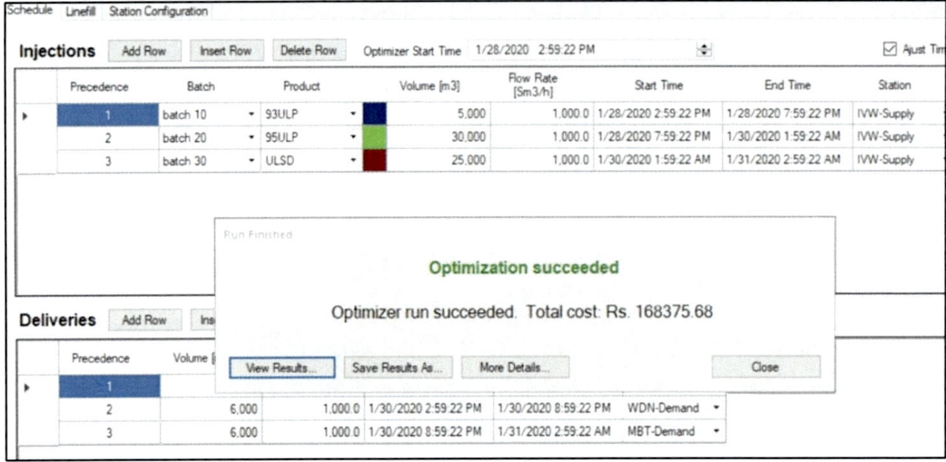

Figure 287 – An Atmos SIM batch schedule optimizer screen

Every product pipeline is scheduled by a responsible team. Despite scheduling being a common task, each team or pipelining company seems to use its own idiosyncratic process and terminology. Oftentimes a scheduler assumes their own set of procedures and terminology are the global way to do it, and expects outsiders to already be familiar with it. As late as the 1990's, product pipelines were scheduled by lining chalkboard erasers up on a table and pushing them along to represent the batches. In the early 2000's, they still used railroad charts,* with the pipeline's length plotted along the horizontal axis and time plotted on the vertical axis. The positions of batch heads are plotted on the chart.

The optimization of a given batch injection schedule involves making decisions about drag reduction agent, heaters and pump speeds. Drag reduction agent deteriorates in a predictable manner as it moves through a pipeline, and there is a trade-off with how much of it to inject and the cost associated with energy consumed by pumps. In addition, there is a cost penalty associated with the inconvenience and wear-and-tear of starting up or shutting down a pump.

---

* Railroad charts are so called because they originated in the railroad industry and scheduling trains is logistically like scheduling of incompressible batches.

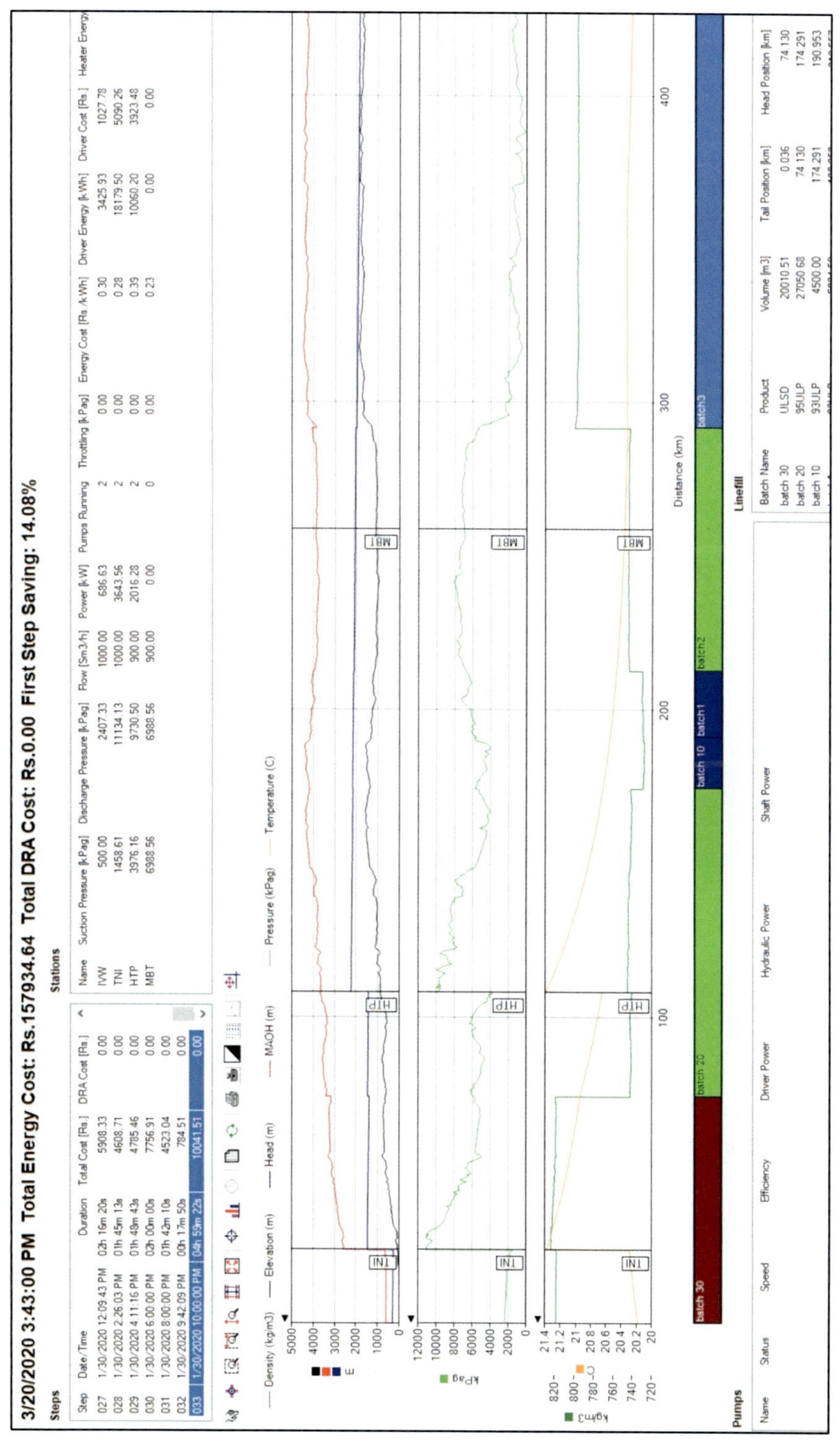

Figure 288 – Detailed hydraulic profiles within Atmos SIM batch schedule optimizer screens

Gas pipeline optimizers

A gas pipeline's throughput capacity is a complicated variable to quantify. It is intertwined with the pipeline's linepack capacity. Both the throughput and the linepack depend on the location of a recipient and the timing of their demand.

In our earlier chapter on gas pipeline operations, we examined the problem of pricing nominations. A procedure of running variations of the same transient scenario in succession, for the purpose of pricing nominations, is one that involves a repetitive task. Computers thrive on repetitive tasks, so surely this lends itself well to automation as a kind of optimizer!

A nominations planning optimizer can quickly optimize packing and unpacking. It runs dynamic simulations, typically with very long time-steps of about 1 hour.

We want to avoid impractical suggestions. For instance, a gas pipeline should not be operated flat out to entirely *"draw down"* its linepack to a minimum level in the first few hours of a scenario, and then sit idle waiting to pack it all back up again, if it could instead be operated in a more consistent, stable manner. In our earlier discussion of look-aheads, we touched upon the *"nose curve"* a gas pipelining team uses when planning their operations to strike an optimal balance between achieving higher flow rates temporarily and maintaining a reasonable inventory as linepack, so flexibility is available both now and later.

The aforementioned consideration alone immediately rules out any optimizer premised on steady states. Optimizing a gas pipeline requires simulation of transient scenarios. * This makes it rather hard to verify a proposed scenario is indeed *"the best"*, so a gas advisor is a more apt descriptor than an optimizer. Despite this caveat, the task of gas optimization is of monumental complexity.

Furthermore, the initial state in a gas pipeline is not at all steady, but transient. This means the best way to advise on a good way to deliver some nominations is to simulate a set of constrained scenarios starting from the most recently saved state, much in the same manner as a look-ahead. The key point to note here is that a look-ahead informs us about the repercussions of operating our pipeline a certain way. This helps us make valuable deductions if we continue to ask the right questions – i.e. if the user analyzes the results of a look-ahead and then sets up another *"better"* look-ahead on that basis, and so forth.

---

* Excepting, perhaps, low pressure gas distribution networks where linepack effects are less important, and their extent and complexity require fast solution.

There might be dozens of compressors and regulators across an extensive gas network. It is not immediately apparent what an efficient and versatile operating philosophy would be. It is difficult to pick apart a gas optimizer's solution into well-defined problems that can be individually implemented. Happily, just providing *recommendations* is useful to a gas pipeline operator. Pipeline simulators making recommendations at the gargantuan scale of a gas pipeline network have the potential to achieve millions in annual savings.

Interestingly, recommendations are almost a different problem from solving for the *best* operating schedule. A true optimization would have to combine every possible way of operating our compressors into an overall map per station. The compressors' surge limits cause this optimization problem to no longer be convex. This prevents us from using many convenient mathematical techniques.

Unless we're careful, an algorithm could inadvertently recommend some quite impractical or ambitious operating strategies that we couldn't reasonably pull off in practice. A solution must be practical, avoiding placing any unreasonable burdens on the operators. How often can a setpoint be adjusted? If we ended up not needing some delivery, would allowable operating limits be exceeded?

It would be revolutionary if such a system were available to gas pipeliners. The challenge of *how* to provide valuable operating recommendations that are optimal in gas pipelines is therefore compounded by the challenge of doing so in a timely fashion. In plain terms: for a gas adviser to fully reach its potential, it must run and analyze hundreds of simulations to reach a best answer within 15 minutes or so.

A gas pipeline operational advisor based on an online state

Atmos SIM has been deployed on a 2 600 km long network supplying power plants, refineries, distribution companies and dozens of city gates in five states, extending to a neighboring country. The network includes a hub: a central pipeline interconnection point at which gas deliveries are divided into branches to reach various regions. Real-time simulation helps operate fifteen compressor stations serving fifty demand points from five supplies. It is known that more efficient operation is possible, while maintaining key parameters such as linepack. Just as look-aheads have revolutionized the way that nominations are planned, an automated gas advisor could provide recommendations at a suitable time to be actionable, achieving big savings when pipelining millions of cubic meters.

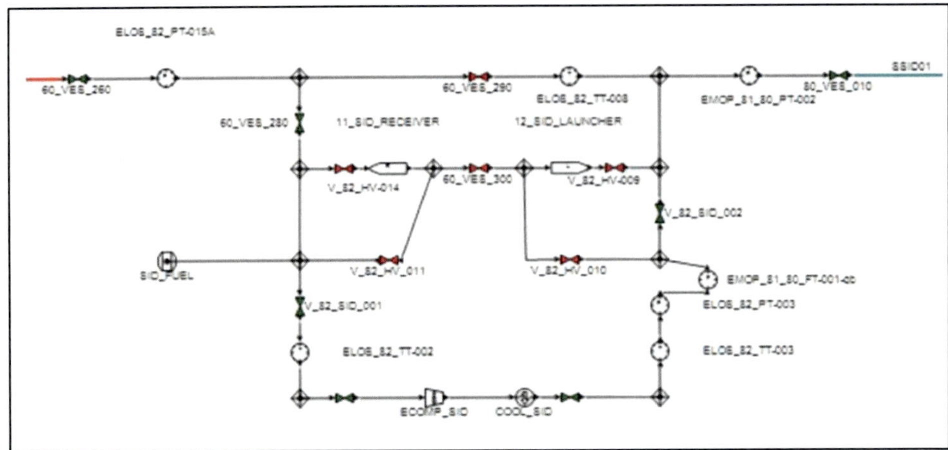

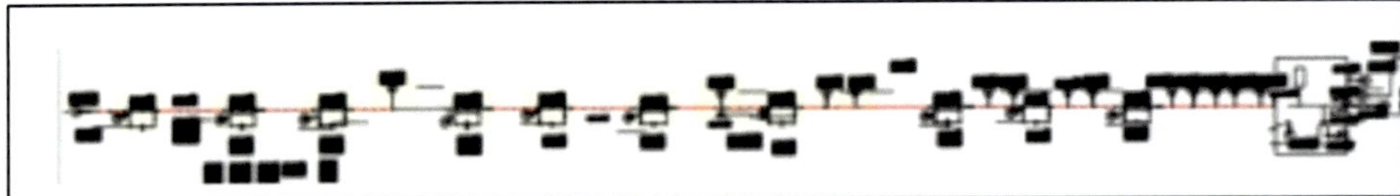

Figure 289 – One of many compressor stations equipped with a fuel offtake and a cooler (top) within a system-wide overview of the extensive gas pipeline network (middle and bottom)

Optimizing hydrogen unit commitment

In electrical power generation, there has long been a dispatching optimization problem around decisions relating to *unit commitment*. The challenge is to select an optimal set of power plants to activate during the right periods. A similar operational question arises in selecting one amongst an available set of possible production and storage facilities to feed hydrogen into a network of gas pipelines. This can be coupled with a second optimization around the decisions about whether to construct a given facility in the first place, an investment analysis coupled with an operational analysis. [88]

Optimizers of this kind have been available standalone for a while, but they neglected changes over the course of each day – such as half-hourly usage patterns – in their models for transport and distribution. [88] This is changing now as state-of-the-art pipeline simulators are harnessing computing power.

In simple terms: the installed capacity for generation and storage at a given set of locations depends on the *linepack* available in a pipeline between those suppliers and their many customers. Those linepacks in turn depend not just on a pipeline's diameter(s), but its operating pressures and flow rates at every given moment in time. An optimizer must be quick enough to yield its results within an acceptable timeframe. These decisions concern installations whose lifespan extends over decades, and whose customers evolve during that time.

Optimizers versus complacency

How do we build a pipeline? The economics of such a project require a huge capital outlay – in plain terms: it's tremendously expensive. A pipeline usually crosses a long distance through many different people's property. Acquiring that property is no mean feat. One owner can delay the whole pipeline's construction by simply demanding some ludicrous sum – say ten thousand times the value of his land! So, if those undertaking the project had to negotiate with everybody on that basis, it would become almost impossible to build a pipeline – in much the same way as a railroad, or any kind of infrastructure.

In many countries, the government will come in if this happens and use a power which the United States calls *eminent domain*. This forces owners to sell their land for its market value before considering the pipeline. But they won't do this for any old private company building a network: the government only provides this aid if the pipeline agrees to be a *common carrier*. This means that even though the pipeline is privately owned, it is akin to a public utility. It must be demonstrably and transparently neutral. For example, the pipeline cannot be excessively profitable: the rates it charges must drop if profits are too high. *

This kills the market for optimizers. If a common carrier pipeline gets an optimizer and reduces fuel costs, next year they must pass those savings onto their customers. But those customers aren't the ones making a decision to buy an optimizer – the pipeliners are! Pipeliners aren't usually directly competing with a rival pipeline in a way that makes them strive to reduce their prices.

One way past this conundrum is for governments to mandate that pipeliners must demonstrate they have indeed reduced their greenhouse gas emissions to minimum viable levels, a task which is now remarkably easy with an optimizer that properly accounts for all the environmental impacts of pipeline operations. Pipelining gas, for example, to displace coal-fired power has a positive impact.

---

* Other distinctions include that common carrier gas pipelines are paid to move an agreed *standard volume*, whereas gas distributors are paid to move *energy*.

13 – BEYOND SINGLE PHASE

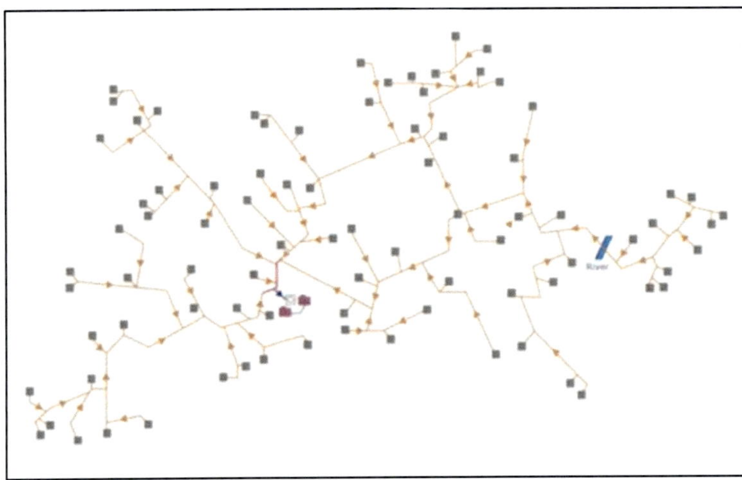

Figure 290 – Multi-phase flows from many wellheads make their way through a complex upstream gathering network to reach a central processing facility.

13.1 – THE CHALLENGE AND THE SCOPE

In this chapter we touch on a number of different flow regimes beyond a single phase. Our focus is on column separation and slack-line flow incorporated in transient physical models. Taking mainstream, off-the-shelf pipeline simulation beyond a single phase is an advance that has been awaited with bated breath. It facilitates leak detection during slack-line operations, and has numerous other benefits.

If flow is not single-phase, experts find it notoriously difficult to estimate even its steady state pressure drop. Some liquid pipelines are intentionally operated in slack-line rather than keeping them tight throughout. On others, column separation only occurs inadvertently, the operating intent being to avoid that condition. In a gas pipeline, liquid drop out is a concern, and despite measures taken to avoid it we find it somewhat inevitable. In some applications, such as transporting natural gas along with natural gas liquids, there are advantages to run at supercritical conditions in what is known as the *"dense phase"*, rather than as a two-phase flow. Navigating all of this for the first time is quite daunting. There is a big family of phenomena in the diverse world of pipelining in which a fluid is not entirely in a single phase!

A multi-phase simulator's capability should be carefully qualified. What kind of flow beyond a single phase are we talking about exactly? Are we only interested in relating steady-state flow rate to pressure drop? Do we need to predict liquid hold-up during stratified flow? Every answer comes with cautious caveats, and the user must think very carefully about the nature of their flow.

Rising to the challenge of flows beyond a single phase demands that we examine each problem on its own merit. We must address each distinct behavior with its own considered approach, avoiding needless complexity and only opting for a physical transient model where it is truly needed to simulate a given pipeline.

13.2 – LIQUID DROP-OUT IN A GAS PIPELINE

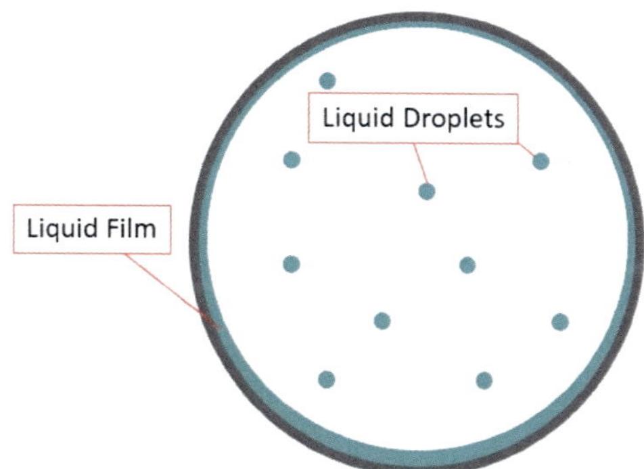

Figure 291 – Cross-section of a gas pipeline with liquid that has dropped out. At high flow rates, it will form spray in the central area of gas, and a film around the circumference. This film is thicker at the bottom of the pipe.

When a typical gas pipeline is stopped during shut-in, it warms up. When it flows, it experiences frictional losses that cause a pressure drop. That in turn causes natural gas to cool down along its route by the Joule-Thomson effect, which results in an equilibrium temperature below ambient. When stopped, heat flows from the ground and the gas temperature gradually rises toward ambient. Its density falls but it occupies no more than the original volume, so pressure rises.

Water and heavier hydrocarbons might *"drop out"* along a natural gas pipeline. A significant change of phase from a vapor to a liquid is a major event for the molecules it's happening to: the volume occupied by a fluid reduces by several orders of magnitude. Moreover, the presence of liquid water in a natural gas pipeline carries a risk of forming hydrates. If slugging occurs, violent transients arise. Liquids drop out of a gas at a different rate during shut-in, affected by the temperatures, and complex phenomena such as retrograde condensation.

Operators of natural gas pipelines give due consideration to liquid drop-out. As rich gas is particularly susceptible to these issues, pipelines from wellheads to a gas processing plant are equipped to handle liquids. *"Dry gas"* pipelines, too, may need to arrange for suitable accommodations to be installed at stations, depending on how *"dry"* it is. Before receiving an incoming pig, we must handle a substantial slug of natural gas liquids (NGL's) preceding it. Slug catchers are installed because equipment designed to handle gas might be ruined if so much liquid enters it. Condensate arrives in quite substantial quantities, so provisions are duly made for its storage and onward transportation to suitable processing facilities.

In a wet gas pipeline, there are many operational conditions in which the fluid might separate into three distinct phases: gas, condensate and water. [*] Instead of a full simulation tracking these elements separately, along with their interactions and mass transfer as pressures and temperatures change, again a single-phase model can sometimes be used for accurate simulation of real-time pressures and flow rates and thereby for leak detection. In this case a method for estimating liquid hold-up is required to be working in tandem, reporting that result via rules of thumb in one special case. [76]

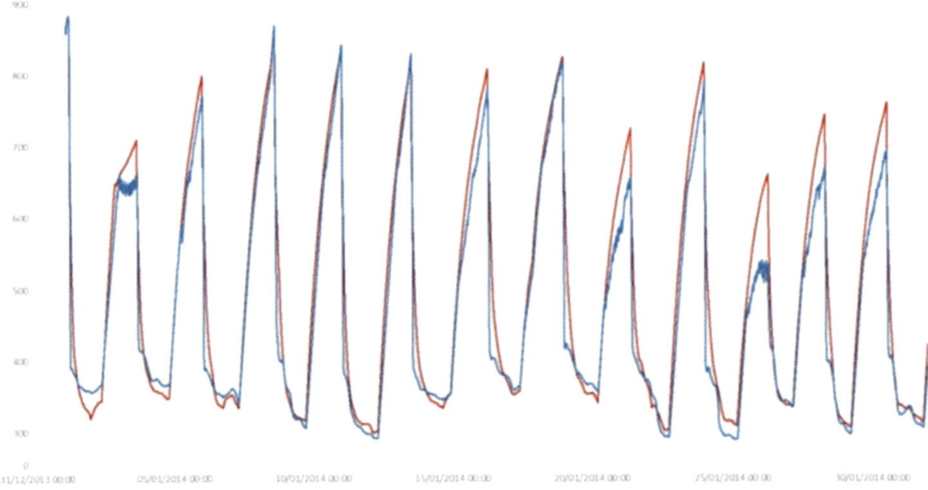

Figure 292 – Liquid volume (m^3) within a wet gas pipeline over time, estimated by a simplified accumulator model (blue) compared against what is expected by a multi-phase model (red). [76]

This approach is only possible with a couple of caveats. Firstly, the liquid must remain stratified – in the case of the pipeline in the reference, it entirely runs downhill and blends into a dry gas pipeline, so there is never slugging or accumulation at low points. And secondly, there must be no sudden changes in operation. If, and only if, both those caveats are met, liquid hold-up can then be estimated without vapor-liquid equilibrium calculations. On the basis of either a pressure drop method or an accumulator method,

[*] If hydrate inhibitors like mono-ethylene glycol (MEG) are added, those will be present in the water liquid phase as well.

each pipeline is considered to be slightly smaller than its nominal internal diameter, learnt efficiency thereby capturing the effect of liquids whenever and wherever they drop out over time. This single-phase model with a tuned parameter performs leak detection well, indicating that it captures the *"important"* physics in that special case. [76]

A good two-phase or three-phase* physical model could potentially perform better in leak detection than a single-phase model with tuning to adjust the pipe diameter. Once a leak happens, particularly a large leak, the physics of a single-phase model from that point onwards would be poor. During the rapid pressure and temperature drop that accompanies the onset of a large leak, much more liquid drops out, making the dynamics of the liquid phase(s) important for accurate simulation. A multi-phase model gives better leak location in that scenario. As some wet gas pipelines *always* operate with a lot of liquid, they really need a multiphase model for any accurate simulation.

In practice, maintaining any true multi-phase model requires such a substantial effort that pipeline enterprises may choose to use the single-phase model instead. For this reason, it is important for pipeline simulators like Atmos SIM to complement its single-phase model to deal with some liquid in a gas pipeline and some gas in a liquid pipeline.

---

* In this context three-phase models track a vapor phase and two liquid phases, one being largely-hydrocarbon, and a separate one being largely-water. This reflects the observed physical behavior of pipelines with high water content.

13.3 – VAPOR-LIQUID EQUILIBRIUM

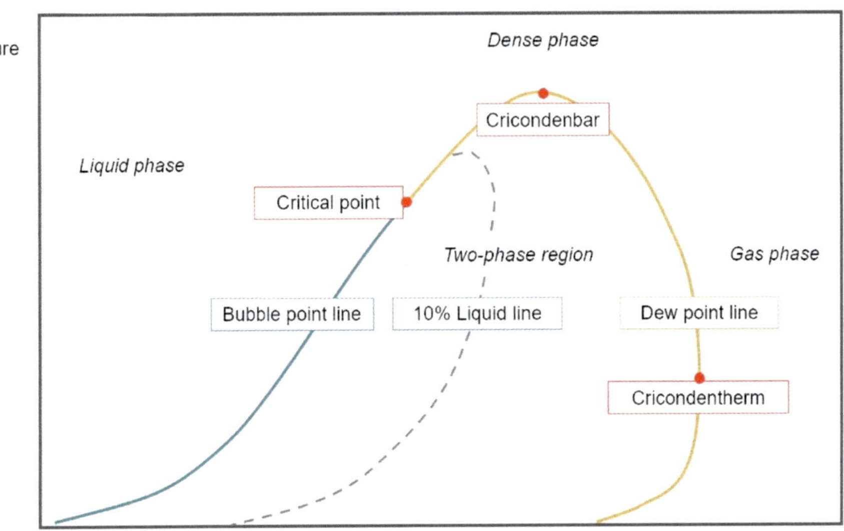

Figure 293 – The phase diagram of a fluid mixture like natural gas includes a liquid phase, a vapor phase (gas phase), a supercritical phase (dense phase), and a *two-phase region* bounded by a *bubble point line* and *dew point line*, together forming the *phase envelope*.

The critical point is a pressure and temperature above which there is only a supercritical phase, rather than a distinct liquid phase and a gas phase. There are pipelines that do operate in the supercritical phase – referred to as the *"dense phase"* by pipeliners, a couple of notable examples being ethylene pipelines and carbon dioxide pipelines. Aside from those exceptions, the vast majority of pipelines operate in the liquid phase, the gas phase or a multi-phase regime.* The cricondenbar at a location is the highest pressure at which distinct dense and vapor phases co-exist. And the cricondentherm at a location is the maximum temperature at which the distinct liquid and vapor phases co-exist.

Slack in liquids evaporates some liquid to the vapor phase. Drop-out in gases condenses some gas to the liquid phase. These two phenomena are similar in that in both cases, a partial change of phase happens when conditions at a point along the pipeline cross the *"phase envelope"* into the two-phase region. However, these two phenomena behave differently enough from one another that we conceptualize and model them separately.

---

* The term *vapor* is used instead of *gas* if a gaseous phase is in equilibrium with a liquid phase, or sometimes just if liquid drop-out is considered likely.

In a liquid pipeline, vapor pressure – i.e. bubble point pressure – is of more interest than bubble point *temperature* because the former is the more likely cause of slack. Slack happens when a liquid's pressure falls below its vapor pressure. For this reason liquid pipeliners speak about *"vapor pressure"* rather than *"bubble point"*. It should be borne in mind that although the vapor pressure of a hydrocarbon liquid at a location depends on its temperature, in a liquid pipeline temperatures don't change all that much over time or space, so the onset of slack is almost always caused by a change in pressure not temperature.

Conversely, in a natural gas pipeline, liquid drop-out (condensation) is unlikely to be due to the dew point pressure. Dew point temperature is of more interest as Joule-Thomson causes that fluid to get progressively cooler along its route. The trajectory of the fluid on the phase diagram as it goes down the pipeline moves significantly on both the pressure and temperature axes.

How much fluid exists as a vapor or as a liquid is determined by a physical calculation of *vapor-liquid equilibrium* (VLE). Some equations of state can be harnessed to perform these full calculations, but pipeline simulators only run these calculations when and where needed, as the task is resource intensive. Detailed VLE calculations are not generally needed in slack models for liquid pipelines because the mass of fluid in the vapor is quite small and it's not important to get its composition exactly right for the sake of its fluid properties. Conversely, detailed VLE calculations are needed for liquid dropout in gas pipelines, where the behavior of the liquid can depend strongly on exactly which components have dropped out.

Scenarios in which vapor-liquid equilibrium calculations are useful include several we considered in the chapter on mimicking leaks. One scenario is a liquid that boils as it escapes, separating into distinct vapor and liquid phases with different compositions. Another scenario is the composition of liquid that drops out of a wet gas as it becomes progressively cooler along its route. A few scenarios are explored in a paper on pressure vessels in the references. [77]

A two-phase system acts as a thermal reservoir, absorbing or releasing heat while remaining at constant temperature. It takes a large amount of heat to boil a liquid; for example, it takes five times more heat to boil water than it does to raise its temperature from freezing (0°C) to boiling point (100°C)! When vapor condenses to liquid, it releases heat, and when liquid evaporates to the vapor phase, it consumes heat. Latent heat of vaporization characterizes the magnitude of this effect, documented in reference books for a fluid at a given set of conditions.

13.4 – SLACK IN A LIQUID PIPELINE

Figure 294 – Cross-section of a part-filled liquid pipeline.

Air pockets in a water pipeline

Air pockets form surprisingly easily in a water pipeline. Taking one's thumb off the edge of a straw to let it drain *mixes* with air. Water is a remarkably excellent solvent, meaning many gases dissolve readily in water at ambient conditions. At some point in its history, water in a pipeline was no doubt in equilibrium with ambient air, so there are bound to be gases dissolved in it. Those gases can come out of solution, a problem that has been known since time immemorial, with well-diggers and those maintaining conduits knowing to test the air enclosed above a body of water before entering. Depending on the operating conditions along a water pipeline, air may be expected to come out of solution in significant quantities, accumulating at high points. Vents are needed there.

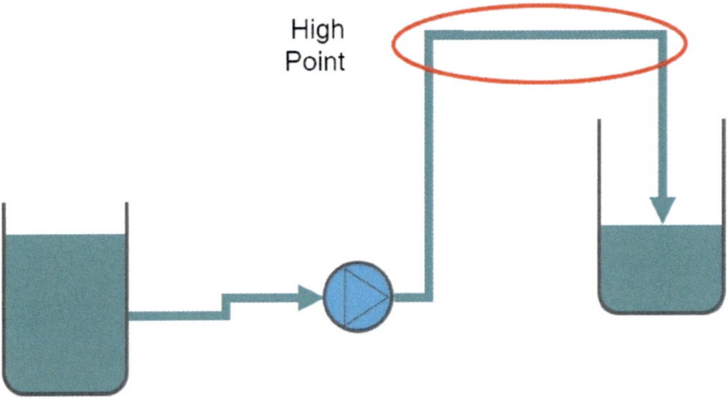

Figure 295 – An isolated high point in a water pipeline is where air is likely to accumulate.

The vapor pressure of a liquid

If a hydrocarbon liquid, such as refined products and crude oil, has been exposed to air since it was extracted from the well, then it will have air dissolved in it and we expect it to behave as air does in water pipelines. There is a further important consideration in hydrocarbon liquid pipelines, namely that the pipeline fluid itself might change phase, partially going to vapor. A hydrocarbon can be characterized by its Reid vapor pressure (RVP) which is the vapor pressure as measured at specific pre-defined conditions. The true vapor pressure (TVP) of the same mixture of liquids might differ because:

- TVP is a function of temperature whereas RVP is measured at 37.8°C.
- TVP depends on vapor-to-liquid ratio whereas RVP is defined at 4:1
- TVP might be defined to exclude dissolved water and air, unlike RVP
- RVP is measured such that some of the sample might have volatized, affecting its composition before it's warmed to the specified 37.8°C

The *Clausius-Clapeyron* relation relates two vapor pressures (P_1, P_2) of a pure liquid at corresponding temperatures (T_1, T_2) to its latent heat of vaporization (Δh_{vap})* by assuming latent heat is independent of temperature:

$$\ln\left[\frac{\mathrm{P}_1}{\mathrm{P}_2}\right] = -\frac{\Delta h_{\text{vap}}}{R} \cdot \left(\frac{1}{T_1} - \frac{1}{T_2}\right)$$

Where R is the universal gas constant. Chemists like the Clausius-Clapeyron equation as it's theoretically elegant. It is good enough for most applications, but makes assumptions which don't always hold true, giving inaccurate results at high pressure or near the critical point. For oil in a pipeline, engineers use a modified form: the *Antoine* equation, fitting three parameters to experimental vapor pressures measured over a restricted temperature range. These *Antoine coefficients* (A, B, C) vary with composition and are specified for each mixture.

$$\log[P] = A - \frac{B}{T + C}$$

---

* This calculated enthalpy of vaporization is mole-basis. The molar volume of a gas is not conveniently constant as it would be for a condensed liquid phase.

For crude oil, coefficient C is typically set to zero, with temperatures then given in absolute units, (i.e. Kelvin or degrees Rankine). * Eliminating by substitution:

$$B = \frac{\log[\mathrm{P}_2] - \log[P_1]}{\left(\frac{1}{T_1} - \frac{1}{T_2}\right)}$$

$$A = \log[P_1] + \frac{B}{T_1}$$

$$P_{\text{vap}} = 10^{\left(A - \frac{B}{T}\right)}$$

This calculates the vapor pressure (P_{vap}) of crude oil at any temperature. If the user doesn't provide reference values, Atmos SIM defaults to take a basis as a crude of 25 °API, having a vapor pressure of $P_1 = 45\ \text{kPa}$ corresponding to a temperature of $T_1 = 300\ \text{K}$, and another at $P_2 = 170\ \text{kPa}$ at $T_2 = 350\ \text{K}$. The Antoine equation is reportedly accurate to a few percent for most volatile liquids encountered in a pipeline whose vapor pressures exceed 0.01 bar. [78]

Historically, Antoine's motivation in his research was avoiding excessive cavitation in naval screw propellers. A liquid whose pressure falls below its vapor pressure forms small bubbles that remain in solution (i.e. cavitation). We don't want excessive cavitation in our pumps for the same reason Antoine didn't want it damaging propellers. Along a streamline a bubble is subjected to increasing pressure. Once pressure rises above the fluid's vapor pressure, the bubble collapses very violently; so violently, in fact, that it can pit steel surfaces.

If bubbles along a pipeline are carried into a station they might damage pumps and other equipment, so liquid pipeline stations are equipped to handle them. The more common scenario in hydrocarbon pipelines is that the bubbles find a high point and accumulate there to form a large contiguous vapor region. † What can happen then is that – unlike air pockets which slowly dissolve if the liquid is saturated – recently-formed hydrocarbon vapor pockets can suddenly collapse, creating a pressure surge which might pose a risk to the pipeline.

---

* This is the August equation. All derived from the Clausius-Clapeyron relation.

† Confusingly, some pipeliners refer to that region itself as a void or an empty space. This is misleading: the volume is occupied by a certain mass of fluid as vapor, albeit negligible compared to the mass of liquid.

Column separation whilst stopped

At lower pressure, a liquid's boiling point is cooler. This is why eggs take longer to cook at higher altitudes: on a mountain or a plateau, as water boils at a lower temperature than the commonly quoted 100°C at sea level, then takes heat out of the cooking pan.

A flowing liquid pipeline is almost always warmer than its surroundings. So if a liquid is unmoving, such as when it sits in shut-in overnight, it typically becomes progressively cooler over time. This fall in temperature causes pressure to fall: the volume (the pipe) and mass (molecules of liquid) stay the same, so density stays the same too (mass divided by volume). Any fluid kept at constant density with falling temperature will have a falling pressure. This effect is quite significant – when a typical crude oil's temperature falls by one degree (1°F) at constant density, its pressure falls by about 7 atmospheres (100 psi)! Although the running temperature of a liquid pipeline is only a few degrees above ambient, a fall of those few degrees during shut-in is often enough to cause a fall from average running pressure to the hydrocarbon's vapor pressure. This in turn might cause liquid to vaporize starting from the high-points where pressures are at their lowest. This vapor pocket gradually grows in both directions down the slope of the hill.

If a liquid line goes slack inadvertently whilst stopped, such as in this scenario it is called *column separation*. The terminology seems woefully inconsistent in the industry, but that is the term pipeliners seem to prefer when temperature is the *"cause"* of vaporization rather than pressure (in contrast to flowing slack).

Three identical lines contain petrol, diesel, and crude oil. Place them in shut-in, and launch a transient simulation, such that temperature falls by 11°C. Does pressure fall?

If the fluid is actually a blend of several components, the lightest components are the most volatile, and so are more liable to vaporize – they go partially into a vapor phase, while all the heavier components remain in the liquid phase. If there is significant air in solution, we expect that to be eager to come out also.

How column separation forms and fills in is well understood. At a high point, whilst the line is stopped, column separation forms an all-vapor region: liquid hold-up suddenly drops from full (100%) to empty (0%). An indicative value of the magnitude of this phenomenon was reported by a pipeliner as an *"empty space"* of around 1 600 m³ along their route, which of course is not empty but rather occupied by product vapor (a *"vapor pocket"*). What happens when that line starts up? How does slack behave when a liquid pipeline is flowing?

Avoiding flowing slack in a tight-line liquid pipeline

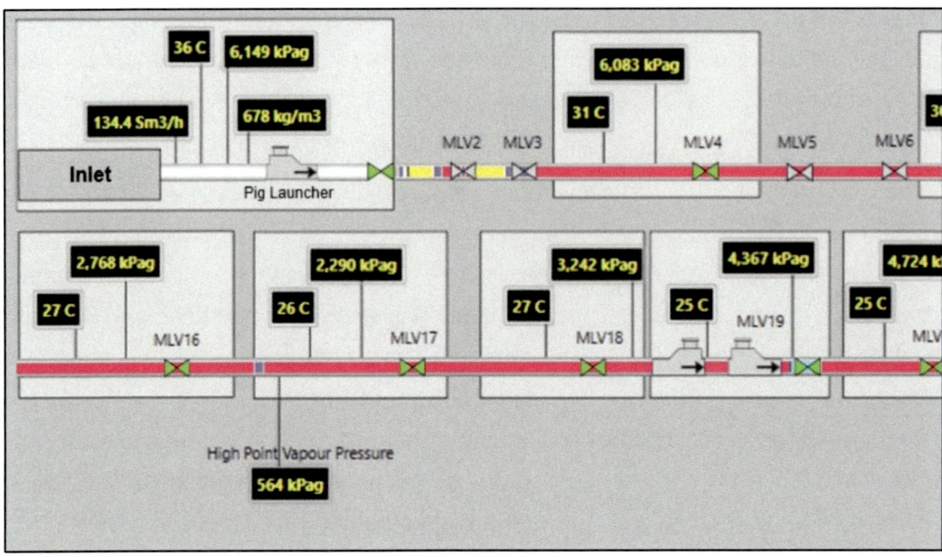

Figure 296 – Atmos SIM calculates the vapour pressure of the fluid batch at a high point and displays it on the Atmos HMI alongside operating conditions and other batch tracking data

Before modern pipeline simulators, a typical liquid pipeline was controlled by operators supervising flow rates and pressures at each end without controlling minimum pressure at intermediate points. It came to be expected that vapor accumulates at high points if pressures there fall below a product's vapor pressure. After flow stopped, liquid flowed from higher points to the lowest points. Every time flowing operation commenced, part of the vapor condensed back into liquid. Before achieving steady flow, operators waited a period of half an hour or longer during which fluid was being pumped in while nothing came out of the other end. If a real-time model does not model slack, no information is available to discern if vapor is compressing and condensing or if a leak is happening (with fluid escaping all the while). For this reason, some liquid pipelines implemented projects to improve safety by avoiding the formation of vapor pockets. Eliminating vapor added an advantage: faster billing of transported batches by improving measurement. A straightforward hydraulic profiler, available in any good pipeline simulator, facilitates the calculation of minimum required pressures at controlled points to maintain a tight-line operation and avoid column separation happening while stopped. [79]

Vapor pockets can form while flowing, not just during shut-in. As a liquid speeds across a hilltop, the local pressure may drop below the minimum required to *"keep the line tight"* (local vapor pressure), causing flowing slack. Flowing slack (*"slack-line"*) presents a number of challenges, complicating leak detection and giving rise to several new surge scenarios. Even while the same pressure is maintained, a slack region might arise in a liquid pipeline as different batches move through in succession. As a new batch enters a sloped pipe, it changes the pressures along the pipeline considerably. The change in vapor pressure with each batch is fairly unimportant; for most pipelined liquids it varies by less than one bar. An exception is liquefied petroleum gas (LPG), whose vapor pressure can be as much as ten bar higher than other liquid products.

Some pipeliners only need a model to tell how to avoid flowing slack. It is possible to predict where slack happens in a steady state without worrying about the physics inside a slack region. This method can't calculate the liquid hold-up, but calculates everything else correctly. It is possible as the non-slack parts of a pipeline are independent enough to solve a steady state model, coupled with the fact pressure inside any slack region is constant at the vapor pressure.* A simple pressure-based online model is relatively accurate for this purpose, adding a slack point at wherever vapor pressure is reached.

In contrast, offline studies involving surge analysis must quantify exactly what takes place when part of a batch flashes to or from the vapor phase. This is a non-trivial physical phenomenon we cannot gloss over by manually *"fudging"* a hydraulic profile. The physics of slack are incredibly complex; there is a feeling any change from tried-and-tested procedure might spell unmitigated disaster. In some cases, slack flow is inevitable. However, its complexity motivates pipeliners to avoid it if at all practical. If a pipeline team decides it is best for a pipeline's operating philosophy to change from slack-line to tight-line, or vice versa, there is a lot more work than merely adjusting a setpoint on a back-pressure control valve. Start-up and shut-down sequences need changing to avoid slack conditions. The system curve will change, increasing power requirements, so pump stations will need to handle new operating points. Pumps may stray away from best efficiency points (BEP), making operating them more costly.

Apart from exceptionally volatile batches like liquefied petroleum gas (LPG), the vapor pressures of all the different liquid batches transported by pipelines differ by less than 1 barg, corresponding to 10 meters of water and 15 meters of a typical hydrocarbon liquid. Yet slack flow may be unavoidable along the route of a given pipeline, being more a consequence of the terrain than of the fluid being pipelined.

---

* This is strictly true for pure substances such as propane. For products or crudes, the contents of a vapor region (and thus its vapor pressure) are expected to very gradually change over time, as the liquid flows past while the vapor remains stuck near the top of the hill. At any given instance in time it is very accurate to say that the vapor pressure and thus the operating pressure are constant over the length of a vapor bubble.

Intentionally operating a flowing slack-line pipeline

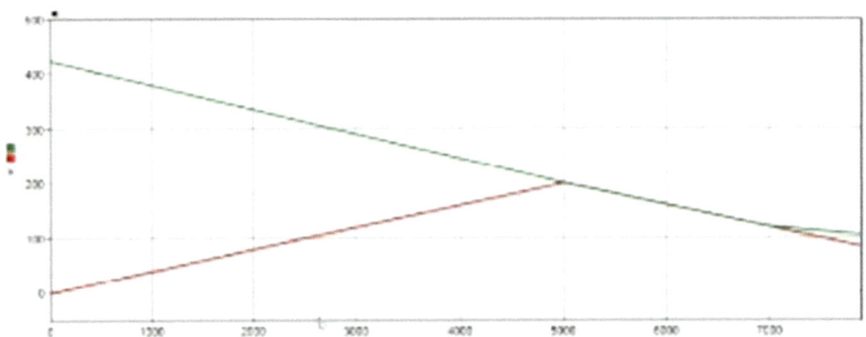

Figure 297 – The dynamic head (green) here is insufficient to have a fully filled cross-section of liquid across the elevation profile (red) so a slack region forms where the profiles coincide

Consider a scenario in which the downstream end of a line is controlled on pressure. A steep downhill pipeline requires a tremendous back-pressure to keep it in tight-line, possibly exceeding maximum allowable operating pressure (MAOP), in which case avoiding slack-line flow is impractical. By adding a throttle valve, the excess pressure could be throttled away on the down slope before it hits MAOP, but this wastes money. As seen in the figure, in a slack region along a flowing liquid pipeline, dynamic head sits exactly on the elevation profile. Fluid partially vaporizes due to local pressure having fallen to its vapor pressure at the local temperature. *

In a single-product pipeline there are no major downsides to slack other than added complication. Indeed, if increasing pressure proves expensive, letting liquid partially vaporize may be *preferable*. Intentional slack-line flow as part of a pipeline's operating philosophy can be seen as an efficient way to operate, especially early on before a pipeline reaches its throughput capacity as designed.

Now let's consider a scenario where a slack region forms in a pipeline controlled on *upstream* pressure. Flow rate through the upstream part is necessarily fixed. Adjusting the downstream pressure control setpoint won't change the flow rate entering at the upstream supply point. The flow delivered to the downstream demand point may change, and that flow imbalance is accounted for by the slack region growing or shrinking by the same amount.

* As opposed to *boiling*, which is due to a high temperature at ambient pressure.

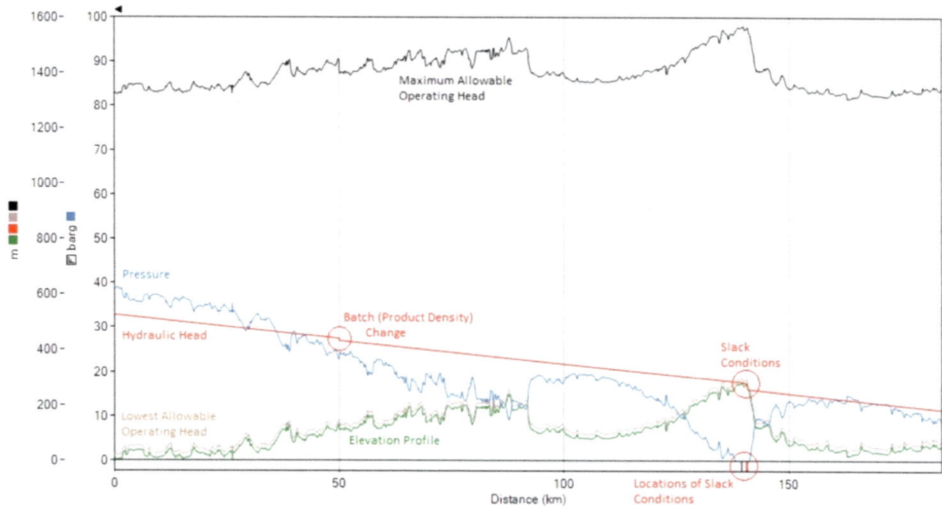

Figure 298 – Slack conditions are shown by the head profile suddenly dropping in the region where a vapor bubble is present

Model performance and stability in slack regions

A handful of simulators attempt steady states that account for slack, but few are capable of capturing transients. They can tell if there isn't any slack, but once there is slack, they give only a vague idea where it might be. To know any more than that, we need a transient model. Describing slack-flow and column separation is more than a simple matter of characterizing a single behavior. The simulator deduces when slack starts and stops, how the ends of slack regions move, and what sort of slack is taking place. This determines the physical equations which are then duly applied.

To remain fast-solving yet accurate at all times for slack scenarios, Atmos SIM incorporates a number of novel features. *Dynamic elevation simplification* stores a very detailed elevation profile, but only uses it for regions near or in slack. And when a slack region disappears, the model returns to using fewer elevation points in that stretch of pipeline, as usual. This is important because slack on a real pipeline manifests as lots of small slack regions rather than one big region – a fact that is not widely appreciated.

If a pipe in stratified two-phase flow becomes more than half-full, it becomes physically unstable: small surface waves grow til they fill the pipe, becoming slugs of liquid. * These surges are real, but simulating them would be complex and somewhat pointless: they are generated in a chaotic fashion, so while the simulator would be correct that

* The actual condition for this is actually a bit more complicated, and every slack region has a more-than-half-full first few meters where it doesn't happen.

there *are* surges, the surges it predicts can never match actual surges in a real pipeline. For this reason Atmos SIM prevents these surges from filling the pipe by cranking up the frictional losses if they become too large. This is another example of spending our effort modelling details that matter for our application, and forgoing details that do not.

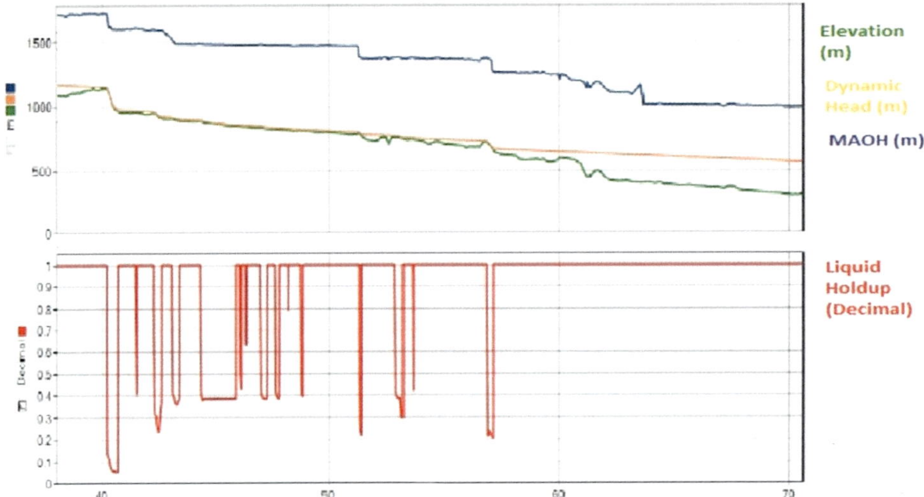

Figure 299 – Before reaching the high point, this pipeline is fully liquid filled. Slack occurs at the high point and extends downhill in the direction of flow. Atmos SIM's slack model reports on the resulting head accurately (top). This corresponds to the vapor pressure as calculated by the principle of vapor-liquid equilibrium. The liquid hold-up is reported as it varies along the length of the pipeline and within the numerous slack regions (bottom).

The impact of simulating slack on a model's performance depends entirely on how much slack is taking place. If a pipeline is flowing in a tight-line operation, the model performance is exactly the same. At the other extreme, if slack is taking place along 30 km of a pipeline with 20 separate slack regions each extending for 1.5 km, then two or three times as many knots may be needed to simulate that, so it might take twice or thrice as long to solve. This situation might arise in a long pipeline across mountainous region particularly if there is stopped flow across numerous high points in the terrain. Atmos SIM takes special measures to maintain numerical stability and model accuracy in a slack region.

One of these is to use a suitable number of knots as needed, as a special exception to user-set closest allowable knot spacing. For example, every slack region, however small, requires a certain number of knots to properly capture the shape of the upstream-end liquid holdup profile, which changes non-linearly over distances of one pipe diameter. This serves both to keep results accurate and avoid giving rise to artificial instability. [80]

Slack region composition and how it collapses

Before considering how a slack region goes away, we must consider what it's made of. There is some question as to the precise composition of vapor in a slack region. If there is air dissolved in the liquid, that will come out first. Enticing air back into solution is a gradual process that is very different to what we can expect to see when a hydrocarbon vapor suddenly condenses into a liquid on being pumped above its vapor pressure.

In most hydrocarbons pipelines, a vapor pocket primarily consists of the light ends of the hydrocarbon mixture: methane (CH_4), ethane (C_2H_6), etc. When slack first forms beside a hydrocarbon batch, a mix of lighter and heavier components comes out of solution. Then, over time, as the bubble and its contents sit on the downstream side of the hilltop while more liquid flows past, more and more light components come out of solution, with some of the heavier components dissolving into the liquid stream and being carried away. This means that a relatively old bubble of flowing slack can be quite difficult to force back into solution, requiring a much higher pressure than was present just before the slack formed originally. The hydrocarbon liquid is no longer in equilibrium with its own vapor, but rather it is in equilibrium with a pocket of gaseous methane or other low-weight alkanes. Forcing those into solution might require fairly high pressures – or allowing a long period of time to pass at moderate pressures.

When a shut-in pipeline with column separation then starts up, the slack region containing that vapor pocket fills in over some period of time. Liquid rises along the upstream pipe. Eventually it crests the top, spilling over the peak into the partially full downstream pipe. In the downstream pipe, liquid runs downhill, filling the slack region from the bottom first, forming a two-phase region where a torrent of liquid runs down the bottom of the cross-section with vapor on top.

Flow rate at the downstream end of a slack region drops discontinuously and suddenly on a profile of flow rate vs distance: more fluid mass can enter a slack region than depart in a sustained imbalance. That extra incoming fluid fills up the partially filled pipe, a packing operation gradually closing the slack region.

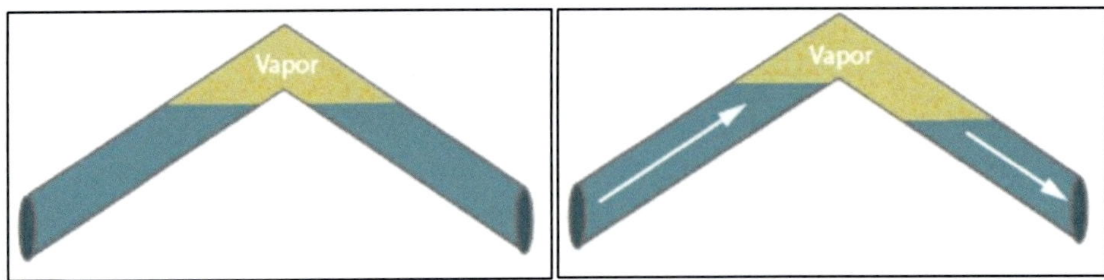

Figure 300 – Column separation in a shut-in pipeline (left) fills in once the line starts up (right). In this example, during start-up the downstream valves were opened, so the downstream-side liquid has begun to drain even before the upstream-side liquid has crested the peak.

A flowing slack region always starts at a local hilltop and extends downhill in the downstream direction. Liquid runs down the bottom of the pipe in open-channel flow, eventually hitting a *"lake"* part-way down the hill – the lake being the completely full pipe at the end of the slack. When pipeline conditions are changing so that the slack goes away, the way this usually happens is the *"lake"* part rises up the hill. When it reaches the top, there's a pressure surge and the slack is gone. Why does this give rise to a pressure surge?

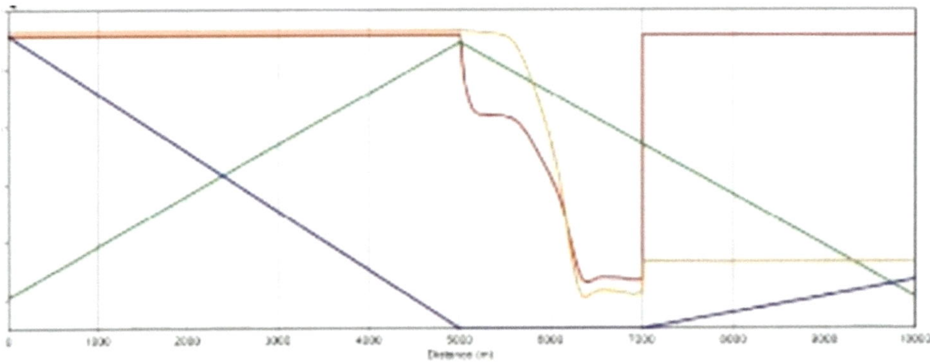

Figure 301 – Pressure (blue) is equal to vapor pressure throughout a slack region near a high point in elevation (green). When flow rate (orange) is further increased, a wave of liquid rushes along the slack region, as visible on the liquid hold-up (red) shown on a profile snapshot.

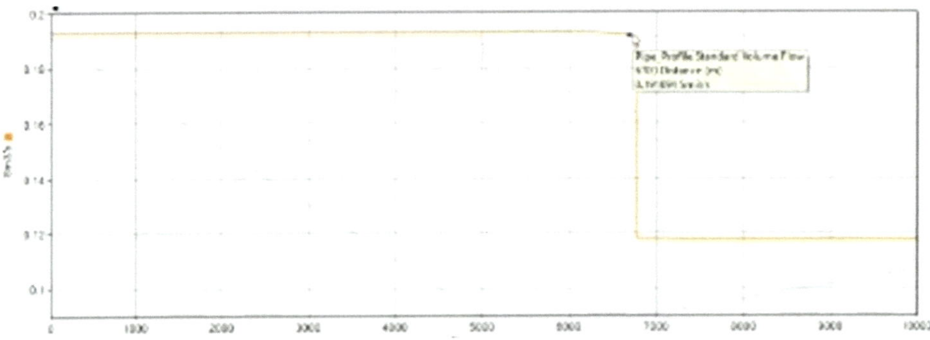

Figure 302 – The flow rate at the downstream end of a slack region drops discontinuously.

If we look at the flow profile versus distance, that explains why: upstream of the slack region, the fluid is moving with one flow rate and associated velocity. Downstream it's moving with a different, lower flow rate and velocity. As long as the slack region exists, this difference in flow rate is compensated by the region filling in. But as the region finishes filling we find that there is a fast-moving column of fluid that just hit a slow-moving column of fluid. The impact produces pressure waves in both directions. The magnitude of the waves can be estimated by the Joukowsky equation, based on each column ending up with the average of the two original velocities.

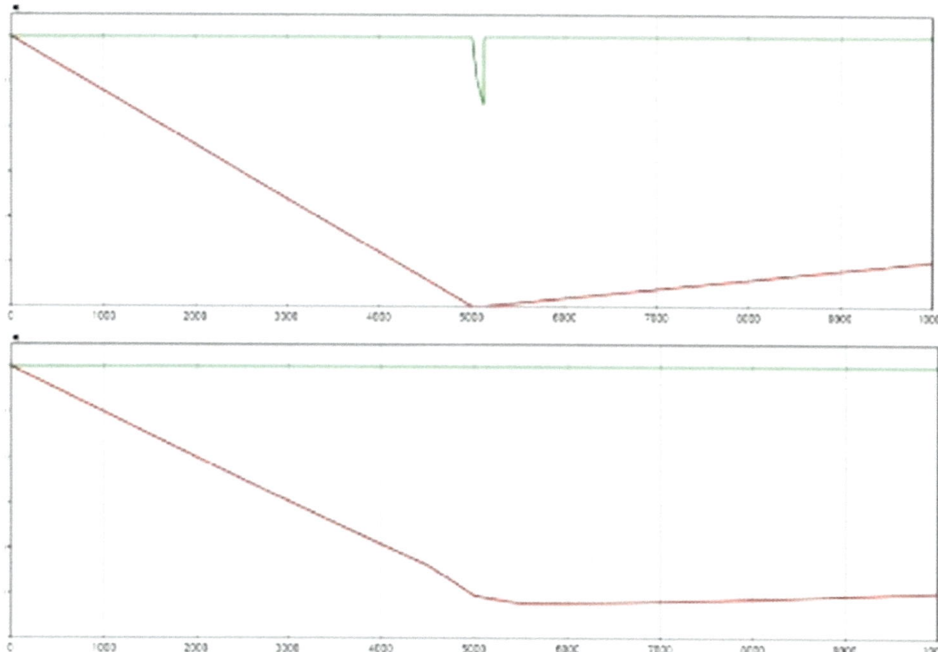

Figure 303 – When a slack region collapses (top), it gives rise to a pressure surge (bottom).

We quantified *water hammer* in the rapid transients chapter – those same principles apply here equally. As liquid is almost incompressible, that pressure surge rapidly propagates throughout the length of the pipeline. Even if the pipe wall at the point of impact is strong enough to contain the pressure excursion, a weak point elsewhere in the system might still rupture as a result. If we run at a short time-step and close knot-spacing, Atmos SIM shows us the surge.

13.5 – STEAM PIPELINES

Two-phase pipeline flow models were first developed in the nuclear power industry, where pipelines carrying steam and water are of utmost importance. It is worth pausing to consider how a pipeline carrying steam works, and how to model it. Steam is a remarkably efficient way of transporting thermal energy. It is useful for district heating, as well as for cooling and air conditioning, helping to offset the electrical demands of buildings and industries in both summer and winter. It has become a ubiquitous utility in process plants and industrial facilities, often provided through a header arrangement. The use of hot water circuits for the purpose of domestic heating is familiar to people living in cold climates. Perhaps less well known are the steam pipelines from geothermal generating plants, some of which extend over dozens of kilometers to reach their users. There are also steam pipelines serving not just industrial complexes but entire cities, one of the largest being the steam system serving New York City. Steam is distributed and fed out to buildings from a large header buried underground.

Modelling the fluid properties of steam is somewhat different to the way we look at regular pipelines, in terms of the temperatures involved, equation of state, and so forth. It is also quite different in that it is not uncommon for large steam lines to routinely carry a portion of the steam that has liquefied, introducing challenges to be met when sizing large steam lines for that *flash steam,* and adding a number of model items not seen in a gas pipeline, for *steam traps* and a diverse array of other specialized equipment.

As water drops out along a steam pipeline, the risk of a potentially dangerous phenomenon known as *steam hammer* becomes a major concern. Steam occupies about 1600 times more volume as a vapor than it does as liquid water. So, when it condenses, it shrinks. In a closed container like a pipe, that collapsing steam can cause water to rush in and fill the vacuum created by condensation, cooling the steam further, a runaway with the violent acceleration and deceleration causing the line to rupture. This situation is one of two major causes of steam hammer, and is due to *thermal shock* – when hot steam comes into contact with a cold pipeline too suddenly during start-up.

Another situation that can give rise to steam hammer happens during the course of operation. Condensed water runs alongside the steam, at a much lower velocity than that of the steam. If condensate is allowed to accumulate into a pool, steam passing over the liquid surface can cause waves. A wave could grow high enough to create a complete liquid seal inside the pipe, with the full pressure of steam behind it. A slug of water accelerates along the pipeline, picking up condensate, eventually slamming into the end. The resulting pressure spike has caused fatal *"steam pipe explosions"*. Engineering steam pipelines – inclinations, steam traps, etc. – is safety-critical. [56]

13.6 – TRUE MULTI-PHASE FLOWS

Multi-phase flow is a specialist topic mostly encountered in upstream gathering lines from wellheads to a central processing facility. It may be 2-phase (oil & water, oil & gas, or water & gas), 3-phase (gas, oil & water), or even 4-phase (gas, oil, water & sand). Modelling a mixture of different fluids flowing together is complicated by the individual characteristics of each product. They may stratify into distinct layers. Dispersed and continuous phase interactions make conservation laws governing mass and energy non-trivial, despite being well-understood. Some specialist pipeline simulators handle all of the physics, tuning the parameters of the interactions against laboratory data. Reactions between fluids can cause pipeline corrosion, risking leaks and ruptures that go unnoticed in a subsea pipeline where no meters are installed. Multi-phase pipelines are likely to need regular pigging to remove the build-up of wax and liquid slugs.

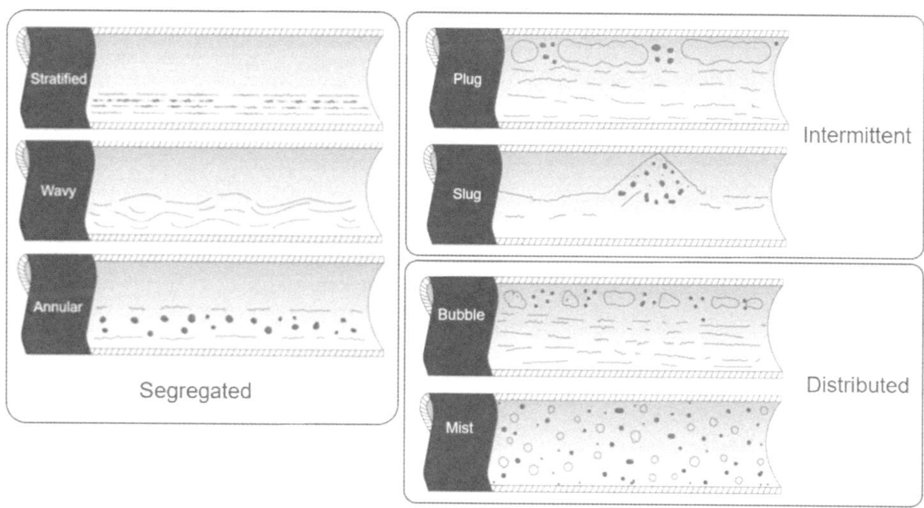

Figure 304 – Multi-phase flow regimes in a horizontal pipeline: segregated (stratified, wavy, annular), intermittent (plug, slug) and distributed (bubble, mist) – adapted from source. [57]

For a model simulating multi-phase flows to be meaningful, it must be obsessively accurate. This includes the thermal model, as the temperature of a slower-moving layer of liquid can differ from the adjacent layer. Multi-phase flow regimes are more difficult to predict than those of a single phase. A gas-liquid interface could be stratified, slugging or bubbly in a horizontal pipe, with more regimes possible in a vertical riser.

Operating a multi-phase pipeline at a higher flow rate might actually cost *less* to pump! To this day, those looking to understand the behavior of a fluid operating in any one of countless multi-phase flow regimes are faced with a choice between building scale-model rigs, or paying for the expense of a niche specialized software package.

SUMMARY

Experts in a domain rarely pause to marvel at the remarkable, working with it routinely that many of us take it for granted. The general public is largely oblivious to reliable pipeline infrastructure. It keeps society running, hidden discretely from view. Behind the scenes are teams of engineers, scientists and specialists striving tirelessly to keep it all in good order, innovating at the forefront of what is possible. Pipeline simulation lies at the heart of their work, harnessing technology to make smarter decisions about design and operation. Pipelines are often the best way to move water, gas and liquid hydrocarbons. Unprecedented transitions are happening within our diverse energy systems, presenting significant challenges for pipeliners to overcome. As renewables gain prominence, gas pipelines are operating differently.

Good models consider only what matters. For us, the pipelines, along with their fluids and equipment, are paramount. Great models intentionally ignore everything that won't affect our system. We make every reasonable assumption about a pipeline, adopting a simplified approach wherever possible for a versatile model to add value throughout the duration of a pipeline's lifecycle. Simulators facilitate many different tasks. One of these is conducting offline studies to help different members of a team solve their problems and make sound decisions about which of several options to move forward with. They explore offline scenarios by adjusting constraints to mimic operations. A constrained simulation lets people run virtual experiments.

A shut-in pipeline at some pressure holds different quantities of compressible fluid at a given instance, depending on its temperature. Equivalently, the same quantity at different temperatures exerts a different pressure. This intertwined relationship between a volume's inventory, its pressure and its temperature is described by equations of state. These are used to find many fluid parameters.

Constrained steady states can expedite the design and selection of equipment. Constrained dynamic simulations can describe operational transients. In liquid these operating scenarios include batching, and fast transients such as surges. In gas pipelines scenarios focus on nominations, throughput and linepack management. Modelling the power and temperature of pumps or compressors helps operate within broader operating envelopes for more versatile pipelines.

A suitable thermal model was overlooked in the past, despite affecting results. Temperature matters. For steady states, it is reasonable to adopt an overall heat transfer coefficient approach. But to capture thermal transients in a buried pipeline, we build a dynamic layer-based model, incorporating heat transfer through the ground. A transient thermal model proves to be essential whenever a simulator needs to estimate arrival time, track composition, or detect leaks.

Similarly, to capture transient hydraulics, we cannot get away with assuming a succession of steady states. A transient scenario lets us accurately model many operations a pipeline undertakes, describing unsteady operations depleting or accumulating both the inventory and the energy. Adaptive knot spacing and time-stepping allow us to run long, complex scenarios quickly and accurately. Switching between idealized control modes mimics a pipeline's operating philosophy by incorporating the effects of its equipment in varying conditions.

Simulators deployed live on site support operations personnel in several ways. Online scenarios interpret incoming raw data from meters to deduce the overall state of the pipeline and its stations, and present all kinds of useful information such as conditions reported at virtual meters, profile trends of the composition along the length of the route, and the inventory held in each section of the line.

When an online and an offline model share an underlying model, we can perform several kinds of time travel. Build-up rapidly brings a live system into focus by sprinting through recent data to reach the transient state that's happening now. Replay runs of historic data let operators train themselves by re-living real scenarios. Saved states from site can be exported to form the basis of offline studies, allowing strategists to assess if something better could have been done. Look-aheads forecast what is to come, anticipating problems and assessing the consequences of possible actions. Look-back validation can check if the look-aheads results were indeed accurate.

Next-generation concurrent look-aheads have opened the door to live optimizers. These harness simulators to go above and beyond, not merely informing us that something is expected to happen, but recommending the best ways to operate our pipelines safely, smoothly and effectively.

Simulators make the fascinating world of fluids more accessible than ever. What used to be excruciating calculations are now performed at a moment's notice. By exploring fluid flow in pipelines, we get to grips with core concepts, and engineers harness physics to implement reliable operations in every pipeline.

This concludes the main volume of the Atmos Book of Pipeline Simulation. The reader is invited to refer to the supplementary text containing dedicated Appendices covering the broader context of today's energy transition, the role of hydrogen in future pipelines, the history of pipeline hydraulics, a basic primer on pipeline essentials, and a quiz.

ABBREVIATIONS

1-D	One-dimensional
AOC	Abnormal operating condition
API	American petroleum institute
ASM	Abnormal situations management
ASV	Anti-surge valve
BC	Boundary condition
BD	Batch detector
BEP	Best efficiency point
BTS	Batch tracking system
BV	Block valve
BWRS	Benedict–Webb–Rubin–Starling
CD	Discharge coefficient
CONCAWE	Conservation of Clean Air and Water in Europe
CV	Flow coefficient, control valve or calorific value
CR	Compression ratio
DN	Nominal diameter
EOB	End of batch
EOS	Equation of state
E-RTTM	Extended real-time transient model
ESD	Emergency shut-down
ESV	Emergency shut-down valve
ETA	Estimated time of arrival
FT	Flow transmitter
GC	Gas chromatograph
GERG	European Gas Research Group
HDPE	High-density polyethylene
HMI	Human-machine interface
ID	Inner diameter
KP	Kilometer post
LAOP	Lowest allowable operating pressure
LDS	Leak detection system
LDU	Leak detection unit
LPG	Liquefied petroleum gas
MAOP	Maximum allowable operating pressure
MCS	Monte Carlo simulation
MLSE	Maximum likelihood state estimator
MOV	Motor-operated valve

MP	Mile-post
NB	Nominal bore
NGL	Natural gas liquids
NIST	National Institute of Standards and Technology
NPS	Nominal pipe size
NPSHa	Net positive suction head available
NPSHr	Net positive suction head required
NTSB	National Transportation Safety Board
OD	Outer diameter
OHTC	Overall heat transfer coefficient
PCV	Pressure control valve
PD	Positive displacement
PDE	Partial differential equation
PFD	Process flow diagram
PHMSA	Pipeline & Hazardous Materials Safety Administration
P&ID	Piping and instrumentation diagram
PID	Proportional integral derivative block
PLC	Programmable logic controller
PT	Pressure transmitter
PTS	Pig tracking system
PR	Peng–Robinson
PR	Pressure ratio
PRV	Pressure relief valve
PS	Pig switch
PV	Process variable
RoC	Rate of change
ROV	Remotely-operated valve
RPM	Revolutions per minute
RTM	Real-time model
RTTM	Real-time transient model
SCADA	Supervisory control and data acquisition
SG	Specific gravity
SOP	Standard operating procedure
SP	Setpoint
SPRT	Sequential probability ratio test
SRK	Soave–Redlich–Kwong
SSS	Succession of steady states
TT	Temperature transmitter
VFD	Variable-frequency drive
VLE	Vapor-liquid equilibrium
VSD	Variable-speed drive

SYMBOLS

Latin symbols

Symbol	Meaning
a	Acceleration
A	Cross-sectional or circumferential area (in context)
A^{std}	Area at standard conditions
B, B_T, B_s	Bulk modulus (isothermal, isentropic)
c, c_P, c_v	Specific heat capacity (isobaric, isochoric)
c, c_s, c_T	Speed of sound (isentropic, adiabatic) = wave speed = sonic velocity
C_D	Discharge coefficient
C_{V}	Flow coefficient, such as for a valve; others use A_V or K_V
D	Inner diameter or outer diameter (in context)
D^{std}	Diameter at standard conditions
e, E	Energy. If lowercase, it is mass-, volume- or mole-specific
Eu	Euler's number
f	Friction factor (Darcy's)
F	Force
F, F^*	Flow rate of some kind in certain pipeline contexts, measured flow rate
g	Acceleration due to gravity
$H, H_{\text{J/kg}}$	Head (height units, specific energy units)
h	Enthalpy (mass- or mole-specific)
h	Convective heat transfer coefficient (in context)
i	Index for spatial knots
j	Joule-Thomson coefficient
k	Thermal conductivity, fluid correction factor, or pipe resistance
L	Length of a pipeline
m_{r}	Relative molar mass (molecular weight)
M	Mass
Ma	Mach number
n	Time-step index, or polytropic index
N	Number of throws in a reciprocating pump or compressor
P, P^{std}, P^*	Pressure, pressure at standard conditions, measured pressure
q	Rate of heat transfer (per unit area or per unit mass)
Q	Actual volumetric flow rate (*"actual flow rate"*)
Q^{std}	Standard volumetric flow rate (*"standard flow rate"*)
Q^{mass}	Mass flow rate

$Q^{\text{net}}, Q^{\text{gross}}$	Energy flow rate (based on net or gross combustion value)
r	Radial co-ordinate; perpendicular to the axis of the pipe
R	Radius, or universal gas constant
Re	Reynolds number
s	Entropy (mass- or mole-specific)
t	Time
T, T^{std}	Temperature, temperature at standard conditions
u	Internal energy (mass- or mole-specific)
U	Overall heat transfer coefficient (OHTC)
v	Flow velocity (averaged over time and over cross section)
V	Volume. We avoid using specific volume
V^{std}	Standard volume (volume at standard conditions)
W, w	Work done. If lowercase, it is mass- or mole-specific
$\dot{W}, \dot{w}$	Power. If lowercase, it is mass- or mole-specific
x	Axis, either along the axis of the pipe or horizontally
y	Generic conserved quantity (mass-, volume- or mole-specific)
Y	Elastic modulus (Young's modulus)
z	Elevation, or vertical axis
Z	Supercompressibility

Greek symbols

Symbol	**Name**	**Meaning**
α	Alpha	Coefficient of thermal expansion
γ	Gamma	Ratio of heat capacities
Γ	Gamma	Specific gravity
δ	Delta	Wall / shell thickness, or leak detection trigger
ε	Epsilon	Strain, or statistical error
$\eta, \eta_p, \eta_s, \eta_m$	Eta	Efficiency (polytropic, isentropic, mechanical)
θ	Theta	Angle of inclination, or variable within a solver
κ	Kappa	Compressibility
λ	Lambda	Statistical measure
Λ_C, Λ_V	Lambda	Clearance ratio, volumetric efficiency
μ	Mu	Dynamic viscosity
ν	Nu	Kinematic viscosity, or Poisson's ratio
ϕ	Phi	Correction factor for pipe restraint
σ	Sigma	Tensile stress, yield strength, standard deviation
ρ, ρ^{std}	Rho	Mass density, standard density
τ	Tau	Shear stress
ψ	Psi	Objective function
ω	Omega	Speed of a pump or compressor

INDEX

REFERENCES

[1] Less is More: Accuracy vs Precision in Modelling, Susan Bachman and Mary Goodreau, PSIG 2000
[2] Dynamics of Physical Systems, Robert Cannon, 2012
[3] Pipeline Rules of Thumb Handbook, McAllister, 2014
[4] Unified Friction Formulation from Laminar to Fully Rough Turbulent Flow, Applied Sciences 2018, Dejan Brkić and Pavel Praks
[5] Advanced Natural Gas Engineering, Xiuli Wang and Michael Economides, 2013
[6] https://www.aboutpipelines.com/en/pipeline-101/why-spills-happen/
[7] What Went Wrong? Trevor Kletz, Elsevier, 1998
[8] Single-phase flow assurance, Ove Bratland, 2013 drbratland.com
[9] thechemicalengineer.com/features/clean-hydrogen-part-2/
[10] Petition for diluted bitumen rulemaking – EPA (accessed online on 24 Aug 2021)
[11] Hazardous Liquid Pipelines - Basics and Issues, Pipeline Safety Trust, 2015
[12] Review of Vapor Cloud Explosion Incidents, UK HSE, 2017
[13] McKay F.F. et al., Hydrocarbon Processing, November 1977, 56(11):487
[14] www.ipcc.ch/site/assets/uploads/2018/02/ipcc_wg3_ar5_chapter7.pdf
[15] Net Zero and Safety, Gilmour, The Chemical Engineer, Dec 2019 / Jan 2020
[16] Natural Gas – Fuel for the 21st Century, Smil, 2015
[17] Oil and gas sector can bring quick climate win by tackling methane emissions, UNEP, 2019
[18] pgjonline.com/news/2020/12-december/eia-natural-gas-venting-flaring-set-records-in-2019
[19] Energy: A Human History, Rhodes, 2018.
[20] Equation of State for Natural Gas Systems (PhD Thesis), Uribe, 1999.
[21] https://tsapps.nist.gov/publication/get_pdf.cfm?pub_id=903432
[22] reutersevents.com/renewables/solar/understanding-energy-density-thermal-storage
[23] Mass and Heat Transfer, Russell, 2019
[24] Crane Technical Paper 410: Flow of Fluids Through Valves, Pipes and Fittings, 1999
[25] SmarterEveryDay on YouTube
[26] Pipeline Leak Detection Handbook, by Morgan E. Henrie, Philip Carpenter, and R. Edward Nicholas
[27] 1984 PSIG talk by Modisette, Nicholas, & Whaley
[28] Preissmann, A. (1961) Propagation des intumescences dans les canaux et rivières. First Congress of the French Association for Computation, Grenoble, 433-442
[29] On Simulation Accuracy, Per Lagoni and Jon Barley, PSIG 0711
[30] Pipeline Variable Uncertainties and Their Effects on Leak Detectability, API 1149, 2nd Ed. (2015)
[31] Pipeline System Automation and Control, Mike Yoon, C. Bruce Warren and Steve Adam, ASME 2007.
[32] Simulation of Rapid Transients in Gas Pipelines for ESD Valve Design, Jason Modisette, Garry Hanmer, PSIG 2021
[33] Aalto M., K. I. Keskinen, J. Aittamaa, and S. Liukkonen, Fluid Phase Equil., 114: 1 (1996)
[34] Tullis, J. Paul. Hydraulics of Pipelines: Pumps, Valves, Cavitation, Transients. United Kingdom: Wiley, 1989.
[35] Managing Liquid Pipeline Slack, Pipeliners Podcast, Jan 2019
[36] An Investigation into how to Optimize the Effect of DRA on the Flow rate & Pressure of Single Phase Fluid Pipelines, Naffaa, PSIG 2019
[37] popularmechanics.com/science/energy/a7480
[38] Design calculations for oil and gas pipelines, Kasch, 2007
[39] Tuning of Subsea Pipeline Models to Optimize Simulation Accuracy, Garry Hanmer, Edward Jackson – Atmos International, Dagfinn Hansen, Ben Velde – Gassco, Svein-Erik Losnegård – Polytec, 2012 (PSIG 1212).
[40] linkedin.com/pulse/cavitation-misunderstandings-other-stories-simon-bradshaw
[41] Chemical Engineers Handbook, Robert Perry and Cecil Chilton, 5.ed 1973
[42] Mark's Standard Handbook for Mechanical Engineers, Eugene Avallone & Theodore Baumeister, 10ed. 1996
[43] Controlling Positive Displacement Pumps, Walter Driedger, 2000, accessed via pumpfundamentals.com
[44] Adapted from a video by WR Training: youtu.be/UmoOXdLLITQ
[45] Hydraulic Design for Business Development Professionals, Pipeliners Podcast, Apr. 2019
[46] Coulson and Richardson's Chemical Engineering Volume 6 (Design), 2ed, by Ray Sinnott, Pergamon Press, 1994
[47] Process Centrifugal Compressors, Lüdtke, 2004
[48] A Working Guide to Process Equipment, Liebermann, 2014
[49] Handbook of Natural Gas Transmission and Processing, Mokhatab 2019
[50] Numerical Simulation of Gas Pipeline Networks, Il'kaev, Seleznev, Aleshin, Klishin, 2005

[51] Fundamentals of Pipeline Engineering, Vincent-Genod, 1984
[52] Jet Propulsion, Cumpsty, 2015
[53] Operation of Centrifugal Compressors in Choke Conditions, Kurz et al, Turbomachinery Symp. 2011
[54] Pipeline Design & Construction, Mohitpour, 2000
[55] Natural Gas Processing, Bahadori, 2019
[56] https://practical.engineering/blog/2018/7/26/what-is-a-steam-hammer
[57] Oil and Gas Pipeline Fundamentals, John L. Kennedy, 1984
[58] Oil and Gas Pipelines in Non-Technical Language, Thomas Miesner and William Leffler, 2nd ed. 2020 PennWell
[59] thetyee.ca/Opinion/2018/11/28/Air-Barrels-Pipeline-System/
[60] *Shock to the System*, US Chemical Safety and Hazard Investigation Board, youtu.be/_icf-5uoZbc
[61] Modelling of Rapid Transients in Gas Pipelines, Hanmer, Mora, de Marco & Lacerda
[62] ntsb.gov/investigations/AccidentReports/Reports/PAR0202.pdf
[63] tandfonline.com/doi/abs/10.1080/1573062X.2020.1729388 Preliminary sizing of surge vessels on pumping mains, Cogent engineering, Nuralhuda Aladdin Jasim, 2020
[64] Adding it Up, Barrett, The Chemical Engineer, Feb. 2019
[65] Simple Method Predicts Gas-line Blowdown Times, Weiss, Botros, Jungowski (NOVA) – O&GJ, 1988
[66] Bhopal Leaves a Lasting Legacy, Kletz, chemicalprocessing.com/articles/2009/238/
[67] https://www.aboutpipelines.com/en/pipeline-101/why-spills-happen/
[68] Protection of piping systems subject to fires and explosions, UK HSE, 2005
[69] pipelinerspodcast.com/episode-27-rupture-detection-giancarlo-milano/
[70] https://www.ntsb.gov/investigations/AccidentReports/Reports/PAR0202.pdf
[71] Recommended Practice API 1540 – Design, Construction, Operation and Maintenance of Aviation Fuelling Facilities, 4th Ed. (2004)
[72] A Sensitivity Analysis of a Computer Model-Based Leak Detection System for Oil Pipelines – Lu, She and Loewen – Energies MDPI
[73] State Estimation of Pipeline Models using the Ensemble Kalman Filter, Jason Modisette, PSIG 2013
[74] Application benefit of integrated simulation software for Gassco's subsea pipeline network, Postvoll Velde (Gassco), Garry Hanmer, James Munro (Atmos International) – PSIG 2013
[75] How Pipelines Make the Oil Market Work – Their Networks, Operation and Regulation, Allegro
[76] Modelling & Simulation of Gassco Wet Gas Pipelines PSIG 1604, Hanmer, Basnett, Issak, Rinde.
[77] Modelling transient leaks from pressure vessels including effects of safety systems, Stene, Harper and Witlox – IChemE Symposium 159, Hazards 24
[78] antoine.frostburg.edu/chem/senese/101/liquids/faq/antoine-vapor-pressure.shtml
[79] Improving Operational Safety in Oil Pipelines, Schraml et. al., Pipeline Technology Conference 2011
[80] A Slack Flow Model With Moving Regime Boundaries, Jason Modisette, PSIG 2018
[81] https://www.iea.org/reports/the-future-of-hydrogen - 2019
[82] thechemicalengineer.com/features/clean-hydrogen-part-1-hydrogen-from-natural-gas-through-cost-effective-co2-capture/
[83] thechemicalengineer.com/features/hynet-demonstrating-an-integrated-hydrogen-economy/
[84] decarbconnect.com/h21-project-testing-the-route-to-domestic-and-industrial-hydrogen-distribution/
[85] www.spiraxsarco.com/learn-about-steam/steam-distribution/pipes-and-pipe-sizing
[86] Technical assessment of Hydrogen transport, compression, processing offshore, North Sea Energy
[87] Transporting Hydrogen-Natural Gas Mixtures, Rainer Kurz et. al., Solar Turbines, PSIG 2110 (2021)
[88] turbomachinerymag.com/view/readying-pipeline-compressor-stations-for-100-hydrogen
[89] The Use Of The Natural-Gas Pipeline Infrastructure For Hydrogen Transport In A Changing Market Structure, Dries Haeseldonck, William D'haeseleer, K.U. Leuven
[90] An integrated solution for planning and operating Power-To-Gas facilities in coupled gas and electricity networks, Adolfo et. al., PSIG 2115 (2021).
[91] Pipeline Transient Optimization for a Gas-Electric Coordination
Decision Support System, Zlotnik et. al., PSIG 1919 (2019).
[92] Multivariable analysis for selection of natural gas compression technology considering the impact of Solar PV power generation, Alvarez et. al. PSIG 1917 (2019)
[93] https://afdc.energy.gov/files/pdfs/hyd_economy_bossel_eliasson.pdf
[94] Tracking Batches Accurately in a Multi-Product Pipeline with large Elevation Changes and Prominent Slack Flow, Giancarlo Milano and David Basnett (Atmos International, Inc.), Nitesh Goyal (Kinder Morgan, Inc.) IPC2018-78715, International Pipeline Conference, 24th to 28th September 2018, Calgary, Alberta, Canada